AF323729

STABILITY CRITERIA FOR FLUID FLOWS

Series on Advances in Mathematics for Applied Sciences – Vol. 81

STABILITY CRITERIA FOR FLUID FLOWS

Adelina Georgescu
Academy of Romanian Scientists, Romania

Lidia Palese
University of Bari, Italy

World Scientific

NEW JERSEY · LONDON · SINGAPORE · BEIJING · SHANGHAI · HONG KONG · TAIPEI · CHENNAI

Published by

World Scientific Publishing Co. Pte. Ltd.

5 Toh Tuck Link, Singapore 596224

USA office: 27 Warren Street, Suite 401-402, Hackensack, NJ 07601

UK office: 57 Shelton Street, Covent Garden, London WC2H 9HE

Library of Congress Cataloging-in-Publication Data
Georgescu, Adelina.
    Stability criteria for fluid flows / by Adelina Georgescu, Lidia Palese.
      p. cm. -- (Series on advances in mathematics for applied sciences ; v. 81)
    Includes bibliographical references.
    ISBN-13: 978-981-4289-56-6 (hardcover : alk. paper)
    ISBN-10: 981-4289-56-6 (hardcover : alk. paper)
  1. Heat--Convection--Mathematics. 2. Fluid mechanics--Mathematics. I. Palese, Lidia. II. Title.

QC327.G46 2009
536'.25--dc22

                                    2009026822

British Library Cataloguing-in-Publication Data
A catalogue record for this book is available from the British Library.

Printed in Singapore by World Scientific Printers

Contents

Introduction

In this volume, the main methods, techniques and tricks used to derive sufficient conditions for fluid flow stability are discussed. In general, nonlinear and linear cases require different treatments, thus we have to differentiate between linear and nonlinear criteria.

With a few exceptions, the treatment is analytical, but connections with the geometric viewpoint of dynamical systems are also outlined.

Inequalities and their use are crucial for finding stability criteria. That is why particular attention is paid to classical or generalized analytical inequalities, especially to those relating integrals of functions and their derivatives. The best constants involved into the last ones can be viewed as extrema of some associated functionals. If the extrema are with constraints, the corresponding inequalities are the so-called isoperimetric inequalities. Further, in order to solve the associated variational problems, direct methods, based on expansions in Fourier series upon total sets of functions, turned out to be among the most efficient. The Fourier series can be introduced directly into the functional or into the corresponding Euler equations, which, in the isoperimetric case, are eigenvalue problems. Moreover, the expansion functions may be chosen to satisfy all boundary conditions of the problem or part of them (especially when even and odd derivatives occur in equations or/and boundary conditions). Finally, in looking for variational principles natural conditions may occur. Several variational aspects related to functional inequalities used to prove stability criteria emerge, to justify the insertion of an entire chapter (3) devoted to variational problems. Algebraic and differential inequalities are summarized in Appendix 1 together with some formulae of tensor analysis.

A great amount of hydrodynamic and hydromagnetic stability criteria exist, and we do not intend to present them all, having chosen to limit ourselves to the founder's criteria, our own results, and a few other results of the Italian and Romanian schools in the field, for mixtures and in the Bénard magnetic case, for free or rigid walls. In addition, we are concerned mainly with convection problems (including temperature; concentration; magnetic, Soret, Dufour, Hall, ion-slip, dielectrophoretic effects in horizontal layers) for viscous fluids or fluid mixtures. Only in a few cases, horizontal convection and other effects are considered.

In most cases, we use variational methods, methods of Hilbert spaces theory, methods based on inequalities (isoperimetric or not), the Fourier series method and a direct method based on the characteristic equation. This is why, with a few exceptions, in the linear cases treated by us, the ordinary differential equations (ode's) have constant coefficients but a higher order. In the nonlinear case, for the sake of simplicity, in order to have a symmetrizable linearized part, we preferred basic equilibria or steady flows described or approximated by affine functions. Consequently, this book considers fluid flows whose stability properties do not depend on local phenomena.

Nowadays, hydrodynamic stability theory is involved in important ecological and industrial problems, requiring a lot of effects, and characteristics of fluids configurations, other than traditional ones, being taken into account. Among them we quote: rotation in spherical and cylindrical configurations; capillarity; heat sources; magneto-elastic effects; compressibility; chemical reactions; temperature-dependent viscosity, nonlinear (*e.g.* cubic) dependence of temperature on the density; porosity; variable gravity; liquid drops; fluids heated rapidly in a time-dependent manner; finite conductivity of the walls; surface films in Bénard convection; concentration and temperature-dependent surface tension; impulsive heating or cooling; heating from above; uniformly accelerated fluids; sound effects in convection; time-dependent basic flows; multi-component fluid layers; physically nonlinear fluids or other continua (second grade fluids, simple fluids, dissociating fluids); fluids with structure (dipolar fluids, micro-polar fluids, ferromagnetic fluids); multiphasic continua and phase transition in continua (mixed phases, melting, mushy zones, frozen zones, phase changing snow packs, ice streets, glaciers directional solidification, solid friction); dynamo. We treat a few of these complex problems in a didactic way, in order to be useful to other similar topics. We do not repeat classical and, by now, simple results of hydrodynamic and hydromagnetic stability theory. They can be found in the basic monographs on the topic: [Lin], [Chan], [DraR], [J76], [Geo85], [Yu2], [Ioo2], [Kos]. We go further instead with more complex but still basic subjects, *e.g.* linearization principle, universal stability criteria, stability spectrum estimates, variational principles, improved energy methods, treatment of problems with intricate boundary conditions.

The treatment in Chapters 1-3 differs from that in Chapters 4-7. Each section in Chapters 4-7 concerns a single physical case and the presented cases are quite complicated. Most of them include several physical effects (*e.g.* mechanical, thermal, magnetic, chemical, porosity) and several mathematical methods (*e.g.* differential and/or integral inequalities, tensorial formulae, variational problems, eigenvalue problems). For each case the physical motivation and conditions are provided. On the contrary, in Chapter 1, a more modern physical basis of mathematical modelling of fluid motions is given. Chapter 2 yields basic mathematical results on the stability involved in the resulted models. Chapter 3 is a brief presentation of needed variational arguments through simple examples whose physical bases are briefly explained. The proofs of the variational results are only sketched. In exchange, comments on their validity are carefully presented.

Moreover, for all worked examples, detailed computations are given.

Throughout the book we try as much as possible to treat the most realistic cases. Firstly, this is related especially to the boundary conditions which "spoil the symmetry" of the mathematical problem, the given problem being reformulated so that elegant functional analytic methods apply to the new setting. Secondly, we bring into actuality some other powerful approaches (*e.g.* Budiansky-DiPrima (B-D) method, backward integration method) frequently used several decades ago, and almost forgotten by now by applied mathematicians. Thirdly, we call the attention of mathematicians interested in fluid flow stability towards powerful methods having an engineering flavor (*e.g.* Joseph's differentiation of parameters approach, B-D method, the direct method).

Minimal prerequisites are: fundamentals of classical calculus of variations (Euler equations, isoperimetric problems, variational principles), linear and nonlinear functional analysis (normed spaces, symmetric operators, generalized derivatives, embedding theorems in Sobolev spaces, quadratic functionals, energy spaces), tensor analysis (flux-divergence formula, representation formulae for solenoidal and poloidal vectors, calculus with differential operators). Their brief presentation can be found in appendices and in Section 3.2.

The book is mainly addressed to applied mathematicians and other researchers in hydrodynamic and hydromagnetic stability. It can also be a companion paper to courses in these fields, *e.g.* [Geo85].

Chapter 1 presents briefly the modelling and the linear and general mathematical models of thermodynamics of fluids in the I. Müller framework. Then the particular classical models used in the book and their corresponding perturbation models around a steady solution are written. The generalized settings of some classical models are provided too.

Chapter 2 deals with geometric reframing of analytical models of hydrodynamic stability theory in dynamical system theory as evolution equations in Banach spaces in view of their use in the linearization principle and energy method in the nonstationary case. Physical and mathematical concepts of energy leading to different definitions of stability are discussed. Improvements of the Prodi estimations for the spectrum of the operator defining the linearized incompressible Navier-Stokes (N-S) equations are obtained, enabling us to derive some universal stability criteria. Some others are obtained by a better use of integral inequalities in a generalized form of the N-S equations. Bounds for linear stability which are also criteria for nonlinear stability are given. Below them, no perturbation amplifies. If these bounds become limits of linear stability, then the existence of subcritical instabilities is excluded.

This is quite a rarely treated case. We considered it in Chapter 4.

Chapter 3, the core of the book, is a systematic presentation of the results of the calculus of variations used in hydrodynamic and hydromagnetic stability theory. First it introduces the two main traditional topics in calculus of variations, *i.e.* variational problems and Euler equations, then shows their relationship revealed by means of the energy method, cases where they are equivalent, and their numerical realization. Due to their notable involvement in deriving stability criteria, the isoperimetric problems, associated isoperimetric inequalities and eigenvalue problems are separately dealt with. Then all these variational methods are exemplified by four two-point problems for a system of three ode's governing the linear stability of some mechanical equilibrium of a fluid: The nonsymmetric matricial differential operators are associated, then they are symmetrized, variational principles for the obtained equations are written, and the use of Fourier series expansions yields stability criteria. This first part provides powerful tools for the investigation of stationary problems. A distinct topic, namely energy method for evolution equations, suited to non-stationary problems, is then dealt with. A bridge between the two tools is represented by isoperimetric inequalities, deduced by stationary tools but used in the non-stationary case.

Chapter 4 deals only with three variants of the non-stationary energy method and their application to concrete nonlinear convection problems governed by pde's. We present only those variants closely related to our own. The first variant is Joseph's method and it was developed by him for a thermal convection problem in the case of a binary fluid mixture, characterized by one vector (velocity) and two scalar (temperature and concentration) unknown functions. His first basic idea was to define a new energy capturing the contribution of some variable-sign terms from the energy relation. This was achieved by introducing as new scalar unknown functions some linear combinations of temperature and concentration. The second idea was to establish a relationship between the undetermined coefficients (parameters) such that the energy relation contains no derivatives of any new scalar functions. The third main idea was to derive the constraint on the parameters in order for the stability bound to be optimal. In Section 4.1 we present our extension of Joseph's method for a convection problem containing an additional interaction between thermal diffusivity and diffusive thermal conductivity. Joseph's lines are followed, but instead of using as unknown parameters the coefficients in the above-mentioned combinations, we use the coefficients in the inverse combinations, yielding a much easier computation. In Section 4.2 the Georgescu-Palese-Redaelli (G-P-R) method is described. It generalizes Joseph's approach translating his ideas in terms of symmetrization of the operators occurring in some equations equivalent to the governing equations and derived from them by suitable scalar multiplications. It is found that the linear and nonlinear stability limits coincide, proving the equivalence of Joseph's and our method. Even more efficient approaches are given in Sections 4.2.2 and 4.2.3. In them we use only the optimality Joseph condition, all other arguments being new. In Section 4.2.4 a horizontal convection is analyzed. For a vertical hydromagnetic convection problem in Section 4.2.5 the G-P-R method is extended to the case of two vector and one scalar unknown functions. The variants based on the energy splitting are sketched in Section 4.3.1. Some of the ideas developed in Section 4.3.1 are applied in Section 4.3.2 to a very complicated hydromagnetic convection problem, whose solution uses our ideas from the linear case (Section 6.2).

Chapters 5 and 6 provide linear stability results mainly obtained by our group using the direct (Chapter 5) or variational (Chapter 6) B-D method. The perturbation fields are taken in the form of normal modes, reducing the problem to a one-dimensional case. Further the B-D method applies: the unknown functions are expanded in Fourier series on sets of sines and cosines which are total in some separable Hilbert subspaces of some Sobolev spaces. In fact, we tried to keep as close to the theory of these spaces as possible in view of avoiding any earlier approximation and its uncontrollable error propagation due to the cumbersome computations involved. In this way, the boundary value problem for a system of ode's becomes an algebraic system in the Fourier coefficients.

Unlike other methods based on Fourier series, in the B-D method the expansion functions are easy to construct and satisfy only part of the boundary conditions of the problem. The other boundary conditions introduce some constraints, to be satisfied by the Fourier coefficients, leading to secular equation defined by a finite-dimensional determinant each entry of which is a converging series. This equation is exact and easier to compute in comparison with the standard secular equation defined by a determinant of infinite order and corresponding to Chandrasekhar-Galerkin methods, where the expansion functions satisfy all boundary conditions. Supplementary simplifications are obtained by splitting the problem into even and odd problems. These advantages show that the B-D method is appropriate to flows subject to several physical effects and to complicated boundary conditions, when the standard methods are practically nonapplicable. In Chapter 5 we illustrate them on complicated cases of convections, involving many unknown functions and parameters, various boundary conditions and high order derivatives which can be of even and odd order. There the B-D method is applied directly to the equations.

In Chapter 6, a variational principle is first shown to hold, and then the series expansions are introduced into the functional. As this functional involves a smaller number of derivatives, the computations are easier the closer is the operator defining the equations to a symmetric operator. Some results obtained by means of the B-D method are the best and, sometimes, the only ones to be found in literature, pleading for the reconsideration of this method. It was applied to stability of elastica in the '40-s and in the '60-s to fluid flow stability. Subsequently, it was almost forgotten until its systematic use by our group.

The transition from the generalized treatment of nonlinear pde's (Chapters 2-4) to the (apparently) classical study of linear ode's (Chapters 5-7) is somehow abrupt. However, the linearized equations are often involved and used in the first step of nonlinear approaches and in their numerical realization based on the eigenvalues of linear problems.

Moreover, in the absence of a quite general linearization principle, comparing nonlinear and linear stability limits is necessary to the physical interpretation of the results. Chapter 5 realizes a "translation" of properties of ode's into those of the associated algebraic system of Fourier coefficients of their solutions. Further, Chapter 6 provides equivalent variational settings for eigenvalue problems defined by matricial differential operators. The direct method applied to several fluid flows in Chapter 7 yields complementary results to linear problems dealt with by methods from Chapters 5 and 6. Among them there are two topics rarely treated in the literature, namely the detection of false neutral manifolds and of the secular equations independent of the boundary conditions. The involved investigation of the multiplicity of the characteristic roots allows one the determination of the true secular equations and, consequently, the detection of false neutral manifolds (occurring in formal direct use of numerical methods). Various approaches to study this multiplicity are shown and the associated bifurcation sets are revealed. The complex cases treated in Chapter 7 by the direct method show its adequacy to real-world problems involving any type of boundary conditions, many parameters and high orders of differentiation. Providing closed-form solutions as a sum of a small number of terms, this method enables one to find secular equations independent of the boundary conditions. In this way, some paradoxical behaviors of certain eigenvalue perturbation problems are explained.

In this book the formulae, theorems, remarks, definitions, lemmas are labeled by three numbers, while in the appendices, by two numbers. The physical quantities occurring in the book are listed at the end of Section 1.1.1.

Apart from section titles, the following abbreviations are frequently used in the text: Budianski-DiPrima (B-D), Navier-Stokes (N-S), Navier-Stokes-Fourier (N-S-F), Oberbeck-Boussinesq (O-B), ordinary differential equation (ode), partial differential equation (pde).

We are deeply indebted to Mihnea Moroianu for reading the entire manuscript and making valuable improvements. We also kindly acknowledge the remarks and suggestions made by our collaborators Catalin-Liviu Bichir, Ioana-Florica Dragomirescu, Arcangelo Labianca and Aldo Redaelli.

The authors

Chapter 1

Mathematical models governing fluid flows stability

Mathematical modelling in thermodynamics is sketched (Section 1.1), following the unitary scheme of Ingo Müller [Müll]. We took this opportunity to show the position of the models used in the book within this general framework.

In Section 1.2 ten classical models used in other chapters are described in the nondimensional form. Their thermodynamic bases are shown and various approximations involved are discussed. The perturbation models are presented in Section 1.3 also in a classical setting. Section 1.4 provides a few models in a generalized setting.

1.1   General mathematical models of thermodynamics

Specific geometric, dynamical, physical and material characterizations of fluid flows occurring in the models used in the book are presented versus the corresponding general ones from thermodynamics. Then the main steps leading to integral and differential models of thermodynamics are shown.

1.1.1   *Physical quantities and their mathematical description*

This section contains the notation for the physical quantities used in the book.

By a *mathematical model of fluid flow* or, simply, *model*, we mean a set of differential equations and initial and/or boundary conditions relating the geometric, dynamic, physical and material properties of fluids in their motion (flow).

An impressive number of such models exists, each one corresponding to a particular mathematical and physical characterization. Among them we quote the models of classical fluid mechanics, concerned with fluids (mostly water and air), supposed as unable to feel the presence of physical influences other than gravity and, in the compressible case, the temperature. Simplifying hypotheses, most of

1

mathematical nature, allow their study in terms of mechanical characteristics only. The prototype is the Navier-Stokes model for viscous fluids, mostly used for the incompressible case.

The tremendous development of various industries, ecology, biology, astrophysics and medicine imposed qualitative changes in the traditional perception of a fluid and its motion. As a result, at the middle of the last century, the thermodynamics of processes far from equilibrium set in.

In the following we call it, shortly, thermodynamics. Its models are very general: they include almost all types of deformations, changes in phase and other physical characteristics, concern traditional or new materials with or without microstructures, in ordinary or unusual situations, in their interactions etc.

Thermodynamic systems range from mechanical to economical. Several schools dealt with the foundations of thermodynamics, from various points of view. For our proposes the most suitable turned out to be Ingo Müller's school: its formalism is quite simple, the physical bases are rigorous and profound, its models are sufficiently general [Müll]. It is in this spirit that in the following we sketch the main steps of the modelling and models in thermodynamics. More exactly, we point out the simplifying assumptions made in the general framework. If one model is studied by us only in a small section, then it will be described in that section.

Geometrically, the fluid body is described by a domain $\Omega \subset \mathbf{R}^n$ of boundary $\partial\Omega$. Each point $\mathbf{x} \in \Omega$, $\mathbf{x} \equiv (x_1, x_2, x_3)$, the components of which are also denoted by x, y, z, is called a material point, being endowed with an inertial characteristic, mass density. With one exception, in our models we do not treat fluids with a micro-polar structure, therefore $\mathbf{x}$ is the single geometric characteristic of the fluid particle centered at it, so we identity the fluid particle with its position $\mathbf{x}$. We assume that during their motion the fluid particles occupy the same domain Ω, therefore the boundary $\partial\Omega$ does not vary in time. However, inside Ω the position $\mathbf{x}$ of a particle depends on the time t and a reference position $\mathbf{x}_0$ at the time $t = 0$. Then filaments, surfaces or subdomains Ω' of fluid particles undergo deformations, *i.e.* the fluid is deformed in its domain of motion Ω. We assume that Ω is a mathematical continuum, hence the fluids are conceived as deformable continua. We deal with those fluids the deformations of which are infinitesimal.

Dynamic (traditionally referred to as kinematic) basic characteristics are the velocity $\mathbf{u}$ and the acceleration $\frac{d\mathbf{u}}{dt}$.

The infinitesimal deformation characteristics are the Fréchet differentials (therefore they are linear in dt) of the displacements, deformations and velocity of deformations and are expressed in terms of $\mathbf{u}$ and the tensor $\nabla\mathbf{u}$. The symmetric part of $\nabla\mathbf{u}$ is referred to as the strain (stretching) tensor and it is denoted by $\mathbf{d}$, where $d_{ij} = \frac{1}{2}\left(\frac{\partial u_i}{\partial x_j} + \frac{\partial u_j}{\partial x_i}\right)$. The skew-symmetric part of $\nabla\mathbf{u}$ is $-\frac{1}{2}(\nabla \times \mathbf{u})\mathbf{R}$, where $\mathbf{R}$ is the Ricci tensor.

Physical (including mechanical) characteristics are fields, *i.e.* functions of time t and space $\mathbf{x}_0$ in the Lagrangian formalism, or functions of t and $\mathbf{x}$ in the Eulerian formalism. The first corresponds to pursuing a particle identified with its initial position $\mathbf{x}_0$ at various instants t. This formalism is most adapted to the mathematical description of the physical laws, in some numerical methods, and in problems where the pressure field is of primary interest. In mathematical studies on fluid flows the Euler formalism is prevailing. Except in this section, we adopt it throughout this book. The main physical quantities occurring in the models we use are: the velocity $\mathbf{u}$, pressure p, mass density ρ, specific body force $\mathbf{F}$, electric field $\mathbf{E}$, magnetic field (called also the potential of current) $\mathbf{H}$, concentration c, density of internal energy ϵ, temperature T, electric charge q, density of momentum $\rho\mathbf{u}$, density of momentum of momentum $\mathbf{x} \times \rho\mathbf{u}$, density of kinetic energy $1/2\rho\mathbf{u}^2$, r specific radiation, $\mathbf{P}$ polarization charge, magnetization current $\mathbf{M}$, total current in the surface $\mathbf{J}$, stress tensor $\mathbf{T}$, strain tensor $\mathbf{d}$, magnetic flux $\mathbf{B} = \mu_0\mathbf{H}$, potential charge called also the dielectric displacement $\mathbf{D} = \epsilon_0\mathbf{E}$. The perturbation velocity, concentration, magnetic field and pressure are denoted by $\mathbf{v}$, θ, γ, $\mathbf{h}$ and p' respectively. The same basic fields are denoted by $\mathbf{U}$ or $\overline{\mathbf{u}}$, T^0, C_0, $\mathbf{H}_0$ and $\overline{p}$ or $\overline{P}$.

Material and other physical characteristics of different kinds of fluids are described by physical parameters like: the coefficient μ of dynamical viscosity, coefficient ν of kinematic viscosity, coefficient μ_b of bulk viscosity, coefficient of thermal conductivity χ, constant of magnetic permeability μ_0, dielectric constant ϵ_0, charge density ρ_e, Hall coefficient β_H, ion-slip coefficient β_I, (β_H and β_I are defined at the end of Section 1.3) barotropic compressibility k_η, isothermal compressibility k_T, specific heat at constant volume c_v, specific heat at constant pressure c_p, coefficient of thermal expansion α, coefficient of concentration expansion β, $\beta_1 = \mid \nabla T \mid$, $\beta_2 = \mid \nabla C \mid$, coefficient of thermoconvection λ, coefficient of thermodiffusive conduction N, coefficient of salt diffusivity (convection) k_c, intensity of gravity acceleration g, σ_e is the coefficient of the electrical conductivity. In the general case, some of these characteristics are not scalars but tensors, *e.g.* χ and σ_e.

In the following we derive the equations satisfied by the basic physical fields. In them the physical parameters occur as coefficients of terms representing various physical effects. This dimensional form of equations is not adequate to a mathematical study, for instance for asymptotic approximation and convergence reasons. This is why we used the nondimensional forms of the governing equations, containing nondimensional parameters. Depending on the characteristic quantities used in the non-dimensionalization, several forms for the nondimensional equations and parameters are obtained. Here are the nondimensional parameters used in our models of interest

$$R_e = \frac{U_\infty L}{\nu}, \text{ the Reynolds number,}$$

$$\mathcal{R}_a = \mathcal{R}^2 = \frac{g\alpha\beta_1 d^4}{\nu k_T}, \text{ the Rayleigh number,}$$

$$\mathcal{C}^2 = \frac{g\beta\beta_2 d^4}{\nu k_c}, \text{ the concentration Rayleigh number,}$$

$$P_r = \nu/k_T, \text{ the Prandtl number,}$$

$$P_m = \nu/\eta, \text{ the magnetic Prandtl number,}$$

$$G_r = \mathcal{R}_a P_r^{-1} \text{ the Grashof number,}$$

$$S_c = \nu/k_c, \text{ the Schmidt number,}$$

$$r_D = k_c/k_T = P_r/S_c = L_e, \text{ Lewis number,}$$

$$Q^2 = \frac{\mu_e H_0^2 d^2}{4\pi\rho\nu\eta}, \text{ the Chandrasekhar number,}$$

$$M^2 = 4\pi Q^2, \text{ the Hartmann number,}$$

where L and d are characteristic lengths, and U_∞ is the characteristic (far-field) velocity, H_0 is the intensity of the (basic) magnetic field, $\eta = (4\pi\mu_0\sigma_e)^{-1}$ is the resistivity (or magnetic viscosity). Other physical characteristics are defined at the end of Section 1.3.

1.1.2 *Global quantities and their integral representation*

In order to form an *integral model* governing a fluid flow, first we must write physical balance laws describing the evolution of the global characteristics of the physical system.

Mathematically *a global quantity* is described by a function of t and a current domain $\Omega'(t) \subset \Omega$ consisting of the same fluid particles in their motion. In fluid mechanics $\Omega'(t)$ is improperly called a *material volume*. It is the analog of the rigid body in classical mechanics. The physical laws are characterizing each material volume and not each point. For instance, we cannot speak of the mass of a fluid point, but of the mass of a material volume $\Omega'(t)$ of the same fluid particles. Denote by $m(t, \Omega'(t))$ the mass of this volume at time t. Hence the mass is a positive-valued function of a real variable and of a set $\Omega'(t)$.

Other global quantities are: the momentum of $\Omega'(t)$, the momentum of momentum of $\Omega'(t)$, the angular momentum of $\Omega'(t)$, the energy of $\Omega'(t)$, the entropy of $\Omega'(t)$ etc. Let us denote them by $G(t, \Omega'(t))$.

Then the most general balance law for G reads

$$\frac{dG(t, \Omega'(t))}{dt} = -{}^G\Phi + {}^G P + {}^G S, \tag{1.1.1}$$

where $-{}^G\Phi$ is the flux through the boundary $\partial\Omega'(t)$, ${}^G P$ is the production of G in $\Omega'(t)$ and ${}^G S$ is the supply of G from $\Omega \setminus \Omega'(t)$. If the production vanishes, then we say that (1.1.1) is a *conservation law*. We write as many balance laws as independent global quantities which characterize the modeled phenomenon. For a mechanical phenomenon we consider the balance of mass, momentum and momentum of momentum. For a thermodynamical phenomenon we take in addition the balance for energy and entropy. For a magneto-thermomechanical phenomenon we consider also the balance of the magnetic field etc.

The law (1.1.1) is very general: it expresses the balance of the rate of change of G and the flux, production and supply. In each specific case these three quantities must be specified for each G. For instance, if G is the momentum, the $-{}^G\Phi$ are the stresses through $\partial\Omega'(t)$, ${}^G P$ are the sources of momentum and ${}^G S$ the body forces.

Let us now introduce the local quantities also called *densities* of the global quantities. As mathematical objects they are fields and, at least at a first glance, they describe quite other physical quantities than their global correspondents. For instance, the mass reads $m(t, \Omega'(t)) = \int_{\Omega'(t)} \rho(t, \mathbf{x}) d\Omega'(t)$, where ρ is the mass density, or, simply, density. In fact, the two descriptions by m or by ρ, of the inertia property are equivalent. Indeed, we can choose $\Omega'(t)$ as close to every $\mathbf{x} \in \Omega$ as we wish. In addition, $\Omega'(t)$ runs over Ω and so does $\mathbf{x}$. Another example: the global momentum $\mathcal{M}$ is expressed in terms of the density of momentum $\rho\mathbf{u}$ in the form $\mathcal{M}(t, \Omega'(t)) = \int_{\Omega'(t)} \rho(t, \mathbf{x})\mathbf{u}(t, \mathbf{x}) d\Omega'(t)$. The integrals over $\Omega'(t)$ are in the sense of Lebesgue. The representation of the mass as an (infinity of) integral(s) follows by the Radon-Nicodym theorem, from the physical properties of the mass (*e.g.* additivity of $\Omega'(t)$), which generated the important mathematical concept of measure.

For the sake of simplicity, we refer to ρ as density, to $\rho\mathbf{u}$ as momentum etc., suppressing the word density in everywhere but in the (mass) density.

A special interest is represented by the global quantity called the stress $\mathcal{T}$. It is the flux of momentum through the surface $\partial\Omega'(t)$. Its integral representation does not follow from measure-theoretical considerations, but from the Cauchy hypothesis that $\mathcal{T}$ depends on a single geometric characteristic of $\partial\Omega'(t)$, namely its outer normal $\mathbf{n}$. Thus $\mathcal{T} = \int_{\partial\Omega'(t)} \mathbf{T} \cdot \mathbf{n} d\sigma(t)$, where $\mathbf{T}$ is the stress tensor. Therefore we deal with a special type of deformable continuum, namely a *Cauchy medium*.

The global body forces read: $\mathcal{F}(t, \Omega'(t)) = \int_{\Omega'(t)} \rho(t, \mathbf{x})\mathbf{F}(t, \mathbf{x}) d\Omega'(t)$, where $\mathbf{F}$ is referred to as the effort or specific force.

In the following, for the sake of simplicity, we omit $(t, \mathbf{x})$ but we recall that the integrands are functions of t and $\mathbf{x}$.

The most complex global quantity is the total internal energy consisting of a sum of two other global energies: internal $\int_{\Omega'(t)} \rho e d\Omega'(t)$, and kinetic $\int_{\Omega'(t)} \rho \frac{\mathbf{u}^2}{2} d\Omega'(t)$.

1.1.3 *Balance equations in integral form*

The substitution of the integral representations of the global quantities in the balance law (1.1.1) leads to the *integral form of the balance laws*. For instance the balance law for momentum is a conservation law and it reads

$$\frac{d}{dt}\int_{\Omega'(t)}\rho\mathbf{u}d\Omega'(t) = \int_{\partial\Omega'(t)}\mathbf{T}\cdot\mathbf{n}d\sigma'(t) + \int_{\Omega'(t)}\rho\mathbf{F}d\Omega'(t), \quad \forall\Omega'(t) \subset \Omega \quad (1.1.2)$$

where the first term in the right-hand side is the momentum flux and the other, the momentum supply. In it the stress tensor is symmetric. Indeed, if into the momentum of momentum balance equation the momentum balance law is taken into account, it follows that the stress tensor is symmetric, *i.e.* $T_{ij} = T_{ji}$.

Since the conservation laws are particular balance laws, in general they are also referred to as balance laws.

In a balance law the significance of the flux, production and supply for some global quantity differs in different situations. For instance, if the material is introduced in an electromagnetic field (and is understood to be influenced by it), then, instead of (1.1.2), we have the so-called mechanical momentum balance law

$$\frac{d}{dt}\int_{\Omega'(t)}\rho\mathbf{u}d\Omega'(t) = \int_{\partial\Omega'(t)}\mathbf{T}\cdot\mathbf{n}d\sigma'(t) + \int_{\Omega'(t)}[q\mathbf{E} + (\mathbf{J}\times\mathbf{B})]d\Omega'(t)$$
$$+ \int_{\Omega'(t)}\rho\mathbf{F}d\Omega'(t) \tag{1.1.3}$$

where the second integral in the right-hand side is the momentum production, therefore this momentum balance law is not a conservation law any longer. Similarly, if the material is not electrically conducting and not magnetizable, the balance law for the total internal energy is a conservation law, *i.e.*

$$\frac{d}{dt}\int_{\Omega'(t)}\rho(\epsilon + \mathbf{u}^2/2)d\Omega'(t) = -\int_{\partial\Omega'(t)}[\mathbf{q} - \mathbf{T}\cdot\mathbf{u}]\cdot\mathbf{n}d\sigma'(t)$$
$$+ \int_{\Omega'(t)}\rho\mathbf{F}\cdot\mathbf{u}d\Omega'(t) + \int_{\Omega'(t)}\rho r d\Omega'(t) \tag{1.1.4}$$

where the first term in the right-hand side is the flux (heat flux minus the power of stress) and the last is the supply due to radiation. We assume that $r = 0$.

If the material is introduced in an electromagnetic field, then, instead of (1.1.4) we have the balance equation of the mechanical energy

$$\frac{d}{dt}\int_{\Omega'(t)}\rho(\epsilon + \mathbf{u}^2/2)d\Omega'(t) = -\int_{\partial\Omega'(t)}(\mathbf{q} - \mathbf{T}\cdot\mathbf{u})\cdot\mathbf{n}d\sigma'(t)$$
$$+ \int_{\Omega'(t)}(\mathbf{J}\cdot\mathbf{E})d\Omega'(t), \tag{1.1.5}$$

where the last term in the right-hand side is the production of mechanical energy, hence (1.1.5) is no longer a conservation law. (The mechanical work was not taken into account.)

However, we can define two other types of momentum and energy the balance laws of which are conservative laws. They read $\int_{\Omega'(t)}(\rho\mathbf{u} + \mathbf{D}\times\mathbf{B})d\Omega'(t)$ and

$$\int_{\Omega'(t)}[\rho(\epsilon + \mathbf{u}^2/2) + (\mathbf{E}\cdot\mathbf{D} + \mathbf{B}\cdot\mathbf{H})/2]d\Omega'(t)$$

and are called the momentum of matter and field and the energy of matter and field respectively, where the concept field has a physical sense as opposed to matter. After some processing, the conservative laws for these global functions read

$$\int_{\Omega'(t)} \left\{ \frac{\partial}{\partial t}(\rho\mathbf{u} + \mathbf{D} \times \mathbf{B}) + \operatorname{div}\left[\rho\mathbf{u} \otimes \mathbf{u} - \mathbf{T} + \frac{1}{2}(\mathbf{E} \cdot \mathbf{D} + \mathbf{B} \cdot \mathbf{H})\mathbf{I} \right.\right. \tag{1.1.6}$$
$$\left.\left. - \mathbf{E} \otimes \mathbf{D} - \mathbf{B} \otimes \mathbf{H}\right] \right\} d\Omega'(t) = 0,$$

$$\int_{\Omega'(t)} \left\{ \frac{\partial}{\partial t}\left[\rho(\epsilon + \mathbf{u}^2/2) + (\mathbf{E} \cdot \mathbf{D} + \mathbf{B} \cdot \mathbf{H})/2\right] \right. \tag{1.1.7}$$
$$\left. + \operatorname{div}\left[\rho(\epsilon + \mathbf{u}^2/2)\mathbf{u} + \mathbf{q} - \mathbf{T} \cdot \mathbf{u} + \mathbf{E} \times \mathbf{H}\right] \right\} d\Omega'(t) = 0,$$

where $\mathbf{I}$ is the identity tensor and $\mathbf{F}$ was considered as absent.

The equalities (1.1.6) and (1.1.7) are not written in the form (1.1.1) because in this case one of the global quantities, *i.e.* the charge q, is expressed as a sum of a volume integral and a surface integral. As we keep this general presentation at the simplest possible form we go on assuming that any global quantity is expressed by a volume integral. However, we remark that, in a most general theory, a global quantity is defined as a sum of volume, surface and line integrals, *e.g.* for elastic shells and roods insulated in a thermally and electrically conducting and magnetizable fluid.

In order to deduce the differential form of the balance laws from their integral form, we apply three theorems: transport theorem [Ser2], flux-divergence theorem (or its scalar variant) and the continuity theorem. The so-called *transport theorem* is an identity expressing the derivative with respect to a parameter of a volume integral depending on a parameter (in our case t). It generalizes the Leibniz-Newton formula

$$\frac{d}{dt}\int_{a(t)}^{b(t)} h(t,s)ds = b'(t)h(t,b(t)) - a'(t)h(t,a(t)) + \int_{a(t)}^{b(t)} \frac{\partial}{\partial t}h(t,s)ds, \tag{1.1.8}$$

which is a transport theorem in one-dimensional case, to the three-dimensional or two-dimensional case. In the three-dimensional case it reads

$$\frac{d}{dt}\int_{\Omega'(t)} \mathbf{g}d\Omega'(t) = \int_{\Omega'(t)} \frac{\partial}{\partial t}\mathbf{g}d\Omega'(t) + \int_{\partial\Omega'(t)} \mathbf{g} \otimes \mathbf{u} \cdot \mathbf{n}d\sigma'(t). \tag{1.1.9}$$

Introducing (1.1.9) in (1.1.2) we obtain an equality containing some integrals over $\Omega'(t)$ and some others over $\partial\Omega'(t)$. In order to write the surface integrals over $\partial\Omega'(t)$ as volume integrals over $\Omega'(t)$ we use the flux-divergence theorem, which is the following identity for a vector quantity

$$\int_{\partial\Omega'(t)} \mathbf{v} \cdot \mathbf{n}d\sigma'(t) = \int_{\Omega'(t)} \operatorname{div}\mathbf{v}d\Omega'(t), \tag{1.1.10}$$

or its corresponding formula for the case of scalar or tensor quantities.

Taking into account (1.1.9) and (1.1.10) in (1.1.2) we obtain

$$\int_{\Omega'(t)} \left[\frac{\partial}{\partial t}(\rho\mathbf{u}) + \operatorname{div}(\rho\mathbf{u} \otimes \mathbf{u}) - \operatorname{div}\mathbf{T} - \rho\mathbf{F}\right] d\Omega'(t) = 0. \quad \forall\Omega'(t) \subset \Omega \tag{1.1.11}$$

1.1.4   *Balance equations in differential form*

Roughly speaking, the continuity theorem in the one-dimensional case states: *given a continuous function $f : [a,b] \to \mathbf{R}$ such that $\int_{a'}^{b'} f(x)dx = 0$ for every subinterval $[a',b'] \subset [a,b]$, then $f(x) \equiv 0$ on $[a,b]$.*

Applying to (1.1.11) an analogous *continuity theorem* for the tree-dimensional case we obtain the *local* or *differential form* of (1.1.2)

$$\frac{\partial}{\partial t}(\rho\mathbf{u})(t,\mathbf{x}) + \operatorname{div}(\rho\mathbf{u}\otimes\mathbf{u})(t,\mathbf{x}) - \operatorname{div}\mathbf{T}(t,\mathbf{x}) - \rho(t,\mathbf{x})\mathbf{F}(t,\mathbf{x}) = 0, \qquad (1.1.12)$$

for all $(t,\mathbf{x}) \in \Omega_{t^*} = (0,t^*) \times \Omega$.

Similarly, if no mass sources, mass flux and mass supply exist, then the integral form of the law of mass conservation reads

$$\int_{\Omega'(t)} \Big(\frac{\partial}{\partial t}\rho + \operatorname{div}(\rho\mathbf{u})\Big)d\Omega'(t) = 0, \quad \forall\Omega'(t) \subset \Omega,$$

and its corresponding differential form is

$$\frac{\partial}{\partial t}\rho + \operatorname{div}(\rho\mathbf{u}) = 0, \quad (t,\mathbf{x}) \in \Omega_{t^*}. \qquad (1.1.13)$$

This equation is referred to as the *continuity equation.*

Taking into account (1.1.13) in (1.1.12) and ignoring the understood dependence on t and $\mathbf{x}$ we have the so-called Cauchy equations

$$\frac{\partial}{\partial t}\mathbf{u} + \mathbf{u}\cdot\operatorname{grad}\mathbf{u} - \frac{1}{\rho}\operatorname{div}\mathbf{T} - \mathbf{F} = \mathbf{0}. \qquad (1.1.14)$$

In general, for a global scalar quantity G of density g, the local form of the balance law reads

$$\frac{\partial}{\partial t}g + \operatorname{div}(g\mathbf{u}) + \operatorname{div}(^G\varphi) - {}^G\mathbf{p} - {}^G\mathbf{s} = \mathbf{0}, \qquad (1.1.15)$$

where $^G\Phi = \int_{\partial\Omega'(t)} {}^G\varphi\cdot\mathbf{n}d\sigma'(\mathbf{t})$ and ^Gp and ^Gs are the density of scalar production and supply of G respectively. Similarly, for a global vector quantity $\mathbf{G}$ of density $\mathbf{g}$ the local form of the balance law reads

$$\frac{\partial}{\partial t}\mathbf{g} + \operatorname{div}(\mathbf{g}\otimes\mathbf{u}) + \operatorname{div}(^G\varphi) - {}^G\mathbf{p} - {}^G\mathbf{s} = \mathbf{0}, \qquad (1.1.16)$$

where $^G\mathbf{\Phi} = \int_{\partial\Omega'(t)} {}^G\varphi\cdot\mathbf{n}d\sigma'(\mathbf{t})$ and $^G\mathbf{p}$ and $^G\mathbf{s}$ are the density of vector production and supply of G respectively.

If the global quantity is expressed as a sum of volume and surface integrals, *e.g.* the global electric charge is

$$\int_{\Omega'(t)} q(t,\mathbf{x})d\Omega'(t) = \int_{\Omega'(t)} q^F(t,\mathbf{x})d\Omega'(t) - \int_{\partial\Omega'(t)} \mathbf{P}\cdot\mathbf{n}d\sigma'(t),$$

then, in order to write the balance equations, we also use a transport and flux-divergence like formula for the functions of t and $\partial\Omega'(t)$. (For details, see [Müll].)

If thermal, electric, magnetic and other phenomena are present, in principle, in each balance law all physical quantities describing these phenomena occur. For instance, in fluids in electromagnetic fields, in the mechanical momentum balance law (1.1.3) the electric and magnetic fields occur, even if the rate of change was expressed only in terms of a mechanical quantity. This is a mathematical expression of the physical interaction of various effects, *e.g.* mechanical, electrical, magnetic. More precisely, (1.1.3) expresses the influence of the nonmechanical effects on the mechanical ones. This is why this balance equation reminds the custom to study an effect by adding other effects but not to relate them structurally. In exchange, equation (1.1.6) represents the balance law for a non purely mechanical quantity in terms of all occurring mechanical and nonmechanical effects. This shows that thermodynamics is not a simple juxtaposition of nonmechanical effects over the mechanical and/or thermal ones. This physical point of view, mathematically means: even if (1.1.3) is reframed into a thermodynamical model, by solving the coupled system of all balance equations (among them being some ones for the electric or magnetic fields) the mechanical quantities are influenced by the presence of non-mechanical ones.

It is understood that among all balance laws which can be written for a specific domain, only the independent one are retained.

1.1.5 *Constitutive equations. State equations*

The examination of the balance equations $(1.1.13) - (1.1.16)$, $(1.1.3)$, $(1.1.7)$, shows that the unknown functions are g, $\mathbf{g}$ and their fluxes, productions and supplies. Therefore the number of unknowns overpasses the number of equations. It is expected that some of these unknowns be basic quantities, the others being expressible in terms of them. Sometimes the experiments suggest which quantities are the basic ones, however the choice of the basic quantities is permanent in view of thermodynamicists. We mention some of the quantities actually accepted as basic: ρ, $\mathbf{u}$, T, $\mathbf{E}$, $\mathbf{H}$, c. Some among the remaining functions (usually productions and supplies) are easier expressible in terms of basic quantities. Special questions rises the problem of expressing fluxes $\mathbf{T}, \mathbf{q}, \mathbf{j}$; ϵ, q; and some other quantities, *e.g.* $\mathbf{P}, \mathbf{M}$, in terms of basic quantities. This is the mathematical point of view.

From physical point of view, the above-written balance equations are valid for general Cauchy continua, *e.g.* fluids, elastica, plastica, fluids in electromagnetic fields, fluid mixtures etc. Whence the necessity to individualize them to specific continua, *e.g.* fluids.

To this aim, the balance equations must be supplemented with relationships between the complex quantities, *e.g.* fluxes; ϵ, $\mathbf{q}$ on one hand and the basic quantities on the other. These relationships are referred to as *constitutive equations*. Each such set of equations define, mathematically, a specific material. Their number and form differ from material to material. Once the basic functions are chosen and the production and supplies are written in terms of them, some quantities, called the *constitutive functions*, remain in the balance equations. The constitutive equations yield expressions of the constitutive functions in terms of the basic functions and/or some of their derivatives. Hence, for different materials the number of constitutive functions and the basic functions on which they depend is different.

For instance, the linear heat-conducting viscous fluid flows are governed by the balance laws for mass (1.1.12), momentum (1.1.14) and energy (1.1.4), (the local form of which in view of (1.1.16) is immediate). There are two possibilities to choose basic quantities: $\rho, \mathbf{u}, T$ or $p, \mathbf{u}, T$. In both cases we have five unknown scalar functions and five scalar equations.

If the second possibility is chosen, then the constitutive functions are ρ, $\mathbf{T} + p\mathbf{I}$, ϵ, $\mathbf{q}$, η (the specific entropy), and $^{\eta}\boldsymbol{\Phi}$ (the entropy flux density), where $p = \frac{1}{3}\mathrm{tr}\mathbf{T}$ in the thermodynamic equilibrium state, *i.e.* $\mathbf{u} = 0$. Indeed, all other functions occurring in the balance laws are the basic functions and their derivatives. The constitutive equations for linear viscous heat conducting fluids read

$$\mathbf{T} + p\mathbf{I} = \mu_b(p, T)(\mathrm{tr}\mathbf{d})\mathbf{I} + 2\mu(p, T)\mathbf{d}, \tag{1.1.17}$$

$$\mathbf{q} = -\chi(p, T)\nabla T, \tag{1.1.18}$$

$$\epsilon = \epsilon(p, T), \tag{1.1.19}$$

$$\rho = \rho(p, T), \tag{1.1.20}$$

$$\eta = \eta(p, T), \tag{1.1.21}$$

$$^{\eta}\varphi = \mathbf{q}/\mathbf{T}, \tag{1.1.22}$$

where $\mathbf{I}$ stands for the unit tensor.

In fact, we should write the balance law for entropy but, since the entropy and its flux, supply and production do not occur explicitly in the balance equations for other quantities, we no longer write the entropy equation and the constitutive equations for η and $^{\eta}\varphi$.

The first and simplest constitutive equation for viscous (non heat-conducting) fluids is a particular case of (1.1.17) and it was deduced by Newton more than three centuries ago. It relates the shear stress T_{12} by the strain d_{12} the proportionality constant being μ. Later Stokes formulated a set of postulates from which a particular case of (1.1.17) followed. Actually, the constitutive equations are deduced by thermodynamic constitutive theory on the basis of three principles: material frame indifference principle, entropy principle and thermodynamic stability principle. Sometimes isotropy and incompressibility properties are assumed too. It is in this way that the second principle of thermodynamics, *i.e.* the entropy balance law contributes to fluid flows models. Indeed, if only the material frame indifference was imposed, the constitutive functions would have been

$$T_{ij} = -\pi\delta_{ij} + 2\mu d_{ij} + c(d^2)_{ij} + eT_{,i}T_{,j} + fT_{,k}d_{k\,(i}T_{,j)} + gT_{,k}(d^2)_{k\,(i}T_{,j)}, \tag{1.1.23}$$

$$q_i = -kT_{,i} + ad_{ij}T_{,j} + b(d^2)_{ij}T_{,j}, \qquad (1.1.24)$$

$$\epsilon = \epsilon(\rho, T, \mathrm{div}\mathbf{d}, (\mathbf{d})^2_{ii}, (\mathbf{d})^3_{ii}, T_{,i}T_{,i}, T_{,i}d_{ij}T_{,j}, T_{,i}(d^2)_{ij}T_{,j}), \qquad (1.1.25)$$

where the coefficients a, b, c, e, f, g and k depend on all functions on which depends ϵ in (1.1.25) and $T_{,i} = \frac{\partial T}{\partial x_i}$, while the brackets to indices stand for the symmetric part of the tensor. The constitutive equations $(1.1.23) - (1.1.25)$ define the most general *viscous heat-conducting fluids*.

When the gradients of the velocity and temperature are small, they are usually neglected, so $(1.1.23) - (1.1.25)$ become

$$\mathbf{T} = -p_{|E}(\rho, T)\mathbf{I} + \mu_b(\rho, T)(\mathrm{tr}\mathbf{d})\mathbf{I} + 2\mu(\rho, T)\mathbf{d}, \qquad (1.1.17')$$

$$\mathbf{q} = -\chi(\rho, T)\nabla T, \qquad (1.1.18')$$

$$\epsilon = \epsilon_{|E}(\rho, T) + \lambda_v(\rho, T)\mathrm{div}\mathbf{u}, \qquad (1.1.19')$$

where $_{|E}$ indicates the equilibrium state (*i.e.* $\mathbf{u} = 0$) and λ_v is shown to be equal to zero. The fluids defined by the constitutive equations $(1.1.17') - (1.1.19')$ are called the *Navier-Stokes-Fourier* (N-S-F) *fluids*. We are concerned only with this type of fluids.

The relations $(1.1.17') - (1.1.19')$ are obtained by linearizing the constitutive equations $(1.1.23) - (1.1.25)$, therefore we deal with the motion of some linear materials. As they are supposed to undergo only linear deformations, it follows that we are interested in *geometrically and materially linear fluids*.

The constitutive equations $(1.1.17')-(1.1.19')$ are equivalent to $(1.1.17)-(1.1.19)$ but the first used $p, \mathbf{u}, T$ as basic fields while the last, the fields $\rho, \mathbf{u}, T$. In both formulations, at any time, the functions p, ρ, ϵ and η are equal to their values in the case of equilibrium. This is a result of the linearization. In $(1.1.17) - (1.1.22)$ it was not necessary to use the subscript $_{|E}$ because from $(1.1.17)$ it follows that p is the value corresponding to $\mathbf{u} = 0$. In this case μ_b and μ are not defined (since the internal friction acts only in moving fluids).

So far the dependence on p and T of the so-called *transport coefficients* μ_b, μ and χ as well as of the functions ϵ, ρ, and η in $(1.1.17) - (1.1.22)$ is unknown. The transport coefficients are supposed to be determined by experiments. Indeed, in applications-oriented papers, curves or tables of values describing the dependence on T of these coefficients are to be found. They also follow from a statistical mechanics viewpoint at a microscopic scale. Throughout this book all constitutive coefficients are supposed to be constant and are referred to as the *material constants*. As far as ϵ, η, and ρ are considered, further study and assumptions must be done. One is the *incompressibility*, which means that all the quoted functions do not depend on p. Usually, in fluid dynamics, two other meanings of this concept are used. Namely, by an incompressible fluid it is understood as an isochoric fluid, *i.e.* a fluid whose deformation leaves unchanged the volumes, consequently, a fluid undergoing neither thermal expansion nor contraction. In fact, for particular fluids, this is a consequence of the Gibbs relations. Supposing, in addition, that ρ does not depend on time, *i.e.* that ρ is constant. This third definition is the most popular. Correspondingly, the constitutive equations $(1.1.17) - (1.1.22)$ become

$$\mathbf{T} + p\mathbf{I} = 2\mu(T)\mathbf{d}, \tag{1.1.26}$$

$$\mathbf{q} = -\chi(T)\nabla T, \tag{1.1.27}$$

$$\epsilon = \epsilon(T), \tag{1.1.28}$$

$$\rho = \text{const} = \rho_0, \tag{1.1.29}$$

$$\eta = \eta(T), \tag{1.1.30}$$

$$^{\eta}\varphi = \mathbf{q}/\mathbf{T}, \tag{1.1.31}$$

where

$$\frac{d\eta}{dT} = \frac{1}{T}\frac{d\epsilon}{dT}. \tag{1.1.32}$$

The fluids possessing the constitutive equations $(1.1.26) - (1.1.31)$ are referred to as *incompressible* N-S-F *fluids*. In fact, by an appropriate formulation of some consequence of the entropy principle, it follows that if ρ does not depend on p, then it does not depend on T either, hence it is necessary to suppose only that ρ does not depend on p and T. This conclusion is not expected to hold for materially nonlinear media.

In addition, assume that μ_b, μ and k are constants. Then $(1.1.26)$ becomes the Stokes law, $(1.1.27)$ becomes the Fourier law and, by rewriting ρ instead of ρ_0 the balance equations for mass and momentum reduce to the equations

$$\text{div}\mathbf{u} = 0 \tag{1.1.33}$$

$$\frac{\partial}{\partial t}\mathbf{u} + \mathbf{u} \cdot \nabla\mathbf{u} = -\frac{1}{\rho}\nabla p + \nu\Delta\mathbf{u} + \mathbf{F}, \tag{1.1.34}$$

which do not depend any longer on T, therefore they are decoupled by the energy equation. Here the pressure is considered a mechanical quantity.

The fluid with the relations $(1.1.26)$, $(1.1.27)$ $(1.1.29)$ where the coefficients μ_b, μ and χ are constants is called an *incompressible* N-S *fluid*.

The balance equations corresponding to a specific fluid bear the same name as the fluid itself. For instance, $(1.1.33)$, $(1.1.34)$ are referred to as *Navier-Stokes equations*. Sometimes this name is assigned only to $(1.1.34)$. Historically these were the first realistic equations in hydrodynamics. As they involve a single constitutive equation, usually in applied hydrodynamics only the constitutive equations for $\mathbf{T}$ is considered.

So far no restriction on state equations and no form of the transport coefficients and the functions $\epsilon(p, T)$, $\rho(p, T)$, $\eta(p, T)$, or $\epsilon(\rho, T)$, $p(\rho, T)$, $\eta(\rho, T)$, was imposed.

These follow from the entropy principle and thermodynamic stability principle. The obtained equations for ϵ, p and η, are called the *equation of classical thermodynamic state, equation of the energy state,* and *equation of the entropic state* respectively.

The entropy principle yields the restrictions

$$\chi \geq 0, \quad \mu \geq 0, \quad \mu_d + \frac{2}{3}\mu \geq 0 \tag{1.1.35}$$

and the Gibbs equation

$$d\eta_{|E} = \frac{1}{T}\big(d\epsilon_{|E} + p_{|E}d(\frac{1}{\rho})\big). \tag{1.1.36}$$

This is of a principal great importance: it enables one to calculate $p_{|E}$, $\epsilon_{|E}$ and $\eta_{|E}$ from experiments. In particular, (1.1.32) follows from (1.1.36) in the case $\rho = \text{const.}$

Particular cases of the equation of the classical thermodynamic state are: (1.1.29); the Gay-Lussac equation, Boyle-Mariotte equation, the van der Waals equation

$$p_{|E}(\rho, t) = \frac{(R/M)T}{(1/\rho) - b} - a\rho^2,$$

where a, b, M are some constants and R is the universal gas constant; the law of perfect gas $p_{|E}(\rho, t) = (R/M)\rho T$ etc. They were deduced experimentally.

In addition, the thermodynamic stability principle implies

$$c_v \equiv \Big(\frac{\partial \epsilon}{\partial T}\Big)_v \geq 0, \;\; k_T \equiv -\frac{1}{v}\Big(\frac{\partial v}{\partial T}\Big)_T \geq 0, \;\; c_p = c_v + vTk_T\Big(\frac{\partial p}{\partial T}\Big)_v^2,$$

$$k_\eta = \frac{c_v}{c_P}k_T, \;\; c_p \geq c_v, \;\; k_\eta \leq k_T, \;\; \alpha^2 \leq \frac{1}{vT}c_p k_T, \tag{1.1.37}$$

where $v = \frac{1}{\rho}$ is the specific material volume.

If, apart from being viscous and heat-conducting, the fluid has additional properties, *e.g.* it is electrically conducting, then, the number of balance equations and constitutive equations increases, the constitutive functions depend on more terms, basic variables and their derivatives, and the constitutive coefficients depend not only on p and T, but also on some other basic functions.

Let us illustrate this for fluids in electromagnetic fields. In some particular situations the basic quantities are ρ, $\mathbf{u}$, T, $\mathbf{E}$, $\mathbf{H}$ and they satisfy the following balance equations of mass (1.1.13), momentum, internal energy, magnetic flux and charge

$$\frac{\partial(\rho\mathbf{u})}{\partial t} + \text{div}(\rho\mathbf{u} \otimes \mathbf{u} - \mathbf{T}) = q\mathcal{E} + \mathcal{J} \times \mathbf{B}, \tag{1.1.38}$$

$$\frac{\partial(\rho\epsilon)}{\partial t} + \text{div}(\rho\epsilon\mathbf{u} + \mathbf{q}) = \mathbf{T} \cdot \nabla\mathbf{u} + \mathcal{J} \cdot \mathcal{E}, \tag{1.1.39}$$

$$\frac{\partial\mathbf{B}}{\partial t} + \nabla \times \mathbf{E} = 0, \quad \text{div}\mathbf{B} = 0, \tag{1.1.40}$$

$$-\frac{\partial \mathbf{D}}{\partial t} + \nabla \times \mathbf{H} = \mathcal{J} + \mathbf{q}\mathbf{u}, \quad \mathrm{div}\mathbf{D} = q, \tag{1.1.41}$$

respectively, where $q = -\mathrm{div}\mathbf{P}$, $\mathcal{J} = \frac{\partial \mathbf{P}}{\partial \mathbf{t}} + \nabla \times \left(\mathbf{P} \times \mathbf{u} + \mathbf{M}\right) + (\mathrm{div}\mathbf{P})\mathbf{u}$, $\mathbf{D} = \epsilon_0 \mathbf{E}$, $\mathbf{H} = \mathbf{B}/\mu_0$, $\mathcal{E} = \mathbf{E} + \mathbf{u} \times \mathbf{B}$, $\mathcal{J}$ is the charge flux (or non-convective current), $\mathbf{J} = \mathcal{J} + \mathbf{q}\mathbf{u}$ is the total current. Equation (1.1.38) is an equivalent local form of (1.1.3) and (1.1.39), of (1.1.5). The constitutive functions are $\mathbf{T}$, q, ϵ, $\mathbf{P}$, $\mathbf{M}$ and, after using the material frame indifference principle and entropy principle, the very intricate constitutive equations for *fluids in electromagnetic fields* are obtained, consisting of the pure mechanical part and an additional electromagnetic part. The electromagnetic part in the stress tensor is a sum of a fourth order polynomial in $\mathcal{E}$, $\mathcal{B}$ and/or their products, and derivatives with respect to ρ of integrals with respect to $\mathcal{E}^2$, $\mathcal{B}^2$ and $(\mathcal{B} \cdot \mathcal{E})^2$. The integrands are sums of $\frac{1}{\rho}$ and $\frac{1}{\rho}$ taken at $\mathcal{E}^2 = \mathbf{0}$, or $\mathcal{B}^2 = \mathbf{0}$, or $\mathcal{B}^2 = \mathbf{0}$ and $\mathcal{E}^2 = \mathbf{0}$. The electromagnetic part in ϵ is similar but here the derivative of the integrals with respect to ρ is absent, while certain terms in the integrands are products of T by derivatives with respect to t of $\frac{1}{\rho}$. Obviously, $\mathbf{P}$ and $\mathbf{M}$ contain only electric and magnetic fields and they are sums of two particular third order monomials in $\mathcal{E}$ and $\mathcal{B}$ [Müll]. The constitutive coefficients depend on ρ, T, $\mathcal{E}^2$, $\mathbf{B}^2$ and $(\mathbf{B}\cdot\mathcal{E})^2$, and $p(\rho, t)$ and $\epsilon(\rho, t)$ are those from the non-electromagnetic case. For $\mathbf{q}$ we have

$$\mathbf{q}_{|E} = -\mathcal{E} \times \mathbf{M}. \tag{1.1.42}$$

We omit to write the entropy in this case too. Hence the presence of the electromagnetic field manifests itself by additional constitutive functions, additional terms with respect to the constitutive functions from the non-electromagnetic case, and additional dependence of the constitutive functions and their constitutive coefficients on $\mathcal{E}^2$, $\mathbf{B}^2$ and $(\mathbf{B} \cdot \mathcal{E})^2$.

Thus the I. Müller scheme allows us to unitarily derive the models by taking into account the constitutive equations into the balance equations and associating with them initial and boundary conditions. Since these conditions differ from one case to another, here we mention only that the initial conditions are the connection of the fluid motion with its history, while the boundary conditions make the fluid in Ω free from the exterior of Ω.

From a mathematical point of view, of primary importance are the *nonlinear terms* in the balance and in the constitutive equations: what kinds of such terms exist and where they appear. Among the nonlinearities we quote: the affinity, (*e.g.* the presence of body forces, radiation and nonhomogeneous conditions), algebraic nonlinearities, *e.g.* powers or products of unknown functions; integral nonlinearities, *e.g.* integrals of products of unknown functions; differential nonlinearities. The most difficult mathematical treatment concerns the last ones and mainly those of the form $aD^{\alpha}b$, *i.e.* products of some unknown functions by their own, or other unknown functions derivatives with respect to $\mathbf{x}$. All constitutive equations in this section contain at most algebraic and /or integral nonlinearities. In these equations, apart from affinity and algebraic nonlinearities, nonlinearities of differential type exist. Some of them come from the rate of change of the global quantities, *e.g.* $\mathrm{div}(g\mathbf{u})$, $\mathrm{div}(\mathbf{g} \otimes \mathbf{u})$, where $g = \rho$, $g = \rho\epsilon$, (in Section 1.2 we have, in addition, $\rho\gamma$, where γ is the concentration) in the balance equations (1.1.13), (1.1.39) and $\mathbf{g} = \rho\mathbf{u}$ in (1.1.38), (1.1.41). (In (1.1.41) the form is somewhat different due to the special type of the global quantity q.) These terms containing the products of the velocity $\mathbf{u}$ and the derivatives of (possibly other) unknown functions are referred to as *advective terms*. Physicists call them convective terms expressing the convection (transport) of a quantity g or $\mathbf{g}$ along the physical trajectories with velocity $\mathbf{u}$. In stability theory, by convection we mean a special type of fluid motion (Section 1.2.2). In addition, in the electromagnetic case, in (1.1.38) and (1.1.39) products of the time derivative of an unknown function by other unknown function occur.

The involved (linear) operators are $\nabla \cdot$ (div), ∇(grad), $\nabla \times$ (rot, curl), Δ(Laplacian) (Appendix 7). By $\otimes$ we mean the tensorial product, the dot between vectors or tensors stands for the scalar product of vectors, the contraction product of a vector by a tensor or a contraction product of two tensors. In the framework of stability theory by a *basic motion* or equilibrium it is understood the motion or the mechanical equilibrium whose stability is investigated.

1.2 Classical mathematical models in thermodynamics of fluids. Basic solutions

We present briefly the nondimensional form of the most important mathematical models used throughout the book.

1.2.1 *Incompressible Navier-Stokes model*

The basic unknown functions are p and $\mathbf{u}$; the balance equations are (1.1.33), (1.1.34); they incorporate the constitutive equations (1.1.26), (1.1.29), where $\mathrm{tr}\,\mathbf{d} = 0$, μ, χ are constants, and the initial and boundary conditions read

$$\lim_{t \to 0} \mathbf{u}(t, \mathbf{x}) = \mathbf{u}_0(\mathbf{x}), \quad \mathbf{x} \in \Omega \tag{1.2.1}$$

$$\lim_{\mathbf{x} \in \Omega \to \mathbf{x}^* \in \partial\Omega} \mathbf{u}(t, \mathbf{x}) = \mathbf{u}_w(\mathbf{x}^*), \quad \mathbf{x}^* \in \partial\Omega, \quad t \in (0, t^*), \tag{1.2.2}$$

where $\Omega_{t^*} = (0, t^*) \times \Omega$. The problem (1.2.1), (1.2.2) for equations (1.1.33), (1.1.34) in the class of functions $\mathbf{u} \in C^2(\Omega) \cap C^0(\overline{\Omega})$ for t fixed in $(0, t^*)$ and $\mathbf{u} \in C^1(0, t^*)$ for $\mathbf{x}$ fixed in Ω and $p \in C^1(\Omega)$ for t fixed in $(0, t^*)$ and $p \in C(0, t^*)$ for $\mathbf{x} \in \Omega$, such that every term of the equations is continuous and the boundary and initial conditions are taken in the sense of limit, is called the *classical incompressible* N-S *model*. All properties of the solutions depend on data: $\dim\Omega$, form of $\partial\Omega$, $\mathbf{u}_0$, $\mathbf{u}_w$, $\mathbf{F}$, t^*. In this model the sources of nonlinearity are: $\mathbf{u}_0$, $\mathbf{u}_w$, $\mathbf{F}$, and $\mathbf{u} \cdot \nabla\mathbf{u}$.

Let L, U_∞, L/U_∞, $\rho U_\infty^2/L$ be the characteristic length, velocity, time and pressure respectively. Then the nondimensional form of equation (1.1.33) is unchanged, while equation (1.1.34) becomes

$$\frac{\partial}{\partial t}\mathbf{u} + (\mathbf{u}\cdot\nabla)\mathbf{u} = -\nabla p + \frac{1}{R_e}\Delta\mathbf{u} + \mathbf{F}. \qquad (1.2.3)$$

The nondimensional quantities are denoted by the same symbols as the dimensional ones. In the book, for the stability purposes, as basic stationary solutions of the model (1.2.3), (1.1.33), (1.2.2) in a bounded domain Ω, we consider a general (unspecified) vector field $\overline{\mathbf{u}}(\mathbf{x})$, assumed to satisfy these equations and the boundary conditions. The pressure was supposed to be eliminated by applying to (1.2.3) the operator $\nabla\times$.

The incompressible N-S model is among the simplest in hydrodynamics. Its traces can be pursued in almost all models governing fluid flows of interest in applications. Its nonlinearity is induced by the presence of $\mathbf{u}_0$, $\mathbf{u}_w$, $\mathbf{F}$ and, mainly, of the advective term $(\mathbf{u}\cdot\nabla)\mathbf{u}$. In stability studies all, but two ($\mathbf{u}_w$, $\mathbf{F}$) sources of nonlinearity and difficulties are inherited by the perturbation models.

1.2.2   *Navier-Stokes-Fourier model in horizontal layers. Oberbeck-Boussinesq approximation*

The basic unknown functions are $\mathbf{u}$, p, T, the local balance equations of mass, momentum and energy are (1.1.13), (1.1.14) and the differential form corresponding to (1.1.4)

$$\frac{\partial}{\partial t}\big(\rho(\epsilon + \mathbf{u}^2/2)\big) + \mathrm{div}\big(\rho(\epsilon + \mathbf{u}^2/2)\mathbf{u}\big) + \mathrm{div}\mathbf{q} = \mathrm{div}(\mathbf{T}\cdot\mathbf{u}) + \rho\mathbf{F}\cdot\mathbf{u}, \qquad (1.2.4)$$

while the constitutive equations are (1.1.17), (1.1.18), (1.1.19) and (1.1.20). Introducing the constitutive equations in (1.1.12) and (1.2.4) and taking into account (1.1.13), the momentum and energy equations become

$$\rho\frac{\partial}{\partial t}\mathbf{u} + \rho(\mathbf{u}\cdot\nabla)\mathbf{u} = -\nabla p + \mu\Delta\mathbf{u} + \rho\mathbf{F}, \qquad (1.2.5)$$

$$\rho\frac{\partial}{\partial t}\epsilon - (\epsilon + \mathbf{u}^2/2)\mathbf{u}\nabla\rho - \rho\mathbf{u}(\mathbf{u}\cdot\nabla\mathbf{u}) + \mathbf{u}\nabla\big(\rho(\epsilon + \mathbf{u}^2/2)\big) - \nabla\cdot(\chi\nabla T)$$

$$= -p\nabla\cdot\mathbf{u} + \big[\mu_b(p, T)(\mathrm{tr}\mathbf{d})\mathbf{I} + 2\mu(p, T)\mathbf{d}\big]\cdot\mathbf{d}, \qquad (1.2.6)$$

where $\mu_b = -\frac{2}{3}\mu$ was chosen (as usual in hydrodynamics). The O-B approximation means to assume for equilibrium density (1.1.20) the expression

$$\rho(T) = \rho_0[1 - \alpha(T - T_0)], \qquad (1.2.7)$$

to take into account (1.2.7) only in the body forces term in the momentum equation (1.2.5) (in all other terms the density variations being neglected), and to ignore the conversion of work to heat in the energy equation. Therefore, the governing N-S-F equations in the O-B approximation are (1.1.33) and

$$\rho_0 \frac{\partial}{\partial t}\mathbf{u} + \rho_0 \mathbf{u} \cdot \nabla \mathbf{u} = -\nabla p + \mu \Delta \mathbf{u} + \rho_0[1 - \alpha(T - T_0)]\mathbf{F}, \qquad (1.2.8)$$

$$\rho \frac{\partial}{\partial t}\epsilon + \rho \mathbf{u} \cdot \nabla \epsilon - \nabla \cdot (\chi \nabla T) = 0. \qquad (1.2.9)$$

In this book χ is a constant, hence in (1.2.9) the diffusion term reads $\chi \Delta T$. The initial conditions read

$$\mathbf{u}(0, \mathbf{x}) = \mathbf{u}_0, \quad T(0, \mathbf{x}) = T_0. \qquad (1.2.10)$$

Consider as Ω the fluid layer situated between the horizontal planes $x = \pm 0.5$, i.e. $\Omega = \{(x, y, z) \in \mathbf{R}^3 \mid z \in (-0.5, 0.5)\}$ and $\partial\Omega = \{(x, y, z) \in \mathbf{R}^3 \mid z = \pm 0.5\}$. The boundary conditions differ according to the nature of the (rigid, impermeable or free) boundary $\partial\Omega$ and to whether there is a heat flux through $\partial\Omega$ or not. They are

$$\mathbf{u} \cdot \mathbf{n} = 0 \qquad \mathbf{n} \times \mathbf{d} \cdot \mathbf{n} = 0, \qquad \text{at} \qquad z = \pm 0.5, \qquad t \geq 0 \qquad (1.2.11)$$

where $\mathbf{n}$ is the outer normal to the layer boundary, if we consider stress-free boundaries.

In the case of the horizontal layer, $\mathbf{n} = \mathbf{k}$ for the upper boundary and $\mathbf{n} = -\mathbf{k}$ for the lower boundary. Therefore, (1.2.11) become

$$\mathbf{u} \cdot \mathbf{k} = 0, \qquad \frac{\partial^2}{\partial z^2}(\mathbf{u} \cdot \mathbf{k}) = 0, \qquad \text{at} \qquad z = \pm 0.5. \qquad (1.2.12)$$

Their derivation uses the solenoidality of $\mathbf{u}$ and can be found in [Chan]. In the case of rigid walls, we have

$$\mathbf{u} = \mathbf{0}, \qquad \text{at} \qquad z = \pm 0.5, \qquad t \geq 0. \qquad (1.2.13)$$

In the same way, in [Chan], taking into account (1.1.33), conditions (1.2.13) read

$$u_3 = \frac{\partial u_1}{\partial z} = \frac{\partial u_2}{\partial z} = 0, \qquad \text{at} \qquad z = \pm 0.5, \qquad t \geq 0. \qquad (1.2.13')$$

If one bounding plane is stress-free and the other rigid, the boundary conditions are changing accordingly. The flow periodicity in the y and z directions observed experimentally or in nature, imposed to use as domain of motion the so-called periodicity cell $V = \{(x, y, z) \in \mathbf{R}^3 \mid x \in (-0.5, 0.5), \, y \in (0, 2\pi/m), z \in (0, 2\pi/n)\}$. On the vertical surfaces of ∂V additional boundary conditions are unnecessary due to the periodicity conditions.

If the planes are rigid and perfectly thermally conducting, then the boundary conditions for the temperature are

$$T = T_0 = \text{const.} \quad \text{at } z = -0.5 \quad \text{and} \quad T = T_1 = \text{const.} \quad \text{at } z = 0.5. \qquad (1.2.14)$$

If some boundary is not thermally conducting, then, on it, we must have

$$(\nabla T) \cdot \mathbf{n} = 0. \qquad (1.2.15)$$

Except in Section 7.11 (where one bounding plane is perfectly thermally conducting and the other is not thermally conducting), throughout the book both bounding planes are perfectly thermally conducting.

In more complex cases, involving additional effects, *e.g.* porosity, chemical reactions, appropriate boundary conditions for T are to be assigned.

The model consisting of (1.2.8), (1.2.9), (1.1.33), (1.2.10) and the suitable boundary conditions of the form $(1.2.13) - (1.2.14)$ has two types of solutions: the mechanical equilibria, *i.e.* for which $\mathbf{u} = \mathbf{0}$, called *conduction states* and the solutions characterized by $\mathbf{u} \neq \mathbf{0}$, called *convections*. The convections are bifurcating from conduction states at those values of the parameters at which these last ones lose their stability due to some perturbations.

Convections treated in Chapter 4, by using the nonlinear equations, are supposed to occur in bounded domains, *e.g.* periodicity cells.

1.3 Classical mathematical models in thermodynamics of electromagnetic fluids

In electrically conducting and non-magnetizable fluids ($\mathbf{M} = \mathbf{0}$), when the free charge and currents are absent, the basic fields are ρ, $\mathbf{u}$, T, $\mathbf{E}$ and $\mathbf{B}$ and the constitutive equations must be written for ϵ, $\mathbf{T}$, $\mathbf{q}$, $\mathbf{P}$. Due to the fact that herein we mainly deal with convections for some particular such fluids, we write only those balance equations used by us and no reference on the approximations performed in the general models to deduce them is made. In addition, for the electromagnetic part we consider only those Maxwell equations which are Galileo (and not Lorentz) invariant. They are [SuS], [Chan],

$$\frac{d}{dt}\mathbf{u} = \frac{1}{\rho}\mathrm{div}\mathbf{T} + \mathbf{F} + \frac{\mu_e}{\rho}\mathbf{J} \times \mathbf{H}, \tag{1.3.1}$$

$$\frac{d}{dt}\rho + \rho\mathrm{div}\mathbf{u} = 0, \tag{1.3.2}$$

$$\rho c_v \frac{d}{dt}\epsilon = \Phi + \mathrm{div}(k_T \mathrm{grad} T), \tag{1.3.3}$$

$$\mathrm{div}\mathbf{E} = \rho_e/\epsilon_0, \tag{1.3.4}$$

$$\mathrm{div}\mathbf{H} = 0, \tag{1.3.5}$$

$$\nabla \times \mathbf{E} = -\mu_0 \frac{\partial}{\partial t}\mathbf{H}, \tag{1.3.6}$$

$$\nabla \times \mathbf{H} = \mathbf{J}, \tag{1.3.7}$$

$$\mathbf{j} = \sigma_e \mathbf{E} - \omega_e \tau_e \mathbf{j} \times \frac{\mathbf{H}}{H} + f^2 \omega_e \tau_e \omega_I \tau_{In} \left[\frac{\mathbf{H}}{H} \left(\frac{\mathbf{H}}{H} \cdot \mathbf{j} \right) - \mathbf{j} \right]. \tag{1.3.8}$$

Equations (1.3.1); (1.3.2); (1.3.3); (1.3.5) and (1.3.6); (1.3.4) and (1.3.7) represent balance equations for momentum, mass, internal energy, electric field, magnetic flux and charge respectively.

Equation (1.3.8) is the constitutive equation for the density current (generalized Ohm's law). The electrical conductivity σ_e is one of the most important rheological parameters in magnetohydrodynamics. If the usual Ohm's law is assumed to hold, then σ_e is a scalar quantity. If (1.3.8) is assumed, then the conductivity becomes a tensor [SuS]. In (1.3.8) σ_e, ω_e, τ_e, ω_I, τ_{In} and f are the scalar electrical conductivity, the Larmor frequency (*i.e.* cyclotron frequency of electrons), the mean electron collision time, the frequency of ions, the (average) time of collision of ions with neutral particles and the mass fraction of atoms which are not ionized, respectively.

Equation (1.3.8) represents the Ohm's law for a partially ionized fluid, *i.e.* when there are neutral particles. In the case of a fully ionized fluid, *i.e.* there are no neutral particles, $f = 0$, and equation (1.3.8) reduces to

$$\mathbf{j} = \sigma_e \mathbf{E} - \omega_e \tau_e \mathbf{j} \times \frac{\mathbf{H}}{H}, \tag{1.3.8'}$$

where the second term in the right-hand side is the Hall current, *i.e.* a transverse current depending on the electron trajectory. This occurs when the ratio $r = \lambda/r_1$ of the electron cyclotron frequency to the electron collision frequency is not negligible, where (only here) r_1 is the average Larmor radius and λ is the mean free path of electrons. It is r which requires the use of (1.3.8).

The last term in the right-hand side of (1.3.8), that is the ion-slip current, becomes important at "small values" of $\omega_e \tau_e$ [SuS] at which the electron-ion collisions dominate the electron motion. For moderate magnetic fields, the Hall current can be neglected, otherwise the tensorial nature of the electrical conductivity must be taken into account. In any case, if $\omega_e \tau_e \gg 1$, transverse conductivities and, therefore, tensorial electrical conductivity (which can be derived from (1.3.8) [Pa], [SuS]) must be considered. In the following we use the notation $\beta_H = \omega_e \tau_e$, $\beta_I = f^2 \omega_e \tau_e \omega_I \tau_{In}$.

The case $\beta_H \neq 0$, *i.e.* when the Hall effect is present, is extensively studied in literature due to its importance in atmospheric physics, astrophysics (especially in the upper layers of the solar atmosphere, due to the high magnetic field). Chapters 5 and 6 of this book are mostly devoted to it.

For other constitutive equations which contain mechanical as well as electromagnetic quantities we quote [Müll]. In our cases of interest, the electric field is eliminated between equations $(1.3.6) - (1.3.8)$.

Apart from the boundary conditions $(1.2.11) - (1.2.15)$ for mechanical (in the sense of non-electromagnetic) quantities $\mathbf{u}$ and T, we impose the following boundary conditions for the magnetic field [Chan]

$$\mathbf{H} = \mathbf{H}^{ex} \qquad \text{and} \qquad J_3 = 0, \tag{1.3.9}$$

$$\mathbf{H} \cdot \mathbf{n} = 0 \qquad \text{and} \qquad E_1 = E_2 = 0, \tag{1.3.10}$$

where $\mathbf{n}$ is the outer normal to the layer, $\mathbf{H}^{ex}$ is a vector field derivable from a potential, (1.3.9) are appropriate for the case of electrically non-conducting boundary, *i.e.* no current can cross the boundary, while, if the boundary is a perfect conductor, *i.e.* no magnetic field can cross the boundary, we must require (1.3.10).

1.4   Classical perturbation models

The basic mechanical equilibria and the corresponding linear and nonlinear perturbation models around the basic state functions are written. The particular cases occurring only in some smaller sections of the book are not presented here. All models are written in Cartesian coordinates only.

1.4.1   *Perturbation models*

A perturbation model is obtained by subtracting the equations and the boundary initial conditions from those of the perturbed motion. With one exception we study only stationary basic flows. The basic state functions are designed by a bar over the quantities or by the same letters as the perturbed quantities but by capital letters. Usually the perturbation quantities are designed by the same symbols as the perturbed quantities but with a prime. In the perturbation models the primes are omitted. The class of the perturbation and the basic state functions are assumed to be the same as for the perturbed flow.

Remark 1.4.1. As we are concerned with the perturbation of the initial conditions only, the boundary conditions for the perturbation quantities are homogeneous and the free terms (*i.e.* which do not depend on the state functions) no longer occur in the perturbed equations. This is why neither the perturbed models nor the model for the basic motion or state can be obtained from the perturbation model only.

1.4.2   *Perturbation incompressible Navier-Stokes model*

In the dimensional form this model reads

$$
\begin{cases}
\dfrac{\partial}{\partial t}\mathbf{v} + \mathbf{v}\cdot\nabla\overline{\mathbf{u}} + \overline{\mathbf{u}}\cdot\nabla\mathbf{v} + \mathbf{v}\cdot\nabla\mathbf{v} = -\dfrac{1}{\rho}\nabla p' + \nu\Delta\mathbf{v}, \\[2mm]
\nabla\cdot\mathbf{v} = 0, \\[2mm]
\mathbf{v}_{|\partial\Omega} = \mathbf{0}, \quad \mathbf{x}^{*}\in\partial\Omega,\ \mathbf{t}\in\mathbf{R}^{+} \\[2mm]
\mathbf{v}_{|t=0} = \mathbf{v}_0(\mathbf{x}), \quad \mathbf{x}\in\Omega,
\end{cases}
\tag{1.4.1}
$$

where the stationary basic state functions are characterized by the velocity $\overline{\mathbf{u}}$ and pressure $\bar{p}$, the perturbation velocity is $\mathbf{v}$ and the perturbation pressure p'. The equations $(1.4.1)_{1,2}$ are taken for $(t, \mathbf{x}) \in \mathbf{R}_+^* \times \Omega$. The basic state functions are not specified here because we are not concerned with its particular forms.

Remark 1.4.2. Unlike in the perturbed N-S model, in the N-S perturbation model the boundary conditions are homogeneous and the body forces no longer occur. This lowers the difficulty of its mathematical investigation.

1.4.3 *Perturbation model for viscous incompressible homogeneous thermoelectrically conducting or nonconducting fluid*

1.4.3.1 *Magnetic case*

In the framework of Cauchy continua and in the domain of validity of the O-B approximation, assume that the fluid is situated in a horizontal layer S bounded by the planes $z = 0$ and $z = d$, which are thermally conductors (characterized by a given temperature T_0 and T_1 respectively) and electrically non-conductors. The basic state is a thermodiffusive mechanical equilibrium m_0

$$m_0 \equiv \{\mathbf{U} = \mathbf{0}, \quad \mathbf{H} = H_0\mathbf{k}, \quad T = -\beta z + T_0, \quad p_0 = p_0(z)\}, \tag{1.4.2}$$

characterized by a constant vertical temperature gradient $-\beta \equiv (T_1 - T_0)/d < 0$ and a uniform vertical magnetic field $\mathbf{H}_0 \equiv H_0\mathbf{k}$ in an orthonormal reference frame $\{O, \mathbf{i}, \mathbf{j}, \mathbf{k}\}$, with $\mathbf{k}$ pointing upwards positive.

Then the nondimensional perturbation equations governing the evolution of the perturbation fields $\mathbf{v}(u, v, w), \mathbf{h}, \theta, p$ follow from $(1.3.1) - (1.3.8)$ and read [Chan], [Pa], [SuS]

$$\begin{cases} \dfrac{\partial}{\partial t}\mathbf{v} = -\mathbf{v} \cdot \nabla\mathbf{v} - \nabla p' + \Delta\mathbf{v} + M^2(\mathbf{H_0} + \mathbf{h}) \cdot \nabla\mathbf{h} + \dfrac{\mathcal{R}}{P_r}\theta\mathbf{k}, \\[2ex] \dfrac{\partial}{\partial t}\mathbf{h} = \nabla \times [\mathbf{v} \times (\mathbf{H_0} + \mathbf{h})] + \dfrac{P_m}{P_r}\Delta\mathbf{h}, \\[2ex] \dfrac{\partial}{\partial t}\theta = -\mathbf{v} \cdot \nabla\theta + \mathbf{v} \cdot \mathbf{k} + \dfrac{1}{P_r}\Delta\theta, \\[2ex] \nabla \cdot \mathbf{v} = 0, \\[1ex] \nabla \cdot \mathbf{h} = 0, \end{cases} \tag{1.4.3}$$

with the boundary conditions (1.2.12)

$$w = \frac{\partial^2}{\partial z^2}w = 0, \quad \text{at} \quad z = \pm 0.5 \tag{1.4.4}$$

for stress-free boundaries, and the boundary conditions (1.2.13), (resp. (1.2.13$'$))

$$u = v = w = 0, \quad \text{at} \quad z = \pm 0.5 \tag{1.4.5}$$

$$w = \frac{\partial u}{\partial z} = \frac{\partial v}{\partial z} = 0, \quad \text{at} \quad z = \pm 0.5, \quad t \geq 0. \tag{1.4.5$'$}$$

for rigid boundaries, respectively and

$$a)\mathbf{h} = \mathbf{0}, \quad \mathbf{k} \cdot \nabla \times \mathbf{h} = \mathbf{0}, \quad \theta = 0 \quad \text{at} \quad z = \pm 0.5, \tag{1.4.6}$$

$$b)\mathbf{k} \cdot \mathbf{h} = \mathbf{0}, \quad \theta = 0 \quad \text{at} \quad z = \pm 0.5, \tag{1.4.7}$$

appropriate to thermally conducting, electrically nonconducting and respectively electrically conducting boundary, where $\mathbf{v}$ is the velocity field, $\mathbf{h}$ is the magnetic field, θ is the temperature and p is the pressure. The four positive coefficients P_r, P_m, M^2 and R are defined at the end of Section 1.1.1.

Condition $(1.4.7)_1$ is the linearization of the condition

$$\left\{ \nabla \times \mathbf{h} + \beta_H [\nabla \times \mathbf{h} \times (\mathbf{H_0} + \mathbf{h})] + \beta_I \left[(\mathbf{H_0} + \mathbf{h}) \times [\nabla \times \mathbf{h} \times (\mathbf{H_0} + \mathbf{h})] \right] \right\} \times \mathbf{n} = \mathbf{0} \tag{1.4.7'}$$

Equations $(1.4.3)_1$, $(1.4.3)_2$-$(1.4.3)_3$ follow, respectively, from $(1.1.38)$, from the Maxwell equations $(1.1.40) - (1.1.41)$ (in the usual hypotheses of the non-relativistic magnetohydrodynamics) and from $(1.2.9)$ if we assume for the internal energy ϵ the validity of the constitutive equation

$$\epsilon = c_v \theta$$

and consider the transport coefficient χ in the Fourier law $(1.1.18')$ as constant.

1.4.3.2   *Perturbation Navier-Stokes-Fourier model in the Oberbeck-Boussinesq approximation*

In the hydrodynamic context this model follows from that one presented in Section 1.4.3.1. We derive it as a particular case from the electromagnetic model because in Chapters 5, 6 and 7 we obtain the mechanical results as some magnetic characteristics vanish.

In the absence of the magnetic field, *i.e.* in the hydrodynamic case, equations $(1.4.3)$, the boundary conditions $(1.4.4)$, $(1.4.6)$ and $(1.4.5)$ become respectively [Chan]

$$\begin{cases} \dfrac{\partial}{\partial t}\mathbf{v} = -\mathbf{v} \cdot \nabla \mathbf{v} - \nabla p' + \Delta \mathbf{v} + \dfrac{\mathcal{R}}{P_r}\theta \mathbf{k}, \\[2mm] \dfrac{\partial}{\partial t}\theta = -\mathbf{v} \cdot \nabla \theta + \mathbf{v} \cdot \mathbf{k} + \dfrac{1}{P_r}\Delta \theta, \\[2mm] \nabla \cdot \mathbf{v} = 0, \end{cases} \tag{1.4.8}$$

$$w = \frac{\partial^2}{\partial z^2}w = \theta = 0, \tag{1.4.9}$$

$$w = \frac{\partial}{\partial z}w = \theta = 0. \tag{1.4.10}$$

They are the perturbations N-S-F equations in the O-B approximation and they correspond to the basic state of the mechanical equilibrium

$$m_0 = \{\mathbf{U} = \mathbf{0}, \, T = -\beta z + T_0, \, p_0 = p_0(z)\}. \tag{1.4.11}$$

We mainly deal with equations (1.4.8) linearized around the null solution (corresponding to the equilibrium (1.4.11)) and assume that the perturbations are normal modes, *i.e.* every unknown scalar function, say f, has the form

$$f(x, y, z, t) = F(z)e^{i(a_x x + a_y y + ct)}. \tag{1.4.12}$$

Then the boundary conditions for perfectly thermally conducting stress-free surfaces (1.4.9) become

$$W = D^2 W = \Theta = 0 \quad \text{at} \quad z = \pm 0.5 \tag{1.4.13}$$

while, taking into account the continuity equation $(1.4.8)_3$, for rigid walls, (1.4.10) read

$$W = DW = \Theta = 0 \quad \text{at} \quad z = \pm 0.5 \tag{1.4.14}$$

where $D = \frac{d}{dz}$. If some bounding surface, say $z = 0$, is not thermally conducting, then the condition $\Theta = 0$ is to be replaced by

$$D\Theta = 0 \quad \text{at} \quad z = 0. \tag{1.4.15}$$

Remark 1.4.3. Observations of convections in nature and experiments reveal some symmetries with respect to the middle plane of the layer. Consequently, the computations simplify if we take the bounding planes at $z = \pm 0.5$. In some other situations it is convenient to make the translation $z' = z + 0.5$, *i.e.* the bounding surfaces previously defined by $z = \pm 0.5$ now are defined by $z' = 0, 1$. Since equations (1.4.8) have constant coefficients, they are invariant to the corresponding translation $z \leftrightarrow z'$. Therefore, we use both definitions at our convenience, of course dropping the $'$. In the case of convections governed by differential equations with variable coefficients the invariance holds any longer and, thus, every particular case must be treated separately.

In the general case, the two-point problems of one of the types (1.4.13), (1.4.14) and (1.4.15) for the linearized equations (1.4.8) have c as an eigenvalue. In all cases treated by us, we assumed that $Re\, c = 0$, implies $Im\, c = 0$, *i.e.* the principle of exchange of stabilities holds.

Apart for physical reasons, for fluids in unbounded domains the normal mode perturbations are chosen for mathematical reasons: in certain situations they form total sets in the space of the nonlinear problems and they satisfy the far-field behavior of the unknown functions. For flows in bounded boxes, no normal modes with arbitrary wave numbers exist. Correspondingly, the form (1.4.12) must be replaced by functions which in the box have a finite number of periods, *i.e.*

$$f(x, y, z, t) = F(z)e^{i(mx + ny + ct)}, \tag{1.4.12'}$$

inside a rectangular box $\{(x, y, z) \in \mathbf{R}^3 \mid x \in [0, a_1], y \in [0, a_2], z \in [-0.5, 0.5]\}$, where $a_1 = L/H$, $a_2 = l/H$, $m = 2m'\pi/a_1$, $n = 2n'\pi/a_2$, L, l and H are the box sizes in the x, y and z direction, respectively and $m' \geq 1$, $n' \geq 1$ are the numbers of cells in the x and y directions respectively. We use them in Sections 4.2.4 and 5.4.4.

In the nonlinear models, the possibility to apply the Hilbert space approaches imposes the hypothesis of the periodicity cells V even in unbounded layers. In this case, due to the periodicity conditions, on the vertical surfaces of ∂V no additional boundary conditions are necessary.

1.4.4   *Perturbation model for viscous incompressible homogeneous thermoelectrically fully ionized conducting fluids*

In this case the nondimensional equations governing the evolution of the perturbation fields $\mathbf{v}, \mathbf{h}, \theta, p$ of the basic thermodiffusive equilibrium m_0 (1.4.2) are derived from $(1.3.1) - (1.3.8)$ and read [Chan], [Pa], [SuS]

$$
\begin{cases}
\dfrac{\partial}{\partial t}\mathbf{v} = -\,\mathbf{v}\cdot\nabla\mathbf{v} - \nabla p + \Delta\mathbf{v} + M^2(\mathbf{H_0}+\mathbf{h})\cdot\nabla\mathbf{h} + \dfrac{\mathcal{R}}{P_r}\theta\mathbf{k}, \\[2ex]
\dfrac{\partial}{\partial t}\mathbf{h} = \nabla\times[\mathbf{v}\times(\mathbf{H_0}+\mathbf{h})] + \dfrac{P_m}{P_r}\Delta\mathbf{h} \\[2ex]
\qquad\quad + \beta_H\dfrac{P_m}{P_r}\nabla\times[(\mathbf{H_0}+\mathbf{h})\times\nabla\times\mathbf{h}], \\[2ex]
\dfrac{\partial}{\partial t}\theta = -\,\mathbf{v}\cdot\nabla\theta + \mathbf{v}\cdot\mathbf{k} + \dfrac{1}{P_r}\Delta\theta, \\[2ex]
\nabla\cdot\mathbf{v} = 0, \\[1ex]
\nabla\cdot\mathbf{h} = 0.
\end{cases}
\tag{1.4.16}
$$

For the null Hall coefficient, *i.e.* $\beta_H = 0$, (1.4.16) reduces to (1.4.3). For other characteristic quantities used in the non-dimensionalization, instead of (1.4.16), we have [Chan], [Pa], [SuS]

$$
\begin{cases}
\dfrac{\partial}{\partial t}\mathbf{v} = -\mathbf{v}\cdot\nabla\mathbf{v} - \nabla p + \Delta\mathbf{v} + P_m M^2(\mathbf{H_0}+\mathbf{h})\cdot\nabla\mathbf{h} + \mathcal{R}\dfrac{P_m^2}{P_r}\theta\mathbf{k}, \\[2ex]
\dfrac{\partial}{\partial t}\mathbf{h} = \nabla\times[\mathbf{v}\times(\mathbf{H_0}+\mathbf{h})] + \Delta\mathbf{h} + \beta_H\nabla\times[(\mathbf{H_0}+\mathbf{h})\times\nabla\times\mathbf{h}], \\[2ex]
\dfrac{\partial}{\partial t}\theta = -\mathbf{v}\cdot\nabla\theta + \mathbf{v}\cdot\mathbf{k} + \dfrac{P_m}{P_r}\Delta\theta, \\[2ex]
\nabla\cdot\mathbf{v} = 0, \\[1ex]
\nabla\cdot\mathbf{h} = 0.
\end{cases}
\tag{1.4.16$'$}
$$

The basic state and the boundary conditions are those from Section 1.4.3. In the linear case, the condition (1.4.7) must be supplemented with the linearized condition (1.4.7$'$) for $\beta_I = 0$.

1.4.5   *Perturbation model for viscous incompressible homogeneous thermoelectrically partially ionized conducting fluid*

We consider the same problem as in Section 1.4.4, but for a partially ionized fluid. Then the nondimensional equations governing the evolution of the perturbation

$\mathbf{v}, \mathbf{h}, \theta, p$ of the basic thermodiffusive equilibrium m_0 (1.4.2) read [Pa], [SuS]

$$
\begin{cases}
\dfrac{\partial}{\partial t}\mathbf{v} = -\,\mathbf{v} \cdot \nabla \mathbf{v} - \nabla p + \Delta \mathbf{v} + M^2 (\mathbf{H_0} + \mathbf{h}) \cdot \nabla \mathbf{h} + \dfrac{\mathcal{R}}{P_r}\theta \mathbf{k}, \\[2ex]
\dfrac{\partial}{\partial t}\mathbf{h} = \nabla \times [\mathbf{v} \times (\mathbf{H_0} + \mathbf{h})] + \dfrac{P_m}{P_r}\Delta \mathbf{h} + \beta_H \dfrac{P_m}{P_r}\nabla \times [(\mathbf{H_0} + \mathbf{h}) \\[2ex]
\qquad \times \nabla \times \mathbf{h}] + \beta_I \dfrac{P_m}{P_r}\nabla \times \{(\mathbf{H_0} + \mathbf{h}) \times [(\mathbf{H_0} + \mathbf{h}) \times \nabla \times \mathbf{h}]\}, \\[2ex]
\dfrac{\partial}{\partial t}\theta = -\,\mathbf{v} \cdot \nabla \theta + \mathbf{v} \cdot \mathbf{k} + \dfrac{1}{P_r}\Delta \theta, \\[2ex]
\nabla \cdot \mathbf{v} = 0, \\[2ex]
\nabla \cdot \mathbf{h} = 0.
\end{cases}
\qquad (1.4.17)
$$

For some other characteristic quantities used in the non-dimensionalization, instead of (1.4.17) we have [Pa], [SuS]

$$
\begin{cases}
\dfrac{\partial}{\partial t}\mathbf{v} = -\,\mathbf{v} \cdot \nabla \mathbf{v} - \nabla p + \Delta \mathbf{v} + M^2 (\mathbf{H_0} + \mathbf{h}) \cdot \nabla \mathbf{h} + \mathcal{R}\dfrac{P_m^2}{P_r}\theta \mathbf{k}, \\[2ex]
\dfrac{\partial}{\partial t}\mathbf{h} = \nabla \times [\mathbf{v} \times (\mathbf{H_0} + \mathbf{h})] + \Delta \mathbf{h} + \beta_H \nabla \times [(\mathbf{H_0} + \mathbf{h}) \times \nabla \times \mathbf{h}] \\[2ex]
\qquad + \beta_I \nabla \times \{(\mathbf{H_0} + \mathbf{h}) \times [(\mathbf{H_0} + \mathbf{h}) \times \nabla \times \mathbf{h}]\}, \\[2ex]
\dfrac{\partial}{\partial t}\theta = -\,\mathbf{v} \cdot \nabla \theta + \mathbf{v} \cdot \mathbf{k} + \dfrac{P_m}{P_r}\Delta \theta, \\[2ex]
\nabla \cdot \mathbf{v} = 0, \\[2ex]
\nabla \cdot \mathbf{h} = 0.
\end{cases}
\qquad (1.4.17')
$$

For the case of stress-free thermally conducting and electrically perfectly conducting walls, for $t \geq 0$ at $z = 0, 1$ the boundary conditions (1.2.11), (1.3.10) and (1.4.7') [SolM] must be imposed.

1.4.6 *Perturbation model for a thermally conducting binary mixture in the presence of the Soret and Dufour effects*

Consider a thermoanisotropic conducting binary mixture in a horizontal layer bounded by the planes $z = 0$ and $z = d$ characterized by assigned temperature and concentration T_0, C_0, and T_d, C_d respectively. The basic state is the mechanical equilibrium

$$
S_0 \equiv \{\mathbf{U} = \mathbf{0}, \quad T = -\beta_1 z + T_0, \quad C = -\beta_2 z + C_0, \quad p_0 = p_0(z)\}, \qquad (1.4.18)
$$

where $-\beta_2 \equiv C_1 - C_0/d > 0$. The nondimensional problem governing the evolution of perturbations of the mechanical equilibrium S_0, in the presence of the Soret and

Dufour effects, reads

$$\frac{\partial \mathbf{v}}{\partial t} + \mathbf{v} \cdot \nabla \mathbf{v} = -\nabla p + (\mathcal{R}\theta - s\mathcal{C}\gamma)\mathbf{k} + \Delta \mathbf{v},$$

$$P_r\left(\frac{\partial \theta}{\partial t} + \mathbf{v} \cdot \nabla \theta\right) = (1 + N\lambda^2\tau^{-1})\Delta\theta + \mathcal{R}\mathbf{v} \cdot \mathbf{k} + N\lambda\sigma\Delta\gamma, \qquad (1.4.19)$$

$$S_c\left(\frac{\partial \gamma}{\partial t} + \mathbf{v} \cdot \nabla \gamma\right) = \Delta\gamma + \lambda\sigma^{-1}\tau^{-1}\Delta\theta + \mathcal{C}\mathbf{v} \cdot \mathbf{k},$$

for $(t, \mathbf{x}) \in (o, \infty) \times V$, in the following subspace of $L_2(V)$

$$\mathcal{N} = \Big\{(\mathbf{v}, \theta, \gamma) \mid \mathbf{v} = 0 \text{ on } \partial V_1, \quad \mathbf{v} \cdot \mathbf{n} = (\mathbf{n} \cdot \mathbf{d}) \times \mathbf{n} = 0 \text{ on } \partial V_2,$$

$$\theta = \gamma = 0 \text{ on } \partial V_1 \cup \partial V_2, \quad \nabla \cdot \mathbf{v} = 0\Big\}. \qquad (1.4.20)$$

The fluid is heated from below and has a larger concentration at the bottom. In (1.4.20) $V \in \mathbf{R}^3$ is the bounded periodicity cell of boundary ∂V. The intersections of ∂V with the rigid and stress-free planes bounding the layer are denoted respectively by ∂V_1, ∂V_2. The unknown functions are $\mathbf{v}$, θ and γ, where γ represents the perturbation concentration field. In (1.4.19) the following eight physical parameters occur: $\mathcal{R}^2$, $\mathcal{C}^2$, λ, N, σ (only in this chapter the ratio of concentrational and thermal expansion coefficients), P_r, S_c, s ($= \pm 1$ if the solute density is greater (less) than the solvent density). The ratios $\tau = S_c/P_r$ and $\alpha = \mathcal{C}/\mathcal{R}$ (only for this model) are used too. Except for s, all other parameters are positive.

The equations governing the evolution of the perturbations of a chemical equilibrium of a thermally conducting two component reactive viscous mixture, situated in a horizontal layer heated from below and experiencing a catalyzed chemical reaction at the bottom plate are (1.4.19) for $N = \lambda = 0$, while the boundary conditions are much more complicated, namely

$$\frac{\partial u}{\partial z} = \frac{\partial v}{\partial z} = w = 0 \quad \text{on} \quad \partial V_2, \quad \mathbf{u} = \mathbf{0} \quad \text{on} \quad \partial V_1, \qquad (1.4.21)$$

$$\theta = \gamma = 0 \quad \text{on} \quad \partial V_2, \quad \frac{\partial \theta}{\partial z} = -s\gamma \quad \frac{\partial \gamma}{\partial z} = r\gamma \quad \text{on} \quad \partial V_1, \qquad (1.4.22)$$

where r and s are some assigned functions.

1.5 Generalized incompressible Navier-Stokes model

Basic ideas for deriving a generalized setting are presented. Then the generalized model for strong solutions is described.

1.5.1 *Generalized models*

For the classical incompressible N-S model (1.1.33), (1.1.34), (1.2.1), (1.2.2), several generalized variants are known [Ler], [Hop], [FoiP], [Lad69], [LadS76], [LiojM],

[LadV]. Their systematic presentation can be found in [Lad69], [Geo85], [Te], [GirR]. In all of them the space of classical solutions is extended. As a consequence, at least the time derivative of velocity $\mathbf{u}$ and/or some of the space derivatives of $\mathbf{u}$ cease to have a classical sense. The derivatives occurring in the generalized formulation have a weak sense, *i.e.* are taken in the sense of distributions. Correspondingly, operators ∇, $\nabla\cdot$, Δ, $\nabla\times$, have a new, generalized meaning.

Formally, in the generalized setting, the derivatives from the classical setting are transferred from $\mathbf{u}$ to the test function φ. Denote by D^α the space derivative of order α. The classical differential operator D^α is $D^\alpha : \mathcal{D}(D^\alpha) = C^{|\alpha|}(\Omega) \to C(\Omega)$. The generalized operator of D^α is the smallest closed extension $\overline{D^\alpha}$ of D^α from $\mathcal{D}(D^\alpha)$ to $\mathcal{D}\overline{(D^\alpha)} \subset L^2(\Omega)$, namely the adjoint of the formally adjoint operator $(\mathcal{D}^\alpha)^+$. Therefore the generalized derivative of order α of $\mathbf{u}$ (if it exists) is an element $\mathbf{g} \in L^2(\Omega)$ such that $(\mathbf{u}, (D^\alpha)^+\varphi) = (\mathbf{g}, \varphi)$, for every test function $\varphi \in \mathbf{C}_0^\infty(\mathbf{\Omega})$. Usually $\mathbf{g}$ is denoted by the same symbol $D^\alpha\mathbf{u}$ as in the classical sense, but this notation is appropriate only if $\mathbf{u} \in W^{|\alpha|,2}(\Omega)$, $| \alpha | \geq 2$. In the same way, the generalized divergence operator $\nabla\cdot$ is defined by $(\mathbf{u}, \nabla \cdot \varphi) = (\mathbf{g}, \varphi)\, \forall\varphi \in \mathbf{C}_0^\infty(\mathbf{\Omega})$, where the element $\mathbf{g} \in L^2(\Omega)$ is denoted abusively by $\nabla\cdot\mathbf{u}$. This notation is correct only for $\mathbf{u} \in W^{1,2}(\Omega)$. The other differential operators are formally defined by integral identities valid for arbitrary test functions φ, where the derivatives of $\mathbf{u}$ are passed over φ.

Similarly, in order to obtain some generalized setting of our model of interest, the classical equations are multiplied by the test function φ and then integrated over the domain of variation of the independent variable. The by-parts integrations are performed until the stage is reached where only the stipulated generalized derivatives on $\mathbf{u}$ occur in the equation. During the integrations Green formulae and the boundary conditions are taken into account. For instance, if we assume that in (1.1.34) $\mathbf{F} = \mathbf{0}$ and $\mathbf{u} \in L^2\big((0, t^*), N^1(\Omega)\big)$, then the weak formulation of (1.1.33), (1.1.34) reads

$$\int_0^{t^*} \big[(\mathbf{u}, \dot\varphi) - ((\mathbf{u} \cdot \nabla)\mathbf{u}, \varphi) - \nu(\nabla\mathbf{u}, \nabla\varphi)\big]\mathbf{dt} = -(\mathbf{u}(0), \varphi(0)), \qquad (1.5.1)$$

for all $\varphi \in \mathbf{C}_0^1\big([\mathbf{0}, \mathbf{t}^*), \mathcal{N}(\mathbf{\Omega})\big)$, where the dot over φ stands for the differentiation with respect to time.

The solenoidality of $\mathbf{u}$, expressed by the continuity equation (1.1.33), was included in $N^1(\Omega)$. If $\mathbf{u} \in W^{1,2}(\Omega_{t^*})$, then $\dot{\mathbf{u}}$ exists as an element of $L^2(\Omega)$ so that instead of $(\mathbf{u}, \dot\varphi)$ we write $-(\dot{\mathbf{u}}, \varphi)$.

We mention that all integrals are in the Lebesgue sense and the existence of a smaller number of derivatives and a weaker sense of them takes us closer to the original balance equation (1.1.1). Increasing the number of generalized derivatives of $\mathbf{u}$ means going away from the sense of (1.1.1) but going closer to the classical model, due to the embedding theorems. For instance if $\mathbf{u} \in W^{3,2}((0, t^*), W^{4,2}(\Omega))$, $\mathbf{u}$ is classical in all generalized models.

Let us call the elements of $L^2(\Omega)$ functions. Almost in all generalized models of fluid dynamics the velocity $\mathbf{u}$ is assumed to be a function of t only and, for any fixed t, $\mathbf{u}$ is an element of $L^2(\Omega)$ or $N(\Omega)$. Hence, in the theory of non-stationary fluid flows, the point of view of dynamical systems theory, and, therefore, of semigroup theory, is adopted. Correspondingly, $L^2(\Omega)$, $N(\Omega)$ or some of their subspaces represent the phase space of the model; we say that the model is defined in $L^2(\Omega)$. For instance, for various generalized models, $\mathbf{u}$ can belong to one among the spaces: $L^2((0, t^*), L^2(\Omega))$, $L^\infty((0, t^*), L^2(\Omega))$, $L^2((0, t^*), N(\Omega))$, $L^2((0, t^*), N^1(\Omega))$, $C((0, t^*), N^1(\Omega))$. In addition, we may have $\dot{\mathbf{u}} \in L^2((0, t^*), N(\Omega))$, $D_t^\gamma \mathbf{u} \in L^2((0, \infty), L^2(\Omega))$, where $D_t^\gamma \mathbf{u}$ is the fractional derivative of order γ, γ being a positive number.

Unlike the classical case, since the possibility to prove existence theorems depends on data, *e.g.* $\mathbf{F}$ and $\mathbf{u}_w$, the types of generalized models, *e.g.* weak, turbulent, strong, are defined in different spaces for different types of data, *e.g.* $\mathbf{u}_w = \mathbf{0}$ or $\neq \mathbf{0}$, $\mathbf{F} = \mathbf{0}$ or $\neq \mathbf{0}$.

1.5.2   *Generalized model for strong solutions*

Among all these models, the closest to the classical model is the Cauchy problem $\mathbf{u}(0) = \mathbf{u}_0$ for the Navier-Stokes equation (1.1.34) in the case $\mathbf{u}_{|\partial\Omega} = 0$, in which $\mathbf{u} \in C((0, t^*), N^1(\Omega)) \cap L^2((0, t^*), N^2(\Omega))$, $\nabla p \in L^2((0, t^*), L^2(\Omega))$, $\dot{\mathbf{u}} \in L^2((0, t^*), N(\Omega))$, $\mathbf{u}_0 \in N^1(\Omega)$, and $\mathbf{F} \in L^2((0, t^*), L^2(\Omega))$. Its solutions $(\mathbf{u}, p)$ are referred to as *strong solutions* of the classical model but they satisfy (1.1.34) in the $L^2(\Omega)$ sense, *i.e.*

$$\int_0^{t^*} \int_\Omega \left(\mathbf{u} + (\mathbf{u} \cdot \nabla)\mathbf{u} - \nu\Delta\mathbf{u} + \frac{1}{\rho}\nabla p - \mathbf{F}\right)\varphi \, \mathbf{dtdx} = \mathbf{0}, \qquad (1.5.2)$$

for all $\varphi \in C_0^1([0, t^*), \mathcal{N}(\Omega))$. The involved operators are taken in a generalized sense. The smoothness properties of the strong solutions depend on data, in particular on $\mathbf{F}$. If $\mathbf{F}$ derives from a potential, and, so, it no longer occurs in (1.5.2), being included in the pressure term, then it can be proved that the strong solution is classical, *i.e.* the elements $\mathbf{u}$ and p of $L^2(\Omega)$ which occur in the strong solution have a representative which satisfies (1.1.34) in the classical sense.

The generalized model for strong solutions $(\mathbf{u}, p)$ of the classical incompressible Navier-Stokes model in the case $\mathbf{u}_{|\partial\Omega} \neq \mathbf{0}$ is the Cauchy problem for (1.1.34), in which

$$\mathbf{u} \in L^2((0, t^*), N(\Omega) \cap W^{2,2}(\Omega)), \qquad \dot{\mathbf{u}} \in L^2((0, t^*), N(\Omega)),$$

$$\mathbf{u} - \overline{\mathbf{u}} \in L^2((0, t^*), N^2(\Omega)), \qquad \nabla p \in L^2((0, t^*), L^2(\Omega)), \qquad (1.5.3)$$

$$\mathbf{F} \in L^2((0, t^*), L^2(\Omega)), \qquad \mathbf{u}_0 \in N^1(\Omega),$$

where $\overline{\mathbf{u}}$ is a stationary solution.

Like in the previous model, the strong solutions in the (1.5.3) class satisfy equation (1.1.34) in the L^2 sense but here the velocity does not vanish on $\partial\Omega$, namely, on $\partial\Omega$, $\mathbf{u}$ and $\overline{\mathbf{u}}$ take the same values. This is ensured by the fact that $\mathbf{u} - \overline{\mathbf{u}}_{|\partial\Omega}$ belongs to $N^2(\Omega)$. Indeed, the embedding $(1.5')$ implies $\mathbf{u} - \overline{\mathbf{u}}_{|\partial\Omega} = 0$ on $\partial\Omega$.

From the stability point of view this model is important because it contains the stationary function $\overline{\mathbf{u}}$, which we take as basic solution. It is constructed by extending $\mathbf{u}_{|\partial\Omega}$ to Ω [Lad69].

The Leray model for turbulent solutions is presented in Section 2.1 and applied in Section 2.7. The weak model of Kiselev and Ladyzhenskaya is shortly used in Section 2.7. For other generalized N-S models we quote [Lad69], [Geo85], [Te], [GirR].

1.5.3 *Perturbation generalized model for strong solutions*

Let us deduce the perturbation model of the generalized incompressible N-S model (1.5.1), (1.5.2) in the class of strong solutions. To this aim we define the perturbation velocity $\mathbf{v} = \mathbf{u} - \overline{\mathbf{u}}$, $\mathbf{v}_0 = \mathbf{u}_0 - \overline{\mathbf{u}}$, therefore the perturbed velocity is $\mathbf{u}$. Introduced in the Cauchy problem for (1.1.34), it yields

$$\frac{\partial}{\partial t}\mathbf{v} - \nu\Delta\mathbf{v} + (\overline{\mathbf{u}}\cdot\nabla)\mathbf{v} + (\mathbf{v}\cdot\nabla)\overline{\mathbf{u}} = -(\mathbf{v}\cdot\nabla)\mathbf{v} - \frac{1}{\rho}\nabla p + \mathbf{F} + \nu\Delta\overline{\mathbf{u}} - (\overline{\mathbf{u}}\cdot\nabla)\overline{\mathbf{u}}.$$

As $\overline{\mathbf{u}}$ is assumed to be a stationary solution of (1.1.34), the expression consisting of the last three terms vanishes and the above equation becomes

$$\frac{\partial}{\partial t}\mathbf{v} - \nu\Delta\mathbf{v} + (\overline{\mathbf{u}}\cdot\nabla)\mathbf{v} + (\mathbf{v}\cdot\nabla)\overline{\mathbf{u}} = -(\mathbf{v}\cdot\nabla)\mathbf{v} - \frac{1}{\rho}\nabla p. \tag{1.5.4}$$

It is valid in the class

$$\mathbf{v} \in C\big((0,t^*), N^1(\Omega)\big) \cap L^2\big((0,t^*), N^2(\Omega)\big), \ \dot{\mathbf{v}} \in L^2\big((0,t^*), N(\Omega)\big),$$
$$\mathbf{v}_0 \in N^1(\Omega), \overline{\mathbf{u}} \in W^{2,2}(\Omega). \tag{1.5.5}$$

Applying to (1.5.4) the projection $P : L^2(\Omega) \to N(\Omega)$ (more precisely $P : N^1(\Omega) \to N^2(\Omega)$), we obtain the evolution equation in $L^2\big((0,t^*), N^2(\Omega)\big)$

$$\frac{d\mathbf{v}}{dt} + A\mathbf{v} + M(\overline{\mathbf{u}}, \mathbf{v}) = R(\mathbf{v}), \tag{1.5.6}$$

where

$$A = -P\nu\Delta, \quad R(\mathbf{v}) = -P\big(\mathbf{v}\cdot\nabla\mathbf{v}\big), \quad M(\overline{\mathbf{u}}, \mathbf{v}) = -P\big[\overline{\mathbf{u}}\cdot\nabla\mathbf{v} + \mathbf{v}\cdot\nabla\overline{\mathbf{u}}\big]. \tag{1.5.7}$$

It follows that the strong solutions in the class (1.5.5) of the classical model satisfy the Cauchy problem

$$\lim_{t\to 0^+} \mathbf{v}(t) = \mathbf{v}_0 \tag{1.5.8}$$

for (1.5.6), where the linear operator A and the nonlinear mapping R are defined by (1.5.7).

We mention that the existence and uniqueness of $\mathbf{v}$ hold and the existence of p follows; here we do not deal with them. In (1.5.6) $\frac{d}{dt}$ is a total derivative with respect to time and $\mathbf{v} \in N^1(\Omega)$, *i.e.* it is a solenoidal vector in the generalized sense. A more detailed presentation of all these questions can be found in [Pro], [Geo85], [Te].

Chapter 2

Incompressible Navier-Stokes fluid. Universal stability criteria. Linearization principle

After a physical and mathematical motivation of the generalized setting of mathematical models of thermodynamics (Section 2.1) the analytical essentials of hydrodynamic and hydromagnetic theory are reformulated in terms of dynamical systems theory (Section 2.2).

2.1 Back to integral setting; involvement of dynamics and bifurcation

In Chapter 1 no reference was made of the physical bases involved into balance, constitutive and state equations other than the formal inclusion into the Müller scheme. Of course, the in-depth presentations to be found in treatises addressed to engineers, physicists and chemists would have helped us to complete the scheme with physical facts. However, we avoided them here due to their not so clear mathematical separation between the unknowns and given data, the confusing mathematical description associated with every physical quantity, *e.g.* parameter, function, functional.

Neither the minutely presentation of the mathematical concepts involved was provided in Chapter 1. This option was motivated by letting the physical scheme to be sufficiently clear and simply exposed. Now, as we proceed with a mathematical analysis, we deal with the mathematical bases lacking in Section 1.1.

The N-S equations were derived about the middle of the XIX century. Then a long period followed when the unknowns were considered as smooth as necessary to permit more or less formal studies. A lot of results were obtained, but the global existence. The cause is threefold.

The first is the use of too strong mathematical properties: in these studies the model was adapted to the existing tools and, so, the analyzed models gone further and further away from the original physical model. Remind: the presence of integral in the Lebesgue sense in the Radon-Nikodym theorem which was not taken into

31

account; the assumption of smoothness properties of the basic unknowns involved into the transport, flux-divergence and continuity theorem ensuring the passage to the local form of the balance equations, to name a few reasons to consider those models as generally unrealistic. As early as 1933 it was Jean Leray that overpassed this standpoint by recommending the use of some other mathematical descriptions of physical quantities, closer to the original integral model. This was the beginning of the generalized settings in differential equations theory. He showed that the bases of hydrodynamics do not require that the unknowns in the balance equations be functions but elements of $L^2(\Omega)$ or some other functionals, involving a generalized meaning of the time and space derivatives of these elements.

In addition, the arbitrariness of $\partial\Omega'(t)$ was replaced by Leray by the arbitrariness of a function φ of a set dense in $L^2(\Omega)$. In this way the continuity theorem was replaced by Theorem 1.2, while the derivative of a function was passed over φ. His resulting model was the first generalized setting of a classical model not only in hydrodynamics but also in mathematical physics. He adapted and created mathematical tools for the physical model, not conversely. We try as much as possible to follow his ideas.

The **Leray model for the turbulent solutions** $\mathbf{u} \in L^2\big((0,t^*), N^1(\Omega)\big) \cap C\big((0,t^*), N(\Omega)\big)$ of the classical incompressible N-S model (1.1.33), (1.1.34), (1.2.1), (1.2.2), with $\mathbf{u}_w = \mathbf{0}$ and body forces deriving from a potential, is the set of integral equations

$$\int_0^{t^*} [(\mathbf{u}, \dot{\varphi}) + ((\mathbf{u} \cdot \nabla)\varphi, \mathbf{u}) + \nu(\mathbf{u}, \Delta\varphi)]dt = -(\mathbf{u}_0, \varphi(0)), \qquad (2.1.1)$$

for every $\varphi \in C_0^1\big([0,T), \mathcal{N}(\Omega)\big)$. Formally (2.1.1) is obtained by projecting (1.1.34) on $N(\Omega)$ and then by taking the integral of the resulting equation. In (1.1.34) p appears through ∇p which belongs to $N^\perp(\Omega)$, therefore the pressure no longer occurs in (2.1.1).

The second cause is the exclusively analytical approach to hydrodynamics. It is only starting with the Hopf generalized setting that the N-S model was associated with an evolution equation in a Banach space and, so, with a semidynamical system. The solutions were considered as functions of time only and, so, they were supposed to belong to $L^2\big((0,t^*), W^{l,p}(\Omega)\big)$. Naturally, the hydrodynamic stability (with respect to perturbations of initial data) became the tubular stability or attractivity of the basic state or motion, (*e.g.* periodic orbit, other standard or strange attractors). In this respect basic contributions are due to Ciprian Foias and Giovanni Prodi [FoiP], Olga A. Ladyzhenskaya [Lad91], V. I. Yudovich [Yu2], Roger Temam [Te], M. I. Vishik [Vis], Vsevolod A. Solonnikov [LadS76], Jean-Louis Lions and E. Magenes [LiojM], Gérard Iooss [Ioo2], Peter Constantin and Ciprian Foias [ConsF], Ciprian Foias and Roger Temam [FoiT], Ciprian Foias, O. Manley and Roger Temam [FoiMT], David Ruelle and Floris Takens [RueT].

The third cause is ignoring the involved bifurcation due to the presence of the physical parameter R_e. As R_e increases, the fluid flows undergo qualitative changes leading to turbulence. The number of stationary solutions increases and their analytical properties worsen such that the existence and uniqueness modify their classical meanings [Sa73], [IooJ], [Ioo2].

2.2 Stability in semidynamical systems: from kinetic energy to Lyapunov energy functionals

With respect to their explicit dependence or independence of time, the solutions of a mathematical model of fluid flow can be classified in stationary (or steady) and non-stationary (unsteady). Here we deal with stability of stationary solutions only. We call them the *basic solutions*. In this book by stability we mean the Lyapunov stability, *i.e.* with respect to perturbations of initial conditions. Therefore the basic solution is supposed to be a function of initial conditions. A corresponding general framework is dynamical system theory.

Let us give a few definitions and results of this theory in a particular case convenient to our study.

Let M be a topological manifold, called the *phase* or *state space*. By definition, a function $\Phi : \mathbf{R} \to M^M$, $\Phi(t) \equiv \phi_t : M \to M$, such that ϕ_t are homeomorphisms of M, $\phi_0 = \mathrm{id}_M$ and $\phi_{t+\tau} = \phi_t \circ \phi_\tau$, $\forall t, \tau \in \mathbf{R}$, is referred to as a *(continuous) dynamical system*. It is denoted by (M, Φ). If instead of $\mathbf{R}$ the function Φ is defined on $\mathbf{R}^+$ then Φ is called the *semidynamical system* (most mathematical models of thermodynamics are semidynamical systems but we first consider the situation $t \in \mathbf{R}$). Let $\mathbf{u}_0$ be an element of M and denote $\phi_t(\mathbf{u}_0) = \mathbf{u}(t, \mathbf{u}_0)$. Since $\phi_0(\mathbf{u}_0) = \mathbf{u}_0$ it follows that $\mathbf{u}(0, \mathbf{u}_0) = \mathbf{u}_0$. This is why we say that at each fixed t, $\mathbf{u}$ is the element of M corresponding to its initial state $\mathbf{u}_0$ and, therefore, $\mathbf{u}$ is the state after t units of time of the initial state $\mathbf{u}_0$. In this way M is the set of all possible initial states and also of all states at some arbitrary instant t. Let $\mathbf{u}_0$ be fixed and let t run over $\mathbf{R}$. The curve described in M by $\mathbf{u}(t, \mathbf{u}_0)$ is called the *phase trajectory* or, simply, *trajectory* through $\mathbf{u}_0$, while the corresponding function $\mathbf{u}(t, \mathbf{u}_0)$ of t is called the *dynamics* or *motion* of $\mathbf{u}_0$.

Consider a set of Cauchy problems for an evolution equation

$$\frac{d}{dt}\mathbf{u} = \mathbf{f}(\mathbf{u}), \tag{2.2.1}$$

$$\mathbf{u}(0) = \mathbf{u}_0, \tag{2.2.2}$$

where $\mathbf{u}_0$ runs over M, and assume that for every $\mathbf{u}_0 \in M$, the solution $\mathbf{u}(t, \mathbf{u}_0)$ of (2.2.1), (2.2.2) exists, is unique and possesses some smoothness and invertibility properties with respect to $\mathbf{u}_0$. Then, with these Cauchy problems we can associate a dynamical or semidynamical system, namely $\phi_t(\mathbf{u}_0) = \mathbf{u}(t, \mathbf{u}_0)$.

Assume that a mathematical model in thermodynamics of fluid flows was written as an initial and/or boundary-value problem for a system of balance equations for the physical quantities p, $\mathbf{u}_{fl}$, T, $\mathbf{H}$, $\mathbf{E}, \ldots$ Here $\mathbf{u}_{fl}$ denotes the fluid velocity. The index $_{fl}$ was put (only in this section) in order to distinguish the velocity from an n-tuple of physical quantities. Denote such an n-tuple by $\mathbf{u}$, *i.e.* $\mathbf{u} = (p, \mathbf{u}_{fl}, T, \mathbf{H}, \mathbf{E}, \ldots)$, and assume that the model was written in the form (2.2.1), (2.2.2). Then $\mathbf{u}_0$ is the physical initial state of the fluid flow, which justifies the mathematical label of state given to $\mathbf{u}_0$.

The concept of state defines the type of phenomenon occurring in a continuous material and, thus, characterizes the continuum. The state function is a vector function $\mathbf{U}$ the components of which are the basic physical quantities, called the *state functions*. For instance, if the state function is $\mathbf{u} = (\mathbf{x}, \dot{\mathbf{x}})$, where $\mathbf{x}$ is the position of a particle and $\dot{\mathbf{x}}$ is its velocity at time t, then the material continuum is degenerated to a point of position $\mathbf{x}(t)$ and velocity $\dot{\mathbf{x}}(t)$, or it is a body which, due to some circumstances, (*e.g.* big distances), can be considered as a material point, *e.g.* Moon, Sun. If $M = \mathbf{R}^2$ then the balance equation governing its mechanical motion is the vector Newton second law written as a system of two ode's in x and $\dot{x}$. A mechanical state is the pair $(x(t), \dot{x}(t))$ for a fixed t. From the dynamical point of view the mechanical motion of $\mathbf{u}_0$ is the function $\mathbf{u}(t, \mathbf{u}_0)$ of t for fixed $\mathbf{u}_0$. Its phase trajectory is the graph of this motion in $\mathbf{R}^2$, *i.e.* it is a curve in the plane $(x, \dot{x})$, while a physical trajectory described by the material point in reality is a curve in the (t, x) plane. The physical motion is the displacement of x_0 along the physical trajectory, while the motion from the dynamical point of view is the evolution of $\mathbf{u}_0(\chi(0), \dot{\chi}(0))$ along the phase trajectory, t being the parameter on this phase trajectory.

Similarly, a thermally and electrically conducting and magnetizable continuous material is a fluid in an electromagnetic field. Its state at a fixed time t is $\mathbf{u} = (p, T, \mathbf{u}_{fl}, \mathbf{E}, \mathbf{H})$ where the index fl indicates the fluid velocity. The phase trajectory of $\mathbf{u}_0$ is a curve in the infinite dimensional space $C^1\big((0, t^*), C^1(\Omega) \cap W^{1,2}(\Omega)\big) \times C^{1,1,1,1}\big((0, t^*), C^2(\Omega) \cap W^{2,2}(\Omega)\big)$, and it is the graph of the motion $\mathbf{u}(t, \mathbf{u}_0)$, where all components of $\mathbf{u}(t, \mathbf{u}_0)$ are functions of t only and their values are functions of $\mathbf{u}_0$ as smooth as necessary for the classical setting of the balance equations having a sense. These equations must be written in the form (2.2.1) if a dynamical study is in view. Unlike the case of mechanics of a finite number of particles or rigid bodies, to transform the balance equations into evolution equations (2.2.1) is a very difficult technical operation; we sketch it in the following section only for the incompressible N-S model. The major difficulty consists in the infinite-dimension of the associated dynamical system, *i.e.* $dim M = \infty$, implying a much more complicated geometry of M, and in nonlinearity of the vector field $\mathbf{f}$. Usually, in (1.2.1), (1.2.2), $\mathbf{u}$ does not have p (pressure) as a component, because the balance equations do not contain the time derivative of p, even if the pressure occurs in them. This represents another difficulty in associating evolution equations with balance equations. Indeed, the pressure occurs in the momentum equation and, possibly, in the state equations for density and internal energy and in the constitutive coefficients. In all cases studied by us p occurred only in the momentum equation, namely through its gradient ∇p. Indeed, except for one case, we assume that the fluid is thermodynamically incompressible and, thus, the constitutive coefficients and the constitutive state equations do not depend on p. The single case of compressible fluid where p would occur in these coefficients and equations was considered in the O-B approximation where ρ depends only on T.

Then, in order to eliminate p from the momentum equation to this equation, we apply the $\nabla \times$ (curl) operator in the classical setting, and some integro-differential orthogonal projection operator P in the generalized case. As a result, the operators and mappings from the momentum equations are composed with P, which complicated the analytical study of these equations, and, so, their writing as an evolution equation. Therefore, in Definitions 2.2.1, 2.2.2 from below, $\mathbf{u}$ belongs no longer to the set M (a current point of which is a tuple of all basic physical state functions) but to a proper subspace $\tilde{M}$ of M corresponding to solenoidal velocities, which are orthogonal to ∇p. In this way, by applying the projection P, in the momentum equation p no longer occurs and the continuity equation is no longer necessary. The space $\tilde{M}$ is a Cartesian product of spaces to which various state functions belong. Therefore, among them $N(\Omega)$ or some of its subspaces $N^l(\Omega)$ is the only one of interest for the case of incompressible fluids. Note that $N(\Omega)$ is a proper subspace of $L^2(\Omega)$. From now on the space of interest for us is $\tilde{M}$ and, for the sake of simplicity, we denote it by M.

Suppose that M is a Banach space and denote by $\| \cdot \|$ its norm.

The natural framework for the Lyapunov stability is dynamical system theory because it assumed that the initial data are arbitrary in a set and this set is the phase space M. We deal only with the stability of stationary solutions $\overline{\mathbf{u}}$ of the classical mathematical models of thermodynamics. They correspond to equilibria of the dynamical system associated with (2.2.1) and satisfy the equation $\mathbf{f}(\overline{\mathbf{u}}) = 0$ and the relation $\overline{\mathbf{u}}(t, \mathbf{u}_0) = \overline{\mathbf{u}}$, *i.e.* at each instant t the equilibrium point $\overline{\mathbf{u}}$ is equal to its initial state, therefore the phase trajectory of $\overline{\mathbf{u}}$ consists of the point $\overline{\mathbf{u}}$ only. Geometrically we say that $\overline{\mathbf{u}}$ is Lyapunov stable if in M we can find a sphere around $\overline{\mathbf{u}}$ such that every phase trajectory initiating at a point of this sphere gives rise to a phase trajectory as close to the trajectory generated by $\overline{\mathbf{u}}$ as we wish, *i.e.* remains in another sphere centered at $\overline{\mathbf{u}}$ and of radius as small as we wish. The formalized definition is

Definition 2.2.1. The equilibrium point $\overline{\mathbf{u}}$ of the dynamical system (M, Φ) generated by (1.2.1), (1.2.2) is *stable* if for every $\epsilon > 0$, there exists $\eta > 0$ such that for every initial point in the ball $S(\eta, \overline{\mathbf{u}}) = \{\mathbf{u}_0 \in M \mid \|\overline{\mathbf{u}} - \mathbf{u}_0\| < \eta\}$ we have that $\mathbf{u}(t, \mathbf{u}_0) \in S(\epsilon, \overline{\mathbf{u}}) = \{\mathbf{u} \in M \mid \|\mathbf{u}(t, \mathbf{u}_0) - \overline{\mathbf{u}}\| < \epsilon\}$.

Denote by $\mathbf{R} \times M$ the *extended phase space*. For $\mathbf{u}_0$ fixed, in $\mathbf{R} \times M$ the graph of the motion $\mathbf{u}(t, \mathbf{u}_0)$ is a curve, called the *integral curve* through $(0, \mathbf{u}_0)$. An equivalent definition of the stability in $\mathbf{R} \times M$ reads

Definition 2.2.2. The equilibrium point $\overline{\mathbf{u}}$ of the dynamical system (M, Φ) generated by (1.2.1) (1.2.2) is *stable* if for every $\epsilon > 0$, there exists $\eta > 0$ such that for every initial point $\mathbf{u}_0 \in S(\eta, \overline{\mathbf{u}})$, $\mathbf{u}(t, \mathbf{u}_0)$ belongs to the cylindrical tube $C(\epsilon, \overline{\mathbf{u}}) = \{(t, \mathbf{u}) \in \mathbf{R} \times M \mid \|\mathbf{u}(t, \mathbf{u}_0) - \overline{\mathbf{u}}\| < \epsilon\}$.

Sometimes, this stability is referred to as *stability in the mean*. The difference $\mathbf{u}(t, \mathbf{u}_0) - \overline{\mathbf{u}}$ is called the *perturbation* of $\overline{\mathbf{u}}$ and it is denoted by $\mathbf{v}$. Then $\mathbf{u}(t, \mathbf{u}_0) = \overline{\mathbf{u}} + \mathbf{v}(t, \mathbf{v}_0)$ is called the *perturbed flow* and $\overline{\mathbf{u}}$ the *unperturbed motion*. Since $\mathbf{u}(0, \mathbf{u}_0) = \mathbf{u}_0$, $\mathbf{v}(0, \mathbf{v}_0) = \mathbf{v}_0$, and $\overline{\mathbf{u}}(t, \overline{\mathbf{u}}) = \overline{\mathbf{u}}(0, \overline{\mathbf{u}}) = \overline{\mathbf{u}}$, we have $\mathbf{u}_0 = \overline{\mathbf{u}} + \mathbf{v}_0$. In addition, if $\mathbf{u}_0 = \overline{\mathbf{u}}$, then $\mathbf{u}(t, \mathbf{u}_0) = \overline{\mathbf{u}}$ (due to the uniqueness) and, consequently, $\mathbf{v}_0 = 0$ and $\mathbf{v}(t, 0) = \mathbf{0}$.

The problem satisfied by $\mathbf{v}(t, \mathbf{v}_0)$ reads

$$\frac{d}{dt}\mathbf{v} = \mathbf{f}(\overline{\mathbf{u}} + \mathbf{v}), \tag{2.2.3}$$

$$\mathbf{v}(0) = \mathbf{v}_0 \tag{2.2.4}$$

and it is called the *perturbation problem*. It contains $\overline{\mathbf{u}}$ and it is obtained by writing the problem (2.2.1), (2.2.2) for the perturbed flow and subtracting from it the problem for the basic motion.

To a perturbed flow $\mathbf{u}$ corresponds a perturbation $\mathbf{v}$ around $\overline{\mathbf{u}}$ and conversely. It follows that to the basic flow $\overline{\mathbf{u}}$ the null perturbation $\mathbf{0}$ corresponds. Hence, to the equilibrium $\overline{\mathbf{u}}$ for (2.2.1), (2.2.2), the equilibrium situated at origin for (2.2.3), (2.2.4), corresponds. In this way, the stability of $\overline{\mathbf{u}}$ for the problem (2.2.1), (2.2.2) can be expressed as the stability of the equilibrium situated at the origin for problem (2.2.3), (2.2.4).

In order to study (2.2.3), (2.2.4) we use some information from the linearized of this problem around $\overline{\mathbf{u}}$, *i.e.*

$$\frac{d}{dt}\mathbf{V} = \mathbf{f}_{\mathbf{u}}(\overline{\mathbf{u}})\mathbf{V}, \tag{2.2.5}$$

$$\mathbf{V}(0) = \mathbf{V}_0, \tag{2.2.6}$$

where $\mathbf{f}_{\mathbf{u}}(\overline{\mathbf{u}})\mathbf{V}$ is the Fréchet differential of $\mathbf{f}$ with respect to $\mathbf{u}$ at $\overline{\mathbf{u}}$ applied to $\mathbf{V}$. In order to distinguish the solutions of (2.2.3), (2.2.4) and (2.2.5), (2.2.6) the standard notation for the solution of (2.2.5), (2.2.6) is $\mathbf{V}$. In the following, instead of $\mathbf{V}$, we use $\mathbf{v}$.

If $\dim M = 2$, then the extended phase space $\mathbf{R} \times M$ is three-dimensional, the integral curve of $\overline{\mathbf{u}}$ is a straight line parallel to the time axis, $S(\eta, \overline{\mathbf{u}}) \in M$ is a two-dimensional disk of radius η centered at $\overline{\mathbf{u}}$ and $C(\epsilon, \overline{\mathbf{u}}) \in \mathbf{R} \times M$ is a usual cylindrical tube of base $S(\eta, \overline{\mathbf{u}})$, having the generatrix parallel to the time axis and the integral curve of $\overline{\mathbf{u}}$ as axis. Therefore $C(\epsilon, \overline{\mathbf{u}})$ is a tube.

In the more general case of unsteady basic flows $\overline{\mathbf{u}}(t, \overline{\mathbf{u}}_0)$ the same definition of stability holds, but $C(\epsilon, \overline{\mathbf{u}}(t, \overline{\mathbf{u}}_0))$ is a tube of radius ϵ having as axis the integral curve corresponding to $\overline{\mathbf{u}}(t, \overline{\mathbf{u}}_0)$, which is no longer a straight line but a space curve. Whence the *tubular Lyapunov stability* in Definitions 2.2.1, 2.2.2.

In the case of the generalized incompressible N-S model, we have $M \subset L^2$, *e.g.* $M = W^{l,2}(\Omega)$, therefore $\|\mathbf{u}(t,\mathbf{x}_0)\|_{l,2} < \infty$. In the classical sense (here denoted by the index c), we could choose $M_c = C^2(\Omega) \cap C(\overline{\Omega})$, which is a normed space with the norm (1.1), *i.e.* $\max_{\mathbf{x} \in \Omega} |\mathbf{u}(t,\mathbf{x})| + \max_{\mathbf{x} \in \Omega} |D\mathbf{u}(t,\mathbf{x})| + \max_{\mathbf{x} \in \Omega} |D^2\mathbf{u}(t,\mathbf{x})|$. We did not make this choice because this space is not complete and the implied limits of sequences of norms do not always exist in M_c. On the other hand, $C^2(\Omega) \cap C(\overline{\Omega}) \subset L^2(\Omega)$ and, so, the generalized setting in $L^2(\Omega)$ applies to the classical case of M_c too; in addition it is simpler.

In general, the definition for pointwise (local, in the small [Geo85]) stability can be formulated in the classical setting of mathematical models defined in some classes of smooth functions. Formally, they are obtained from Definitions 2.2.1, 2.2.2 by simply replacing $\|\cdot\|_{l,p}$ by $|\cdot|_c$, if the generalized setting has $M = W^{l,p}$, $l \geq 2$. The results from the generalized setting are weaker because $|\mathbf{u}|_c > \|\mathbf{u}\|_{l,p}$. However, they are more appropriate to the fluid flows in laminar- turbulent regime and in some special conditions when the smoothness of physical quantities is not justified experimentally.

In Section 1.1 we defined three types of energies: *kinetic*, $E_c(t,\Omega'(t)) = \frac{1}{2}\int_{\Omega'(t)} \rho \mathbf{u}_{fl}^2 d\Omega'(t)$, *internal* $E_i(t,\Omega'(t)) = \int_{\Omega'(t)} \rho \epsilon d\Omega'(t)$, and *total internal* $E_{ti}(t,\Omega'(t)) = E_i(t,\Omega'(t)) + E_c(t,\Omega'(t))$, were $\mathbf{u}_{fl}$ was the velocity field.

From the physical point of view the stability of some stationary (and not only) flows must be expressed in terms of two among these physical energies (because only two are linearly independent), while from the mathematical point of view the stability of equilibrium $\overline{\mathbf{u}}$ involves other quantities, namely the norm of the perturbation $\|\mathbf{u}(t,\mathbf{u}_0) - \overline{\mathbf{u}}\|$, where $\mathbf{u}$ is the state function vector. Among the components of this $\mathbf{u}$ there is only one which defines a physical energy, namely the velocity, denoted only in this paragraph by $\mathbf{u}_{fl}$. In the models studied by us ρ and ϵ are functions of T only and in the particular case of the incompressible N-S model ρ is a constant and ϵ does not occur in the equations of the model. Consequently, it is only in this model that the kinetic energy coincides with the norm of the unknown state vector, up to a constant. Moreover, in this case, up to a constant, the kinetic energy is the L^2 norm. But, even in this case, if M is $W^{l,2}$ with $l \neq 2$, then the mathematical energy contains $\mathbf{u}$ as well as the derivative $D\mathbf{u}$, therefore the physical kinetic energy is not equal to the norm of the state vector.

Therefore, in general, our mathematical definition of stability does not involve the physical energy. In addition, sometimes, in applications, from this definition it is not easy to obtain estimations for the norm of perturbation but rather for the norm of some linear combinations of the state functions and their derivatives. Consequently, in fact, instead of estimating the physical energy, in hydrodynamic and hydromagnetic stability we estimate some Lyapunov functional. Then, by Lyapunov second method (criterion), the stability as in Definitions 2.2.1 and 2.2.2 follows. This is why, instead of the Lyapunov functional we say the "energy". The Lyapunov criterion states: let $\overline{\mathbf{x}}$ be an equilibrium point for a dynamical system defined in a finite-dimensional phase space M. Assume that M is a normed space and denote by $S_{\overline{\mathbf{x}}}$ some neighborhood of $\overline{\mathbf{x}}$ in M. In these conditions, if in $S_{\overline{\mathbf{x}}}$ there exists a Lyapunov functional $V : M \to \mathbf{R}$, then $\overline{\mathbf{x}}$ is stable. We remind that V has the following properties: in $S_{\overline{\mathbf{x}}}$, V is continuous; in $S_{\overline{\mathbf{x}}} \setminus \{\overline{\mathbf{x}}\}$, V is differentiable; V has a local minimum at $\overline{\mathbf{x}}$, *i.e.* it can be assumed that, in $S_{\overline{\mathbf{x}}}$, $V(\overline{\mathbf{x}}) = 0$, $V(\mathbf{x}) > 0$ if $\mathbf{x} \neq \overline{\mathbf{x}}$; V is a monotonous nonincreasing function along any phase trajectory $\mathbf{x}(t, \mathbf{x}_0)$, where $\mathbf{x}_0 \neq \overline{\mathbf{x}}$, $\mathbf{x}_0 \in S_{\overline{\mathbf{x}}}$, *i.e.* for every $t \geq 0$ we have $\dot{V}(\mathbf{x}(t, \mathbf{x}_0)) \leq 0$ [Fab].

From the perturbation problem we cannot derive an energy equation but an *energy relation*, because, besides the energy, this equality contains some other functionals in the form of integrals of products of various state functions and their derivatives.

Usually, the energy relation is obtained by multiplying the balance equations or/and some space derivatives of these equations by scalars and some state functions such that in the resulting equation the derivative with respect to time, *i.e.* the rate of change, of the energy occurs (*e.g.* the Lyapunov function). Recently, [GeoPalR00], [GeoPalR01], better stability results by first modifying the balance equations and then applying the above mentioned procedure were obtained (Sections 4.1, 4.2).

As usual with the Lyapunov functional V, there is no *a priori* guiding idea of the multiplications and differentiations of the governing equations in order to form V. In exchange, we are limited in our attempts by the possibility to obtain the positivity of V, the negativity of $\frac{dV}{dt}$ as required in the Lyapunov criterion. As a consequence, a lot of stability problems of interest in applications are not yet solved and are simplified by mathematical reasons, leading to Lyapunov functionals for which the hypotheses of the Lyapunov principle hold. However, there are treatments overcoming these difficulties, *e.g.* in [J70b], [Rio88a], [RioM88c], [MuloR94], [Pal05], by means of some guidelines [Rio88a], [Mulo06] to find V (Section 4.3).

Another big difference between the physical and mathematical energy, comes from the domain over which the integral in their definition is taken: in the physical case it is the arbitrary material domain $\Omega'(t)$, while in the mathematical case it is the entire domain of motion Ω. As a consequence, the mathematical definition is appropriate to the fluid flows which have relevant global (*i.e.* at the scale of Ω) properties, *e.g.* convections and rotating flows. In the case of fluid flows with relevant local (*i.e.* strongly depending on $\Omega'(t)$) physical properties, *e.g.* boundary layers, the stability results following from Definitions 2.2.1, 2.2.2 no longer hold. Thus, in the case of fluid flows with relevant local properties the definition of stability in the small must be used.

2.3 Perturbations; asymptotic stability; linear stability

The class of perturbations is essentially the same as the class of solutions, *i.e.* M, with the difference that on $\partial\Omega$ the perturbations vanish. The perturbations must satisfy the equation (2.2.3), which does not contain free terms, *e.g.* body forces $\mathbf{F}$ independent of $\mathbf{u}$, but is much more complicated than (2.2.1). Remind that stability theory provides qualitative results on $\overline{\mathbf{u}}$ without knowing the solutions corresponding to other initial data, and, so, perturbations. At this high degree of generality only few studies exist and they concern the linearization principle.

A more reduced degree of generality corresponds to perturbations taken in the class of difference motions. (Recall that a *difference motion* satisfies the boundary conditions (2.2.4) and solenoidality conditions of some state functions, has the same smoothness as these ones, but does not necessarily satisfy (2.2.3).) This is the class where universal stability criteria and more general results obtained by the non-stationary energy method (Section 3.2.7) are deduced. In particular, in order to establish the universal stability criteria, the rate of change of perturbation energy (more precisely Lyapunov functional) provided by the energy relation is estimated by means of isoperimetric inequalities of the Poincaré type and by using some other integral inequalities in the class of difference motions.

Even more particular but stronger results can be obtained if the perturbations are looked for in smaller subclasses of difference motions, *e.g.* axisymmetric, two-dimensional, normal mode perturbations.

For applications, a concept more interesting than the stable equilibrium or motion is the attractor. In general, the attractor is a subset of M with a simple or complicated geometric structure [Mil], [Geo92]. For our purposes the following definition will suffice.

Definition 2.3.1. An *attractor* is an equilibrium $\overline{\mathbf{u}} \in M$ for which a neighborhood $\mathcal{B} \subset M$ of $\overline{\mathbf{u}}$ exists, such that for every $\mathbf{u}_0 \in \mathcal{B}$ we have

$$\lim_{t\to\infty} \mathbf{u}(t,\mathbf{u}_0) = \overline{\mathbf{u}}, \qquad (2.3.1)$$

or, equivalently, there exists a corresponding neighborhood $\mathcal{B}_1 \subset M$ of $\mathbf{0}$ such that for every $\mathbf{v}_0 \in \mathcal{B}_1$ we have $\lim_{t\to\infty} \mathbf{v}(t,\mathbf{v}_0) = \mathbf{0}$.

This means that $\overline{\mathbf{u}}$ attracts all phase trajectories started from $\mathcal{B}$. The largest neighborhood $\mathcal{B}$ the points of which are attracted by $\overline{\mathbf{u}}$ is called the *basin of attraction* of $\overline{\mathbf{u}}$.

Definition 2.3.2. Let $\mathcal{B}$ be the basin of attraction of $\overline{\mathbf{u}}$. If $\mathcal{B} = M$, then $\overline{\mathbf{u}}$ is called a *global attractor*, while if $\mathcal{B}$ is strictly included in M, $\overline{\mathbf{u}}$ is called a *local attractor*.

There are concrete examples showing that an attractor is not necessary stable. Whence

Definition 2.3.3. The equilibrium $\overline{\mathbf{u}}$ is asymptotically stable if it is stable and is an attractor.

In applications, asymptotic stability is always understood even if it is referred to as stability. Sometimes, the term stable equilibrium means attractor. Therefore, since the label may be inadequate, the reader must carefully look at its meaning in each paper.

In [J76], the *conditional stability* is a stability for initial energy from some strict subset of M and *unconditional stability*, for every initial energy from M. In the case of asymptotic stability these concepts correspond to the local or global attractivity of $\overline{\mathbf{u}}$, *i.e.* to $\mathcal{B} \subset M$ or $\mathcal{B} = M$. Global attractivity is expected to hold if $\overline{\mathbf{u}}$ is the unique attractor, while as the turbulent regime is approaching and several stationary states are plausible too, as a result of bifurcation, local attractivity is expected.

It is understood that attractivity and attractors are defined only if a global existence theorem for the solutions of problem (2.2.1), (2.2.2) can be proved. In this case $t^* = \infty$. In addition, in order to associate a dynamical or semidynamical system, we suppose from the very beginning $t \in \mathbf{R}$ or $t \in \mathbf{R}^+$ respectively. This is why we did not mention this earlier.

In differential equations theory, a global result concerns every t and data (initial and boundary conditions, every value of the parameters). In dynamical systems theory globality is associated with attractivity for the entire phase space M.

Finally, in hydrodynamic and hydromagnetic stability theory, "global" can refer to all possible perturbations. This is why, every time that we characterize a notion as global we must say with respect to t, data, domain of attraction or perturbation.

Although related to stability question, the existence results are not the concern of stability theory and are not presented in this book.

As standard in hydrodynamic and hydromagnetic stability theory, let R_G be the *limit of global stability* of $\overline{\mathbf{u}}$ with respect to all perturbations of M. Let R_L be the *limit of linear stability* of $\overline{\mathbf{u}}$, *i.e.* for perturbations satisfying (2.2.5), (2.2.6). Our general aim is to derive bounds for R_G and R_L and to study some situations in which $R_L = R_G$. In the linear case we assume that the perturbations belong to some classes which are physically plausible or experimentally observed and lead to feasible mathematical treatments by means of the existing tools. In most cases they are chosen to be normal modes.

2.4   Linear stability

The representation formulae for the solutions of linear ode's (Section 2.4.1) or pde's (Section 2.4.2) in terms of the eigenvectors of the operator defining these equations are presented. In the nonlinear case the contribution of the linearized operators is minutely shown. In the finite-dimensional case the algebraic and geometric multiplicities of the eigenfunctions are largely exploited.

2.4.1 *Finite-dimensional case*

Definition 2.4.1. An equilibrium $\bar{\mathbf{u}}$ of the dynamical system associated with (2.2.1), (2.2.2) is *linearly stable* if the equilibrium from the origin of M of the linearized problem (2.2.5), (2.2.6) is stable.

In order to study this stability problem we first recall a representation formula for the solution $\mathbf{u}$ of the Cauchy problem (2.2.2) for the linear evolution equation in a finite-dimensional Hilbert space $\mathbf{R}^n$

$$\frac{d}{dt}\mathbf{u} = \mathbf{A}\mathbf{u}, \tag{2.4.1}$$

where $\mathbf{u} = (u_1, \ldots, u_n)$, $u_i : [a, b] \to \mathbf{R}$, $u_i = u_i(t)$, $i = 1, \ldots, n$, and $\mathbf{A}$ is an $n \times n$ matrix.

Denote by $\lambda_1, \ldots, \lambda_l$ the eigenvalues of $\mathbf{A}$ and let $\mathbf{v}_1, \ldots, \mathbf{v}_l$ stand for l corresponding eigenvectors. The eigenvalues λ_k, $k = 1, \ldots, l$ are the roots of the characteristic equation $\det(\mathbf{A} - \lambda\mathbf{I}) = 0$, therefore, in general, they have various algebraic $m_{a_k}(\lambda_k)$ and geometric $m_{g_k}(\lambda_k)$ multiplicities. To different eigenvalues, linearly-independent eigenvectors correspond. Thus, $\mathbf{v}_1, \mathbf{v}_2, \ldots, \mathbf{v}_l$ are supposed to be linearly independent. If $\mathbf{A}$ has n mutually distinct eigenvalues, then the corresponding eigenvectors form a basis of $\mathbf{R}^n$ or of $\mathbf{C}^n$ (if some of the eigenvalues are complex, then instead of (2.4.1) we consider the corresponding complexified form). Then, for any fixed t, $\mathbf{u}$ is a linear combination of n linearly independent eigenvectors $\mathbf{v}_i$, while if we take into account the time dependence, then $\mathbf{u}$ is a linear combination of $\mathbf{v}_i e^{\lambda_i t}$. These vectors form a basis for $\mathcal{R}(A)$, where, for any fixed t, A is a linear operator defined by the matrix $\mathbf{A}$. If at least some λ_j is multiple, the number of linearly independent eigenvectors can be smaller than n and, in order to form a basis of $\mathbf{R}^n$ or, in general, of $\mathbf{C}^n$, some of the associated eigenvectors of λ_j must be considered, *i.e.* we obtain $\mathbf{C}^n = \cup_j \mathcal{N}(\mathbf{A} - \lambda_j\mathbf{I})^{m_{a_j}}$. This was natural because the operator corresponding to $\mathbf{A}$ is compact (Appendix 2) and $\sum_j m_{a_j} = n$. Then the representation of $\mathbf{u}$ in this basis changes accordingly. For the sake of an easier presentation, from now on we consider only (2.4.1) in $\mathbf{R}^n$.

For stability purposed we are interested in two aspects: how to construct the eigenvectors and the associate eigenvectors, and, especially, which is the dependence on the time t of their coefficients. If all eigenvalues are simple, then $\mathbf{v}_1, \ldots, \mathbf{v}_n$ were constructed as minima of the Rayleigh quotient functional in successive spaces of vectors orthogonal to the already constructed eigenvectors. This procedure holds in finite as well as in infinite-dimensional Hilbert spaces, for symmetric operators for which the eigenvalues are real and, hence, these minima are real numbers too. In the complex case (at least in the finite dimensional case) more or less similar procedures were imagined.

In order to determine the coefficients, functions of t, of the solution $\mathbf{u}$, the knowledge of algebraic and geometric multiplicity must be taken into account. In addition, the representation of the solution $\mathbf{u}$ must be introduced in the equation (2.4.1) and in the condition (2.2.2) in order to obtain the Cauchy problems for the coefficients. Several cases must be considered.

• $\lambda_i \neq \lambda_j$ *for* $i \neq j$ $i, j = 1, \ldots, n$. In this case the eigenvectors $\mathbf{v}_1, \ldots, \mathbf{v}_n$ form a basis of $\mathbf{R}^n$ such that for every fixed t, $\mathbf{u}(t)$ is a linear combination of them, *i.e.* $\mathbf{u}(t) = \sum_{i=1}^n C_i(t)\mathbf{v}_i$. Substituting this expression into (2.4.1), denoting $\cdot \equiv \frac{d}{dt}$ and taking into account that $\mathbf{A}\mathbf{v}_i = \lambda_i \mathbf{v}_i$ for every $i = 1, \ldots, n$, we obtain

$$(\dot{C}_1 - \lambda_1 C_1)\mathbf{v}_1 + \ldots + (\dot{C}_n - \lambda_n C_n)\mathbf{v}_n = 0.$$

Then the linear independence of $\mathbf{v}_1, \ldots, \mathbf{v}_n$ implies for the coefficients $C_i(t)$ the following equations $(\dot{C}_i - \lambda_i C_i) = 0$, $\forall i = 1, \ldots, n$ and the solutions of these equations are $C_i(t) = C_{i0}e^{\lambda_i t}$, where C_{i0} are constants. These constants are uniquely determined in terms of the coordinates of the initial state $\mathbf{u}_0$ from the system of n affine equations in C_{i0} corresponding to the vector equation $\mathbf{u}_0 = \sum_{i=1}^n C_{i0}\mathbf{v}_i$. The uniqueness follows from the fact that the Kramer determinant is $\det(v_{ij}) \neq 0$, where v_{ij} is the j-th component of $\mathbf{v}_i$, and taking into account that $\mathbf{v}_i$ are linearly independent. Therefore we obtain the following representation for the solution $\mathbf{u}$ of (2.2.2), (2.4.1)

$$\mathbf{u}(t) = \sum_{i=1}^n C_{i0}e^{\lambda_i t}\mathbf{v}_i. \tag{2.4.2}$$

Remark 2.4.1. Formula (2.4.2) holds also if some eigenvalues are multiple roots of the characteristic equation, but their geometric multiplicity is equal to the algebraic multiplicity (*e.g.* if $\mathbf{A}$ is symmetric).

In the following we assume that the system consisting in the eigenvectors and the associated eigenvectors of $\mathbf{A}$, in each algebraic eigenspace is orthonormal. The orthonormalization is carried out only after the representation formula (2.4.2) is written in terms of n linearly independent combinations of the elements of the algebraic eigenspace.

• $\lambda_1 = \lambda_2 = \lambda_0$, $\lambda_i \neq \lambda_j$ *for* $i \neq j$, $i, j = 2, \ldots, n$. By Remark 2.4.1, only the case $m_a \neq m_g$ must be considered. In this case $m_a = 2$ and $m_g = 1$. Let $\mathbf{v}_1$ be the eigenvector of $\mathbf{A}$ corresponding to λ_0 and denote by $\mathbf{w}$ the associate eigenvector of order one of λ_0, the other eigenvalues of $\mathbf{A}$ being as in the first case. We have $(\mathbf{A} - \lambda_0\mathbf{I})^2\mathbf{w} = 0$, implying $(\mathbf{A} - \lambda_0\mathbf{I})[(\mathbf{A} - \lambda_0\mathbf{I})\mathbf{w}] = 0$, and, thus, $(\mathbf{A} - \lambda_0\mathbf{I})\mathbf{w} = \mathbf{v}_1$, or, equivalently,

$$\mathbf{A}\mathbf{w} = \lambda_0\mathbf{w} + \mathbf{v}_1.$$

A basis for $\mathbf{C}^n$ is $\mathbf{v}_1, \mathbf{w}, \mathbf{v}_3, \ldots, \mathbf{v}_n$ and, therefore, a general solution of (2.2.2), (2.4.1) reads $\mathbf{u} = C_1(t)\mathbf{v}_1 + C_2(t)\mathbf{w} + \sum_{i=3}^n C_i(t)\mathbf{v}_i$. Substituting this expression into (2.4.1), following the same procedure and taking into account the relation

$\mathbf{A}\mathbf{w} = \lambda_0\mathbf{w} + \mathbf{v}_1$, we obtain

$$(\dot{C}_1 - \lambda_0 C_1)\mathbf{v}_1 + \dot{C}_2\mathbf{w} - (\lambda_0\mathbf{w} + \mathbf{v}_1)C_2 + \sum_{i=3}^{n}(\dot{C}_i - \lambda_i C_i)\mathbf{v}_i = 0,$$

implying $C_i = C_{i0}e^{\lambda_i t}\mathbf{v}_i$, $i = 2, \ldots, n$ and $\dot{C}_1 - \lambda_0 C_1 = C_2$, whence $C_1(t) = (C_{10} + C_{20}t)e^{\lambda_0 t}$, so the representation of the solution $\mathbf{u}$ reads

$$\mathbf{u}(t) = (C_{10} + C_{20}t)e^{\lambda_0 t}\mathbf{v}_1 + C_{20}e^{\lambda_0 t}\mathbf{w} + \sum_{i=3}^{n}C_{i0}e^{\lambda_i t}\mathbf{v}_i. \qquad (2.4.3)$$

Remark 2.4.2. The multiplicity of λ_0 does not influence the part of the solution representation containing the other eigenvalues. The vectors $\mathbf{w}$ and $\mathbf{v}_1$ are linearly independent, but their coefficients in the representation of $\mathbf{u}$ are related, namely the coefficient of t in the coefficient of $\mathbf{v}_1$ is the same as the coefficient of $\mathbf{w}$.

Remark 2.4.3. In order to derive some formulae briefly presented in [Keld], we use a somewhat nonstandard notation (Appendix 2). Recall that an associate eigenvector of order k of an operator $\mathbf{A}$ (corresponding to the matrix $\mathbf{A}$) satisfies the equation $(\mathbf{A} - \lambda\mathbf{I})^{k+1}\mathbf{w}_k = 0$. Then $\mathbf{w}_0$ is an eigenvector. An appropriate decomposition of the operator $(\mathbf{A} - \lambda\mathbf{I})^{k+1}$ shows that the associate vectors of order $l \leq k$ are associate vectors of order $k + 1$ too, while $\mathbf{w}_k$ is not an associate vector of order $l < k$. For instance, $\mathbf{w}_0$ is an eigenvector and it is an associate eigenvector of an arbitrary order. However, $\mathbf{w}_k$ does not satisfy any equation $(\mathbf{A} - \lambda\mathbf{I})^l$, for $l \leq k$. If to $\mathbf{w}_k$ we add linear combinations of $\mathbf{w}_{k-1}, \ldots, \mathbf{w}_0$, which are associated eigenvectors of order $k - 1, k - 2, \ldots, 0$ respectively, the sum is still an associate eigenvector of order k. It is this property that determine us to do the orthonormalization only after knowing which among the associated eigenvectors occur in the representation formulae.

 • $\lambda_1 = \lambda_2 = \lambda_3 = \lambda_0$, $\lambda_i \neq \lambda_j$ for $i \neq j$, $i, j = 3, \ldots, n$. In this case $m_a = 3$ and two subcases may occur. In the first, there are two orthogonal eigenvectors $\mathbf{v}_1$ and $\mathbf{v}_2$ corresponding to λ_0, i.e. $m_g(\lambda_0) = 2$. Then, as in the previous case, if $\mathbf{w}$ stands for an associate eigenvector of λ_0, then $\mathbf{v}_1, \mathbf{v}_2, \mathbf{w}, \mathbf{v}_3, \ldots, \mathbf{v}_n$ is a basis of $\mathbf{R}^n$. In this subcase $A\mathbf{w} = \lambda_0\mathbf{w} + \mathbf{v}_1 + a\mathbf{v}_2$, where a is an arbitrary constant, and the representation for $\mathbf{u}$ reads

$$\mathbf{u}(t) = \left[(C_{10} + C_{30}t)\mathbf{v}_1 + (C_{20} + aC_{30}t)\mathbf{v}_2 + C_{30}\mathbf{w}\right]e^{\lambda_0 t} + \sum_{i=1}^{4}C_{i0}e^{\lambda_i t}\mathbf{v}_i. \qquad (2.4.4)$$

Since, by Remark 2.4.3, $C_{10}\mathbf{v}_1 + C_{20}\mathbf{v}_2 + C_{30}\mathbf{w} = \tilde{\mathbf{w}}_1$ is an associate eigenvector of order 1 and $C_{30}(\mathbf{v}_1 + a\mathbf{v}_2) = \tilde{\mathbf{w}}_0$ is still an eigenvector, the term in $e^{\lambda_0 t}$ reads $(\tilde{\mathbf{w}}_1 + t\tilde{\mathbf{w}}_0)e^{\lambda_0 t}$, where $\tilde{\mathbf{w}}_0$ and $\tilde{\mathbf{w}}_1$ are linearly independent.

In the second subcase, there is only one eigenvector $\mathbf{v}_1$ corresponding to λ_0, therefore the basis of $\mathbf{R}^n$ has the orthogonal vectors $\mathbf{v}_1, \mathbf{v}_4, \ldots, \mathbf{v}_n$. The other two can be taken from the associated eigenvectors satisfying $(\mathbf{A} - \lambda_0\mathbf{I})^2\mathbf{w} = \mathbf{0}$, and $\mathbf{z}$, satisfying $(\mathbf{A} - \lambda_0\mathbf{I})^3\mathbf{z} = 0$, where $\mathbf{A}\mathbf{w} = \lambda_0\mathbf{w} + \mathbf{v}_1$ and $\mathbf{A}\mathbf{z} = \lambda_0\mathbf{z} + \mathbf{w}$. Thus $\mathbf{u}(t) = C_1(t)\mathbf{v} + C_2(t)\mathbf{w} + C_3(t)\mathbf{z} + \sum_{i=4}^{n} C_i(t)\mathbf{v}_i$, implying $\dot{C}_1 - C_1\lambda_0 - C_2 = 0$, $\dot{C}_2 - C_2\lambda_0 - C_3 = 0$, $\dot{C}_3 - C_3\lambda_0 = 0$ $\dot{C}_i - C_i\lambda_0 = 0$, $i = 4,\ldots,n$, leading to the representation

$$\mathbf{u}(t) = (C_{10} + C_{20}t + C_{30}\frac{t^2}{2})e^{\lambda_0 t}\mathbf{v}_1 + (C_{20} + C_{30}t)e^{\lambda_0 t}\mathbf{w}$$
$$+ C_{30}e^{\lambda_0 t}\mathbf{z} + \sum_{i=1}^{4} C_{i0}e^{\lambda_i t}\mathbf{v}_i, \tag{2.4.5}$$

where $\mathbf{v}_1$, $\mathbf{w}$ and $\mathbf{z}$ are linearly independent. Denote $C_{10}\mathbf{v} + C_{20}\mathbf{w} + C_{30}\mathbf{z} = \tilde{\mathbf{w}}_2$, $C_{20}\mathbf{v}_1 + C_{30}\mathbf{w} = \tilde{\mathbf{w}}_1$, and $C_{30}/2\mathbf{v}_1 = \tilde{\mathbf{w}}_0$. Then the terms in $e^{\lambda_0 t}$ read $(\tilde{\mathbf{w}}_2 + t\tilde{\mathbf{w}}_1 + \frac{t^2}{2}\tilde{\mathbf{w}}_0)e^{\lambda_0 t}$.

Here we assumed that the equation $(\mathbf{A} - \lambda_0\mathbf{I})\mathbf{w} = \mathbf{v}_1$ has a single solution $\mathbf{w}$. Supposing that there are two orthogonal solutions $\mathbf{w}_1$ and $\mathbf{w}_2$ such that $\mathbf{A}\mathbf{w}_1 = \lambda_0\mathbf{w}_1 + \mathbf{v}_1$ and $\mathbf{A}\mathbf{w}_2 = \lambda_0\mathbf{w}_2 + \mathbf{v}_1$ we obtain the representation

$$\mathbf{u}(t) = \left[C_{10} + (C_{20} + C_{30})t\right]e^{\lambda_0 t}\mathbf{v}_1 + C_{20}e^{\lambda_0 t}\mathbf{w}_1 + C_{30}e^{\lambda_0 t}\mathbf{w}_2 + \sum_{i=1}^{4} C_{i0}e^{\lambda_i t}\mathbf{v}_i. \tag{2.4.6}$$

Denoting $C_{10}\mathbf{v}_1 + C_{20}\mathbf{w}_1 + C_{30}\mathbf{w}_2 = \tilde{\mathbf{w}}_1$, $(C_{20} + C_{30})\mathbf{v}_1 = \tilde{\mathbf{w}}_0$, the term in $e^{\lambda_0 t}$ reads $(\tilde{\mathbf{w}}_1 + t\tilde{\mathbf{w}}_0)e^{\lambda_0 t}$. Therefore, so far we studied the only case when the eigenvalue λ_1 (denoted by λ_0) was multiple.

• Consider the general case when the matrix $\mathbf{A}$ in (2.4.1) has $s < n$ distinct eigenvalues and denote them by $\mu_1, \mu_2, \ldots, \mu_s$, $\mu_i \neq \mu_j$ for every $i, j = 1, \ldots, s$. Assume that μ_r has eigenvectors $\mathbf{w}_0^{(r)}$, associate eigenvectors of order one $\mathbf{w}_1^{(r)}$, and so on up to the associated eigenvectors $\mathbf{w}_{k_r}^{(r)}$ of order k_r. Hence, the solution $\mathbf{u}$ of (2.4.1) has the representation [Keld]

$$\mathbf{u}(t) = e^{\mu_1 t}\left[\mathbf{w}_{k_1}^{(1)} + \mathbf{w}_{k_1-1}^{(1)}t + \ldots + \frac{t^{k_1}}{k_1!}\mathbf{w}_0^{(1)}\right] + e^{\mu_2 t}\left[\mathbf{w}_{k_2}^{(2)} + \mathbf{w}_{k_2-1}^{(2)}t + \ldots + \frac{t^{k_2}}{k_2!}\mathbf{w}_0^{(2)}\right]$$
$$+ \ldots + e^{\mu_s t}\left[\mathbf{w}_{k_s}^{(s)} + \mathbf{w}_{k_s-1}^{(s)}t + \ldots + \frac{t^{k_s}}{k_s!}\mathbf{w}_0^{(s)}\right],$$
$$\tag{2.4.7}$$

such that $k_1 + \ldots + k_s = n$. By construction, each of the eigenvectors or associated eigenvectors $\mathbf{w}_{k_r-q_r}^{(r)}$, $r = 1, \ldots, s$, $q_r = 0, \ldots, k_r$ can be uniquely determined in terms of $\mathbf{u}_0$.

From the dynamical point of view, another form of (2.4.7) is more convenient. In order to write it let us examine the particular representation of $\mathbf{u}$, assuming that the Gram-Schmidt procedure was applied such that the algebraic eigenspace is orthonormal. Then taking the inner product of (2.4.2), written for $t = 0$, by $\mathbf{v}_k$ we have $C_{i0} = (\mathbf{u}_0, \mathbf{v}_i)$. Similarly, from (2.4.3) at $t = 0$ we have $C_{10} = (\mathbf{u}_0, \mathbf{v}_1)$, $C_{20} = (\mathbf{u}_0, \mathbf{w})$, $C_{i0} = (\mathbf{u}_0, \mathbf{v}_i)$, for $i = 3,\ldots,n$, in (2.4.4) at $t = 0$ we have $C_{10} = (\mathbf{u}_0, \mathbf{v}_1) = u_{01}$, $C_{20} = (\mathbf{u}_0, \mathbf{v}_2) = u_{02}$, $C_{30} = (\mathbf{u}_0, \mathbf{w}) = u_{03}$, $C_{i0} = (\mathbf{u}_0, \mathbf{v}_i) = u_{0i}$, for

$i = 4, \ldots, n$. Then (2.4.2) reads

$$\mathbf{u}(t, \mathbf{u}_0) = \sum_{i=1}^{n} u_{0i} e^{\lambda_i t} \mathbf{v}_i = \begin{pmatrix} u_{01} e^{\lambda_1 t} \mathbf{v}_1 \\ u_{02} e^{\lambda_2 t} \mathbf{v}_2 \\ \cdots \\ u_{0n} e^{\lambda_n t} \mathbf{v}_n \end{pmatrix} = e^{\lambda_1 t} \begin{pmatrix} v_{11} & 0 & \cdots & 0 \\ v_{12} & & & \\ & \cdots & & \\ v_{1n} & 0 & \cdots & 0 \end{pmatrix} \cdot \begin{pmatrix} u_{01} \\ u_{02} \\ \cdots \\ u_{0n} \end{pmatrix}$$

$$+ e^{\lambda_2 t} \begin{pmatrix} 0 & v_{21} & 0 & \cdots & 0 \\ 0 & v_{22} & & & \\ & \cdots & & & \\ 0 & v_{2n} & 0 & \cdots & 0 \end{pmatrix} \cdot \begin{pmatrix} u_{01} \\ u_{02} \\ \cdots \\ u_{0n} \end{pmatrix} + \cdots + e^{\lambda_n t} \begin{pmatrix} 0 & \cdots & 0 & v_{n1} \\ 0 & \cdots & 0 & v_{n2} \\ & \cdots & & \\ 0 & \cdots & 0 & v_{nn} \end{pmatrix} \cdot \begin{pmatrix} u_{01} \\ u_{02} \\ \cdots \\ u_{0n} \end{pmatrix}$$

$$= \left(\mathbf{P}_{01} e^{\lambda_1 t} + \cdot + \mathbf{P}_{0n} e^{\lambda_n t} \right) \mathbf{u}_0,$$

where $\mathbf{P}_{0i}$ are matrices the entries of which are constants, *i.e.* polynomials of order zero in t. Similarly, letting λ_1 instead of λ_0, (2.4.3) reads

$$\mathbf{u}(t, \mathbf{u}_0) = \mathbf{u}_{01} e^{\lambda_1 t} \mathbf{v}_1 + \mathbf{u}_{02} e^{\lambda_1 t} \mathbf{w} + \mathbf{u}_{02} t e^{\lambda_1 t} \mathbf{v}_1 + \sum_{i=3}^{n} \mathbf{P}_{0i} e^{\lambda_i t} \mathbf{u}_0$$

$$= e^{\lambda_1 t} \begin{pmatrix} v_{11} & v_{11} t + w_1 & 0 & \cdots & 0 \\ v_{12} & v_{12} t + w_2 & 0 & \cdots & 0 \\ & \cdots & & & \\ v_{1n} & v_{1n} t + w_n & 0 & \cdots & 0 \end{pmatrix} \cdot \begin{pmatrix} u_{01} \\ u_{02} \\ \cdots \\ u_{0n} \end{pmatrix} + \sum_{i=3}^{n} \mathbf{P}_{0i} e^{\lambda_i t} \mathbf{u}_0$$

$$= \left(\mathbf{P}_{11} e^{\lambda_1 t} + \sum_{i=3}^{n} \mathbf{P}_{0i} e^{\lambda_i t} \right) \mathbf{u}_0,$$

where $\mathbf{P}_{11}$ is an $n \times n$ matrix the entries of which are polynomials of order one in t. In the same way (2.4.5) can be written as

$$\mathbf{u}(t, \mathbf{u}_0) = e^{\lambda_1 t} \begin{pmatrix} v_1 & v_1 t + w_1 & v_1 \frac{t^2}{2} + w_1 t + z_1 & 0 & \cdots & 0 \\ v_2 & v_2 t + w_2 & v_2 \frac{t^2}{2} + w_2 t + z_2 & 0 & \cdots & 0 \\ & \cdots & & & & \\ v_n & v_n t + w_n & v_n \frac{t^2}{2} + w_n t + z_n & 0 & \cdots & 0 \end{pmatrix} \cdot \begin{pmatrix} u_{01} \\ u_{02} \\ \cdots \\ u_{0n} \end{pmatrix}$$

$$+ \sum_{i=4}^{n} \mathbf{P}_{0i} e^{\lambda_i t} \mathbf{u}_0 = \left(\mathbf{P}_{21} e^{\lambda_1 t} + \sum_{i=4}^{n} \mathbf{P}_{0i} e^{\lambda_i t} \right) \mathbf{u}_0.$$

In this way (2.4.7) reads

$$\mathbf{u}(t, \mathbf{u}_0) = \sum_{i=1}^{s} \mathbf{P}_i e^{\lambda_i t} \mathbf{u}_0, \tag{2.4.8}$$

where P_i are $n \times n$ matrices the entries of which are polynomials in t of degree at most k_i (we recall that $\mathbf{w}_{k_j - l}^{(j)}$ can be expressed in terms of $(\mathbf{u}_0, \mathbf{x}_k)$, where $\mathbf{x}_k$ is a notation for eigenvectors and associated eigenvectors).

Remark 2.4.4. In a standard reasoning $\mathbf{w}^{(j)}_{k_j-l}$ have the simplest possible form $\tilde{\mathbf{w}}^{(j)}_{k_j-l}$ and in the formula corresponding to (2.4.7) they are multiplied by some constants $A_{k_j l}$, determined from the boundary conditions (2.2.2). Introduce the notation $e^{\mathbf{A}t} = \mathbf{I} + \mathbf{A}t + \mathbf{A}^2\frac{t^2}{2!} + \mathbf{A}^3\frac{t^3}{3!} + \ldots$ and take into account that the (generalized) eigenspace $\{\tilde{\mathbf{w}}^{(j)}_{k_j-l}\}$ is a basis for $\mathbf{R}^n$. Therefore $\mathbf{u}_0 = \sum_{j,l} A_{k_j l}\tilde{\mathbf{w}}^{(j)}_{k_j-l}$. Moreover $\mathbf{A}^s\tilde{\mathbf{w}}^{(j)}_0 = \lambda^s_j\tilde{\mathbf{w}}^{(j)}_0$, implying

$$
\begin{aligned}
e^{\mathbf{A}t}\tilde{\mathbf{w}}^{(j)}_0 &= (\mathbf{I} + \mathbf{A}t + \mathbf{A}^2\frac{t^2}{2!} + \ldots)\tilde{\mathbf{w}}^{(j)}_0 \\
&= \tilde{\mathbf{w}}^{(j)}_0 + \lambda_j t\tilde{\mathbf{w}}^{(j)}_0 + \lambda^2_j\frac{t^2}{2!} + \lambda^3_j\frac{t^3}{3!}\tilde{\mathbf{w}}^{(j)}_0 + \ldots \\
&= e^{\lambda_j t}\tilde{\mathbf{w}}^{(j)}_0, \\
e^{\mathbf{A}t}\tilde{\mathbf{w}}^{(j)}_1 &= e^{\lambda_j t}[\tilde{\mathbf{w}}^{(j)}_1 + \tilde{\mathbf{w}}^{(j)}_0 t].
\end{aligned}
$$

Similar equalities hold for $\tilde{\mathbf{w}}^{(j)}_{k_j-l}$, so (2.4.8) has the following equivalent form

$$
\mathbf{u}(t, \mathbf{u}_0) = e^{\mathbf{A}t}\mathbf{u}_0. \tag{2.4.9}
$$

For each fixed t, $e^{\mathbf{A}t} : \mathbf{R}^n \to \mathbf{R}^n$ is a linear operator A and it plays the role of an evolution operator ϕ_t for the dynamical system Φ associated with (2.4.1), (2.2.2). Due to the fact that Φ is also defined as the group $\{\phi_t\}$ of its values ϕ_t, it follows that $e^{\mathbf{A}t}$ generates this dynamical system and it is called the infinitesimal generator of Φ. The operators $e^{\mathbf{A}t}$ are compact (Appendix 2), therefore they have a finite number of eigenvalues $e^{\lambda_j t}$ corresponding to the eigenvectors $\tilde{\mathbf{w}}^{(j)}_0$ and associate eigenvectors $\mathbf{w}^{(j)}_{k_j-l}$.

Assume that (2.4.1) represents the linearized equation of the nonlinear vector equation (2.2.1) in $M = \mathbf{R}^n$. Writing (2.2.1) as

$$
\frac{d}{dt}\mathbf{u} = \mathbf{A}\mathbf{u} + \mathbf{f}(\mathbf{u}) - \mathbf{f_u}(\overline{\mathbf{u}})\mathbf{u}, \tag{2.2.1$'$}
$$

where $\mathbf{A} = \mathbf{f_u}(\overline{\mathbf{u}})$, around some equilibrium $\overline{\mathbf{u}}$ or, equivalently, $\frac{d}{dt}\mathbf{u} = \mathbf{A}\mathbf{u} + \mathbf{h}(\mathbf{u})$, where $\mathbf{h_u}(\overline{\mathbf{u}}) = \mathbf{0}$, by Lagrange method of variation of coefficients, the solution of (2.2.1) formally reads

$$
\mathbf{u}(t, \mathbf{u}_0) = e^{\mathbf{A}t}\mathbf{u}_0 + \int_0^t \mathbf{h}(\mathbf{u}(s))e^{-\mathbf{A}(s-t)}ds, \tag{2.4.9$'$}
$$

revealing a suitable decomposition of the solution space of (2.2.1$'$) into a direct sum of a finite-dimensional part, *i.e.* the eigenspace of the linearized operator and an infinite-dimensional complementary subspace. In fact, (2.4.9$'$) is not the closed-form of the solution of (2.2.1$'$), but only an integral form of Equation (2.2.1$'$).

The form (2.4.8) shows that if all eigenvalues have a negative real part, then $\mathbf{u}$ decreases to zero as $t \to \infty$. This means that the null solution of (2.4.1), (2.2.2) is (linearly) asymptotically stable. In addition, the Lyapunov theorem upon the first approximation states: *if $\mathcal{R}e\lambda_i < 0$, $i = 1,\ldots,n$, then $\overline{\mathbf{u}}$ is (nonlinearly) asymptotically stable* for (2.2.1).

Suppose that at least one eigenvalue λ_j of $\mathbf{A}$ has $\mathcal{R}e\lambda_j > 0$, while the remained eigenvalues have $\mathcal{R}e\lambda_i < 0$, $i = 1,\ldots,j-i,j+i,\ldots,n$. In this case the representation formula (2.4.8) shows that $\overline{\mathbf{u}}$ is linearly asymptotically unstable. In addition, the Perron theorem states: *if $\mathcal{R}e\lambda_j > 0$, $\mathcal{R}e\lambda_i < 0$, $i = 1,\ldots,j-i,j+i,\ldots,n$, then $\overline{\mathbf{u}}$ is (nonlinearly) asymptotically unstable* for (2.2.1).

These two theorems are the two parts of a linearization principle for (2.2.1). In general, by a *linearization principle* we mean a proposition asserting that a property valid for a linearized equation holds for the nonlinear equation too. We mention that the linearization principles for stability and those for bifurcation are some of the most important theorems in nonlinear analysis.

The Lyapunov and Perron theorems are criteria for asymptotic stability or instability respectively. They are valid for small perturbations $\mathbf{u}$ and only for those equilibria $\overline{\mathbf{u}}$ which are hyperbolic, *i.e.* the corresponding $\mathbf{A}$ has no eigenvalues with null real part. In addition, these theorems are valid for small neighborhood of $\mathbf{u}$, hence these results concern the conditional asymptotic stability. The proofs use the expansion of $\mathbf{f}$ in a Taylor (asymptotic) series involving bounded higher order Fréchet derivatives of f at $\overline{\mathbf{u}}$ multiplied by $(\mathbf{u} - \overline{\mathbf{u}})^k$, $k \geq 2$, enabling one to neglect such terms as irrelevant for the stability. The remaining terms are the linear ones. In this reasoning the splitting of the space of the nonlinear equation (2.2.1) as a direct sum of a finite-dimensional eigenspace of the linearized equation (2.1.5) and an infinite-dimensional complementary subspace is understood.

2.4.2 *Infinite-dimensional case*

Assume that in (2.2.1), (2.2.2) $\mathbf{u} \in \mathcal{B}$, where $\mathcal{B}$ is a Banach space, the mapping is such that the global existence and uniqueness theorem holds and let $\overline{\mathbf{u}}$ be an equilibrium point of the associated dynamical system. The linearized equation about $\overline{\mathbf{u}}$ is (2.2.5). Let us summarize the main characteristics of $\mathbf{A}$, the main steps followed and the main assumptions done in Section 2.4.1 in order to obtain the representation formula (2.4.8) and then the linearization principle. They are: $\mathbf{A}$ has constant entries, the corresponding operator is compact, the point spectrum is discrete, it has a finite number of eigenvalues of finite geometric and algebraic multiplicities, the mapping defining the nonlinear equation (2.4.9) is differentiable.

Looking at the spectra of various classes of operators in Appendix 2, we see that the most general convenient framework for the evolution equations (2.2.1) and (2.2.5) is a complex infinite-dimensional Hilbert space. In addition, the linear operator of the linearized perturbation equation (2.2.5) must be positive definite, compact and must possess a discrete spectrum, (in the sense of Mikhlin).

We might encounter difficulty induced by the nonconstant coefficients in the operator in (2.2.5).

The various mathematical models in Section 1.2 are nonlinear due to: the supplies, *e.g.* body forces, initial and boundary conditions and the advective terms from the rate of change of densities of global quantities. The presence of the linear terms defined by the Laplacian operator (Δ) is encouraging since $-\Delta$ is a positive definite operator in an appropriate Hilbert space. A complication can occur because in order to get rid of the pressure and of the equations expressing the solenoidal vector fields (velocity, magnetic field), a suitable projection P must be applied. In this way, instead of studying the operator $-\Delta$ we must investigate the operator $-P\Delta$. As it is only positive definite, we must extend it up to the selfadjoint operator $\tilde{A}$ which has real eigenvalues. The compactness for some related operators ensures the boundedness from below of spectra. This is essential for proving a linearization principle.

Consider the Cauchy problem $u(0) = u_0$ for a one-dimensional evolution equation, namely the first-order affine ode in $\mathbf{R}$

$$\dot{u} + a(t)u + h(t) = 0. \tag{2.4.10}$$

Assume that the functions a and h are of class $C^1(\mathbf{R})$, ensuring the existence and uniqueness of the solution $u(t, u_0)$. Similarly to (2.4.9'), its closed-form reads

$$u(t, u_0) = e^{-\int_0^t a(r)dr} u_0 - \int_0^t h(s) e^{-\int_s^t a(r)dr} ds \tag{2.4.11}$$

or, equivalently,

$$u(t, u_0) = e^{-\int_0^t a(r)dr} u_0 - \int_0^t h(t - \tau) e^{-\int_{t-\tau}^t a(r)dr} d\tau \tag{2.4.12}$$

or,

$$u(t, u_0) = e^{-\int_0^t a(r)dr} u_0 - \int_0^t h(t - \tau) e^{\int_0^{t-\tau} a(r)dr} e^{-\int_0^t a(r)dr} d\tau. \tag{2.4.13}$$

For $a = \text{const.}$ and $h(t) \equiv 0$, (2.4.11) becomes

$$u(t, u_0) = e^{-at} u_0 \tag{2.4.14}$$

and can also be written as

$$u(t, u_0) = \phi_t(u_0) \tag{2.4.14}$$

where $\phi_t = e^{-at}$ is the evolution operator at time t, $\phi_t : \mathbf{R} \to \mathbf{R}$, $\phi_t = \phi_t(u_0)$, and the corresponding $\Phi : \mathbf{R} \to \mathbf{R}^n$ is the associated dynamical system. If instead of imposing $t \in \mathbf{R}$ we require $t \in \mathbf{R}^+$, then ϕ_t is an operator of the semigroup $\{\phi_t\}_{t \geq 0}$. In this case the usual notation is $S_{(t)}$ instead of ϕ_t.

If $a = \text{const.}$ but $h(t)$ a function of t, then (2.4.13) yields

$$u(t, u_0) = S_{(t)} u_0 - \int_0^t h(t - \tau) S_{(\tau-t)} \circ S_{(t)} d\tau = S_{(t)}(u_0) - \int_0^t h(t - \tau) S_{(\tau)} d\tau,$$

where the semigroup property $S_{(\tau-t)} S_{(t)} = S_{(\tau-t+t)} = S_{(\tau)}$ was used. Therefore, the closed form of the solution of a nonlinear ode of affine type was expressed by means of the operators of this semigroup.

If h depends not on t but on u, *i.e.* instead of equation (2.4.10) we have the equation

$$\dot{u} + a(t)u + h(u) = 0, \qquad (2.4.15)$$

then (2.4.13) becomes

$$u(t, u_0) = S_t(u_0) - \int_0^t h(u(t - \tau))S_\tau d\tau, \qquad (2.4.16)$$

where $S_t = e^{-\int_0^t a(r)dr}$. Of course, (2.4.16) is no longer a closed-form solution of (2.4.15) but an integral form of (2.4.15).

Consider now an evolution linear equation in $\mathbf{R}^n$ defined by an $n \times n$ matrix $\mathbf{A}$ and having a nonlinear part $R(\mathbf{u})$

$$\frac{d}{dt}\mathbf{u} + \mathbf{A}\mathbf{u} = R(\mathbf{u}). \qquad (2.4.17)$$

Then, an expression similar to (2.4.16) reads

$$\mathbf{u}(t, \mathbf{u}_0) = S_t(\mathbf{u}_0) + \int_0^t R(\mathbf{u}(t - \tau))S_\tau d\tau, \qquad (2.4.18)$$

where $S_t = e^{-\mathbf{A}t} = \mathbf{I} - \mathbf{A}t + \frac{\mathbf{A}^2}{2!}t^2 - \ldots$ is a matrix depending on t. In general, if the entries of $\mathbf{A}$ depend on t, this is a series. If $\mathbf{A}$ has constant entries, then this sum is finite and its form depends on the algebraic and geometric multiplicity of the eigenvalues of $\mathbf{A}$. In fact, if $R \equiv 0$, (2.4.17) becomes identical to (2.4.1), and (2.4.8) to (2.4.18).

Similar reasonings can be done in an infinite-dimensional Hilbert space H for the Cauchy problem $\mathbf{u}(0) = \mathbf{u}_0$ for an evolution equation of the form (2.4.17). In this case instead of (2.4.18) we have

$$\mathbf{u}(t, \mathbf{u}_0) = exp\{-tA\}\mathbf{u}_0 + \int_0^t R(\mathbf{u}(t - \tau))exp(-\tau A)d\tau. \qquad (2.4.19)$$

An operatorial calculus proceeds similarly, but here as t runs over $\mathbf{R}^+$, $exp\{-tA\}$ generates a semigroup of operators, whose infinitesimal generator is the (linear) operator A. Like in the finite-dimensional case, if R is not a nonlinear mapping of $\mathbf{u}$ but a function of t, then instead of equation (2.4.19) we have the closed form solution of (2.4.17)

$$\mathbf{u}(t, \mathbf{u}_0) = exp\{-tA\}\mathbf{u}_0 + \int_0^t \mathbf{h}(t - \tau)exp(-\tau A)d\tau. \qquad (2.4.20)$$

Let us now discuss not equations but inequalities. Thus, if instead of an equation (2.4.10) or (2.4.15) in $\mathbf{R}$ we have inequalities

$$\dot{u} \leq -a(t)u - h(t), \qquad (2.4.21)$$

$$\dot{u} \leq -a(t)u - h(u), \qquad (2.4.22)$$

where u is a nonnegative absolutely continuous function and $a(t)$ and $h(t)$ are nonnegative Lebesgue integrable functions on $[0, T]$. Then the following Gronwall lemmas (inequalities)

$$u(t) \leq e^{-\int_0^t a(r)dr}\Big[u_0 - \int_0^t h(s)e^{\int_0^s a(r)dr}ds\Big], \qquad (2.4.23)$$

$$u(t) \leq e^{-\int_0^t a(r)dr}\Big[u_0 - \int_0^t h(u(s))e^{\int_0^s a(r)dr}ds\Big] \qquad (2.4.24)$$

hold. Formally, (2.4.23) and (2.4.24) are obtained by integrating equations (2.4.21) and (2.4.22) and, then, instead of writing that the solution $u(t)$ is equal to the found expression, we write that it is smaller or equal. The rigorous justification is: multiply (2.4.21) by $e^{\int_0^t a(r)dr}$ to obtain $\dfrac{d[u(t)e^{\int_0^t a(r)dr}]}{dt} - u(t)a(t)e^{\int_0^t a(r)dr} \leq - u(t)a(t)e^{\int_0^t a(r)dr} - h(t)e^{\int_0^t a(r)dr}$, implying $\dfrac{d[u(t)e^{\int_0^t a(r)dr}]}{dt} \leq -\dfrac{d}{dt}\int_0^t h(s)e^{\int_0^s a(r)dr}ds$ or, equivalently,

$$\frac{d}{dt}\Big[u(t)e^{\int_0^t a(r)dr} + \int_0^t h(s)e^{\int_0^s a(r)dr}ds\Big] \leq 0,$$

hence the function in the square bracket is decreasing from $u(0)$, whence (2.4.23). The inequality (2.4.24) follows similarly.

2.5   Prodi's linearization principle

In order to establish such a principle we start with the Cauchy problem (1.5.8) for the perturbation evolution equation (1.5.6) in N^1,

$$\frac{d}{dt}\mathbf{v} + \tilde{\mathbf{A}}\mathbf{v} = R(\mathbf{v}), \qquad (2.5.1)$$

where

$$\tilde{\mathbf{A}}\mathbf{v} = \mathbf{A}\mathbf{v} + M_{\bar{\mathbf{u}}}(\mathbf{v}), \qquad (2.5.2)$$

therefore ν is taken equal to 1. Then it is shown that (2.5.1), (1.5.8) can be written in the form (2.4.17), where A is replaced by $\tilde{A}$ and R, by its concrete form (1.5.7). At the end, from (2.4.17), the bound of the energy $\|\mathbf{v}(t, \mathbf{v}_0)\|$ in terms of $\|\mathbf{v}_0\|$ and time t gives us the criterion for the decay of the perturbation energy of the nonlinear equation.

This criterion uses the *fundamental hypothesis of Prodi*: let λ be a spectral value of $\tilde{A}$. Then there exists $\delta > 0$ such that

$$Re\{\lambda\} > \delta, \quad n = 1, 2, \dots. \qquad (2.5.3)$$

Prodi supposes that $\mathbf{u}$ is a hyperbolic equilibrium of the semigroup of operators which has $\tilde{A}$ as its infinitesimal generator.

Prodi's theorem 2.5.1 [Pro]. *If $\Omega \subset \mathbf{R}^2$ or $\mathbf{R}^3$ is bounded, $\partial\Omega$ is of class C^2 and the linearized operator $\tilde{A}$ has the spectrum bounded as in (2.5.3), then the basic solution $\bar{\mathbf{u}}$ of the strong incompressible N-S model (1.5.8), (1.5.6) is (nonlinearly) stable for small initial perturbations $\mathbf{v}_0$.*

Since inequality (2.5.3) is a sufficient condition for linear stability, Prodi's theorem states that the same inequality represents a nonlinear stability criterion too. This is the first part of the linearization principle and it was given in 1962. The second part was given in 1965 by Yudovich [Yu1]. Subsequently, other linearization principles were obtained for other physical situations. Those related to our research are those by David H. Sattinger [Sa], Jean-Pierre Guiraud and Gérard Iooss [GuI], Gérard Iooss [Ioo1], Klaus Kirchgässner and Peter Sorger, V. I. Yudovich [Yu1], Daniel D. Joseph [J70a] [J65], [J76] and [J66], Peter Herfort (see [Geo85]) and by Bruno Carbonaro [Carba1], [Carba2], Lidia Palese, Adelina Georgescu, Aldo Redaelli [GeoPalR96c], [GeoPalR00], Olga A. Ladyzhenskaya and Vsevolod A. Solonnikov [LadS67], Giuseppe Mulone and Franco Salemi [MuloS85], Salvatore Rionero [Rio78], Salvatore Rionero, Giovanni Paolo Galdi [RioG79], Giovanni Paolo Galdi and Brian Straughan [GaldS82]. Other details on this topics can be found in [Yu2], [Geo85], [EbS1], [EbS2], [GeoPalR96c].

A complete form of the linearization principle is a criterion for the equality $R_L = R_G$ of the two limits of linear and nonlinear stability. This is why this equality was recently taken as a definition of this principle.

Proof (sketch). Let λ be a real number $\lambda < 0$. Then, by (2.5.3), the resolvent $(\tilde{A} - \lambda I)^{-1}$ exists and, for $-\lambda > c_1^2/4$, we have $\|(\tilde{A} - \lambda I)^{-1}\|_{N \to N} \leq (-\lambda - c_1^2/4)^{-1}$. This last inequality follows by taking the inner product in N by $\mathbf{v}$ of the relation $\tilde{A}\mathbf{v} - \lambda\mathbf{v} = \mathbf{g}$, then using the inequality

$$|M_{\bar{\mathbf{u}}}\mathbf{v}| \leq c_1\|\mathbf{v}\| \tag{2.5.4}$$

for $\mathbf{v} \in N^1$, where $|\cdot|$ and $\|\cdot\|$ are the norms in $L^2(\Omega)$ and $N^1(\Omega)$ respectively, and the Young inequality. Since $-\tilde{A} : \mathcal{D}(-\tilde{A}) = N^2 \to N^1$ is a closed operator with $\overline{\mathcal{D}(-\tilde{A})} = N$, by Hille-Yosida theorem [Yos], it follows that the operator $-\tilde{A}$ is a generator of a semigroup of operators $exp\{-\tilde{A}t\}$ strongly continuous in N. Similarly, by multiplying $\tilde{A}\mathbf{v} - \lambda\mathbf{v} = \mathbf{g}$ by $\mathbf{v} \in \tilde{N} = \{\mathbf{v} \in N^2 \mid \tilde{A}\mathbf{v} \in N^1\}$, in the same condition for λ it follows that the restriction of $-\tilde{A}$ to $\tilde{N}$ is a generator of a semigroup of maps $exp\{-\tilde{A}t\}$ strongly continuous in N^1 satisfying the inequality $\|exp\{-\tilde{A}t\}\|_{N^1 \to N^1} \leq exp(-c_1^2 t/4)$.

This result allows us to argue that in the class (1.5.5) the solution of the affine equation corresponding to (2.5.1)

$$\frac{d}{dt}\mathbf{v} + \tilde{\mathbf{A}}\mathbf{v} = \mathbf{R}(t), \tag{2.5.5}$$

has an expression of the form (2.4.20), namely

$$\mathbf{v}(t) = exp\{-t\tilde{A}\}\mathbf{v}_0 + \int_0^t exp(-\tau\tilde{A})\mathbf{R}(t - \tau)d\tau,$$

where $exp\{-t\tilde{A}\}$ is the semigroup generated by $-\tilde{A}$ in N^1. Since, in particular, $\mathbf{R}(t)$ can be $R(\mathbf{u}(t))$, this yields the first part of the proof, namely the representation of the solution of (2.5.5) and (2.5.1). Now we must derive bounds for $\|\mathbf{v}\|$.

Remark that for the Cauchy problem (1.5.8), (2.5.5) the superposition principle holds, *i.e.* $\mathbf{v} = \mathbf{v}_1 + \mathbf{v}_2$, where $\mathbf{v}_1$ is the solution of the linear equation (2.5.5) where $\mathbf{R} = \mathbf{0}$, *i.e.*

$$\frac{d}{dt}\mathbf{v} + \tilde{\mathbf{A}}\mathbf{v} = \mathbf{0}, \tag{2.5.6}$$

and satisfies the nonhomogeneous initial conditions (1.5.8), while $\mathbf{v}_2$ is the solution of the affine equation (2.5.5) with the initial condition $\mathbf{v}_0 = \mathbf{0}$. Then, in order to obtain bounds for $\|\mathbf{v}_1\|$ equation (2.5.6) is multiplied by $\mathbf{v}$ in $N(\Omega)$, (2.5.4) and Gronwall lemma (2.4.23) are used, yielding

$$\|\mathbf{v}_1\| = \|exp\{-\tilde{A}t\}\|_{N\to N^1} \leq c_2/t, \tag{2.5.7}$$

where $\mathbf{v} = exp(-t\tilde{A})\mathbf{v}_0$.

Then multiplying (2.5.5) by $A\mathbf{v}$, applying the Young inequality and using (2.4.23) we get

$$\|\mathbf{v}_2\| \leq c_5 \int_0^t |\mathbf{R}(\tau)|^2 d\tau, \quad 0 \leq t \leq 1. \tag{2.5.8}$$

Using all these results and many standard inequalities from Hilbert spaces theory (*e.g.* Hölder inequality), calculus and semigroup theory, Prodi obtained the following estimation for $t > 0$ for the solution of (2.5.1), (1.5.8)

$$\|\mathbf{v}\|^2 + \int_0^t |A\mathbf{v}(\tau)|^2 d\tau \leq c_6 \{\|\mathbf{v}_0\|^2 + \int_0^t |\mathbf{R}(\tau)|^2 d\tau. \tag{2.5.9}$$

Furthermore, the assumption (2.5.3) concerning the location of the spectrum of $\tilde{A}$ in $\mathbf{C}$ is used to define a new function $\mathbf{w}(t) = \mathbf{v}(t)exp\{\delta t\}$, (we used this trick in proving (2.4.23)), which satisfies the equation

$$\frac{d}{dt}\mathbf{w} + \tilde{\mathbf{A}}_1\mathbf{w} = \mathbf{R}_1(t), \tag{2.5.10}$$

deduced from (2.5.5), where $\tilde{A}_1 = \tilde{A} - \delta I$, $\mathbf{R}_1(t) = exp(\delta t)\mathbf{R}(t)$. Then the spectrum of $\tilde{A}_1$ is situated in the right halfplane of the complex plane. Then, (2.5.9) for (2.5.10) implies

$$\|\mathbf{v}\|^2 + \int_0^t exp(-2\delta(t - \tau))|A\mathbf{v}(\tau)|^2 d\tau \leq c^* \{exp(-2\delta t)\|\mathbf{v}_0\|^2$$
$$+ \int_0^t exp(-2\delta(t - \tau))|\mathbf{R}(\tau)|^2 d\tau\}. \tag{2.5.11}$$

Since $\mathbf{R}_1(t)$ can be in particular $R(\mathbf{v})$, this means that for the solution of (2.5.1) we have the estimation (2.5.11), where $\mathbf{R}(\tau)$ is replaced by $R(\mathbf{v}(\tau))$. If in it we take into account the inequality

$$\|R(\mathbf{v})\|^2 \leq k_3 \|\mathbf{v}\|^3 |A\mathbf{v}|, \quad \mathbf{v} \in N^2(\Omega) \tag{2.5.12}$$

we obtain

$$\|\mathbf{v}(t)\|^2 + \int_0^t exp(-2\delta(t-\tau))|A\mathbf{v}(\tau)|^2 d\tau \leq c^* \{exp(-2\delta t)\|\mathbf{v}_0\|^2$$

$$+ k_3 \int_0^t exp(-2\delta(t-\tau))|A\mathbf{v}(\tau)|\|\mathbf{v}(\tau)\|^3 d\tau\}. \tag{2.5.13}$$

Denoting

$$\eta_1 = \left(\int_0^t exp(-2\delta(t-\tau))|A\mathbf{v}(\tau)|^2 d\tau\right)^{1/2}, \qquad \eta_2 = \int_0^t exp(-2\delta(t-\tau))\|\mathbf{v}(\tau)\|^6 d\tau,$$

using the Hölder inequality (1.11) for $u = \eta_1$ and $v = \eta_2$ and the Young inequality (1.12) in the form $\epsilon\eta_1\eta_2 \leq \eta_1^2 + \frac{\epsilon^2}{4}\eta_2^2$ for $\epsilon = k_3 c^*$, we get

$$c^* k_3 \int_0^t exp(-2\delta(t-\tau))|A\mathbf{v}(\tau)|\|\mathbf{v}(\tau)\|^3 d\tau \leq c^* k_3 \eta_1 \eta_2 \leq \eta_1^2 + \frac{k_3^2 c^{*2}}{4}\eta_2^2$$

$$= \int_0^t exp(-2\delta(t-\tau))|A\mathbf{v}(\tau)|^2 d\tau + \frac{k_3^2 c^{*2}}{4}\int_0^t exp(-2\delta(t-\tau))\|\mathbf{v}(\tau)\|^6 d\tau.$$

In this way, in inequality (2.5.13) the terms containing $A\mathbf{v}(\tau)$ cancel out, yielding

$$\|\mathbf{v}(t)\|^2 \leq c^* exp(-2\delta t)\|\mathbf{v}_0\|^2 + \frac{k_3^2 c^{*2}}{4}\int_0^t exp(-2\delta(t-\tau))\|\mathbf{v}(\tau)\|^6 d\tau. \tag{2.5.14}$$

Introducing the notation $\phi(t) = \|\mathbf{v}\|^2 exp(2\delta t)$, the inequality (2.5.14) becomes

$$\phi(t) \leq c^* \|\mathbf{v}_0\|^2 + \frac{k_3^2 c^{*2}}{4}\int_0^t exp(-4\delta\tau)\phi^3(\tau)d\tau, \tag{2.5.15}$$

where $\phi(0) = \|\mathbf{v}_0\|^2$. For every $t \geq 0$, $\phi(t) \leq \psi(t)$, where $\psi(t)$ is the solution of the integral equation

$$\psi(t) = c^* \|\mathbf{v}_0\|^2 + \frac{k_3^2 c^{*2}}{4}\int_0^t exp(-4\delta\tau)\psi^3(\tau)d\tau, \tag{2.5.16}$$

corresponding to the initial value $\psi(0) = c^* \|\mathbf{v}_0\|^2$. The differential form of (2.5.16) is

$$\frac{d}{dt}\psi = \frac{k_3^2 c^{*2}}{4}exp(-4\delta t)\psi^3(t)$$

and has the solution $\psi(t) = \dfrac{c^* \|\mathbf{v}_0\|^2}{\sqrt{1-\beta(1-e^{-4\delta t})}}$, where $\beta = \dfrac{c^{*4} k_3^2}{8\delta}\|\mathbf{v}_0\|^4$. The function ψ is real for every $t \geq 0$ if

$$\|\mathbf{v}_0\|^4 < \frac{8\delta}{c^{*4} k_3^2}. \tag{2.5.17}$$

Then $\phi(t) \leq \psi(t)$ reads

$$\|\mathbf{v}(t)\|^2 \leq \frac{c^* \|\mathbf{v}_0\|^2 e^{-2\delta t}}{\sqrt{1 - \frac{c^{*4} k_3^2}{8\delta}(1 - e^{-4\delta t})\|\mathbf{v}_0\|^4}}, \qquad 0 \leq t < \infty. \tag{2.5.18}$$

The estimation (2.5.18) shows that if (2.5.17) holds, then $\mathbf{u}$ is nonlinearly stable, which finishes the proof of Prodi's theorem.

The proof of the linearization principle imposed the theory of the N-S equations, as one of the mostly investigated theories of differential operators [Te], [Yu2].

2.6 Estimates for the spectrum of $\tilde{A}$

In order to keep the expressions as simple as possible, in Sections 2.4 and 2.5 the presence of some physical parameters was ignored. Moreover, the properties of $\tilde{A}$ were quite good and the hypothesis (2.5.3) seemed improbable for general fluid flows in bounded Ω. We expect that for various basic flows $\overline{\mathbf{u}}$ the corresponding operators $\tilde{A}$, which depend on $\overline{\mathbf{u}}$ and define the linearized N-S equations around $\overline{\mathbf{u}}$, have different properties. Indeed, in Appendix 2 the main characteristics of spectra of a few classes of operators were presented, showing that the geometry, and the related multiplicity of eigenvalues, can be very different and complicated.

A supplementary complication can arise from the dependence of $\tilde{A}$, and therefore of its spectrum, on the physical parameter, in our case ν or R_e. As a consequence, we can expect that the spectrum $\sigma(\tilde{A})$ be a subset of $\mathbf{C}$ with a complicated geometric structure and this set changes its form, continuous part of it becomes discrete sets, others coalesce etc.

A simplified image is that of a spectrum consisting in discrete points, namely eigenvalues, describing in $\mathbf{C}$ some curves as the parameter increases. It is possible that at some R_e several such curves intersect each other, the "moving" eigenvalues change their direction, for some small R_e all eigenvalues have positive real parts, then some eigenvalues arrive at the imaginary axis in $\mathbf{C}$. If at some critical Reynolds number R_{eL} a single eigenvalue arrives at this axis, cuts it at the origin and then passes in the other halfplane, then $\overline{\mathbf{u}}$ becomes linearly unstable and the principle of exchange of stabilities holds. If at R_{eL} two complex-conjugate eigenvalues cut the imaginary axis, then at R_{eL} a Hopf bifurcation sets in [IooJ], [Geo85].

Hence, essential facts of hydrodynamic stability and bifurcation depend on the geometric characteristics of $\sigma(\tilde{A})$ in $\mathbf{C}$. This is why the contribution of G. Prodi in these fields is fundamental in the theory of N-S equations: in 1962 he established that $\sigma(\tilde{A})$ is bounded by a parabola

$$Re(\sigma) = \frac{\nu}{4c_1^2} Im^2(\sigma) - \frac{c_1^2}{\nu}, \qquad (2.6.1)$$

where c_1 is the constant in (2.5.4) and σ is a point of $\sigma(\tilde{A})$, such that $Re(\sigma) > 0$ implies linear stability (in fact Prodi took $\nu = 1$).

As c_1 depends on $\overline{\mathbf{u}}$ and (2.6.1) depends on ν (or $\mathcal{R}_e$), the position of this parabola depends on $\overline{\mathbf{u}}$ and ν too: for small $\mathcal{R}_e$ (and large ν) it is situated in the right halfplane of $\mathbf{C}$ and as $\mathcal{R}_e$ increases (while ν decreases) this parabola reaches the imaginary axis and passes in the left halfplane.

Equation (2.6.1) was deduced by taking into account the relation

$$(A\mathbf{u}, \mathbf{u}) = ((\mathbf{u}, \mathbf{u})) \qquad (2.6.2)$$

for the operator $A = -P\Delta : N^2(\Omega) \to N^1(\Omega)$, where $(\cdot, \cdot)$ and $((\cdot, \cdot))$ are the scalar products in $L^2(\Omega)$ and $N^1(\Omega)$ respectively. Since the domain of motion Ω is bounded in $\mathbf{R}^2$ or $\mathbf{R}^3$, it follows that $((\mathbf{u}, \mathbf{v})) = (D\mathbf{u}, D\mathbf{v})$, where D is the differentiation operator with respect to the space variable $\mathbf{x}$. The relation (2.6.1) justified the hypothesis (2.5.3) used in proving the linearization principle. The location of this parabola depends on R_e; the better positioning of it leads to improved stability criteria. In this section we present our estimations for $\sigma(\tilde{A})$, which improve the Prodi's one [GeoPal95], [GeoPal97], [Geo77].

2.6.1 *Necessary conditions for belonging to* $\sigma(-\tilde{A})$

Consider the strong incompressible N-S model (2.5.1), (1.5.8) such that the stability spectrum is the spectrum $\sigma(-\tilde{A})$ of the operator $-\tilde{A}$. For the properties of $-\tilde{A}$ and its spectrum we quote [Mikh3] [Mikh5], [Pro], [Lad69], [Geo85], [IooJ]. Let $\rho(-\tilde{A})$ stand for the resolvent set of $-\tilde{A}$, which consists of $\sigma \in \mathbf{C}$ for which the resolvent operator $(-\tilde{A} - \sigma I)^{-1}$ exists and is a densely defined and bounded operator. So, if for σ lying in some set of $\mathbf{C}$ we have an estimation $\|(-\tilde{A} - \sigma I)^{-1}\| < \infty$, then σ must belong to the set complementary to $\sigma(-\tilde{A})$. Correspondingly, in order to determine bounds of the region where $\sigma(-\tilde{A})$ is located, we consider the equation

$$-\tilde{A}\mathbf{v} - \sigma\mathbf{v} = \mathbf{g}, \tag{2.6.3}$$

where $\mathbf{g} \in L^2(\Omega)$, and look for sufficient conditions for $|\mathbf{v}|/|\mathbf{g}| < \infty$. Indeed, in this case, we have

$$\|(-\tilde{A} - \sigma I)^{-1}\| = \sup_{\mathbf{g} \in \rho(-\tilde{A})} |(-\tilde{A} - \sigma I)^{-1}\mathbf{g}|/|\mathbf{g}| = \sup_{\mathbf{g} \in \rho(-\tilde{A})} |\mathbf{v}|/|\mathbf{g}| < \infty$$

and, therefore, these conditions are sufficient for σ to belong to $\rho(-\tilde{A})$.

Let $\sigma = \sigma_r + i\sigma_i$, where σ_r and σ_i stand for the real and imaginary parts of σ. If $\sigma_r < 0$ for all $\sigma \in \sigma(-\tilde{A})$, the perturbations $\mathbf{v}$, satisfying the linear equation corresponding to (2.5.1) (formally obtained for $R(\mathbf{v}) = 0$), damps out and, correspondingly, $\bar{\mathbf{u}}$ is linearly stable.

Let us put $\sigma = -\lambda$ and let us rename $-\mathbf{g}$ by $\mathbf{g}$. Then (2.6.3) reads

$$\tilde{A}\mathbf{v} - \lambda\mathbf{v} = \mathbf{g} \tag{2.6.3$'$}$$

and, by scalar multiplication by $\mathbf{v}$ in N, it implies

$$(\tilde{A}\mathbf{v}, \mathbf{v}) - \lambda|\mathbf{v}|^2 = (\mathbf{g}, \mathbf{v}). \tag{2.6.4}$$

Taking into account (2.6.2) and (2.5.4), where $c_1 > 0$ is a constant, (2.6.4) can be written in the equivalent form

$$\nu\|\mathbf{v}\|^2 - \lambda|\mathbf{v}|^2 = (\mathbf{g}, \mathbf{v}) - (M_{\bar{\mathbf{u}}}\mathbf{v}, \mathbf{v}), \tag{2.6.4$'$}$$

or, equivalently,

$$\nu\|\mathbf{v}\|^2 - \lambda_r|\mathbf{v}|^2 = Re\{(\mathbf{g}, \mathbf{v})\} - Re\{(M_{\bar{\mathbf{u}}}\mathbf{v}, \mathbf{v})\}, \tag{2.6.5}$$

$$-\lambda_i|\mathbf{v}|^2 = Im\{(\mathbf{g}, \mathbf{v})\} - Im\{(M_{\bar{\mathbf{u}}}\mathbf{v}, \mathbf{v})\}, \tag{2.6.6}$$

$$\lambda_i |\mathbf{v}|^2 = -Im\{(\mathbf{g},\mathbf{v})\} + Im\{(M_{\bar{\mathbf{u}}}\mathbf{v},\mathbf{v})\}, \tag{2.6.7}$$

where λ_r and λ_i represent the real and imaginary parts of λ. These are the fundamental relations which permit us to bound the spectrum $\sigma(-\tilde{A})$. To this aim we use the Schwarz inequality, (2.5.4) and Young inequality, then we introduce arbitrary positive constants and determine them such that the region where we locate the spectrum be the narrowest.

Using the Schwarz inequality, and inequality (2.5.4) we have

$$Re\{(\mathbf{g},\mathbf{v})\} - Re\{(M_{\bar{\mathbf{u}}}\mathbf{v},\mathbf{v})\} \leq |(\mathbf{g},\mathbf{v})| + |(M_{\bar{\mathbf{u}}}\mathbf{v},\mathbf{v})| \leq |\mathbf{g}||\mathbf{v}| + |M_{\bar{\mathbf{u}}}\mathbf{v}||\mathbf{v}|$$
$$\leq |\mathbf{g}||\mathbf{v}| + c_1 \|\mathbf{v}\||\mathbf{v}|.$$

Now let us use the Young inequality

$$c_1 \|\mathbf{v}\||\mathbf{v}| \leq \nu b \|\mathbf{v}\|^2 + c_1^2 |\mathbf{v}|^2/(4\nu b), \tag{2.5.4$'$}$$

where $b > 0$ is an arbitrary constant. Then from (2.6.5) we get

$$\nu(1-b)\|\mathbf{v}\|^2 - \lambda_r |\mathbf{v}|^2 \leq |\mathbf{g}||\mathbf{v}| + c_1^2 |\mathbf{v}|^2/(4\nu b) \tag{2.6.5$'$}$$

and, similarly,

$$-\lambda_i |\mathbf{v}|^2 \leq |\mathbf{g}||\mathbf{v}| + \nu e \|\mathbf{v}\|^2 + c_1^2 |\mathbf{v}|^2/(4\nu e), \tag{2.6.6$'$}$$

$$\lambda_i |\mathbf{v}|^2 \leq |\mathbf{g}||\mathbf{v}| + \nu e \|\mathbf{v}\|^2 + c_1^2 |\mathbf{v}|^2/(4\nu e), \tag{2.6.7$'$}$$

where $e > 0$ is another arbitrary constant. Multiplying (2.6.6$'$) by an arbitrary constant $c \geq 0$ and adding the result to (2.6.5$'$), we obtain

$$\nu(1-b-ce)\|\mathbf{v}\|^2 - \left[\lambda_r + c\lambda_i + \frac{c_1^2}{4\nu}\left(\frac{1}{b}+\frac{c}{e}\right)\right]|\mathbf{v}|^2 \leq |\mathbf{g}||\mathbf{v}|(1+c). \tag{2.6.8}$$

Assuming $1-b-ce \geq 0$ and taking into account the dimensional Poincaré inequality $\alpha d^{-2}|\mathbf{v}|^2 \leq \|\mathbf{v}\|^2$, $\alpha, d > 0$, where d is the diameter of Ω, we have

$$-\left[\lambda_r + c\lambda_i + \frac{c_1^2}{4\nu}\left(\frac{1}{b}+\frac{c}{e}\right) - \frac{\nu\alpha}{d^2}(1-b-ce)\right]|\mathbf{v}|^2 \leq |\mathbf{g}||\mathbf{v}|(1+c). \tag{2.6.9}$$

If $\lambda_r + c\lambda_i \leq -\frac{c_1^2}{4\nu}\left(\frac{1}{b}+\frac{c}{e}\right) + \frac{\nu\alpha}{d^2}(1-b-ce)$, then $\sigma = -\lambda \in \rho(-\tilde{A})$, therefore a necessary condition for σ (and, so $-\lambda$) to belong to the spectrum of $-\tilde{A}$ is

$$\lambda_r + c\lambda_i \geq -\frac{c_1^2}{4\nu}\left(\frac{1}{b}+\frac{c}{e}\right) + \frac{\nu\alpha}{d^2}(1-b-ce). \tag{2.6.10}$$

If instead of (2.6.6$'$) we use (2.6.7$'$), then we obtain

$$\lambda_r - c\lambda_i \geq -\frac{c_1^2}{4\nu}\left(\frac{1}{b}+\frac{c}{e}\right) + \frac{\nu\alpha}{d^2}(1-b-ce). \tag{2.6.11}$$

Hence, the stability spectrum of $-\tilde{A}$ (expressed in terms of λ) lies in the angle formed by the straight lines

$$\lambda_r \pm c\lambda_i = -\frac{c_1^2}{4\nu}\left(\frac{1}{b}+\frac{c}{e}\right) + \frac{\nu\alpha}{d^2}(1-b-ce) \tag{2.6.12}$$

containing the origin of the complex plane $\mathbf{C}$.

2.6.2 *Spectrum bounds based on straight lines*

Let us now choose those straight lines corresponding to the narrower region containing the spectrum. To this purpose we distinguish several cases.

Case c = 0. Hence we are interested only in bounds for λ_r. If $b = 1$, we obtain the Prodi bound (he took $\nu = 1$) for λ_r

$$\lambda_r \geq - c_1^2/(4\nu), \tag{2.6.13}$$

whereas if $0 < b < 1$, we have the estimate

$$\lambda_r \geq \frac{\nu}{d^2 b} \left[\alpha(1 - b)b - \frac{c_1^2 d^2}{4\nu^2} \right], \tag{2.6.13'}$$

which improves (2.6.13) if we have the constraint $\frac{c_1^2}{4\nu} < \nu\alpha b(1 - b)/d^2$. The best choice of b, corresponding to $\max[b(1 - b)]$ is $b = 1/2$.

Let us introduce the Reynolds number $R = c_1 d/\nu$. Then, if

$$R < \sqrt{\alpha}, \tag{2.6.14}$$

we have $\lambda_r > 0$, hence $\overline{\mathbf{u}}$ is linearly stable. If $c_1 = \sqrt{2}\|\overline{\mathbf{u}}\|$, this represents Ladyzhenskaya's universal (nonlinear) stability criterion [Lad69].

Inequality (2.6.14) satisfied the quoted constraint if $b \geq 1/4$, therefore (2.6.13') improves (2.6.13).

Case c > 0, 1 − b − ce = 0. Then (2.6.12) implies

$$\lambda_r \pm c\lambda_i \geq - \frac{c_1^2}{4\nu} \left(\frac{1}{1 - ce} + \frac{c}{e} \right). \tag{2.6.15}$$

Among the straight lines bounding the spectrum according to (2.6.15) we quote the straight lines $S_\pm$, corresponding to $c = 1$, $e = b = 1/2$,

$$\lambda_r \pm \lambda_i = - \frac{c_1^2}{\nu}. \tag{2.6.16}$$

Since the reasonings are analogous for $\lambda_i > 0$, we restrict ourselves to the case $\lambda_i < 0$. Then the corresponding straight line S_+ cuts the λ_i-axis at the point $B = (0, -c_1^2/\nu)$.

Let $-c_1^2 \eta/(4\nu)$ be the ordinate of the point Q where the straight line

$$\lambda_r + c\lambda_i = - \frac{c_1^2}{4\nu} \left(\frac{1}{1 - ce} + \frac{c}{e} \right) \tag{2.6.17}$$

cuts the λ_i-axis. Let O be the origin of the system of coordinates in the (λ_r, λ_i) plane. The point Q cannot be situated between O and B, *i.e.*, we cannot have $\eta < 4$, where the parameter η is, by definition,

$$\eta = \frac{1}{c(1 - ce)} + \frac{1}{e}. \tag{2.6.18}$$

Indeed, (2.6.18) can be thought of as an equation in c

$$c^2(e - e^2\eta) - c(1 - e\eta) - e = 0, \tag{2.6.18'}$$

which has real solutions if and only if $(1 - e\eta)^2 + 4e(e - e^2\eta) \geq 0$. The last inequality implies $(1 - e\eta)(1 - e\eta + 4e^2) \geq 0$. As (2.6.18) shows that $1/e < \eta$, it follows that we must have

$$1 - e\eta + 4e^2 \leq 0, \tag{2.6.19}$$

which is true if and only if $\eta > 4$ and $e \in (e_1, e_2)$ where $e_{1,2} = \frac{\eta \pm \sqrt{\eta^2 - 16}}{8}$.

For each η fixed, (2.6.17) represents a fascicle of straight lines passing through Q and it is situated inside the angle formed by the extreme straight lines $S_1(\eta)$ and $S_2(\eta)$ corresponding to the slopes $-1/\tilde{c}_1$ and $-1/\tilde{c}_2$, where

$$\begin{aligned}
\tilde{c}_1 = c(e_1) = \frac{1}{2e_1} = 2e_2 = \frac{\eta - \sqrt{\eta^2 - 16}}{4}, \\
\tilde{c}_2 = c(e_2) = \frac{1}{2e_2} = 2e_1 = \frac{\eta + \sqrt{\eta^2 - 16}}{4}.
\end{aligned} \tag{2.6.20}$$

If $\eta = 4$ we have $e = 1/2$, $\tilde{c}_1 = \tilde{c}_2 = 1$, hence the fascicle reduces to the straight line S_+. For all $\eta > 4$ there is a subfascicle situated between S_+ and the λ_i-axis and the other is situated on the other part of S_+. The first subfascicle cuts the λ_r-axis at a point situated between $E = \left(-c_1^2/(2\nu), 0\right)$ and $D = \left(-c_1^2/\nu, 0\right)$. The second subfascicle cuts the λ_r-axis at a point situated far away from D.

Each straight line of the fascicle realizes a bound of the spectrum. For η fixed, the best bounds are provided by $S_1(\eta)$ for the first subfascicle and $S_2(\eta)$ for the second one. The straight line $S_1(\eta)$ represents a bound for the spectrum lying in the halfplane $\lambda_r < 0$, whereas the straight line $S_2(\eta)$ represents a bound for the spectrum with $\lambda_r > 0$. It follows that, as η runs from 4 to ∞, the best bound is provided by the two envelopes S_1 and S_2 of the two subfascicles. In order to deduce S_1, we must eliminate η between the relation

$$\lambda_r + \tilde{c}_1\lambda_i = -\frac{c_1^2}{4\nu}\tilde{c}_1\eta \tag{2.6.21}$$

and the relation obtained by differentiating (2.6.21) with respect to η

$$\tilde{c}_1'\lambda_i = -\frac{c_1^2}{4\nu}[\tilde{c}_1 + \eta\tilde{c}_1'], \tag{2.6.22}$$

where prime indicates this differentiation. So, taking into account (2.6.20), we have $\tilde{c}_1' = -\tilde{c}_1/\sqrt{\eta^2 - 16}$, such that (2.6.22) implies $-\lambda_i/\sqrt{\eta^2 - 16} = -\frac{c_1^2}{4\nu} + \eta\frac{c_1^2}{4\nu}/\sqrt{\eta^2 - 16}$ or, taking into account (2.6.20), we have $-\lambda_i = \frac{c_1^2}{\nu}\tilde{c}_1$ and, therefore, $\tilde{c}_1 = -\frac{\nu}{c_1^2}\lambda_i$. Substituting $\tilde{c}_1$ in (2.6.21) by this expression we obtain $\eta = 4\frac{\lambda_r}{\lambda_i} - 4\frac{\nu}{c_1^2}\lambda_i$ and introducing in (2.6.20), written in the form $-\frac{\nu}{c_1^2}\lambda_i = \frac{\eta - \sqrt{\eta^2 - 16}}{4}$, we get the part of the parabola

$$\lambda_r = -\frac{c_1^2}{2\nu} + \frac{\nu}{2c_1^2}\lambda_i^2, \tag{2.6.23}$$

for which λ_r, $\lambda_i < 0$. This portion of parabola is S_1 and it cuts the λ_r and λ_i axes at E and B respectively. Similar reasonings show that S_2 is the portion of parabola (2.6.23) corresponding to $\lambda_r > 0$, $\lambda_i < 0$. S_1 and S_2 meet at B.

If instead of (2.6.17) we take

$$\lambda_r - c\lambda_i = -\frac{c_1^2}{4\nu}\left(\frac{1}{1-ce} + \frac{c}{e}\right) \tag{2.6.17'}$$

all the above reasonings hold if λ_i is replaced by $-\lambda_i$, therefore if we consider positive λ_i. Correspondingly, we obtain that in the upper halfplane $\lambda_i > 0$, the best bound for the spectrum, found with the aid of straight lines, is the upper branch of (2.6.23). In this way we proved the following

Theorem 2.6.1. *The spectrum of the N-S problem linearized around a stationary solution, in the class of Leray turbulent solutions, is situated inside the parabola* (2.6.23) *and for* $\lambda_r > -\frac{c_1^2}{4\nu}$.

Corollary. *The spectrum in Theorem* 2.6.1 *is situated in the angular region determined by the straight lines* (2.6.16) *in the complex plane* **C**.

This last assertion follows from the fact that the straight lines $S_\pm$ are tangent to the parabola (2.6.23) at B and B', where B' is the symmetric of B with respect to the λ_r-axis.

Case c > 0, $1 - b - ce > 0$. Then (2.6.12) implies

$$\lambda_r \pm c\lambda_i \geq \frac{\nu}{d^2 be}\left[\alpha be(1 - b - ce) - \frac{c_1^2 d^2}{4\nu^2}(e + bc)\right], \tag{2.6.15'}$$

which improves (2.6.15) if we have the constraint $\frac{c_1^2}{4\nu} < \alpha\nu b(1 - ce)/d^2$.

The best choice of b, corresponding to $\max \dfrac{be(1 - b - ce)}{e + bc}$, is

$$b = \frac{e(1 - ce)}{e + \sqrt{e^2 + ec - c^2 e^2}}$$

and it leads to

$$\max \frac{be(1 - b - ce)}{e + bc} = \left[\frac{e(1 - ce)}{e + \sqrt{e^2 + ec - c^2 e^2}}\right]^2.$$

Accordingly,

$$R < 2\sqrt{\alpha}\,\frac{e(1 - ce)}{e + \sqrt{e^2 + ec - c^2 e^2}} \tag{2.6.14'}$$

represents a universal stability criterion.

The above constraint is fulfilled if $\sqrt{b} < \dfrac{e\sqrt{1 - ce}}{e + \sqrt{e^2 + ec - c^2 e^2}}$. The right hand-side of (2.6.14') is a decreasing function of ec, where $0 < ec < 1$.

Hence, the best criterion obtained by the straight lines (2.6.12) remains (2.6.14).

2.6.3   *Spectrum bounds based on parabolas*

If in (2.6.6) and (2.6.7) we apply the Schwarz inequality, we have

$$-\lambda_i |\mathbf{v}|^2 \leq |\mathbf{g}||\mathbf{v}| + c_1 \|\mathbf{v}\||\mathbf{v}|, \tag{2.6.24}$$

$$\lambda_i |\mathbf{v}|^2 \leq |\mathbf{g}||\mathbf{v}| + c_1 \|\mathbf{v}\||\mathbf{v}|, \tag{2.6.25}$$

and, simplifying by $|\mathbf{v}|$, we obtain

$$-\lambda_i |\mathbf{v}| \leq |\mathbf{g}| + c_1 \|\mathbf{v}\|, \tag{2.6.24'}$$

$$\lambda_i |\mathbf{v}| \leq |\mathbf{g}| + c_1 \|\mathbf{v}\|, \tag{2.6.25'}$$

whence $\lambda_i^2 |\mathbf{v}|^2 \leq |\mathbf{g}|^2 + 2c_1 |\mathbf{g}|\|\mathbf{v}\| + c_1^2 \|\mathbf{v}\|^2$ and, by Schwarz inequality, we have

$$\lambda_i^2 |\mathbf{v}|^2 \leq |\mathbf{g}|^2 \left(1 + \frac{1}{\nu e}\right) + c_1^2 (1 + \nu e)\|\mathbf{v}\|^2, \tag{2.6.26}$$

where $e > 0$. Applying the Young inequality in (2.6.5') we obtain

$$-\lambda_r |\mathbf{v}|^2 + \nu(1 - b)\|\mathbf{v}\|^2 \leq \frac{a\nu}{c_1^2}|\mathbf{g}|^2 + |\mathbf{v}|^2 \frac{c_1^2}{4\nu}\left(\frac{1}{a} + \frac{1}{b}\right), \tag{2.6.27}$$

where $a > 0$. Adding (2.6.27) to (2.6.26) multiplied by $c \geq 0$, we have

$$\nu\left[1 - b - cc_1^2\left(e + \frac{1}{\nu}\right)\right]\|\mathbf{v}\|^2 - \left[\lambda_r - c\lambda_i^2 + \frac{c_1^2}{4\nu}\left(\frac{1}{a} + \frac{1}{b}\right)\right]|\mathbf{v}|^2 \leq |\mathbf{g}|^2 \left(\frac{a\nu}{c_1^2} + c + \frac{c}{\nu e}\right). \tag{2.6.28}$$

Case c = 0. Denoting $\lambda_r + \frac{c_1^2}{4\nu a} = \tilde{\lambda}_r$, we obtain the estimates (2.6.13) and (2.6.13'), where λ_r is replaced by $\tilde{\lambda}_r$. Next, letting $a \to \infty$, we get just (2.6.13) and the criterion (2.6.14). Hence the best results remain those of Section 2.6.2.

Case $1 - \mathbf{b} - \mathbf{cc}_1^2(\mathbf{e} + 1/\nu) = 0 = 0$. The spectrum is situated inside the parabola

$$\lambda_r = -\frac{c_1^2}{4\nu}\left(\frac{1}{a} + \frac{1}{b}\right) + \frac{(1 - b)\nu}{(1 + e\nu)c_1^2}\lambda_i^2, \tag{2.6.29}$$

which, for $a = b = 1/2$, $e = 1/\nu$, becomes Prodi's parabola

$$\lambda_r = -\frac{c_1^2}{\nu} + \frac{\nu}{4c_1^2}\lambda_i^2. \tag{2.6.30}$$

In fact, Prodi took $\nu = 1$, so that, instead of (2.6.30), he had

$$\lambda_r = -c_1^2 + \frac{1}{4c_1^2}\lambda_i^2. \tag{2.6.30'}$$

The starting point in our investigation was the remark that Prodi's bound for the real part of the spectrum obtained from (2.6.30') was only $\lambda_r = -c_1^2$, while when considered λ as real he obtained the better bound $\lambda = -c_1^2/4$, corresponding to (2.6.13). In fact, in deducing (2.6.13), no information on λ_i was used (we took $c = 0$), so we might assume from the beginning that our reasonings concern real λ.

The best parabola (2.6.29) would correspond to $\min\left(\frac{1}{a} + \frac{1}{b}\right)$ and $\max\frac{(1-b)}{(1+e\nu)}$. Since a and e are arbitrary positive numbers, the best choice for them is $a \to \infty$, $e \to 0$. However, the requirements for b are contradictory, because $1/b$ is minimum as $b \to 1$, while this situation leads to $1-b \to 0$. It follows that we cannot obtain the narrowest parabola which, at the same time, has the smallest range for λ_r. In fact, even for $a \to \infty$ and $b \to 1$, we cannot have an estimate better than $\lambda_r > -c_1^2/4\nu$. So, we prefer to have a narrower parabola, because the spectrum will be contained in the intersection of parabolas (2.6.29) and (2.6.23); therefore, in any case, the region near the vertex of parabola (2.6.29) does not count. The narrowest parabola (2.6.29) is obtained for $\min\frac{1}{b(1-b)} = 4$ (attained for $b = 1/2$) and it has the equation

$$\tilde{\lambda}_r = -\frac{c_1^2}{2\nu} + \frac{\nu}{2c_1^2}\tilde{\lambda}_i^2, \tag{2.6.31}$$

where $\lambda_r = \tilde{\lambda}_r$, $\tilde{\lambda}_i^2 = \lambda_i^2/(1 + e\nu)$, $c_1^2/(2\nu a) << 1$, $e\nu << 1$. This parabola, at the limit as $a \to \infty$ and $e \to 0$, is just (2.6.23).

The best bounds are represented by the envelope of this family of parabolas (2.6.29) written in the form

$$x = -\frac{1}{b} + (1 - b)y, \tag{2.6.29'}$$

where $x = \frac{\tilde{\lambda}_r}{B}$, $y = \tilde{\lambda}_i^2/(4B^2)$, $B = c_1^2/(4\nu)$. This envelope has the equation $x = y - 2\sqrt{y}$, where $y \geq 1$ and, correspondingly, $x \geq -1$. It reads, equivalently,

$$\tilde{\lambda}_r = -|\tilde{\lambda}_i| + \frac{\nu}{c_1^2}\tilde{\lambda}_i^2, \quad |\tilde{\lambda}_i| > \frac{c_1^2}{2\nu}, \quad \tilde{\lambda}_r > -\frac{c_1^2}{4\nu}, \tag{2.6.32}$$

therefore consists of two parabolas branches E_+ and E_-. The portions situated between the straight lines $\tilde{\lambda}_r = -\frac{c_1^2}{4\nu}$ and $\tilde{\lambda}_r = 0$ are tangent at the points $(\tilde{\lambda}_r, \tilde{\lambda}_i) = \left(\frac{c_1^2(1-2b)}{4\nu b^2}, \pm\frac{c_1^2}{2\nu b}\right)$ to (and, therefore are the envelopes of) the parabolas (2.6.29) for $1/2 < b < 1$. The limit parabola (2.6.29) corresponding to $b = 1/2$ is (2.6.31) and it is tangent to the envelope (2.6.29) at the point $\left(0, \pm\frac{c_1^2}{\nu}\right)$. The degenerate limit parabola (2.6.29) corresponding to $b = 1$, is the straight line $\tilde{\lambda}_r = -1$ and it is tangent to the envelope (2.6.32) at the points $\left(-\frac{c_1^2}{4\nu}, \pm\frac{c_1^2}{2\nu}\right)$, *i.e.* the vertices of the two branches of the envelope. The portions of the curves (2.6.32), situated in the halfplane $\tilde{\lambda}_r > 0$, are tangent at the points $(\tilde{\lambda}_r, \tilde{\lambda}_i) = \left(\frac{c_1^2(1-2b)}{4\nu b^2}, \pm\frac{c_1^2}{2\nu b}\right)$ to (and, are the envelopes of) the parabolas (2.6.29) for $0 < b < 1/2$.

Taking into account Theorem 2.6.1 and these reasonings, we have the following best result

Theorem 2.6.2 [GeoPal95,97], [Geo77]. *The spectrum of the linearized N-S equation in the class of Leray turbulent solutions is situated in the region containing the origin and delimited by the straight line $\lambda_r = -\frac{c_1^2}{4\nu}$ and the outer branches of parabolas $\lambda_r = -|\lambda_i| + \lambda_i^2\nu/c_1^2$.*

Case $1 - b - cc_1^2(e + 1/\nu) > 0$. The spectrum is situated in the region

$$\tilde{\lambda}_r - c\tilde{\lambda}_i^2 \geq \frac{\nu}{bd^2}\left\{\alpha b(1 - b) - \frac{c_1^2 d^2}{4\nu^2}\left[1 + 4\alpha\nu cd^{-2}(1 + \nu e)b\right]\right\}. \tag{2.6.33}$$

Let $k = 4\alpha\nu cd^{-2}(1+\nu e)$. The best choice for b, corresponding to $\max\{b(1-b)/(1+kb)\}$, is $b = \left[\sqrt{1+k} - 1\right]/k$ and gives $\max[b(1 - \alpha)/(1 + kb)] = \left[\sqrt{1+k} - 1\right]^2/k^2$. Accordingly,

$$R < 2\sqrt{\alpha}\left[\sqrt{1 + k} - 1\right]/k \tag{2.6.34}$$

represents a universal stability criterion, which is worse than (2.6.14).

Summing up, there is a fivefold source of improvements in our stability spectrum estimates: 1) the better handling with Young inequality in inequalities for the real λ_r, and imaginary parts λ_i of the eigenvalues; 2) the introduction of additional parameters by applying different Young inequalities for the same product in λ_i and λ_r; 3) the introduction of an additional parameter with which one must multiply the inequality for λ_i; 4) the use of envelopes of families of straight lines and parabolas bounding the spectrum; 5) in addition to Prodi's type inequalities for λ_r and λ_i^2, we used an inequality for λ_i. This allowed us to bound the spectrum by families of straight lines, nonexistent to Prodi. Their envelope is (2.6.23). The advantage of having many parameters was exploited in Section 2.6.3 in deriving families of bounding parabolas (2.6.29). Their envelope proved to be the best bound for $\lambda_r \geq -\frac{c_1^2}{4\nu}$. Since for $\lambda_r < -\frac{c_1^2}{4\nu}$ there is no eigenvalue, it follows that the envelope $\lambda_r = \pm\lambda_i + \lambda_i^2\nu/c_1^2$ of parabolas (2.6.29) provides the best bound.

2.7   Universal stability criteria

Improvements of some classical stability criteria realized by one of the authors (A.G) by appropriate combinations of Hölder inequality of three indices and Young inequality are presented by following [Geo76]. We also present the results in [GeoPal95,97] concerning criteria for linear stability obtained by $\tilde{A}$ spectrum estimates (Section 2.6). When compared with (nonlinear) universal stability criteria, sufficient conditions for nonexistence of subcritical stability type follow.

2.7.1   *Energy relation*

Let $\mathbf{u}$ stand for a vector whose components are the state functions other than the pressure, and assume that the fluid is incompressible. Then, for each such state function, a balance equation occurs in the mathematical model governing the fluid flow. Denote by H the Hilbert phase space of the model. Let $\bar{\mathbf{u}}$ be a basic stationary solution of the model and denote by $\mathbf{u}$ the perturbed solution and by $\mathbf{v} = \mathbf{u} - \bar{\mathbf{u}}$ the perturbation. Assume that H splits as $H = H_1 \oplus H_2$, where H_1 is a Cartesian product of Hilbert spaces, some of which consist of solenoidal vectors. Let P stand for the projection operator $P : H \to H_1$; in this way the equations expressing the solenoidality no longer exist. In this context we take the norm $\|\mathbf{v}\|$ of $\mathbf{v}$ in H (Section 2.2) by the perturbation energy.

Remind that $\|\mathbf{v}\|$ is a function of t only. Indeed, H is the phase space, therefore its norm is taken with respect to the space variable $\mathbf{x}$ only.

Definition 2.7.1. *Universal stability criteria* are sufficient conditions ensuring the decay of $\|\mathbf{v}\|$ as $t \to \infty$, which are valid for an arbitrary domain of motion Ω and basic fluid flow $\overline{\mathbf{u}}$.

The starting point in deriving such a criterion is the determination of the *energy relation*.

Looking at the balance equations and constitutive equations in Section 1.1, we find that the energy equality, obtained by projecting the balance equations on H_1 by means of P, has the form

$$\frac{d}{dt}\|\mathbf{v}\|^2 = \int_{\Omega} f(\mathbf{v}, D\mathbf{v}, D^2\mathbf{v}, \overline{\mathbf{u}}, D\overline{\mathbf{u}})d\mathbf{x}, \tag{2.7.1}$$

where the integral depends on $\mathbf{v}$ and its derivatives of order one and two and on $\overline{\mathbf{u}}$ and its first derivative with respect to $\mathbf{x}$. As a consequence, the equation governing the (time) evolution of $\|\mathbf{v}\|$ contains not only terms in $\|\mathbf{v}\|$, and, so, the label of the energy equation for (2.7.1) is inadequate.

The absence of an energy equation is the source of many subsequent approximations. Indeed, from the energy relation we must deduce the *energy inequality*

$$\frac{d}{dt}\|\mathbf{v}\|^2 \leq -c\|\mathbf{v}\|^2, \tag{2.7.2}$$

where $c > 0$ is a constant independent of $\mathbf{v}$. Taking into account (2.4.21) and (2.4.23) it follows that $\frac{\|\mathbf{v}\|^2}{2} \leq e^{-ct}\frac{\|\mathbf{v}_0\|^2}{2}$, *i.e.* $\overline{\mathbf{u}}$ is exponentially asymptotically stable in the Lyapunov sense.

All criteria in Section 2.7 concern this type of stability.

The inequality (2.7.2) is deduced from the energy relation by means of various tricks and types of inequalities: in general pre-Hilbertian spaces (*e.g.* Schwarz inequality), of integral type, (*e.g.* between L^p and L^q spaces), Hölder inequality (1.11), integro-differential inequalities (*e.g.* the Poincaré inequality, embeddings inequalities $(1.5) - (1.10)$), of algebraic type (*e.g.* the Young inequality (1.12)).

In addition, inequalities specific to the operators involved into the mathematical models governing the fluid flows are frequently used. The most important concern the *advective nonlinear terms*

$$b(\mathbf{a}, \mathbf{b}, \mathbf{c}) \equiv \int_{\Omega} \mathbf{a} \cdot \nabla \mathbf{b} \cdot \mathbf{c}\,d\mathbf{x} \leq \|\mathbf{a}\|_4\|\mathbf{b}\|_2\|\mathbf{c}\|_4, \quad \mathbf{a},\, \mathbf{c} \in L^4(\Omega), \quad \mathbf{b} \in N^1(\Omega) \tag{2.7.3}$$

and the following inequalities [Lad69] for bounded $\Omega \subset \mathbf{R}^n$

$$\|\mathbf{u}\|_4^4 \leq \left(\frac{4}{3}\right)^{\frac{3}{4}}|\mathbf{u}|\|\mathbf{u}\|^3, \quad \mathbf{u} \in N^1(\Omega) \subset L^4(\Omega), \quad \Omega \subset \mathbf{R}^3, \tag{2.7.4}$$

$$\|\mathbf{u}\|_4^4 \leq 2|\mathbf{u}|^2\|\mathbf{u}\|^2, \quad \mathbf{u} \in N^1(\Omega) \subset L^4(\Omega), \quad \Omega \subset \mathbf{R}^2. \tag{2.7.5}$$

In these inequalities and until the end of Section 2.7 $|\cdot|$ stands for the norm in $L^2(\Omega)$ and $\|\cdot\|$ for the norm in $N^1(\Omega)$. In this way, $\frac{|\overline{\mathbf{u}}|^2}{2}$ is the kinetic energy of the basic state and $\nu\frac{\|\overline{\mathbf{u}}\|^2}{2} = \frac{\nu}{2}|\nabla\overline{\mathbf{u}}|^2 = \frac{\nu}{2}|D\overline{\mathbf{u}}|^2 = \frac{\nu}{2}\int_\Omega \nabla\overline{\mathbf{u}} : \nabla\overline{\mathbf{u}}d\mathbf{x}$ is the energy of the shear (viscous)stress of the basic state. Loosely speaking, $|\overline{\mathbf{u}}|^2$ and $\|\overline{\mathbf{u}}\|^2$ are these energies.

In dimensional setting the Poincaré (isoperimetric) inequality reads

$$\alpha d^{-2}|\mathbf{u}|^2 \leq \|\mathbf{u}\|^2, \quad \mathbf{u} \in N^1(\Omega), \tag{2.7.6}$$

where d is the diameter of Ω and the best constant αd^{-2} is the minimum of the functional $\frac{\|\mathbf{u}\|^2}{|\mathbf{u}|^2}$ in $N^1(\Omega)$. It is associated with the eigenvalue problem (3.2.58).

We also mention the identities

$$b(\mathbf{a}, \mathbf{b}, \mathbf{b}) = 0, \quad b(\mathbf{b}, \mathbf{b}, \mathbf{b}) = 0, \quad b(\mathbf{a}, \mathbf{b}, \mathbf{c}) = -b(\mathbf{a}, \mathbf{c}, \mathbf{b}). \tag{2.7.7}$$

Each inequality introduces an approximation, and, so, weakens the criterion. For some purposes, *e.g.* for the linearization principle, even poorer approximations were useful. For instance, the imprecise bounds (2.5.3) and the resulting criterion (2.5.17) proved to be of an exceptional theoretical value. However, a stability criterion like (2.5.17) cannot be used in applications. It contains a lot of imprecisely defined positive constants δ, c^*, k_3 and all those constants on which δ, c^* and k_3 depend.

Whence, the idea that in order for a stability criterion to be important not only theoretically but also for applications, all the used inequalities of the type $\leq$ must be the best possible, *e.g.* of isoperimetric type, or as closed to the equality as possible. Moreover, since for a given model governing a fluid flow the operators and mappings are once for ever defined by the balance equations, there are a few types of inequalities we may use. Therefore the tricks used in this section, as well as some others in Chapter 4 could be of a real value. Another notable approximation introduced in passing from the energy equality to the energy inequality is the use of universal inequalities. Recall that the perturbations must satisfy the smoothness properties, the evolution equations and the boundary and initial conditions while in the used inequalities, *e.g.* of isoperimetric type, usually only the smoothness properties, boundary conditions and sometimes the solenoidality restriction, are required.

The idea of using universal criteria to obtain *a priori* inequalities for the solutions of N-S equations comes from Leray [Ler]. The *a priori* estimates permit one to establish the boundedness of some operators and mappings in the balance equations, leading to the solution existence via compactness theorems, the Riesz representation theorem, some fixed point theorem, Lax-Milgram lemma, or the like.

Such an estimate is the so-called *Leray energy inequality*

$$\frac{1}{2}|\mathbf{u}|^2 + \nu\int_0^T |D\mathbf{u}|^2 dt \leq \frac{1}{2}|\mathbf{u}_0|^2, \tag{2.7.8}$$

where $|\cdot|$ is the norm in $L^2(\Omega)$. This inequality is not of the form (2.7.2) and seems inadequate to yield criteria for asymptotic stability because it does not contain the time t in an explicit form. By including (2.7.8) in the definition of the turbulent solution for the N-S equations (2.1.1), Leray proved the existence theorem for them in the class

$$\mathbf{u} \in L^2\big((0,t^*), N^1(\Omega)\big) \cap L^\infty\big((0,t^*), N(\Omega)\big). \tag{2.7.9}$$

However, (2.7.8) suggests that as T increases, the viscous dissipation term $\nu \cdot \int_0^T |D\mathbf{u}|^2(t)dt$ increases and, so, $|\mathbf{u}|$ decreases as fast as some powers of t. That this case really occurs is shown in [Pro] for some convenient T. In the next two sections we show that even an exponential decrease of $\|\mathbf{u}\|$ takes place if $|\mathbf{u}_0|$ is sufficiently small.

Before doing this, recall the further contribution of Ratip Berker [Berk] and Jean Kampé de Férié [Kam], in emphasizing the importance of the energy inequality [Ser1].

The first to deduce universal stability criteria for fluid flows in the classical framework was James Serrin in his pioneering paper [Ser1] of 1959. Then Ladyzhenskaya [Lad69] determined their corresponding generalized variants for the N-S fluid and for a more general fluid [Lad67]. By using the Galerkin-Faedo-Hopf method, Foias and Prodi [FoiP] deduced for the N-S model (under the Leray turbulent form) some criterion, which if applied to the Ladyzhenskaya case was weaker, in spite of the fact that their energy relation was better. This gave the motivation of the paper [Geo76], where all these results were shown to be particular or limiting cases of new improved criteria. The improvement was due to the use of additional Hölder inequalities and their more adequate correlation with the Young inequality. Namely, the newly introduced parameters were chosen such that the best bounds be obtained.

Take d (the diameter of Ω) as the characteristic length, let $\Omega \subset \mathbf{R}^n$ be a bounded domain of motion, where $n = 2, 3$, assume that $\partial\Omega \in C^2$, let $\overline{\mathbf{u}}$ be the stationary basic solution of the model. In the classical case the criteria are expressed in terms of the Reynolds number $R_{e_1} = \frac{|\overline{\mathbf{u}}_{max}|d}{\nu}$, where $|\overline{\mathbf{u}}_{max}|^2 = \max_{x\in\Omega} |\overline{\mathbf{u}}(\mathbf{x})| \cdot |\overline{\mathbf{u}}(\mathbf{x})|$, while in the generalized setting the generalized Reynolds numbers involved were expressed either in terms of the rate of viscous shear $\|\overline{\mathbf{u}}\|$, namely $R_{e_2} = \frac{\|\overline{\mathbf{u}}\|d^{\frac{1}{2}}}{\nu}$ (for $n = 3$), $R'_{e_2} = \frac{\|\overline{\mathbf{u}}\|d}{\nu}$ (for $n = 2$), or in terms of both kinetic energy $|\overline{\mathbf{u}}|^2$ and rate of the shear $\|\overline{\mathbf{u}}\|^2$ of the basic flow, namely $R_{e_3} = \frac{|\overline{\mathbf{u}}|^{\frac{1}{4}}\|\overline{\mathbf{u}}\|^{\frac{3}{4}}d^{\frac{1}{4}}}{\nu}$ (for $n = 3$) and $R'_{e_3} = \frac{|\overline{\mathbf{u}}|^{\frac{1}{2}}\|\overline{\mathbf{u}}\|^{\frac{1}{2}}d^{\frac{1}{2}}}{\nu}$ (for $n = 2$).

2.7.2 *Three-dimensional case*

It is worth noting that no universal stability criteria in terms of $|\overline{\mathbf{u}}|^2$ only were obtained. For that case the appropriate Reynolds numbers would have been $R_{e_4} = \frac{|\overline{\mathbf{u}}|d^{-1/2}}{\nu}$ (for $n = 3$) and $R'_{e_4} = \frac{\|\overline{\mathbf{u}}\|}{\nu}$ (for $n = 2$).

2.7.2.1 *Incompressible Navier-Stokes fluid*

Criteria in terms of $\overline{u}_{max}$. Consider the Leray model (2.1.1) in the class (2.7.9). Leray proved that in this class the solution exists and is strong [Ler]. Therefore it satisfies the same equations (1.1.33) and (1.1.34) as the classical solutions but, instead of being continuous, each of their terms belongs to $L^2(\Omega)$. Then the corresponding perturbation equations around $\overline{\mathbf{u}}$ is $(1.4.1)_1$, $(1.4.1)_2$. Projecting $(1.4.1)_1$ on $N^1(\Omega)$ we obtain the energy relation

$$\frac{1}{2}\frac{d}{dt}|\mathbf{v}|^2 = -b(\mathbf{v}, |\overline{\mathbf{u}}|, \mathbf{v}) - \nu\|\mathbf{v}\|^2, \tag{2.7.10}$$

or, by (2.7.7), its equivalent form

$$\frac{1}{2}\frac{d}{dt}|\mathbf{v}|^2 = b(\mathbf{v}, \mathbf{v}, |\overline{\mathbf{u}}|) - \nu\|\mathbf{v}\|^2. \tag{2.7.11}$$

By Young inequality (1.12) and taking the maximum, Serrin [Ser1] deduced successively

$$(\mathbf{v} \cdot \nabla\mathbf{v}) \cdot \overline{\mathbf{u}} \leq \frac{1}{2}\Big(\nu\nabla\mathbf{v} : \nabla\mathbf{v} + \frac{\mathbf{v}^2|\overline{\mathbf{u}}|^2}{\nu}\Big), \tag{2.7.12}$$

$$|b(\mathbf{v}, \mathbf{v}, \overline{\mathbf{u}})| \leq \frac{\nu}{2}\|\mathbf{v}\|^2 + \frac{\overline{u}_{max}^2}{\nu}\frac{|\mathbf{v}|^2}{2}, \tag{2.7.13}$$

whence the energy relation (2.7.11) implied the following kinetic energy inequality

$$\frac{1}{2}\frac{d}{dt}|\mathbf{v}|^2 \leq -\frac{\nu}{2}\|\mathbf{v}\|^2 + \frac{\overline{u}_{max}^2}{\nu}\frac{|\mathbf{v}|^2}{2} \tag{2.7.14}$$

and, by Poincaré inequality (2.7.6), (2.7.14) took the form (2.7.2), where $c = \nu\alpha d^{-2} - \overline{u}_{max}^2\nu^{-1}$, namely

$$\frac{1}{2}\frac{d}{dt}|\mathbf{v}|^2 \leq \Big(\frac{\overline{u}_{max}^2}{\nu} - \nu\alpha d^{-2}\Big)\frac{|\mathbf{v}|^2}{2}, \tag{2.7.15}$$

implying the asymptotic Lyapunov stability of $\overline{\mathbf{u}}$ if the universal criterion

$$R_{e_1} \leq \sqrt{\alpha} \tag{2.7.16}$$

holds.

The inequality (2.7.12) is very strong. Indeed, in [Geo76] all classical universal criteria were improved but the Serrin's one (2.7.16). For instance, by Hölder inequality, the Poincaré inequality (2.7.6) and taking the maximum we have

$$|b(\mathbf{v}, \mathbf{v}, |\overline{\mathbf{u}}|)| \leq \|\mathbf{v}\|\Big(\int_\Omega (\mathbf{v} \cdot \overline{\mathbf{u}})^2 d\mathbf{x}\Big)^{\frac{1}{2}} \leq \|\mathbf{v}\|\overline{u}_{max}|\mathbf{v}| \leq \frac{\|\mathbf{v}\|^2\overline{u}_{max}d}{\sqrt{\alpha}}, \tag{2.7.17}$$

so (2.7.11) yields the energy inequality

$$\frac{1}{2}\frac{d}{dt}|\mathbf{v}|^2 \le 2\Big(\frac{\overline{u}_{max}d}{\sqrt{\alpha}} - \nu\Big)\frac{\|\mathbf{v}\|^2}{2}$$

implying (2.7.16).

Remark 2.7.1. If in (2.7.17), instead of applying the Poincaré inequality we would have used the Young inequality such that the term in $\|\mathbf{v}\|^2$ be equal to that in (2.7.11), namely

$$|b(\mathbf{v},\mathbf{v},\overline{\mathbf{u}})| \le \|\mathbf{v}\|\overline{u}_{max}|\mathbf{v}| \le \|\mathbf{v}\|^2 + \frac{\overline{u}_{max}^2}{\nu}\frac{|\mathbf{v}|^2}{4},$$

then the corresponding energy inequality following from (2.7.11) would have been

$$\frac{1}{2}\frac{d}{dt}|\mathbf{v}|^2 \le \frac{\overline{u}_{max}^2}{\nu}\frac{|\mathbf{v}|^2}{4},$$

which is of no use from the stability point of view. The cause is that the single term in the energy relation, which is negative for all $\mathbf{x}$ and t is $-\nu\|\mathbf{v}\|^2$. It gives the only sure contribution to the stability. Consequently, a method is more successful, the larger is the absolute value of the $\|\mathbf{v}\|^2$ left in the energy inequality. On the other hand, as in (2.7.11) we consider only the absolute value of $b(\mathbf{v},\mathbf{v},\overline{\mathbf{u}})$, the smaller its estimations, the better criteria follow.

Criterion in terms of $|\overline{\mathbf{u}}|$ and $\|\overline{\mathbf{u}}\|$. By Hölder inequality and using (2.7.4) and (2.7.6) in the energy relation (2.7.11) we have

$$|b(\mathbf{v},\mathbf{v},\overline{\mathbf{u}})| \le \|\mathbf{v}\|\Big(\int_\Omega (\mathbf{v}\cdot\overline{\mathbf{u}})^2 d\mathbf{x}\Big)^{\frac{1}{2}} \le \|\mathbf{v}\|\big(\|\mathbf{v}\|_4^4\|\overline{\mathbf{u}}\|_4^4\big)^{\frac{1}{4}}$$
$$\le \Big(\frac{4}{3}\Big)^{\frac{3}{4}}\|\mathbf{v}\|^{\frac{7}{4}}|\mathbf{v}|^{\frac{1}{4}}|\overline{\mathbf{u}}|^{\frac{1}{4}}\|\overline{\mathbf{u}}\|^{\frac{3}{4}} \le \Big(\frac{4}{3}\Big)^{\frac{3}{4}}\|\mathbf{v}\|^2|\overline{\mathbf{u}}|^{\frac{1}{4}}\|\overline{\mathbf{u}}\|^{\frac{3}{4}}\alpha^{-\frac{1}{8}}d^{\frac{1}{4}}$$

(2.7.18)

leading to the energy inequality

$$\frac{1}{2}\frac{d}{dt}|\mathbf{v}|^2 \le \Big[\Big(\frac{4}{3}\Big)^{\frac{3}{4}}|\overline{\mathbf{u}}|^{\frac{1}{4}}\|\overline{\mathbf{u}}\|^{\frac{3}{4}}\alpha^{-\frac{1}{8}}d^{\frac{1}{4}} - \nu\Big]\|\mathbf{v}\|^2. \tag{2.7.19}$$

If the universal criterion [Geo76]

$$R_{e_3} < \Big(\frac{3}{4}\Big)^{\frac{3}{4}}\alpha^{\frac{1}{8}} \tag{2.7.20}$$

holds, then (2.7.19) implies

$$\frac{1}{2}\frac{d}{dt}|\mathbf{v}|^2 \le \Big[\Big(\frac{4}{3}\Big)^{\frac{3}{4}}|\overline{\mathbf{u}}|^{\frac{1}{4}}\|\overline{\mathbf{u}}\|^{\frac{3}{4}}\alpha^{-\frac{1}{8}}d^{\frac{1}{4}} - \nu\Big]2\alpha d^{-2}\frac{|\mathbf{v}|^2}{2}, \tag{2.7.21}$$

which is (2.7.2) where $c = -2\alpha d^{-2}\big[\big(\frac{4}{3}\big)^{\frac{3}{4}}|\overline{\mathbf{u}}|^{\frac{1}{4}}\|\overline{\mathbf{u}}\|^{\frac{3}{4}}\alpha^{-\frac{1}{8}}d^{\frac{1}{4}} - \nu\big]$. Consequently $\overline{\mathbf{u}}$ is exponentially asymptotically stable.

Remark 2.7.2. Since $R_{e_3} = \frac{|\overline{\mathbf{u}}|^{\frac{1}{4}}\|\overline{\mathbf{u}}\|^{\frac{3}{4}}d^{\frac{1}{4}}}{\nu} \ge \alpha^{\frac{3}{8}}\frac{|\overline{\mathbf{u}}|d^{-\frac{1}{2}}}{\nu} = \alpha^{\frac{3}{8}}R_{e_4}$, we are tempted to state that (2.7.21) implies $R_{e_4} < \big(\frac{3}{4}\big)^{\frac{3}{4}}\alpha^{-\frac{1}{4}}$. Similarly, $R_{e_3}/n < R_{e_3}$ and, so, (2.7.21) would imply $R_{e_3} < n\big(\frac{3}{4}\big)^{\frac{3}{4}}\alpha^{\frac{1}{8}}$ for every $n > 0$. Therefore the same reasoning would lead to $R_{e_3} < \infty$, so the domain of stability would be the entire real line. In these reasonings the wrong hypothesis is that every $R_{e_3}/n < R_{e_3}$ satisfies (2.7.19). In fact, the square bracket in (2.7.19) is not derived from R_{e_3}/n, since R_{e_3}/n is not related to the inequality (2.7.19). In exchange, (2.7.19) can be continued as follows

$$\frac{1}{2}\frac{d}{dt}|\mathbf{v}|^2 \le \left[\left(\frac{4}{3}\right)^{\frac{3}{4}}|\overline{\mathbf{u}}|^{\frac{1}{4}}\|\overline{\mathbf{u}}\|^{\frac{3}{4}}\alpha^{-\frac{1}{8}}d^{\frac{1}{4}} - \nu\right]\|\mathbf{v}\|^2 \le \left[n\left(\frac{4}{3}\right)^{\frac{3}{4}}|\overline{\mathbf{u}}|^{\frac{1}{4}}\|\overline{\mathbf{u}}\|^{\frac{3}{4}}\alpha^{-\frac{1}{8}}d^{\frac{1}{4}} - \nu\right]\|\mathbf{v}\|^2,$$

leading to the weaker criterion $R_{e_3} < \frac{1}{n}\left(\frac{3}{4}\right)^{\frac{3}{4}}\alpha^{\frac{1}{8}}$.

Criterion in terms of $\|\overline{\mathbf{u}}\|$. A further application of (2.7.6) to (2.7.18) gives

$$|b(\mathbf{v},\mathbf{v},\overline{\mathbf{u}})| \le \left(\frac{4}{3}\right)^{\frac{3}{4}}\|\mathbf{v}\|^2|\overline{\mathbf{u}}|^{\frac{1}{4}}\|\overline{\mathbf{u}}\|^{\frac{3}{4}}\alpha^{-\frac{1}{8}}d^{\frac{1}{4}} \le \left(\frac{4}{3}\right)^{\frac{3}{4}}\alpha^{-\frac{1}{4}}d^{\frac{1}{2}}\|\overline{\mathbf{u}}\|^2\|\mathbf{v}\|^2, \qquad (2.7.21')$$

leading to the energy inequality

$$\frac{1}{2}\frac{d}{dt}|\mathbf{v}|^2 \le \left[\left(\frac{4}{3}\right)^{\frac{3}{4}}\alpha^{-\frac{1}{4}}d^{\frac{1}{2}}\|\overline{\mathbf{u}}\|^2 - \nu\right]\|\mathbf{v}\|^2. \qquad (2.7.22)$$

If the hypothesis of the universal criterion [Geo76]

$$R_{e_2} \le \left(\frac{3}{4}\right)^{\frac{3}{4}}\alpha^{\frac{1}{4}} \qquad (2.7.23)$$

holds, by applying (2.7.6) to (2.7.22), we obtain the energy inequality of the form (2.7.2), where $c = 2\alpha d^{-2}\left[\left(\frac{4}{3}\right)^{\frac{3}{4}}\alpha^{-\frac{1}{4}}d^{\frac{1}{2}}\|\overline{\mathbf{u}}\|^2 - \nu\right]$, whence, again, the exponential asymptotic stability of $\overline{\mathbf{u}}$. It improves the criterion from [Lad67].

Instead of (2.7.11), let us use the energy relation (2.7.10). Thus, using Hölder inequality, (2.7.6) and (2.7.4) we successively get

$$|b(\mathbf{v},\overline{\mathbf{u}},\mathbf{v})| \le \|\overline{\mathbf{u}}\|\left(\int_\Omega (\mathbf{v}^4 d\mathbf{x}\right)^{\frac{1}{2}} \le \|\overline{\mathbf{u}}\|\left(\frac{4}{3}\right)^{\frac{3}{4}}|\mathbf{v}|^{\frac{1}{2}}\|\mathbf{v}\|^{\frac{3}{2}} \le \left(\frac{4}{3}\right)^{\frac{3}{4}}\alpha^{-\frac{1}{4}}d^{\frac{1}{2}}\|\overline{\mathbf{u}}\|^2\|\mathbf{v}\|^2$$

$$(2.7.24)$$

leading to the energy inequality

$$\frac{1}{2}\frac{d}{dt}|\mathbf{v}|^2 \le \left[\left(\frac{4}{3}\right)^{\frac{3}{4}}\alpha^{-\frac{1}{4}}d^{\frac{1}{2}}\|\overline{\mathbf{u}}\| - \nu\right]\|\mathbf{v}\|^2,$$

implying the same criterion (2.7.23).

2.7.2.2   *Materially nonlinear fluid*

In [Lad67] Ladyzhenskaya proposed some new equations for the description of fluid motions. Instead of N-S equations she took

$$\frac{\partial}{\partial t}\mathbf{u} - \frac{\partial}{\partial x_i}\left[A(\nabla\mathbf{u})\frac{\partial\mathbf{u}}{\partial x_i}\right] + (\mathbf{u}\cdot\nabla)\mathbf{u} = -\nabla p + \mathbf{F}(t,\mathbf{x}) \qquad (2.7.25)$$

where, by definition,

$$A(\nabla\mathbf{u}) = \nu_0 + \nu_1|\nabla\mathbf{u}|^{2\mu}, \qquad (2.7.26)$$

ν_0 and ν_1 being positive constants. The second term in (2.7.25) corresponds to the momentum flux and is defined by a stress tensor $\mathbf{T}$, given by (2.7.26). Comparing (2.7.26) with (1.1.24) it follows that the fluid in [Lad67] is a nonlinear material of other type than those considered by [Müll].

A generalized solution of (2.7.25) is defined as satisfying the following integral equation

$$\int_{\Omega\times[0,T]}\left[(\frac{\partial}{\partial t}\mathbf{u}+(\mathbf{u}\cdot\nabla)\mathbf{u})\boldsymbol{\Phi}+(\nu_0+\nu_1|\nabla\mathbf{u}|^{2\mu})\nabla\mathbf{u}\nabla\boldsymbol{\Phi}\right]d\mathbf{x}dt=\int_{\Omega\times[0,T]}\mathbf{f}\boldsymbol{\Phi}d\mathbf{x}dt$$

$$(2.7.27)$$

in some subspace of $L^2\big((0,t)\times\Omega\big)$ defined in [Lad67]. The test function was supposed to belong to the same space. Ladyzhenskaya took into consideration the action of the body forces. As they have no importance for our study we shall neglect them.

The energy relation obtained by letting $\boldsymbol{\Phi}$ be the velocity perturbation is identical to (2.7.10). Following the method based on (2.2.24) we obtain the criterion (2.2.23). The two viscosity constants ν_0 and ν_1 enter the definition of the basic solution $\overline{\mathbf{u}}$.

2.7.3 *Two-dimensional case*

2.7.3.1 *Leray setting*

Using the methods from Section 2.7.2.1 we present three criteria first deduced in [Geo76]. In this case the Poincaré inequality preserves its form but the value of α is different [Bic]. Another difference from the three-dimensional case consists in the fact that, instead of (2.7.4), (2.7.5) must be used.

As in (2.7.5) the coefficients and the powers of the norm differ from those in (2.7.4), the criteria change accordingly but the proofs are the same. This is why we present them briefly.

For these reasons *criterion in terms of* $\overline{\mathbf{u}}_{max}$ is (2.7.16).

Criterion in terms of $|\overline{\mathbf{u}}|$ *and* $\|\overline{\mathbf{u}}\|$. Similar computations as for (2.7.18) lead to

$$|b(\mathbf{v},\mathbf{v},\overline{\mathbf{u}})|\leq\sqrt{2}|\overline{\mathbf{u}}|^{\frac{1}{2}}\|\overline{\mathbf{u}}\|^{\frac{1}{2}}\alpha^{-\frac{1}{4}}d^{\frac{1}{2}}\|\mathbf{v}\|^2 \tag{2.7.28}$$

leading to the universal stability criterion

$$R'_{e_3}<\alpha^{\frac{1}{4}}/\sqrt{2}. \tag{2.7.29}$$

Criterion in terms of $\|\overline{\mathbf{u}}\|$. Further use of (2.7.6) in (2.7.28) yields

$$|b(\mathbf{v},\mathbf{v},\overline{\mathbf{u}})|\leq\sqrt{2/\alpha}\|\overline{\mathbf{u}}\|\|\mathbf{v}\|^2 \tag{2.7.30}$$

implying the criterion

$$R'_{e_2}\leq\sqrt{\alpha/2}. \tag{2.7.31}$$

2.7.3.2 *Weak setting*

Here we show an alternative way to derive criterion (2.7.31) for the N-S model. Let $\Omega_T = (0, T) \times \Omega$. The generalized (weak) solutions of this model belong to the class $\mathbf{u} \in V_2(\Omega_T) \cap L_{r,q}(\Omega_T) \cap J(\Omega_T)$. They are determined by the body forces $\mathbf{f} \in L_{1,2}(\Omega_T)$ and the initial velocity field $\mathbf{u}_0 \in J^{\circ(1)}(\Omega) \cap W^{2,2}(\Omega)$ and satisfy the relation

$$\int_{\Omega_T} \left(-\mathbf{u}\frac{\partial}{\partial t}\mathbf{\Phi} + \nu \nabla \mathbf{u} : \nabla \mathbf{\Phi} - u_k \mathbf{u}\frac{\partial}{\partial x_k}\mathbf{\Phi} \right) dt dx$$
$$+ \int_\Omega \mathbf{u}\mathbf{\Phi}\,|_{t=T}\, dx - \int_\Omega \mathbf{u}_0\mathbf{\Phi}\,|_{t=0}\, dx = \int_{\Omega_T} \mathbf{F}\mathbf{\Phi} dt dx, \tag{2.7.32}$$

for every test function $\mathbf{\Phi} \in W^{\circ 1,1;2}(\Omega_T) \cap J^\circ(\Omega_T)$, where $J(\Omega_T)$ is the subset of $C_\infty^0(\Omega_T)$ consisting of solenoidal vectors and $J^\circ(\Omega_T)$ is its closure in the $L_2(\Omega_T)$ norm. The space $J^{\circ(1)}(\Omega)$ is the closure of $J(\Omega)$ in the $W^{1,2}(\Omega)$ norm, while $V_2(\Omega_T)$ is a subspace of $L_2(\Omega_T)$ [Lad69].

The compact notation $L_{1,2}$, $L_{q,r}$ and $W^{\circ 1,1;2}$ stands for spaces of vectors functions having the indicated properties by the first index as functions of t for fixed $\mathbf{x}$ and the properties indicated by the second index as functions of $\mathbf{x}$ for fixed t.

The weak setting (2.7.32) is appropriate to an analytical rather than geometrical study. Taking $\mathbf{\Phi} = \mathbf{v} = \mathbf{u} - \bar{\mathbf{u}}$, where, as in the above, $\mathbf{v}$ stands for perturbation velocity and $\bar{\mathbf{u}}$ for the basic velocity, in the standard way, *i.e.* subtracting from (2.7.32) the equation for $\bar{\mathbf{u}}$, (2.7.10) is obtained. This explains why in [Lad69] the same criterion (2.7.31) was obtained (it was written as $\sqrt{2}C_\Omega^* \|u_x\|\nu^{-1} < 1$, where $C_\Omega^* = d\alpha^{-1}$).

2.7.3.3 *A variant of Leray setting. Method based on orthogonal projections*

Foias and Prodi [FoiP] considered a special type of asymptotic stability of the basic solution $\bar{\mathbf{u}}$, on which the perturbation $\mathbf{v}$ giving rise to the perturbed velocity $\mathbf{u}$ is superposed. They derived the criterion: *if for n, λ_{n+1} satisfy the inequality*

$$\lambda_{n+1} > \left(\frac{4}{\nu}\right)^2 c^2 \tag{2.7.33}$$

then $\|\mathbf{u}\| \to \|\bar{\mathbf{u}}\|$ *as* $t \to \infty$ *in the sense* $\mathrm{pr}_n\mathbf{u}(t) - \mathrm{pr}_n\bar{\mathbf{u}}(t) \to 0$ *in* $\mathbf{R}^n$ *for* $t \to \infty$. Here λ_n are the eigenvalues of the following problem (related to problem (3.2.58), (3.2.59), (3.2.57))

$$\Delta \mathbf{w}_n + \nabla q = -\lambda_n \mathbf{w}_n, \quad \nabla \cdot \mathbf{w}_n = 0, \quad \mathbf{w}_{n|\partial\Omega} = 0$$

(in particular $\lambda_1 = \alpha d^{-2}$, in our notation), and, by definition, $\mathrm{pr}_0\mathbf{v} = 0$, $\mathrm{pr}_n\mathbf{v} = ((\mathbf{v}, \mathbf{w}_1), \ldots, (\mathbf{v}, \mathbf{w}_n)) \in \mathbf{R}^n$ and $\mathcal{C} = \lim_{\sigma \to \infty} \mathcal{C}_\sigma$, ($\|\bar{\mathbf{u}}(t)\| \leq \mathcal{C}_\sigma$, $\|\mathbf{u}(t)\| \leq \mathcal{C}_\sigma$ for $\sigma \leq t \leq \infty$).

Foias and Prodi criterion expresses the fact that the asymptotic behavior of the solution of the N-S equations depends on the asymptotic behavior of its projection on a finite number of orthogonal vectors $\mathbf{w}_1, \ldots, \mathbf{w}_n$, which are the eigenvectors corresponding to the first n eigenvalues $\lambda_1, \ldots, \lambda_n$ of the eigenvalue problem (3.2.58), (3.2.59), defined by the positive definite operator $\tilde{A} = -P\Delta$ (Section 1.3). This operator is positive and selfadjoint, therefore the eigenvalues are $\infty \leftarrow \cdots \geq \lambda_n, \cdots, \geq \lambda_2 \geq \lambda_1 > 0$ (Appendix 2). The corresponding eigenvectors $\mathbf{w}_k$ form a total set in N^1.

The criterion (2.7.33) is better the larger is n. The weakest inequality (2.7.33) is the one corresponding to $n = 0$ which can be written as

$$\frac{\mathcal{C}}{\nu\sqrt{\lambda_1}} < \frac{1}{4} \tag{2.7.33$'$}$$

or, in notation of Section 2.7.3.2, $\dfrac{\|\overline{\mathbf{u}}\|d}{\nu} < \dfrac{\sqrt{\alpha}}{4}$. It represents a criterion of the type (2.7.31) but weaker than that of Ladyzhenskaya [Lad69], in spite of the fact that the energy relation used by Foias and Prodi was stronger than that of Ladyzhenskaya. That is why in [Geo76] the paper [FoiP] was analyzed and some inequality stronger than that leading to (2.7.33) was derived.

So, together with Foias and Prodi, we say that a function of t, $\overline{\mathbf{u}}(t)$, is a weak basic non-stationary solution of the incompressible N-S problem (we do not take into account body forces) if

$$\overline{\mathbf{u}}(t) \in L^2_{loc}\big((T,\infty); N^1(\Omega)\big) \cap L^\infty_{loc}\big((T,\infty); N(\Omega)\big) \quad \lim_{t\to T+0} \overline{\mathbf{u}}(t) = \mathbf{u}_0 \in N$$

and it satisfies a Leray-like model

$$\int_T^\infty \left[-\left(\overline{\mathbf{u}}(t), \frac{\partial}{\partial t}\boldsymbol{\Phi}(t)\right) + \nu\big((\overline{\mathbf{u}}(t), \boldsymbol{\Phi}(t))\big) + b(\overline{\mathbf{u}}, \overline{\mathbf{u}}, \boldsymbol{\Phi})\right] dt = 0 \tag{2.7.34}$$

for every function $\boldsymbol{\Phi}(t)$ such that, $\boldsymbol{\Phi} \in C\big((T,\infty); N^1(\Omega)\big)$, $\boldsymbol{\Phi}$ is differentiable (in $N(\Omega)$) and $\dot{\boldsymbol{\Phi}} \in L^2_{loc}\big((T,\infty); N(\Omega)\big)$, $\boldsymbol{\Phi}(t)$ has a compact support in (T,∞). Note that $((\cdot,\cdot))$ is the inner product in N^1. They proved the existence and uniqueness of the turbulent solution of this problem by Galerkin-Faedo-Hopf method, *i.e.* by projecting the equations on some finite-dimensional spaces.

The same interesting idea was used in [FoiP] in order to derive the energy relation. This derivation was standard (from the perturbed equation the equation for the basic state was subtracted) but they took $\boldsymbol{\Phi} = B(\mathbf{u} - \overline{\mathbf{u}})$, where $B \geq 0$, $B = -AP_n$, is a linear bounded operator in N, commuting with some operator $A = -P\Delta$ (Section 1.3). In addition, $P_n : N^1 \to V^n$ is a projection on the linear space V^n determined by $\mathbf{w}_1, \ldots, \mathbf{w}_n$. Thus they obtained the so-called generalized energy relation in $L^2(\Omega)$. Then they allowed for B to be equal to E (an orthogonal projection in N), so the energy relation became

$$\frac{1}{2}\frac{d|E\mathbf{v}(t)|^2}{dt} + \nu\|E\mathbf{v}(t)\|^2 + b(\mathbf{u}, \mathbf{v}, E\mathbf{v}) + b(\mathbf{v}, \overline{\mathbf{u}}, E\mathbf{v}) = 0. \tag{2.7.35}$$

For $E = I$ (the identity operator) (2.7.35) reduces to (2.7.10). In the same way as in Section 2.7.3.1, we have

$$|b(\mathbf{u}, \mathbf{v}, E\mathbf{v})| = |b(\mathbf{u}, (I - E)\mathbf{v}, E\mathbf{v})| = |b(\mathbf{u}, E\mathbf{v}, (I - E)\mathbf{v})|$$

$$\leq \sqrt{2}|\mathbf{u}|^{\frac{1}{2}}\|\mathbf{u}\|^{\frac{1}{2}}\|E\mathbf{v}\||(I - E)\mathbf{v}|^{\frac{1}{2}}\|(I - E)\mathbf{v}\|^{\frac{1}{2}},$$

and

$$|b(\mathbf{v}, \overline{\mathbf{u}}, E\mathbf{v})| \leq |b(E\mathbf{v}, \overline{\mathbf{u}}, E\mathbf{v})| + |b((I - E)\mathbf{v}, \overline{\mathbf{u}}, E\mathbf{v})|$$

$$\leq \sqrt{2}\|\overline{\mathbf{u}}\|\|E\mathbf{v}\|\|E\mathbf{u}\| + \sqrt{2}|\overline{\mathbf{u}}|^{\frac{1}{2}}\|\overline{\mathbf{u}}\|^{\frac{1}{2}}\|E\mathbf{v}\||(I - E)\mathbf{v}|^{\frac{1}{2}}\|(I - E)\mathbf{v}\|^{\frac{1}{2}}$$

whence [FoiP], for $t \geq \sigma$,

$$\frac{1}{2}\frac{d|E\mathbf{v}|^2}{dt} + \nu\|E\mathbf{v}\|^2 \leq 2\sqrt{2C_\Omega}\mathcal{C}_\sigma|(I - E)\mathbf{v}|^{\frac{1}{2}}\|(I - E)\mathbf{v}\|^{\frac{1}{2}}\|E\mathbf{v}\| + \sqrt{2}\mathcal{C}_\sigma|E\mathbf{v}|\|E\mathbf{v}\|, \tag{2.7.36}$$

where $C_\Omega = \alpha^{-1/2}d$ in our notation. Using the following Young inequalities

$$2\sqrt{2C_\Omega}\mathcal{C}_\sigma|(I-E)\mathbf{v}|^{\frac{1}{2}}\|(I-E)\mathbf{v}\|^{\frac{1}{2}}\|E\mathbf{v}\| \leq \frac{8C_\Omega\mathcal{C}_\sigma^2}{\nu}x^2|(I-E)\mathbf{v}|\|(I-E)\mathbf{v}\| + \frac{\nu}{4x^2}\|E\mathbf{v}\|^2, \tag{2.7.37}$$

and

$$\sqrt{2}\mathcal{C}_\sigma|E\mathbf{v}|\|E\mathbf{v}\| \leq \frac{4\mathcal{C}_\sigma^2}{\nu}y^2|E\mathbf{v}|^2 + \frac{\nu}{8y^2}\|E\mathbf{v}\|^2, \tag{2.7.38}$$

in (2.7.36), instead of the Foias and Prodi relationship (2.7.36), in [Geo76] it was deduced the inequality

$$\frac{1}{2}\frac{d|E\mathbf{v}|^2}{dt} + \nu\|E\mathbf{v}\|^2 \leq \frac{8C_\Omega\mathcal{C}_\sigma^2}{\nu}x^2|(I - E)\mathbf{v}|\|(I - E)\mathbf{v}\|$$
$$+ \frac{4\mathcal{C}_\sigma^2}{\nu}y^2|E\mathbf{v}|^2 + \left(\frac{\nu}{4x^2} + \frac{\nu}{8y^2}\right)\|E\mathbf{v}\|^2. \tag{2.7.39}$$

Taking $E = E_n$, where $E_n\mathbf{u} = (\mathbf{u}, \mathbf{w}_{n+1})\mathbf{w}_{n+1} + \ldots + (\mathbf{u}, \mathbf{w}_{n+2})\mathbf{w}_{n+2} + \ldots$, (2.7.39) becomes the energy inequality

$$\frac{1}{2}\frac{d|E_n\mathbf{v}|^2}{dt} + \left[\frac{\nu\lambda_{n+1}(8x^2y^2 - 2y^2 - x^2)}{8x^2y^2} - \frac{4y^2\mathcal{C}_\sigma^2}{\nu}\right]|E_n\mathbf{v}|^2 \leq \frac{8\alpha^{-1/2}d\mathcal{C}_\sigma^2\sqrt{\lambda_n}}{\nu}|pr_n\mathbf{v}|^2x^2 \tag{2.7.40}$$

implying $|\mathbf{v}(t)|^2 \to 0$ as $t \to \infty$ if the square bracket in (2.7.40) is positive. If we allow for the positive numbers x, y (so far arbitrary) to be equal to 1, we obtain the coefficient $\frac{5\nu}{8}$ for $\|E\mathbf{v}\|^2$ on the left-hand side of (2.7.39) and it is better than $\frac{\nu}{4}$ given in [FoiP]. From (2.7.39) in [Geo76] it was obtained the stability criterion

$$\lambda_{n+1} > \frac{32\mathcal{C}_\sigma^2}{\nu^2}\frac{y^4x^2}{8x^2y^2 - 2y^2 - x^2}, \tag{2.7.41}$$

which takes its best form for $x \to \infty$ and $y = \frac{1}{2}$, namely

$$\lambda_{n+1} > \frac{2\mathcal{C}_\sigma^2}{\nu^2}. \tag{2.7.42}$$

For $n = 0$ (2.3.42) becomes just the Ladyzhenskaya criterion which is better than (2.7.33'). For larger n better criteria are obtained and this is due to the suitable choice of the Young inequalities (2.7.37), (2.7.38).

In spite of the fact that for $x \to \infty$ (2.7.37) loses its sense, since in (2.3.41) we have a strict inequality, we can consider x as large as we want and all the above considerations still hold.

The same type of improvement was obtained in [Geo85] for a slightly more general case, including perturbations of body forces and boundary conditions.

2.7.3.4 *Sufficient criteria for nonexistence of subcritical instabilities*

We assume that Ω is bounded and $\partial\Omega$ is of class C^2.

Three-dimensional case. In Section 2.6 we derived the linear stability criterion (2.6.14) asserting: for $c_1 d/\nu < \sqrt{\alpha}$ the stability spectrum $\{\sigma\}$ is situated in the left complex halfplane and the basic stationary motion $\overline{\mathbf{u}}$ is linearly stable with respect to all kinds of perturbations. Note that c_1 is defined by (2.5.4), *i.e.* $|M_{\overline{\mathbf{u}}}\mathbf{v}| \leq c_1\|\mathbf{v}\|$. Hence c_1 depends on $\overline{\mathbf{u}}$.

Mainly, the spectrum estimates were obtained by using the following successive inequalities

$$|(M_{\overline{\mathbf{u}}}\mathbf{v}, \mathbf{v})| \leq c_1\|\mathbf{v}\|\,|\mathbf{v}| \leq c_1|\mathbf{v}|^2/(4b\nu) + \nu b\|\mathbf{v}\|^2 \leq c_1|\mathbf{v}|^2/(4b\nu) + \nu\|\mathbf{v}\|^2 + \nu(b-1)\|\mathbf{v}\|^2$$

$$\leq c_1|\mathbf{v}|^2/(4b\nu) + \nu\|\mathbf{v}\|^2 + \nu(b-1)|\mathbf{v}|^2 d^2\alpha^{-1} \tag{2.7.43}$$

and by imposing to $b(1-b)$ to be maximal.

In Section 2.7.2.1 we started with the energy relation (2.7.11) and proved three types of (nonlinear) universal criteria. The best ones were (2.7.16), (2.7.20), (2.7.23), asserting: if $\mathcal{R}e_1 \equiv \overline{u}_{\max}d/\nu \leq \sqrt{\alpha}$, $\mathcal{R}e_3 \equiv |\overline{\mathbf{u}}|^{\frac{1}{4}}\|\overline{\mathbf{u}}\|^{\frac{3}{4}}d^{\frac{1}{4}}/\nu < \left(\frac{3}{4}\right)^{\frac{3}{4}}\alpha^{\frac{1}{8}}$ and $\mathcal{R}e_2 \equiv \|\overline{\mathbf{u}}\|d^{\frac{1}{2}}/\nu \leq \left(\frac{3}{4}\right)^{\frac{3}{4}}\alpha$ respectively, then the solution $\overline{\mathbf{u}}$ is (exponentially asymptotically nonlinearly) stable with respect to every kind of perturbations.

Recall that

$$|(M_{\overline{\mathbf{u}}}\mathbf{v}, \mathbf{v})| = |b(\mathbf{v}, \overline{\mathbf{u}}, \mathbf{v}) + b(\overline{\mathbf{u}}, \mathbf{v}, \mathbf{v})| = |b(\mathbf{v}, \overline{\mathbf{u}}, \mathbf{v})|.$$

Moreover, in (2.7.11) the nonlinear (advective) terms were absent because $b(\mathbf{v}, \mathbf{v}, \mathbf{v}) = (R(\mathbf{v}), \mathbf{v}) = 0$. Hence in Section 2.7 all our computations concern the linearized equations only and they were defined by $|b(\mathbf{v}, \overline{\mathbf{u}}, \mathbf{v})|$, hence the same term was involved in the linear criterion. These are the premises that the two approaches, used in the linear and nonlinear cases, give the same criterion for linear and nonlinear stability. Such a criterion is also a criterion for the nonexistence of subcritical instability, more exactly no perturbation amplifies in this range. Here the subcritical means for $R_e < R_{e_L}$, where R_{e_L} denotes the linear stability limit.

A common range of validity of the linear and nonlinear criteria (2.6.14) and (2.7.23) corresponds to $c_1 = (4/3)^3/4d^{-1/2}\alpha^{1/4}\|\overline{\mathbf{u}}\|$.

A criterion improving (2.7.23) is obtained by the same method as in Section 2.7.2.1, but taking into account the sequence of inequalities [GeoPal95], [GeoPal97]

$$|b(\mathbf{v},\mathbf{v},\overline{\mathbf{u}})|\leq|\mathbf{v}|_4\|\mathbf{v}\|\|\overline{\mathbf{u}}|_4\leq(4/3)^{3/4}|\mathbf{v}|^{1/4}\|\mathbf{v}\|^{3/4}\|\mathbf{v}\|\|\overline{\mathbf{u}}|^{1/4}\|\overline{\mathbf{u}}\|^{3/4}$$

$$\leq(4/3)^{3/4}d^{1/4}\alpha^{-1/8}\|\mathbf{v}\|^2|\overline{\mathbf{u}}|^{1/4-\gamma}\|\overline{\mathbf{u}}\|^{3/4}|\overline{\mathbf{u}}|^\gamma$$

$$\leq(4/3)^{3/4}d^{1/4}\alpha^{-1/8}\|\mathbf{v}\|^2|\overline{\mathbf{u}}|^{1/4-\gamma}\|\overline{\mathbf{u}}\|^{3/4}d^\gamma\alpha^{-\gamma/2}\|\overline{\mathbf{u}}\|^\gamma$$

$$=(4/3)^{3/4}d^{1/4+\gamma}\alpha^{-1/8-\gamma/2}|\overline{\mathbf{u}}|^{1/4-\gamma}\|\overline{\mathbf{u}}\|^{3/4+\gamma}\|\mathbf{v}\|^2. \tag{2.7.43}*$$

This criterion states the nonlinear stability for

$$\widehat{R}_{e_\gamma} < (3/4)^{3/4}\alpha^{1/8+\gamma/2}, \tag{2.7.44}$$

where $0\leq\gamma\leq1/4$ and $R_{e_\gamma} \equiv d^{1/4+\gamma}|\overline{\mathbf{u}}|^{1/4-\gamma}\|\overline{\mathbf{u}}\|^{3/4+\gamma}/\nu$. We have $R_{e_3} = \widehat{R}_{e_{1/4}}$ and $R_{e_2} = \widehat{R}_{e_0}$.

Then, (2.6.14) and (2.7.44) give the same criterion if

$$c_1 = (4/3)^{3/4}d^{\gamma-3/4}\alpha^{3/8-\gamma/2}|\overline{\mathbf{u}}|^{1/4-\gamma}\|\overline{\mathbf{u}}\|^{3/4+\gamma}. \tag{2.7.45}$$

Theorem 2.7.1 [GeoPal95], [GeoPal97]. *The best linear and nonlinear stability criterion is (2.6.14) where c_1 is defined by (2.7.45). It is a sufficient condition for nonexistence of any amplified perturbation.*

Two-dimensional case. All comments from the three-dimensional case hold except for (2.7.43)*, which is now

$$|b(\mathbf{v},\mathbf{v},\overline{\mathbf{u}})|\leq|\mathbf{v}|_4\|\mathbf{v}\|\|\overline{\mathbf{u}}|_4\leq2^{1/2}|\mathbf{v}|^{1/2}\|\mathbf{v}\|^{1/2}\|\mathbf{v}\|\|\overline{\mathbf{u}}|^{1/2}\|\overline{\mathbf{u}}\|^{1/2}$$

$$\leq2^{1/2}d^{1/2}\alpha^{-1/4}\|\mathbf{v}\|^2|\overline{\mathbf{u}}|^{1/2-\gamma}\|\overline{\mathbf{u}}\|^{1/2}|\overline{\mathbf{u}}|^\gamma$$

$$\leq2^{1/2}d^{1/2}\alpha^{-1/4}\|\mathbf{v}\|^2|\overline{\mathbf{u}}|^{1/2-\gamma}\|\overline{\mathbf{u}}\|^{1/2}d^\gamma\alpha^{-\gamma/2}\|\overline{\mathbf{u}}\|^\gamma$$

$$=2^{1/2}d^{1/2+\gamma}\alpha^{-1/4-\gamma/2}|\overline{\mathbf{u}}|^{1/2-\gamma}\|\overline{\mathbf{u}}\|^{1/2+\gamma}\|\mathbf{v}\|^2.$$

Then a criterion for nonlinear stability reads

$$R'_{e_\gamma} < \alpha^{1/4+\gamma/2}/\sqrt{2}, \tag{2.7.46}$$

where $R'_{e_\gamma} \equiv d^{1/2+\gamma}|\overline{\mathbf{u}}|^{1/2-\gamma}\|\overline{\mathbf{u}}\|^{1/2+\gamma}/\nu$. The criteria (2.7.46) and (2.6.14) are the same if

$$c_1 = \sqrt{2}d^{\gamma-1/2}|\overline{\mathbf{u}}|^{1/2-\gamma}\|\overline{\mathbf{u}}\|^{1/2+\gamma}. \tag{2.7.47}$$

Theorem 2.7.2 [GeoPal95], [GeoPal97]. *The best linear and nonlinear stability criterion is (2.6.14) where c_1 is defined by (2.7.47). It is a sufficient condition for nonexistence of any amplified perturbation.*

Chapter 3

Elements of calculus of variations

After positioning calculus of variations among other branches of mathematics in its historical evolution (Section 3.1), the main topics, concepts and results in this field are presented
(Section 3.2). Some of these questions are exemplified on four thermoelectrical convection problems involving matricial ordinary differential operators (Sections 3.3, 3.4 and
3.5). Similar questions are examined for the case of fluid flow stability problems involving
partial differential operators in Sections 3.2.6 and 3.2.7. Among them the four most important ideas are presented throughout this chapter. The role of symmetrization of operators
in simplifying computations is shown and more realistic stability criteria are obtained.
Another important topic concerns the variational principles enabling one to replace a variational problem with the equivalent boundary-value problem for a differential equation or
conversely. Isoperimetric inequalities and their equivalent eigenvalue problems are closely
analyzed. Finally, direct methods based on Fourier series expansions in calculus of variations are minutely described. For ordinary differential operators we use extensively the
results in [LavLy] while for the functional analytic backgrounds of calculus of variations
we are mainly based on treatises by V. I. Smirnov, S. L. Sobolev, S. G. Mikhlin, L. A.
Lyusternik, M. A. Lavrentiev, V. I. Levin, C. Foias, O. A. Ladyzhenskaya, P. M. Prenter,
H. F. Weinberger.

For the Fourier series we mainly use the Fikhtengoltz's treatise, while for the expansion
functions which do not satisfy all boundary conditions of the problem we are inspired from
the papers by B. Budiansky and R. C. DiPrima.

Our presentation of the energy method for evolution equations was influenced by the
contributions of J. Leray, R. Berker, J. Serrin, S. Chandrasekhar and D. D. Joseph and
his collaborators.

3.1 Generalities

Historical facts and main topics in calculus of variations are briefly presented. The con-

75

nection of this field with other domains of pure and applied mathematics are shown.

Calculus of variations deals with minimization of functionals. In particular, the functionals can be expressed by integrals or/and may represent various geometrical or physical quantities. Together with asymptotics and dynamical systems (including bifurcation), calculus of variations provides some of the most powerful mathematical tools to investigate nonlinear problems in science and engineering.

The origins of calculus of variations are to be looked for in the ancient Greek mathematics and its history is strongly connected with geometry and analytical mechanics [Goldsti], [Lec]; in the long run this calculus gained its autonomy evolving further as a genuine mathematical discipline. For the time being the calculus of variations tends to be included in a unitary theory of optimization which contains the optimal control.

The non-Eulerian calculus of variations before Lagrange begins, perhaps, with the Pappus problem (A. D. 290): *Among all plane curves of given length and passing through two given points P_1 and P_2, find that curve such that the area of geometric figure formed between the curve and the cord $P_1 P_2$ be maximum.* This is an isoperimetric problem.

Another famous problem of the calculus of variations is the brachistochrone problem (in Greek *brachistos*=the shortest, *chronos*=time), proposed in 1696 by Johann Bernoulli and solved (by means of special artifices) by himself and, among others, by Newton, Leibniz, l'Hospital, James: *Determine the trajectory of a material point starting at point P_1 with the velocity v_0 and falling down, under the gravitational field, at point P_2, during a minimum time.* It reduces to minimize the time functional.

Finally, let us mention the Maupertuis principle which, in the Jacobi form, is expressed as a variational problem with the Lagrangian $L = 2\sqrt{(h - U)T}$ where U is the potential energy, T stands for the kinematic velocity and h is a constant.

The Euler algorithm (elaborated in 1744) is the first to allow a unitary and quite general treatment of variational problems for functionals defined on spaces of smooth functions, by reducing the variational problem to the solution of an associated differential equation. It permits to find the second term of the asymptotic expansion in powers of a small parameter of the functional in the variational problem by using the asymptotic expansion of the argument function truncated also to two terms. The vanishing of this second term leads to the Euler equation which is a necessary condition of extremum. The application of the same algorithm by Euler (1707-1783) and the successors to the case of various types of functionals expressed in the form of integrals, provided Euler-type equations for functionals whose Lagrangian depends also on higher order derivatives and/or on many argument functions and/or is defined by a line-, surface-, volume- or general multiple-integrals.

The Euler method is still (mainly) applied by non-mathematicians.

Lagrange's variations. In 1759, in connection with its investigations on Taylor series of real functions defined on R^n on one hand, and with his research in mechanics on the other hand, Lagrange (1736-1813) elaborates an algorithm to find necessary conditions of extremum on the basis of the concept of *variation* (whence the name of calculus of variations). This notion will remain for a long time a nebulous mathematical object. However, it extends to functions of a more general nature (*i.e.* functionals) the notion of the differential of a function of a real argument. Lagrange extends to the case of functionals the principle according to which the differential vanishes at the point of extremum. By means of variations, Lagrange tried to avoid the use of infinitesimals; in fact he defined a differential whose main property, besides its order of magnitude, was its linearity. The two centuries that passed showed that this way of Lagrange opened the premises of a natural generalization to more complicated situations like Fréchet and Gateaux differentiable functionals, the Lagrange variation of a functional being, in fact, the Fréchet or Gateaux differential of that functional. This is why, as a rule, Lagrange is considered to be the founder of the calculus of variations, although its name is often associated with that of Euler.

For a long time, subsequent to Lagrange, the calculus of variations has dealt with: the search of necessary and sufficient conditions of extremum for various functionals defined on spaces of smooth or piecewise smooth functions of one or many real variables and with particular variational problems of special interest for applications, among which a first and permanent place was occupied by problems of analytic mechanics and its various formalisms (*i.e.* mathematical models describing the motion of material points): Newtonian, Lagrangian, Hamiltonian and, more recently, Kirchhoffian. A proof of the strong impregnation of the calculus of variations with analytic mechanics is, for instance, the assertion made by Carathéodory in the *Preface* to his famous book [Cara] of 1935 that he presents the theory of the calculus of variations as a servant to mechanics.

Along with its 250 years of existence the *post-Lagrangian classical calculus of variations* dealt with necessary and sufficient conditions for (local and global) maxima and minima of functionals. Cases of positively defined Lagrangians and quadratic functionals were investigated for many types of spaces of admissible functions, classified upon the smoothness of their functions, the boundary conditions they satisfy and upon various constraints. All these lead to many important particular problems which are by now distinct chapters of the *classical analytic calculus of variations*: isoperimetric problems (where the constraint is expressed by means of an integral), the general Lagrange problem (where the constraint is a differential equation), the Mayer problem (where the Lagrangian is linear in the derivative of the argument), problems with conditioned extremum, discontinuous problems [LavLy] etc. Spaces of functions which on part of the boundary of their domain of definition have not an assigned limit have also been considered. Direct methods have been used in the calculus of variations where the extremum problem is no longer reduced to a differential equation. Much attention was paid to the connection between isoperimetric and eigenvalue problems. Because the functions defined in R^n are particular functionals, their maxima and minima are studied also in the framework of the calculus of the variations (as a part of the tensor analysis in $\mathbf{R}^n$); this case occupied a large space in the Lagrange studies themselves. On the same line Tonelli's contributions must also be mentioned.

The modern analytic calculus of variations associates with an operatorial equation a variational formulation with the aid of some functional. The most simple case is : *Let $A : H \to H$ be a positive definite operator in a Hilbert space H. Then the equation $Au = f$, $f \in H$ has at most one solution. If this solution exists, it realizes the minimum of the functional $F(u) = (Au, u) - (u, f) - (f, u)$, where $(\cdot, \cdot)$ is the scalar product on H. Conversely, the element $u_0 \in H$ which realizes the minimum of the functional F is a solution of the given equation.* These last assertions, (developed first by Mikhlin and his collaborators) form a variational principle, that is a proposition which establishes an equivalence between a boundary-value problem and an associated variational problem. The case of quadratic functionals associated with affine equations is treated extensively. The problem of derivation of variational principles not only for positive operators is old. As early as 1853, in the geometric optics, Hamilton observed the correspondence between first-order partial differential equations and the associated variational problems; ten years later this fact was remarked in a more general case by Jacobi. Today the derivation of variational principles plays an important role in the abstract calculus of variations.

The existence of such a principle may lead to generalized solutions of a problem, *i.e.* to a mathematical object which realizes the minimum of the attached functional but not necessary to the given problem. In order to find the functional for the classically formulated problem, one associated to it (linear or nonlinear) operator which is continued up to an operator whose argument is a distribution; if this operator has certain compactness properties, it may be associated with a functional. Hence this formulation no longer gives a variational alternative equivalent to the given problem (but to an intermediate problem). In exchange, it yields a new-type of solution, nonexistent in the initial setting and often more useful for applications than the classical solution. In this context the simplest example is represented by discontinuous variational problems whose extremals belong to a class larger than that of the problem. The calculus of variations on spaces of distributions is related to mechanics of continua which operates with infinite dimensional dynamical systems.

The problems with a free (*i.e.* unknown) boundary are associated with variational problems of other types, expressed as variational or quasivariational inequalities [Kin], [KinS], [Mosc], NanP], [Aus], [BaC], [AthMR] and they are alternative formulations of the principle of the virtual powers. In a variational inequality the involved convex set does not depend on the solution [AsE], [Aub2], [EkT], [EkG], [GruW], [Rab], [Roc], [RocW], [SmitP], [Weinste], [Wi]. If it is not the case, it is referred to as a quasivariational inequality. The calculus of variations occurs in inequalities of Sobolev spaces. Nowadays the calculus of variations tends to unite to the theory of nonlinear monotonous operators in Banach spaces. Finally, the generalized variational framework often proves to be the natural framework for the formulation of physical laws (*e.g.* the case of N-S equations for viscous fluid flows which admit a generalized formulation but do not have a classical variational setting. On the contrary, Euler equations governing the flows of ideal fluids do admit a classical variational setting).

The geometrical calculus of variations is an alternative in differential geometry or topology, to problems of extremum, formulated by means of suitable geometrical objects. In this setting the extremals are looked for in classes of curves and surfaces; when the last ones are parametrized (and, thus, are defined by functions independent of the systems of coordinates), extremals are found which were not admitted by the analytic study which depends on Cartesian coordinates. This shows that the geometrical setting of variational problems leads to generalizations specific to the geometrical point of view. At present, the analytic mechanics still requires new ideas of geometrical calculus of variations because, the last few years, after the basic contributions by R. Abraham, J. E. Marsden and V. I. Arnold, it appeared that the geometrical bases of this mechanics are still unsatisfactory.

Let us also remark that in this context the Lagrange variation, in the form of the Fréchet differential defined on the tangent space of the manifold, is the key mathematical object.

A much investigated branch of classical geometrical calculus of variations is that of closed extremal curves (geodesics) on closed two-dimensional surfaces, subject of great interest for problems of mechanical stability. This calculus was also extended to three-dimensional manifolds, where the variational problem must take into account the topology of the manifold. For instance, Carathéodory showed that the closed extremals which realize an absolute minimum in the analytical sense do not always give a minimum if they are considered as fibers of a fiber bundle. The appropriate definition of the variation in differential geometry enabled the application of the calculus of variations to the study of geodesics of Riemannian manifolds. We mention the *calculus of variations in the large* (of Morse, Milnor etc.), where the property of extremization is valid throughout the domain of definition of the parameter on the curve, and the critical points of functions defined on differentiable manifolds are studied. In the three-dimensional calculus of variations much attention was paid to the Plateau problem. Other important problems of the classical geometrical calculus of variations is the search of invariants connected with any curve or surface of the space where the variational problem is considered and on which a pointwise transformation is defined. This problem, of special interest for analytic mechanics, was assumed very much among others by Emmy Noether and P. Finsler (this one using the spaces bearing his name).

Remark also the existence of *computation algorithms* based on concepts of calculus of variations (*e.g.* spline functions and finite elements). In this respect we mention the existence of a variational finite element theory. However, a non-variational finite elements' analysis also exists. Numerous applications based on finite elements were made to diffusion convection problems but they mainly concern the computation of the fluid flows and, consequently, are not of interest in this book.

Among all the quoted topics in the calculus of variations we shall be concerned with a few closely connected to criteria of fluid flow stability: isoperimetric problems and inequalities, direct methods based on Fourier expansions upon total sets, variational principles for matricial ordinary differential equations and other related topics, *e.g.* eigenvalue problems, bifurcation sets for characteristic equations, symmetrization of matricial differential equations, energy method for evolution equations.

The bibliography on analytic calculus of variations is extremely rich, *e.g.* [Ak], [Bl], [Bol], [ButH], [El], [FucNS], [Ces2], [Cont], [Cou2], [GelF], [Gi1], [Gi2], [GiMS], [GiH], [Gou], [Gra], [IofRS1], [IofRS2], [IofT], [Kou], [Lav], [LavLy], [Lebe], [Mau2], [Mikh1-5], [Morr], [Nir], [Rot], [Smir], [Sob2], [Tik], [Tuc], [Va], [We1], [We2], [Weinsto], [Ze]. To the reader especially interested in applications we recommend [Bal], [AraL], [Berg], [Dre], [FrB], [FrAS], [Berd], [Kup], [Lan], [Lau], [OdR], [Pr], [RedV], [Rek], [Sche], [StriA], [Stru], [Your], [Wan].

3.2 Direct and inverse problems of calculus of variations

The main concepts, methods and results in calculus of variations for the analytical setting are presented. Technicalities are avoided as much as possible.

3.2.1 *Variational problems in classical, generalized and abstract setting*

Let $\Omega \subset \mathbf{R}^n$ be a domain and denote by $\partial\Omega$ its boundary. Consider a functional $\mathcal{F}$ and denote by $\mathcal{D}(\mathcal{F})$ its domain of definition, referred to as *the class of admissible functions*. The set $\mathcal{D}(\mathcal{F})$ is a space of functions u, where $u : \Omega \to \mathbf{R}$ and u possesses certain smoothness properties, satisfies certain constraints in Ω and assumes, eventually, together with some of its derivatives, certain values on $\partial\Omega$.

Let $\delta\mathcal{F}(u_0)$ be the Fréchet differential of $\mathcal{F}$ at u_0, where δ is the Lagrange symbol of differentiation with respect to u. It commutes with the symbol D of differentiation with respect to x. If $\delta\mathcal{F}(u_0) = 0$, then u_0 is called a *stationary point* of $\mathcal{F}$. In particular, u_0 can be a *point of minimum (maximum)* of $\mathcal{F}$ if the value $\mathcal{F}(u_0)$ is a minimum (maximum) in the range $\mathcal{R}(\mathcal{F})$ of $\mathcal{F}$. The points of minima or maxima are called *the extremals* of $\mathcal{F}$, while the corresponding values of $\mathcal{F}$ are referred to as the *extrema*, namely *minima* and *maxima* of $\mathcal{F}$.

Definition 3.2.1. The *variational problem* for $\mathcal{F}$ in $\mathcal{D}(\mathcal{F})$ is the problem of finding its extremals and extrema.

Since a point of maximum for $\mathcal{F}$ is a point of minimum for $\frac{1}{\mathcal{F}}$ and in order to compute a minimum, it is necessary to know the point of minimum, usually the following definition is used

Definition 3.2.1$'$. The *variational problem* for $\mathcal{F}$ in $\mathcal{D}(\mathcal{F})$ is the problem of finding its minimum and we write

$$\min_{u \in \mathcal{D}(\mathcal{F})} \mathcal{F}(u). \tag{3.2.1}$$

Calculus of variations extends the calculus in $\mathbf{R}^n$ concerning the stationarity of functions of several real variables, from the case of a finite number of variables to an infinite number of variables. Indeed, $\mathcal{DF}$ is an infinite-dimensional set. Similarly to $\mathbf{R}^n$, the following definition is of interest in applications.

Definition 3.2.1$''$. The *variational problem* for $\mathcal{F}$ in $\mathcal{D}(\mathcal{F})$ is the problem of finding the stationary points of $\mathcal{F}$, *i.e.* the functions u_0 for which the Fréchet differential of $\mathcal{F}$ vanishes: $\delta\mathcal{F}(u_0) = 0$. In older papers, the variational problems are also referred to as the *direct variational problems*.

Like in $\mathbf{R}^n$, the extremals can be relative, *i.e.* local, or absolute, *i.e.* global. Unlike $\mathbf{R}^n$, $\mathcal{DF}$ involves in its definition certain regularity properties of u. Consequently, the definition of a minimum value can have different meanings, corresponding to the sense of the norm and, thus, of the neighborhood defined in $\mathcal{D}(\mathcal{F})$. More exactly, in non-degenerate cases, the stationary points of $\mathcal{F}$ are isolated points of $\mathcal{D}(\mathcal{F})$. If the stationary point is not unique in $\mathcal{D}(\mathcal{F})$, then it is called a *relative stationary point*, and an *absolute stationary point* otherwise. In particular, there exist relative and absolute minima and maxima. In order to have a unique extremal, instead of $\mathcal{D}(\mathcal{F})$ it is considered an appropriate subspace of $\mathcal{D}(\mathcal{F})$. This is why the theory assumes that in $\mathcal{D}(\mathcal{F})$ there exists a single extremal and it is a point of minimum for $\mathcal{F}$.

Every variational problem involves two basic components: the functional $\mathcal{F}$ and its domain of definition $\mathcal{D}(\mathcal{F})$. As already mentioned, the class of admissible functions, $\mathcal{D}(\mathcal{F})$, is defined by: the smoothness of the functions u of $\mathcal{D}(\mathcal{F})$; their boundary conditions; other constraints on u. The change in each of these three characteristics implies significant changes in the solution of the problem. For instance, if u are smooth functions we are in the framework of classical calculus of variations, otherwise we are in the generalized case. Different types of constraints (*e.g.* integral, differential) correspond to different types of variational problems, called the *isoperimetric problems*.

A variational problem depends also strongly on the dimension of Ω and the shape of $\partial\Omega$.

In the classical theory, $\mathcal{F}$ is an integral of the form $\mathcal{F}(u) = \int_\Omega F(x, u, D^\alpha u)dx$, where the function F, called *the Lagrangian*, depends on u and its derivatives $D^\alpha u$ of order $|\alpha|$, with $0 \leq |\alpha| \leq m$. If Ω is one-dimensional, *i.e.* $x \in \mathbf{R}$, then $\alpha \in \mathbf{N}$

and D^α stands simply for the differentiation of order α. If Ω is n-dimensional, *i.e.* $x \in \mathbf{R}^n$, then α is a multi-index $\alpha = (\alpha_1, \ldots, \alpha_n)$, $\alpha_i \geq 0$, $\mid \alpha \mid = \alpha_1 + \ldots + \alpha_n$, and $D^\alpha u = \frac{\partial^{|\alpha|} u}{\partial x_1^{\alpha_1} \ldots \partial x_n^{\alpha_n}}$. In this case $u \in \mathcal{DF} \subset C^m(\Omega)$.

In the generalized framework of Sobolev spaces $W^{l,p}(\Omega)$, *i.e.* $u \in W^{l,p}(\Omega)$, the integral defining $\mathcal{F}$ is taken in the Lebesgue sense and it is understood that the integrand F is an m-order differential mapping A. The derivatives $D^\alpha u$ are taken in the sense of distributions and they are regular distributions. If the derivatives are allowed to be singular distributions, then the integral stands for the dual pairing parentheses. More generally, A need not be a differential mapping, but an abstract mapping. As a consequence, the generalized and abstract theory of calculus of variations can be viewed as special chapter of nonlinear functional analysis.

The generalized setting cannot be avoided. Indeed, in some cases, classical extremals do not exist. In exchange, in these cases, extremals can exist which do not belong to $C^m(\Omega)$. Thus, $C^m(\Omega)$ must be enlarged up to $W^{l,p}(\Omega)$ or a larger function space. In enlarging $\mathcal{D}(\mathcal{F})$, the meaning of derivatives weakens, therefore the generalized setting is more natural.

Remark 3.2.1. Since there are classical variational problems which have no solution while the generalized corresponding ones do, the generalized setting can be viewed as an extension intended to ensure the existence of a solution (even if in a non-classical sense), *i.e.* to achieve solvability. (Similarly, there are polynomial algebraic equations in $\mathbf{R}$ the solvability of which does not take place in $\mathbf{R}$ but is always possible in $\mathbf{C}$, where $\mathbf{R}$ is embedded.)

This confers to calculus of variations a central place in the theory of differential equations, offering its natural framework.

A still more natural setting is the geometrical one. For instance, if $\Omega \subset \mathbf{R}$ and the boundary conditions specify that all admissible functions assume given values on $\partial\Omega$, geometrically this means that $\mathcal{D}(\mathcal{F})$ consists of curves $u = u(x)$ passing through the same two points if $\Omega \subset \mathbf{R}^1$ or through the same curve if $\Omega \subset \mathbf{R}^2$ etc.

In general, the boundary conditions are easy to express in geometrical terms; in a concrete problem they express physical requirements.

The extremals can be calculated directly, for instance expanding u in Fourier series and introducing it into the functional. In this way, the stationarity of the functional is transformed into the stationarity of the obtained function of the Fourier coefficients. These methods are called *direct methods* of calculus of variations.

Sometimes, in order to solve a variational problem, it is more convenient to reduce it to an associated boundary-value problem, generically called the *Euler equation*. In this case the methods do not apply directly to the given variational problem but to the boundary-value problem for the Euler equation. These methods are referred to as *indirect methods* of calculus of variations.

The converse situation can be of interest too: starting with a given boundary-value problem, it is required to find the associated variational problem. Thus the difficult question whether or not these two associated problems are equivalent, *i.e.* have the same solutions, arises.

Definition 3.2.2. A proposition asserting the equivalence of a variational problem and some boundary-value problem for the associated Euler equation is referred to as a *variational principle* [Moi], [Kup], [Berd], [FrA], [Lan], [Your], [Fil], [För].

3.2.2 Construction of the boundary-value problem associated with a variational problem. Necessary conditions for extremum

First, let us show how to construct the Euler equation starting with the given functional $\mathcal{F}$. Most textbooks of calculus of variations are concerned with just this problem, namely with proving that the stationary points of $\mathcal{F}$ are solutions of some boundary-value problem for the Euler equation. Usually, this name is bore by the simplest case of one-dimensional x and of a single derivative occurring in F, *i.e.* $m = 1$. In some other cases, the name of Euler-Lagrange, Euler-Poisson etc. is used. If u is a vector function, then the Euler equation is a vector equation too. Formally, the Euler equation can be obtained by the Euler procedure, which mimics the method to derive the minima, maxima and inflexion points of real functions. Let us sketch it in the generalized framework, the single one appropriate to variational problems.

Unlike in theoretical studies, in applications these problems have a classical setting, involving continuous derivatives. Therefore, the admissible functions satisfy some boundary conditions and belong to some $C'^k(\overline{\Omega})$, *i.e.* the space of k times continuously differentiable functions on Ω and which can be continued up to continuous functions on $\overline{\Omega}$. Without this possibility of continuation, the boundary condition could not always be imposed. As $\mathcal{D}(\mathcal{F}) \subset C'^k(\overline{\Omega})$, in order to rephrase the classical variational problem into a generalized sense, we must perform some embeddings. Thus, let $\mathcal{X}$ be a Banach space and embed $\mathcal{D}(\mathcal{F})$ in $\mathcal{X}$. As usual in nonlinear functional analysis, denote also by $\mathcal{D}(\mathcal{F})$ the subset of $\mathcal{X}$ isomorphic to the given $\mathcal{D}(\mathcal{F})$ from the classical setting. Correspondingly, integrable will mean integrable in the Lebesgue sense.

Convention *In the following we adopt this twofold notation and sense of integrability without any further specification.*

Let $\overline{u} \in \mathcal{D}(\mathcal{F})$ be a fixed element of $\mathcal{D}(\mathcal{F})$ and denote by $\mathcal{M}$ a linear subset of $\mathcal{X}$. Write every element u of $\mathcal{D}(\mathcal{F})$ as $u = \overline{u} + \eta$, where $\eta \in \mathcal{M}$; then $\mathcal{D}(\mathcal{F})$ is a linear manifold of $\mathcal{X}$. Remark that $\mathcal{D}(\mathcal{F})$ is not a linear set and that η satisfies the same type of boundary conditions on $\partial\Omega$ as u, but they are homogeneous. The space $\mathcal{M}$ is constructed once $\mathcal{D}(\mathcal{F})$ is given. Assume that if η belongs to a finite-dimensional subspace $\mathcal{M}_1$ of $\mathcal{M}$ then $\mathcal{F}(u) = \mathcal{F}(\overline{u} + \eta)$ is a sufficiently smooth functional. In addition, suppose that $\mathcal{D}(\mathcal{F})$ is dense in $\mathcal{X}$, which is equivalent to say that $\mathcal{M}$ is dense in $\mathcal{X}$. In these conditions define

$$\delta\mathcal{F}(u,v) = \lim_{\epsilon\to 0} \frac{d\mathcal{F}(u+\epsilon v)}{d\epsilon} \tag{3.2.2}$$

and refer to $\delta\mathcal{F}(u,v)$ as the first variation of $\mathcal{F}$ with respect to u applied to v. The modern usual notation for $\delta\mathcal{F}$ is $\mathcal{F}'_u v$ because it can be proved that $\delta\mathcal{F}(u,v)$ is a linear functional of v.

(In fact, (3.2.2) is the Gateaux differential and the method based on it is referred to as the *Euler method* in calculus of variations. The linear part of $\mathcal{F}(u+\delta u) - \mathcal{F}(u)$ proves to be the first Fréchet differential and it coincides with the Gateaux differential in our cases of interest. Here the δu is the Lagrange variation, *i.e.* the Fréchet differential of u, where I is the identity operator.)

Assume that $\delta\mathcal{F}$ is also bounded (more exactly we consider a subspace $\mathcal{D}_1(\mathcal{F})$ of $\mathcal{D}(\mathcal{F})$ on which $\delta\mathcal{F}$ is bounded). Then, for each u fixed, $\delta\mathcal{F}(u) \in \mathcal{X}^*$, where $\mathcal{X}^*$ is the space adjoint to $\mathcal{X}$. It follows that $\delta\mathcal{F}(u,v) =< \delta\mathcal{F}(u), v >$ where $< \cdot, \cdot >$ stands for the duality pairing relation. As u is running over $\mathcal{D}(\mathcal{F})$, $\delta\mathcal{F}(u)$ generates a nonlinear mapping denoted by $\mathrm{grad}\mathcal{F}(u)$ and called the *gradient of* $\mathcal{F}$, where $\mathrm{grad}\mathcal{F}(u) : X \to X^*$, *i.e.* $< \delta\mathcal{F}(u), v >=< \mathrm{grad}\mathcal{F}(u), v >$. In order to avoid the technical discussion related to the possibility of splitting the Banach space X for $\eta \in \mathcal{M}_1$, we assume that $\mathcal{X} = \mathcal{H}$, where $\mathcal{H}$ is a Hilbert space. Then the duality pairing becomes the scalar product in $\mathcal{H}$ and, due to the arbitrariness of $v \in \mathcal{M}$, and of the density of $\mathcal{M}$ in $\mathcal{X}$, by Theorem 1.2, $\delta\mathcal{F}(u,v) = 0$ implies the Euler equation [Mikh 5]

$$\mathrm{grad}\mathcal{F}(u) = 0. \tag{3.2.3}$$

In the classical case the last assertion follows from the following two lemmas:

Paul du Bois-Raymond lemma [LavLy]. *Let $f : [a,b] \to \mathbf{R}$ be a continuous function and let $v : [a,b] \to \mathbf{R}$ an arbitrary continuous function with continuous derivatives and $v(a) = v(b) = 0$. Suppose that $\int_a^b f(x)v'(x)dx = 0, \ \forall v \in C^1([a,b], \mathbf{R})$. Then f is a constant function on $[a,b]$.*

Lagrange lemma [LavLy]. *In the same conditions as in the previous lemma, assume that $\int_a^b f(x)v(x)dx = 0, \quad \forall v \in C^1([a,b], \mathbf{R})$. Then $f(x) \equiv 0$ on $[a,b]$.*

Remark 3.2.2. Euler equation (3.2.3) is only a necessary condition of stationarity of $\mathcal{F}$. This follows from the fact that, as a function of ϵ, $\mathcal{F}(u+\epsilon v)$ is stationary if $\lim_{\epsilon\to 0} \frac{d}{d\epsilon}\mathcal{F}(u+\epsilon v) = 0$. Therefore, every stationary function of $\mathcal{F}$, in particular every extremal of the variational problem (3.2.1), is a solution of the Euler equation (3.2.3) and it belongs to $\mathcal{D}(\mathrm{grad}\mathcal{F})$. In other words, with the variational problem (3.2.1) we associated the boundary-value problem

$$u \in \mathcal{D}(\mathrm{grad}\mathcal{F}) \tag{3.2.4}$$

for the Euler equation (3.2.3).

Remark 3.2.3. In $\mathcal{D}(grad\mathcal{F})$ the smoothness as well as the boundary conditions for u are specified. The smoothness is the primary concern in theory, while the boundary conditions are the first concern in applications.

Remark 3.2.4. Sufficient conditions for minimum involve the second Lagrange variation, *i.e.* the second-order Fréchet differential. They are not given here.

The form of the Euler equation depends on the Banach space $\mathcal{X}$ in which $\mathcal{D}(\mathcal{F})$ is embedded. However, the solution of the Euler equation does not depend on the choice of $\mathcal{X}$. This property is known as the *invariance of the Euler equation*.

Remark 3.2.5. In general, as specific examples show, $grad\mathcal{F}$ is a nonlinear mapping whose domain of definition, $\mathcal{D}(grad\mathcal{F})$, is contained in $\mathcal{D}(\mathcal{F})$. In these cases smoothness requirements ensuring the boundedness of the functional $\delta\mathcal{F}(u)$ must be imposed on $\mathcal{F}$. For example, if $C^1[a, b]$ is embedded not in $L^2(a, b)$ but in $W^{\circ 1,2}(a, b)$, then $\mathcal{D}(\mathcal{F})$ becomes equal to $\mathcal{D}(grad\mathcal{F})$. The fact that, in general, $\mathcal{D}(\mathcal{F}) \neq \mathcal{D}(grad\mathcal{F})$, *i.e.* the class of admissible functions differs from the domain of definition of the mapping defining the associated Euler equation, is the main source of difficulty in applying inverse methods in calculus of variations.

3.2.3   *Classical Euler equations associated with variational problems for particular functionals*

Consider the particular case of a functional $\mathcal{F}$ expressed by an integral, the integrand of which is denoted by F. Assume that $\mathcal{D}(\mathcal{F}) \subset \mathcal{H}$, where the Hilbert space $\mathcal{H}$ is a function space consisting of vector functions $\mathbf{u} : \mathbf{R}^n \to \mathbf{R}^s$, $\mathbf{u} = \mathbf{u}(\mathbf{x})$. In this section we sketch the main types of variational problems for the case of a Lagrangian F depending on x and on u and its derivatives up to the r-th order. Their detailed presentation can be found in the basic treatises on analytical calculus of variations: [Mikh5] and [LavLy] for readers who are interested in mathematics and applications respectively. In the case $n = r = s = 1$ we follow [Mikh5], while for the other cases we follow [LavLy].

n = r = s = 1. This simplest problem in calculus of variation concerns the functional $\mathcal{F}(u) = \int_a^b F\big(x, u(x), u'(x)\big)dx$, $\mathcal{D}(\mathcal{F}) = \{u \in C^1[a, b] \mid u(a) = \alpha_1, \quad u(b) = \beta_1\}$, where the prime stands for the derivative, α_1 and α_2 are given constants, $\mathcal{M} = \{v \in C^1[a, b] \mid v(a) = v(b) = 0\}$, F is continuous in x, u and u' and has continuous partial derivatives F_u $F_{u'}$ for $a \leq x \leq b$, $-\infty < u < +\infty$, $-\infty < u' < +\infty$. Choosing $\mathcal{X} = L^2(a, b)$, we embed $\mathcal{D}(\mathcal{F})$ in $L^2(a, b)$ and denote also by $\mathcal{D}(\mathcal{F})$ the subset of $L^2(a, b)$ isometric to it. Then $\mathcal{D}(grad\mathcal{F}) = \{u \in \mathcal{D}(\mathcal{F}) \mid F_{u'}(x, u, u')$ is absolutely continuous on $[a, b]$ and has a square integrable derivative with respect to $x\}$ and $grad\mathcal{F} = F_u - \frac{d}{dx}F_{u'}$. Therefore, in the generalized sense, the Euler equation reads

$$F_u - \frac{d}{dx}F_{u'} = 0, \quad x \in (a, b), \quad u \in \mathcal{D}(grad\mathcal{F}). \tag{3.2.5}$$

In additional assumptions on $\mathcal{F}$, another form for this equation can be obtained. For instance, if all partial derivatives F_x, F_u, $F_{u'}$, $F_{xu'}$, $F_{uu'}$, $F_{u'u'}$ are continuous and $F_{u'u'} \neq 0$, then it can be proved that the solution u of the Euler equation has the derivative u'' continuous on $[a, b]$. In this case (3.2.5) becomes

$$u'' = \{F_u - F_{xu'} - F_{uu'}u'\}/F_{u'u'}, \quad x \in [a, b], \quad u \in \tilde{\mathcal{D}}(\mathcal{F}) \tag{3.2.6}$$

where $\tilde{\mathcal{D}}(\mathcal{F}) = \{u \in C^2[a, b] \mid u(a) = \alpha_1, \quad u(b) = \beta_1\}$.

A formal derivation of (3.2.5) by the Euler method is immediate. Indeed,

$$\lim_{\epsilon \to 0} \frac{\mathcal{F}(u + \epsilon v) - \mathcal{F}(u)}{\epsilon} = \lim_{\epsilon \to 0} \int_a^b \frac{\mathcal{F}(x, u + \epsilon v, u' + \epsilon v') - \mathcal{F}(x, u, u')}{\epsilon}dx$$

$$= \lim_{\epsilon \to 0} \int_a^b \frac{1}{\epsilon}\left[F_u\epsilon v + F_{u'}\epsilon v'\right]dx = \int_a^b \left[F_u + \frac{d}{dx}(F_{u'}v) - v\frac{d}{dx}F_{u'}\right]dx$$

$$= [F_{u'}v]_a^b + \int_a^b (F_u - \frac{d}{dx}F_{u'})vdx.$$

Since $v(a) = v(b) = 0$, the free term vanishes. Then, the Paul du-Bois-Raymond and Lagrange lemmas imply that the integrand (which is just the left-hand side of (3.2.5)) vanishes too.

Remark 3.2.6 [Mikh2]. These lemmas solve the question of the class of the solution involved in deriving (3.2.5) in the case of one independent variable. This question proves to be more difficult for deriving (3.2.3) in the case of several independent variables.

In calculus of variations there are two main methods enabling the derivation of the Euler equations from a variational problem: *Euler method* (Section 3.2.1) and *Lagrange method*. By Lagrange method [GiH] extremals are stationarity points for $\mathcal{F}$, *i.e.* for them the linear part of the difference $\mathcal{F}(u + \delta u) - \mathcal{F}(u)$ must vanish. Here δu is the Lagrange variation of u. Let us exemplify this method for the simplest variational problem. We have

$$\mathcal{F}(u + \delta u) - \mathcal{F}(u) = \int_a^b [F(x, u + \delta u, u' + \delta u') - F(x, u, u')]dx.$$

The linear (in δu) part of this difference, denoted by $\delta\mathcal{F}$, is

$$\delta\mathcal{F} = \int_a^b (F_u\delta u + F_{u'}\delta u')dx$$

$$= \int_a^b [F_u\delta u + F_{u'}\frac{d}{dx}(\delta u)]dx = \int_a^b [F_u\delta u + \frac{d}{dx}(F_{u'}\delta u) - \frac{d}{dx}(F_{u'})\delta u]dx.$$

Here we used the commutativity of δ and $\frac{d}{dx}$ too. Since, in $\mathcal{DF}$, $\delta u(a) = \delta u(b) = 0$, it follows that $\int_a^b \frac{d}{dx}(F_{u'}\delta u)dx = [F_{u'}\delta u]_a^b = 0$. Therefore $\delta\mathcal{F} = \int_a^b(F_u - \frac{d}{dx}F_{u'})\delta u dx$ and, consequently, by the above quoted lemmas, $\delta\mathcal{F} = 0$ implies (3.2.5).

We add that, just as $v(a)$ and $v(b)$ in the Euler method, $\delta u(a)$ and $\delta u(b)$ vanish only in the class of curves passing through the two fixed points, whence a reason for Mikhlin's Remark 3.2.6.

Note that the case $n = r = s = 1$ concerns one-dimensional functions u depending on a single independent variable x; the Lagrangian F depends on x, u and a single derivative of u, namely u'; in the (x, u) plane, the graphs of the admissible functions, *i.e.* from $\mathcal{D}(\mathcal{F})$, are C^1 curves passing through two fixed points (a, α_1), (b, β_1).

If $n = 1$, $s \in \mathbf{N}^*$, $s > 1$, $r = 1$, F depends on x, on s functions $u_1(x), \ldots, u_s(x)$ and on their first derivatives. In other words,

$$\mathcal{F}(\mathbf{u}) = \int_a^b F(x, u_1, \ldots u_s, u_1', \ldots u_s')dx,$$

where $\mathbf{u} = (u_1, \ldots, u_s)$. The boundary-value problem associated with the variational problem for $\mathcal{F}$ reads

$$F_{u_i} - \frac{d}{dx}F_{u_i'} = 0, \quad i = 1, \ldots, s \quad x \in (a, b), \quad u \in \mathcal{D}(grad\mathcal{F}) \tag{3.2.7}$$

$$u_1(a) = \alpha_1, \ldots, u_s(a) = \alpha_s \quad u_1(b) = \beta_1, \ldots, u_s(b) = \beta_s, \tag{3.2.8}$$

where (3.2.7) is known as the *Euler-Lagrange* equation and the condition (3.2.8) means that the admissible functions $\mathbf{u}$ pass through two admissible points from $\mathbf{R}^s$.

If $n = 1$, $s = 1$, $r \in \mathbf{N}^*$, $r > 1$, F depends on a single real variable x, on a function $u(x)$ and on its first r derivatives with respect to x. Therefore

$$\mathcal{F}(\mathbf{u}) = \int_a^b F(x, u, u', \ldots, u^{(r)})dx.$$

The boundary-value problem associated with the variational problem for $\mathcal{F}$ reads

$$F_u - \frac{d}{dx}F_{u'} + \frac{d^2}{dx^2}F_{u''} - \ldots + (-1)^r\frac{d^r}{dx^r}F_{u^{(r)}} = 0, \quad x \in (a, b), \quad u \in \mathcal{D}(grad\mathcal{F}) \tag{3.2.9}$$

$$\begin{aligned}
u(a) = \alpha_0, u'(a) = \alpha_1, \ldots, u^{(r-1)}(a) = \alpha_{r-1}, \\
u(b) = \beta_0, u'(b) = \beta_1, \ldots, u^{(r-1)}(b) = \beta_{r-1}.
\end{aligned} \tag{3.2.10}$$

Equation (3.2.9) is the *Euler-Poisson equation*. The domain of definition of $\mathcal{F}$ is $\mathcal{D}(\mathcal{F}) = \{\mathbf{u} \in C^1([a, b], \mathbf{R}) \mid \mathbf{u} \text{ satisfies (3.2.10)}\}$ and consists in vectors $\mathbf{u} = (u, u', \ldots, u^{(r)})$, functions of x. In the space $(x, \mathbf{u})$, the graphs of all admissible functions pass through two points, namely $\mathbf{u}(a)$ and $\mathbf{u}(b)$.

For the case of admissible functions the graphs of which are running over two parallel straight lines, we recommend [LavLy].

If $n \in \mathbf{N}^*$, $n > 1$, $s = 1$, $r = 1$, the admissible functions u are one-dimensional but they depend on n independent variables $x_1, \ldots, x_n$ therefore, $u : \Omega \to \mathbf{R}$ where $\Omega \subset \mathbf{R}^n$. Denote $\mathbf{x} = (x_1, \ldots, x_n)$. The functional $\mathcal{F}$ is $\mathcal{F}(u) = \int_\Omega F(\mathbf{x}, u, \frac{\partial u}{\partial x_i}) d\mathbf{x}$ and $\mathcal{D}(\mathcal{F}) = \{u \in C^1(\Omega) \mid u_{|\partial\Omega} = u_w \text{ assigned functions}\}$. The boundary-value problem associated with the variational problem for $\mathcal{F}$ is

$$F_u - \sum_{i=1}^{n} \frac{\partial}{\partial x_i} F_{\frac{\partial u}{\partial x_i}} = 0, \quad \text{for} \quad \mathbf{x} \in \Omega, \tag{3.2.11}$$

$$u = u_w \quad \text{for} \quad \mathbf{x} \in \partial\Omega, \tag{3.2.12}$$

where (3.2.11) is called the *Euler-Ostrogradski* equation.

In all these problems the admissible functions satisfy some boundary conditions and have the smoothness required by the classical setting of the associated boundary-value problems. In the following variational problem, the admissible functions are subject to some *integral constraints*. This supplementary constraint imposed on admissible functions has important consequences: instead of an associated boundary-value problem, we have an associated eigenvalue boundary-value problem.

Isoperimetric problems with one integral constraint. Here we consider the simplest case $\mathcal{F}(u) = \int_a^b F(x, u, u')dx$, $\mathcal{D}(\mathcal{F}) = \{u \in C^2([a, b], \mathbf{R}) \mid u(a) = \alpha_1, u(b) = \beta_1\}$ but, in addition, u must satisfy the integral constraint

$$\mathcal{G}(u) = \int_a^b G(x, u, u')dx = \mathcal{G}_0, \tag{3.2.13}$$

where $\mathcal{G}_0$ is an assigned real number.

The variational problem for $\mathcal{F}$ with the constraint (3.2.11) is equivalent to the variational problem without constraint for the functional $\mathcal{F} + \mu\mathcal{G}$ and leads to the following *Euler eigenvalue problem*

$$F_u - \frac{d}{dx}F_{u'} + \mu(G_u - \frac{d}{dx}G_{u'}) = 0 \tag{3.2.14}$$

$$u(a) = \alpha_1, \qquad u(b) = \beta_1 \tag{3.2.15}$$

where $\mu \neq 0$ is, by now, an arbitrary constant, *i.e.* it is a parameter, called *the Lagrange multiplier*.

Remark 3.2.7. Here, by an *eigenvalue problem* it means a boundary-value problem $H(\mu, u) = 0$, or in particular, $A(u) - \mu B(u) = 0$, where A and B are some mappings, containing a physical vector parameter, usually denoted by R, such that this problem possesses nontrivial solutions, called eigenfunctions, only if μ takes certain values, called eigenvalues. Before the foundation of bifurcation theory, A and B were assumed to be linear operators [Mikh3]. In our case this concept coincides with the classical concept used in [Mikh3] only when $G_u - \frac{d}{dx}G_{u'} = \mathcal{I}$ where $\mathcal{I}$ is the identity operator, and $F_u - \frac{d}{dx}F_{u'}$ is a linear operator of u, of when one of these two operators is invertible.

The requirement that $F_u - \frac{d}{dx}F_{u'}$ be linear is fulfilled only for homogeneous quadratic functionals $\mathcal{F}$. In this case, we determine the μ-depending characteristic values and eigensolutions $u(\mu, R)$ of (3.2.14) and replace them in the boundary conditions (3.2.15). Then we solve the obtained (secular) equation in μ. The roots μ_i are subsequently substituted in the eigensolutions and compute the value of $\mathcal{F}$ for all these eigensolutions and values μ_i. As a result, the extremals and extrema are found. The physical parameter R is present in the secular equation, hence μ_i depend on R and, so is the minimum of $\mathcal{F}$.

For non-quadratic functionals, (3.2.14), (3.2.15) is a *bifurcation problem*. Since, depending on the possible values of μ, a general bifurcation problem can have one or several or none solutions, the same holds for the problem (3.2.14), (3.2.15). As already stated, we refer to the existing nontrivial solutions of (3.2.14), (3.2.15) as eigensolutions and to the corresponding values of the parameters, as eigenvalues and they depend on R. Then, replacing the eigenvalues and eigenvectors of (3.2.14), (3.2.15) in (3.2.13) we obtain an algebraic bifurcation problem in R. If this problem possesses nontrivial roots, then the extremals are to be found among the eigensolutions as in the case of homogeneous quadratic functionals.

Example 3.2.1. Consider the variational problem $\min_{u \in \mathcal{D}(\mathcal{F})} \mathcal{F}(u)$ where

$$\mathcal{F}(u) = \int_0^1 (u'(x))^2 dx \Big/ \int_0^1 u^2(x)dx, \tag{3.2.16}$$

$$\mathcal{D}(\mathcal{F}) = \{u \in C^1(0,1) \mid u(0) = u(1) = 0,\ u \neq 0\}$$

Since this problem is very difficult as it stands, let us introduce the new function

$$v(x) = \frac{u(x)}{\sqrt{\int_0^1 u^2(x)dx}}.$$

Obviously,

$$\mathcal{G}(v) = \int_0^1 v^2(x)dx = 1. \tag{3.2.17}$$

Then $\mathcal{F}(u)$ becomes $\mathcal{F}_1(v)$ where

$$\mathcal{F}_1(v) = \int_0^1 (v'(x))^2 dx, \tag{3.2.18}$$

$$\mathcal{D}_1(\mathcal{F}_1) = \{v \in C^1(0,1) \mid v(0) = v(1) = 0,\ v \text{ satisfies } (3.2.17)\}$$

Since $\mathcal{F}(u) = \mathcal{F}_1(v)$, it follows that $\min_{u \in \mathcal{D}(\mathcal{F})} \mathcal{F}(u) = \min_{v \in \mathcal{D}_1(\mathcal{F}_1)} \mathcal{F}_1(v)$. Therefore, the variational problem without constraints for $\mathcal{F}(u)$ has been reduced to the isoperimetric problem (3.2.18) for $\mathcal{F}_1(v)$. In its turn, the isoperimetric problem (3.2.18) for $\mathcal{F}_1(v)$ is reduced to a simpler variational problem without constraints, but involving the parameter μ, for the functional $\mathcal{F}_1(v) + \mu\mathcal{G}_1(v)$, which leads to the Euler eigenvalue two-point problem

$$v'' - \mu v = 0, \quad v(0) = v(1) = 0. \tag{3.2.19}$$

The eigensolutions have the general form $v(x) = C_1 e^{\lambda x} + C_2 e^{-\lambda x}$ where C_1 and C_2 are arbitrary constants. The corresponding characteristic equation $\lambda^2 - \mu = 0$ has the roots $\lambda_{1,2} = \pm\sqrt{\mu}$ for $\mu > 0$ and $\lambda_{1,2} = \pm i\sqrt{-\mu}$ for $\mu < 0$.

For $\mu < 0$, the general solution of $(3.2.19)_1$ is $v(x) = C_1 \cos(\sqrt{-\mu}x) + C_2 \sin(\sqrt{-\mu}x)$. Imposing this solution to satisfy the boundary conditions $(3.2.19)_{2,3}$, we are led to the secular equation, which is an algebraic one-parameter bifurcation equation,

$$\begin{vmatrix} 1 & 0 \\ \cos\sqrt{-\mu} & \sin\sqrt{-\mu} \end{vmatrix} = 0,$$

whence $\mu = -k^2\pi^2$, $k \in \mathbf{Z}$, and $C_1 = 0$. Hence, $v_k(x) = C_2 \sin(k\pi x)$. Introducing these eigenfunctions in the constraint (3.2.17) we get $C_2 = \sqrt{2}$, therefore $v_k(x) = \sqrt{2}\sin(k\pi x)$ and $\mathcal{F}_1(v_k) = k^2\pi^2$. Consequently, $\mathcal{F}_1$ has a minimum of π^2 corresponding to the point of minimum $v_1(x) = \sqrt{2}\sin(\pi x)$.

For $\mu > 0$, the general solution of $(3.2.19)_1$ reads $v(x) = C_1 \cosh(\sqrt{\mu}x) + C_2 \sinh(\sqrt{\mu}x)$ and leads to the secular equation $\sinh\sqrt{\mu} = 0$, implying $\mu = 0$, which is not convenient. Hence, the single extremal of $\mathcal{F}_1$ remains v_1.

The fact that $\mathcal{F}_1$ attains its minimum and not its maximum can be checked directly, *e.g.* for $v_1 = x(x-1)$ we have $\mathcal{F}_1(v_1) = 10$, while $\mathcal{F}_1(v_1) = \pi^2 < 10$. The rigorous proof is provided by a sufficient condition of minimum fulfilled by F, namely $F_{v'v'} > 0$ (in our case $F_{v'v'} = 2$) [LavLy].

Instead of (3.2.18), let us take $\mathcal{F}_1(v) = R\int_0^1 (v'(x))^2 dx$, where $R > 0$ is a physical parameter. Then (3.2.19) reads $Rv'' - \mu v = 0$. The secular equation becomes an algebraic two-parameter bifurcation problem and yields $-\mu = Rk^2\pi^2$. The eigensolutions have the same form as in Example 3.2.1 because their dependence on μ/R is not changed. The resulting $\mathcal{F}(v_n)$ preserves the previous form but now, as expected, it is multiplied by R. So is its minimum.

Example 3.2.1′. Take $\mathcal{F}_1(v) = \int_0^1 (v''(x))^2 dx + R\int_0^1 (v'(x))^2 dx$, assume that the constraint (3.2.17) holds, and impose the additional boundary conditions $v'(0) = v'(1) = 0$. Then instead of (3.2.19) we have $v^{IV} - Rv'' + \mu v = 0$, leading to the bifurcation secular equation $2\sqrt{\mu}(1 - \cosh r_1 \cosh r_2) + R\sinh r_1 \sinh r_2 = 0$, where $r_{1,2}$ and $r_{3,4} = -r_{1,2}$ are the roots of the characteristic equation, *i.e.* $r_{1,2} = \sqrt{\dfrac{R\pm\sqrt{R^2-4\mu}}{2}}$. This bifurcation equation is entitled to serve as secular equation only when the characteristic roots are mutually distinct, *i.e.* for $\mu \neq R^2/4$, $\mu = 0$.

If $\mu = 0$ the isoperimetric problem becomes a variational problem without constraints, hence we assume $\mu \neq 0$.

For $0 < \mu < R^2/4$ we have

$$r_{1,2} = \frac{\sqrt{R + 2\sqrt{\mu}} \pm \sqrt{R - 2\sqrt{\mu}}}{2} = a \pm b,$$

for $\mu < 0$ we have

$$r_{1,3} = \sqrt{\frac{R + \sqrt{R^2 - 4\mu}}{2}}, \qquad r_{2,4} = \pm i\sqrt{\frac{\sqrt{R^2 - 4\mu} - R}{2}} = ia,$$

while for $\mu > R^2/4$ we have

$$r_{1,2} = \sqrt{\frac{R \pm i\sqrt{4\mu - R^2}}{2}} = \frac{\sqrt{2\sqrt{\mu} + R}}{2} \pm i\frac{\sqrt{2\sqrt{\mu} - R}}{2} = a \pm ib,$$

where $a, b \in \mathbf{R}$.

In the case $0 < \mu < R^2/4$ the secular equation reads $-2\sqrt{\mu}(\sinh^2 a + \sinh^2 b) + R(\sinh^2 a - \sinh^2 b) = 0$, or, equivalently, $b^2 \sinh^2 a - a^2 \sinh^2 b = 0$. Since $\sinh a/a$ is a strictly monotonous function, increasing from 1 (for $a = 0$) to ∞ (for $a \to \infty$), and $\sinh^2 b/b^2$ decreases from 1 to 0, the secular equation has no root other than in the unacceptable situation $\mu = 0$.

In the case $\mu > R^2/4$ the secular equation becomes $2\sqrt{\mu}(\sin^2 b - \sinh^2 a) + R(\sin^2 b + \sinh^2 a) = 0$, or, equivalently, $a^2 \sin^2 b - b^2 \sinh^2 a = 0$. As $\sinh a > a$ and $\sin b < b$ it follows that this equation implies the inconvenient case $\mu = 0$.

For $\mu < 0$, we have $r_1 > 0$, $r_2 = ia$, the secular equation reads $2\sqrt{-\mu}(1 - \cosh r_1 \cos a) - R\sinh r_1 \sin a = 0$ and it has an infinity of solutions.

Finally, the case $R = 0$ is contained here as a particular case.

Thus, on the lines of Example 3.2.1, $\min\mathcal{F}$ can be computed. From the bifurcation point of view we must study also the cases $\mu = R^2/4$; $\mu = 0$, $R \neq 0$; and $\mu = R = 0$. Irrespective of the fact that in these cases the corresponding (other) secular equation possesses or not solutions, $\min\mathcal{F}_1$ can be calculated along the lines of Example 3.2.1 separately for each case and then decide which is the absolute minimum for μ, $R \neq 0$.

The treatment of this example is difficult due to the presence, in an affine way, of the physical parameter in the secular equation. Its difficulty would increase if more than one parameter occur. The form of the eigensolutions contains a single coefficient, since in the secular determinant there is a third order non-vanishing minor. So, the second bifurcation equation, *i.e.* that obtained by introducing the eigensolutions in the constraint, is easily solved, which is not the case if all third-order minors are vanishing or if the Euler equation (3.2.20) is nonlinear. Whence

Remark 3.2.8. Due to the fact that the solution of the two quoted bifurcation equations is much more difficult if a physical parameter is present, in hydrodynamic and hydromagnetic stability theory mostly the so-called *universal isoperimetric inequalities, i.e.* corresponding to isoperimetric problems independent of the physical parameters, are used. As a consequence, the choice of the most appropriate existing (universal) isoperimetric inequalities becomes crucial. More precisely, one of our guiding idea in stability studies was to transform the given problem into an equivalent form, enabling us the most efficient use of these inequalities.

Remark 3.2.9. The introduction of the Lagrange multiplier μ, makes the variational problem (3.2.16) free of the constraint (3.2.17). This is obtained by allowing the parameter μ to be arbitrary and subsequently determining it such that the constraint be fulfilled. This profound idea of Lagrange is encountered and proves to be fruitful also at least in two other basic situations: the variation of coefficient method in the theory of ode's and in analytic mechanics. Of course, the Lagrange method in calculus of variations (described by us for the case $r = s = n = 1$) is in the same line of thoughts and makes Lagrange's contribution to mathematics among the most important throughout its history.

All these types of arbitrariness involved into: parameters in the isoperimetric problems; variable coefficients in the general solutions of affine ode's; virtual displacements in analytic mechanics; and the variation from calculus of variations respectively, define, in fact, some classes of mathematical objects rigorously formalized more than 200 years later.

Remark 3.2.10. Isoperimetric problems are related to the so-called *isoperimetric inequalities*. For instance, in the case of the isoperimetric problem (3.2.18) we have $\mathcal{F}_1(v) > \mathcal{F}_1(v_1) = \pi^2$ for every $v \in \mathcal{D}_1(\mathcal{F}_1)$. This implies the particular Poincaré isoperimetric inequality in u

$$\int_0^1 (u'(x))^2 dx \geq \pi^2 \int_0^1 u^2(x)dx. \tag{3.2.20}$$

In Chapters 2 and 4 the use of isoperimetric inequalities to prove linearization principle and to derive the energy inequality and, so, nonlinear stability criteria, is crucial. It is only by using the best constants in them that some positivity of coefficients leading to fine results can be obtained.

Due to the importance in the study of stability of fluids flows, the related topics of isoperimetric problems and inequalities and eigenvalue problems are treated in much more detail and in a more general context in Section 3.2.6.

3.2.4   *Construction of the variational problem associated with an Euler equation: energy method. Quadratic functionals associated with affine or linear equations*

In Section 3.2.2 with a variational problem (3.2.1) we associated an Euler equation (3.2.3). In Section 3.2.3, for particular functionals, we showed the corresponding Euler equations, expressed as boundary-value problems for some differential equations. In this section we deal with the inverse topic: we present a method in order to construct a variational problem, *i.e.* a functional $\mathcal{F}$, associated with a given Euler equation. This method is referred to as the *energy method* [Mikh1], [Mikh2]. It is also referred to as the *minimization of a functional.*

The derivation of the Euler equation and the minimization of a functional are the two main topics in calculus of variations. This section concerns particular

Euler equations, namely those corresponding to associated quadratic functionals. In fact, since to the properties of the operators defining the Euler equations some properties of the functionals in the associated variational problems correspond, we present variational principles rather than the energy method alone.

First, we show that the Euler equations corresponding to particular quadratic functionals are affine, or linear.

The relationship between the affinity of the Euler equation and the quadraticity of the associated functional is further shown in a more general case in the form of the variational principle 3.2.2. Then we show how the existence of the Friedrichs extension of the functional leads to the solvability of the Euler equation. Next we consider two-point problems for equations defined by symmetric or nonsymmetric operators in $L^2(a, b)$.

In all cases, in this section, given the affine or linear equations defined by a symmetric operator A

$$Au = g \quad \text{or} \quad Au = 0$$

in a (real or complex) Hilbert space H, the functional $\mathcal{F}$ defining the associated variational problem is constructed by the same formulae

$$\mathcal{F}(u) = (Au, u) - (u, g) - (g, u) \quad \text{and} \quad \mathcal{F}(u) = (Au, u)$$

respectively, like for positive definite operators in Theorem 3.2.2 (even if A possesses only part of the properties of the operator in this theorem). If the given linear equations are defined by a nonsymmetric operator A, we construct the sesquilinear form starting with the inner product (Au, u^*) and then integrate by-parts. In this case the associated equations are the given equation and its adjoint.

3.2.4.1 *Equation* **Au** $= 0$, *where* **A** *is a symmetric differential operator in* $L^2(a,b)$

Consider a variational problem (3.2.1), the associated Euler equation of which is (3.2.3). Denote the nonlinear mapping $grad\mathcal{F}$ by T. Then (3.2.3) reads

$$Tu = 0, \quad u \in \mathcal{D}(T). \tag{3.2.3'}$$

Remark 3.2.11. For theory, as well as for applications, the most important variational problems are those defined by *quadratic functionals $\mathcal{F}$ because their associated equations are affine*. In this case equation (3.2.3') becomes

$$Au = g, \tag{3.2.21}$$

where A is a linear operator and g is a given function of $\mathcal{R}(T)$. This enables one to use the results of linear functional analysis and, mainly, of spectral theory of linear operators. On the other hand, in mechanics of continua, *e.g.* hydrodynamics, elasticity, hydromagnetics, the major part of the encountered problems falls into this situation.

If the quadratic functional $\mathcal{F}$ is an integral and the corresponding Lagrangian is a differential mapping $F(x, D^\alpha u)$, then, formally, T is obtained by transporting all derivatives involving v in $F(x, D^\alpha(u + \epsilon v))$ on u and then letting $\epsilon = 0$.

Consider, in addition, that $\mathcal{X} = \mathcal{H}$, where $\mathcal{H}$ is a Hilbert space, and $\mathcal{F}$ is a symmetric quadratic functional of the form

$$\mathcal{F}(u) = \int_\Omega \sum_{|\alpha|=1}^n a_\alpha (D^\alpha u)^2 dx, \tag{3.2.22}$$

where a_α are constant coefficients. Assume that this functional is obtained from the inner product (Au, u), where A is a linear operator on $\mathcal{H}$, transporting on u the derivatives up to half of the highest order. More exactly,

$$A = \sum_{|\alpha|=1}^n (-1)^{|\alpha|} a_\alpha (D^{2\alpha} u)(x) dx. \tag{3.2.23}$$

Let us compute the operator T in the Euler equation $(3.2.3')$ corresponding to the minimum problem for the functional $\mathcal{F}$ defined by (3.2.22). By the Euler method we have

$$\delta\mathcal{F}(u, v) = \lim_{\epsilon \to 0} \frac{d\mathcal{F}(u + \epsilon v)}{d\epsilon} = \lim_{\epsilon \to 0} \int_\Omega \left\{ \sum_{|\alpha|=1}^n a_\alpha \left[(D^\alpha u)^2 + 2\epsilon(D^\alpha u)(D^\alpha v) \right. \right.$$

$$\left. \left. + \epsilon^2 (D^\alpha v)^2 \right] \right\} dx = 2 \sum_{|\alpha|=1}^n a_\alpha \int_\Omega (D^\alpha u)(D^\alpha v) dx,$$

where $v \in \mathcal{M}$. Assume that A is a densely defined symmetric operator. Then, by construction, $\mathcal{D}(\mathcal{F}) = \mathcal{D}(A) = \mathcal{D}(grad\mathcal{F})$ and the symmetry of A implies $\mathcal{D}(A^*) = \mathcal{M}$. Consequently, the last expression of $\delta\mathcal{F}$ is equal to (Au, v), namely

$$\sum_{|\alpha|=1}^n a_\alpha \int_\Omega (D^\alpha u)(D^\alpha v) dx = 2(Au, v). \tag{3.2.24}$$

Due to the fact that v is arbitrary in $\mathcal{M}$ and $\mathcal{M}$ is dense in H, it follows that $\delta\mathcal{F} = 0$ implies

$$Au = 0, \quad u \in \mathcal{D}(A). \tag{3.2.25}$$

This means that (3.2.25), which is the Euler equation (3.2.22), is linear and it has the form (3.2.21), where $g = 0$ and $T = A$ is a linear operator.

Consequently, if we start with the linear equation (3.2.25), where A has the above-specified definition and properties, by performing the inner product (Au, u) and by defining the functional

$$\mathcal{F}(u) = (Au, u), \quad u \in \mathcal{D}(\mathcal{F}) = \mathcal{D}(A), \tag{3.2.26}$$

it follows that (3.2.25) is the Euler equation for the functional $\mathcal{F}$ in $\mathcal{D}(A)$. All these show the validity of the following variational principle.

Theorem 3.2.1. *If A defined by (3.2.23) and $\mathcal{F}$ by (3.2.22) and A and $\mathcal{F}$ have the above properties then, in $\mathcal{D}(A)$, $\delta\mathcal{F} = 0$ if and only if $Au = 0$.*

If $a_\alpha > 0$, from (3.2.24) it follows that $(Au, u) = \sum_{|\alpha|=1}^{n} a_\alpha \int_\Omega (D^\alpha u)^2 dx$. Using appropriate integro-differential inequalities of the form

$$\int_\Omega (D^\alpha u)^2 dx \geq \gamma^2 \int_\Omega u^2 dx \tag{3.2.27}$$

which prove to be isoperimetric inequalities of Poincaré type (Section 3.2.8), we have that A is a particular positive definite operator.

3.2.4.2 Equation $Au = 0$, where A is a positive definite or merely positive operator

The variational principle expressed by Theorem 3.2.1 holds for general positive definite operators as follows.

Theorem 3.2.2 of a minimum of a functional [Mikh2]. *Let H be a complex Hilbert space and let $A : \mathcal{D}(A) \subset H \to H$ be a positive definite operator. If the equation*

$$Au = g, \quad u \in \mathcal{D}(A) \tag{3.2.28}$$

has a solution u_g, then the functional

$$\mathcal{F}(u) = (Au, u) - (u, g) - (g, u) \tag{3.2.29}$$

assumes its minimum value for this solution. Conversely, if at u_g, $\mathcal{F}(u)$ attains its minimum, then u_g is the solution of (3.2.28). In (3.2.28) g is a given function from H (which does not depend on u).

Proof. Assume that (3.2.28) has a solution u_g. Therefore, $Au_g = g$. Substituting this expression for g in (3.2.29) we have $\mathcal{F}(u) = (Au, u) - (u, Au_g) - (Au_g, u) = (A(u - u_g), u - u_g) - (Au_g, u_g)$. Obviously, $\mathcal{F}$ is minimal for $u = u_g$ since, by the positive definiteness of A we have $(Au_g, u_g) > 0$. The minimum of $\mathcal{F}$ is

$$\min_{u \in \mathcal{D}A} \mathcal{F}(u) = -(Au_g, u_g) = -(g, u_g) = (u_g, g). \tag{3.2.30}$$

Conversely, assuming that $min\mathcal{F}$ is attained for u_g, we must have $\delta\mathcal{F}(u_g, v) = \lim_{\epsilon \to 0} \frac{d}{d\epsilon} \mathcal{F}(u_g + \epsilon v) = 0$, $\forall v \in \mathcal{M}$, where $\mathcal{M}$ is the linear manifold corresponding to $\mathcal{D}(\mathcal{F}) \subset H$. Substituting in this equality the expression (3.2.29) and taking into account that A is symmetric we obtain the equality $\mathcal{R}e(Au_g - f, v) = 0$. Replacing v by iv it follows $\mathcal{I}m(Au_g - f, v) = 0$, therefore $(Au_g - f, v) = 0$, whence, since $\overline{\mathcal{D}(\mathcal{F})} = H$, (3.2.28).

Remark that (3.2.29) can also be written as

$$\mathcal{F}(u) = (Au, u) - 2Re(u, g). \tag{3.2.29'}$$

Thus, if $\mathcal{H}$ is a real Hilbert space, the corresponding real version of Theorem 3.2.2 holds, where (3.2.29') is replaced by

$$\mathcal{F}(u) = (Au, u) - 2(u, g). \tag{3.2.29''}$$

Note that if A is merely a positive operator, the following result holds.

Theorem 3.2.3 [Mikh3]. *Let H be a complex Hilbert space and let $A : H \to H$ be a positive operator. Then equation (3.2.25) has at most one solution u_g.*

Theorem 3.2.3 implies the uniqueness of the solution of (3.2.28) and, correspondingly, for the minimum of $\mathcal{F}$ defined by (3.2.29).

In general, (3.2.28) has no solution in $\mathcal{D}(\mathcal{A})$ and $\mathcal{F}$, no minimum. In order to achieve the existence of the solution of (3.2.28), we first extend the meaning of the associated variational problem for $\mathcal{F}$ and only then we define a corresponding generalized meaning for the solution of (3.2.28). To this aim we first extend $\mathcal{D}(\mathcal{F}) = \mathcal{D}(A)$ up to a certain Hilbert space H_A defined by means of A; it is only in such way that the extended functional can have a minimum on H_A. Simultaneously, in passing from $\mathcal{D}(A)$ to H_A the operator A extends to a selfadjoint operator $\tilde{A}$. The following construction shows that *a positive definite operator A can always be extended to a particular selfadjoint operator $\tilde{A}$, called the Friedrichs extension of A* [Fri1].

Thus, define [Mikh 2] the new inner product $[u, v]_A = (Au, v)$, $\forall u, v \in \mathcal{D}(\mathcal{A})$ and denote by H_A the corresponding pre-Hilbertian space, *i.e.* $H_A = \big(\mathcal{D}(A), [\cdot, \cdot]_A\big)$. Since, in general, H_A is not complete, we complete it with respect to the norm $|\cdot|_A$, defined by $[\cdot, \cdot]_A$, up to the Hilbert space $\overline{H}_A$, where $\overline{H}_A = H_A{}^{|\cdot|_A}$.

To this aim we use the relation

$$\| \cdot \|_H < \frac{1}{\gamma} | \cdot |_A, \tag{3.2.31}$$

the fact that A is densely defined, where γ^2 is the constant from the definition of positive definite operators (Appendix 2), and the following fundamental result

Theorem (Friedrichs) [Fri1]. *Every element of $\overline{H}_A$ belongs to H and H_A can be completed by means of elements of H.*

Hence, $\mathcal{D}(\mathcal{A}) \subset \overline{H}_A \subset H$, the first symbol $\subset$ represents immersion and the second, inclusion.

The functional

$$\mathcal{F}_A(u) = [u, u] - [u, u_g] - [u_g, u], \quad u \in \overline{H}_A \tag{3.2.32}$$

is the extension of $\mathcal{F}$ from $\mathcal{D}(A)$ to $\overline{H}_A$. Here u_g is the unique element of $\overline{H}_A$ such that

$$(u, g) = [u, u_g], \quad \forall \ u \in \overline{H}_A. \tag{3.2.33}$$

Since (3.2.32) equivalently reads

$$\mathcal{F}_A(u) = | u - u_g |_A^2 - | u_g |_A^2, \tag{3.2.32'}$$

it follows that the minimum of $\mathcal{F}_A$ is u_g defined by (3.2.33).

Definition 3.2.3. When such a point of minimum of $\mathcal{F}_A$, u_g, exists, it is referred to as the *generalized (variational) solution* of (3.2.28).

In general, $u_g \notin \mathcal{D}(\mathcal{A})$ and, so it is not a solution of (3.2.28), but satisfies the equation

$$\tilde{A}u = g \tag{3.2.28'}$$

in the generalized (weak) sense, *i.e.* $(g, v) = (u, Av)$, $\forall v \in \mathcal{M}$. Whence, (cf. Remark 3.2.1) the contribution of an associated variational problem to the solvability of a given non-variational problem.

We have $\mathcal{D}(\tilde{A}) = \overline{H}_A$ and $\tilde{A}$ is selfadjoint. More exactly, it can be proved that $\overline{H}_A$ is the domain of definition of the positive square root operator $\tilde{A}^{1/2}$ (see (3.2.24)). It is very important in numerical computations.

Assume that H is a separable Hilbert space and so is $\overline{H}_A$. Let $\{e_n(x)\}$ be a total orthonormal set of $\overline{H}_A$, therefore, $u_g = \sum_{n=1}^{\infty}[u_g, e_n]e_n$. Take in (3.2.33) $u = e_n$ to obtain $(e_n, g) = [e_n, u_g]$, *i.e.* $[u_g, e_n] = (g, e_n)$, implying $u_g = \sum_{n=1}^{\infty}(g, e_n)e_n$. Consequently, in the case of separable H, the generalized solution of (3.2.28) can be computed numerically by using this formula, where the involved series is convergent.

Remark 3.2.12. The uniqueness of u_g in (3.2.33) shows the existence of an operator G such that $u_g = Gg$. This is very much exploited in proving Theorem 3.2.2 and many results in the spectral theory of positive definite operators. Let A be a positive definite operator and let $\mathcal{F}(u) = (Au, u)/(u, u)$ be its Rayleigh quotient functional. It also reads $\mathcal{F}(u) = \|u\|_A^2/\|u\|^2$. By Theorem 3.2.2, it is proved [Mikh5] that $\inf_{u \in \mathcal{D}(A)} \mathcal{F}(u) = \inf_{u \in \overline{H}_A} F(u) = \gamma^2$, where γ^2 is the constant from the definition of the positive definiteness of A. If, in addition, A is completely continuous and H is separable, then A has at most a countable set of eigenvalues (in H_A) and the least, λ_1, is just γ^2 (Appendix 2). Therefore,

$$\lambda_1 = \min_{\mathbf{u} \in \overline{H}_A} \frac{(A\mathbf{u}, \mathbf{u})}{(\mathbf{u}, \mathbf{u})}. \tag{3.2.34}$$

Moreover, the corresponding eigenvector $\mathbf{u}_1$ can be determined by using a sequence (called a minimizing one, cf. Section 3.2.5) which, converging weakly, converges in the norm too (since A is compact). Therefore the approximations for λ_1 and $\mathbf{u}_1$ satisfy the corresponding equations

$$(u_1^{(k)}, u_1^{(k)}) = \lambda_1^{(k)}(Au_1^{(k)}, u_1^{(k)}). \tag{3.2.34'}$$

If in (3.2.28) A is only a positive operator, then (3.2.28) may have no solution at all or may have a generalized solution belonging to $\overline{H}_A$ but not to H [Mikh2]. This time $\overline{H}_A \not\subset H$. In addition, A can be extended up to a selfadjoint operator. The positive square root operator $\tilde{A}^{1/2}$ can be defined too, but only on $\overline{H}_A \cap H$.

The first variational principle (of Dirichlet) is a particular case of Theorem 3.2.2: *Let $\Omega \in \mathbf{R}^2$ be a domain and let $\mathcal{D}(\mathcal{F})$ be the set of functions assuming specified values on the boundary $\partial\Omega$ of Ω. The function for which the Dirichlet functional*

$\mathcal{F}(u) = \int_\Omega \left[(\frac{\partial u}{\partial x})^2 + (\frac{\partial u}{\partial y})^2 \right] dx$ *takes the minimum value is the solution of the Laplace equation* $\Delta u = 0$ *in* Ω. Inappropriate choices of $\mathcal{D}(\mathcal{F})$ lead to the apparently counterexamples of Weierstrass and Hadamard. They put into an unfavorable light the association of variational problems with differential equations. It is only in 1908 that this situation was reconsidered: Ritz published his method for approximate solution of variational problems. Moreover, starting with the invention of generalized solutions by J. Leray [Ler] for a hydrodynamic problem and by S. L. Sobolev [Sob2] for general differential operators in the years '30 and '40 of the past century, it was proved that it is due only to their variational setting that the differential equations can be suitably studied. This re-established completely the position of variational principles and problems.

In mechanics of continua boundary-value problems for an equation of the form (3.2.28) occur, where A is a differential operator. The corresponding functional (3.2.29) represents the potential *energy* of the modeled phenomenon. Then, $[\cdot, \cdot]_A$, $| \cdot |_A$ and $\overline{H}_A$ are called the *energy inner product, energy norm* and *energy space* respectively. Whence the name of *energy method*. Sometimes, [Mikh3], instead of energy method the label of *variational method* is used.

The most extensive presentation of the energy method is done in [Mikh1] and [Mikh3]. In [Mikh3] it is adopted an abstract setting in a Hilbert space $\mathcal{H}$ for the Euler equation (3.2.28). Then applications to equations of elliptic type and to elasticity are given. In [Mikh1] $\mathcal{H}$ is a function space and the operator involved into the Euler equation is differential. Moreover, it is understood that $\mathcal{H}$ is $L^2(\Omega)$ or a Hilbert subspace of $L^2(\Omega)$. The problem (3.2.28) is studied in these spaces for differential operators A.

For equations (3.2.28) where A is a linear operator which is nonsymmetric, but can be symmetrized by means of some auxiliary operator, or A is a nonlinear mapping with such Gateaux differential, variational principles generalizing Theorem 3.2.2 can be found in [Fil]. In these cases the Friedrichs extension is defined accordingly for the nonlinear mappings and the corresponding theory is developed in spaces more general than Sobolev spaces.

3.2.4.3   *Equations* $\mathbf{Au = g}$ *and* $\mathbf{Au = 0}$ *in* $L^2(a, b)$ *where* $\mathbf{A}$ *is not a positive definite operator. Natural boundary conditions. Principal boundary conditions*

The energy method presented above concerns particular cases of given equations (3.2.28) and corresponding operators, while for all other cases it is silent. A lot of worked examples show that the class of equations (3.2.28) defined by positive definite or merely positive operators is not the single one with which some functional can be associated such that a variational principle holds for it. The situation is similar to that of functions in $\mathbf{R}$ which are not differentiable at some points at which they achieve their minimum (*e.g.* the function $u : [-1, 1] \to \mathbf{R}$, $u(x) =$

$\sqrt{1 - x^2}$ is not differentiable at $x = \pm 1$, but at each of these points u achieves a minimum). A general theory as that one for differentiable functions cannot be elaborated. However, in each specific case of a non-differentiable function it is possible to decide if the minimum exists or not. Exactly in the same way, even the simplest boundary-value problems in linear hydrodynamic and hydromagnetic stability are modeled by equations defined by non-symmetrizable, or non-selfadjoint, or non-positive operators (Section 3.3.3, 3.3.4). In addition, the symmetry for operators with variable coefficients, in most cases, is not to be expected to occur. However, for part of these cases, variational principles do occur (Section 3.3.5). In Appendix 7 of [Geo85], the validity of variational principles for some equations which are not in the (3.2.28) class was checked by B-D method. On particular problems the equivalence of direct and variational methods [Geo77], [GeoPo], [GeoC], [Geo87], [GeoS], [GeoOP], was proved using Fourier series expansions.

Example 3.2.2. Consider the quadratic functional in $C^2(a, b) \cap C[a, b] \subset L^2[a, b]$

$$\mathcal{F}(u) = \int_a^b [c_1(x)u^2 + c_2(x)u'^2 + c_3(x)uu' + c_4(x)u + c_5(x)u' + c_6(x)]dx, \quad (3.2.35)$$

where $c_1, \ldots, c_5$ are continuously differentiable functions on $[a, b]$ and $c_2 \neq 0$ on $[a, b]$, and the associated Euler equation written in the form of a two-point problem

$$Au = g, \quad (3.2.36)$$

$$u(a) = u(b) = 0, \quad (3.2.37)$$

where $Au = -2c_2u'' - 2c_2'u' + (2c_1 - c_3')u$, $g(x) = c_5' - c_4$. Inspired by (3.2.29') let us construct the functional

$$\mathcal{J}(u) = \frac{1}{2}[(Au, u) - 2(u, g)]. \quad (3.2.38)$$

Integrating by parts and taking into account (3.2.33), it follows that $\mathcal{J}(u) = \mathcal{F}(u) + \int_a^b c_6(x)dx$. But the last integral is a constant, which, like for functions in $\mathbf{R}$, does not modify the point of minimum. Due to the uniqueness of the solution of (3.2.36), (3.2.37), the variational principle follows: $\delta F = 0$ *if and only if u is a solution of* (3.2.36), (3.2.37). However, there exist particular functions $c_1, \ldots, c_5$ for which A is not positive, *e.g.* $c_1 = x$, $c_2 = -(x^2 + 1)$, for $a = -0.5$, $b = 0.5$. Indeed, $(Au, u) = 2\int_a^b[c_2(x)u'^2(x) + c_1(x)u^2(x)]dx$. Then $\int_{-0.5}^{0.5} c_2(x)u'^2(x)dx < \max_{x \in [-0.5, 0.5]} c_2(x) \cdot \int_{-0.5}^{0.5} u'^2(x)dx = -\int_{-0.5}^{0.5} u'^2(x)dx$ and, by the Poincaré inequality (3.2.20) we obtain $(Au, u) < \left[-\pi^2 + \max x_{x \in [-0.5, 0.5]}\right] \int_{-0.5}^{0.5} u^2(x)dx = (-\pi^2 + 1) \int_{-0.5}^{0.5} u^2(x)dx < 0$. On the other hand, requiring the positivity of A could lead to physical non-realistic situation. For instance, assume that (3.2.36), where $a = -0.5$, $b = 0.5$, $c_1 = R_1 + R_2$, $c_2 = -R_3$ and R_1, R_2 and R_3 are physical parameters, govern some fluid flow stability. Imposing the positivity of A, it follows the constraint $(R_1 + R_2)\pi^2 > R_3$, which does not occur among the physical hypotheses.

Often in hydrodynamic and hydromagnetic stability, problems in the real Hilbert subspace of $L^2(a, b)$, like those in the following example, occur.

Example 3.2.3. Consider the two-point problem

$$[B_k(u)]_a^b = 0, \quad k = 1, \ldots, n \tag{3.2.39}$$

for the n-th order ode with constant coefficients

$$Au = 0, \tag{3.2.40}$$

where $A : \mathcal{D}(A) \subset L^2(a, b)$, $\mathcal{D}(A) = \{u \in C^n(a, b) \cap C^{n-1}[a, b] \subset L^2(a, b) \mid u \,\text{satisfy} \,(3.2.29)\}$, $Au = \sum_{k=0}^n a_k D^k u$, $u : (a, b) \to \mathbf{R}$, $u = u(x)$, $D = \frac{d}{dx}$, B_k are $n - 1$-th order ordinary differential operators with constant coefficients given in (3.2). By Consequence 3.1, the solutions of (3.2.39), (3.2.40) belong to $C^\infty([a, b], \mathbf{R})$, so re-defining $\mathcal{D}(A) = \{u \in C^\infty([a, b], \mathbf{R}) \mid u \,\text{satisfies}\,(3.2.29)\}$ we can construct (Remark 3.2) the Hilbert subspace H of $L^2(a, b)$ and redefine $\mathcal{D}(A) = H$.

Assume that A is symmetric and the boundary conditions (3.2.39) are such that (3.16) holds. Then, (3.16) defines the functional $\mathcal{F}(u) = (Au, u)$

$$\mathcal{F}(u) = (Au, u) = \sum_{k=0}^{n/2} (-1)^k a_k \int_a^b (D^k u)^2(x)dx = (A^{1/2}u, A^{1/2}u). \tag{3.2.41}$$

By construction, $\mathcal{D}(F) = \mathcal{D}(A)$. Similarly to Section 3.2.4.1, it can be proved that the variational problem for $\mathcal{F}$ and the problem (3.2.39), (3.2.40) are associated, even if, in general, A is merely symmetric.

In fact, in (3.2.41) the notation $A^{1/2}$ is abusive. It is justified only if A is positive and, for positive definite operators A too. We use this notation because the associated sesquilinear form $\mathcal{F}(u, v) \equiv (Au, v)$, *i.e.* (3.15), contains products of the same type of derivatives of u and v, just as in $\tilde{A}^{1/2}$ for positive operators.

For symmetric matricial differential operators A, the associated form $\mathcal{F}(u, v)$ can contain products of various type of derivatives of the components of u and v (Example 3.4.1). This follows from Remark 4.7 stating that if A is symmetric it can contain even - as well as odd - order derivatives.

Even if $A^{1/2}$ is not positive or does not exist, the form (3.2.41), involving only derivatives of order smaller than or equal to $n/2$, is very important for applications. Indeed, by using numerical methods based on Fourier series to solve the problem of the minimum of the functional (3.2.38) associated with problem (3.2.36), (3.2.37) for $g = 0$, this form reduces to half the necessary computations. Whence *the special interest in symmetrization of operators*.

If in (3.2.41) all coefficients $(-1)^k a_k$ are positive, then A is positive definite and the variational principle 3.2.2 applies. A can be positive also in the case when some coefficients are negative provided that, after applying Poincaré type inequalities, the resulting coefficients of (u, u) be positive. This is exactly the case in Example 3.2.2. Moreover, exactly as in Example 3.2.2, A can be non-positive but the variational principle holds.

Assume that A is not symmetric. In this case, heuristic treatments are prevailing over rigorous proofs and they are carried out by those interested in concrete applications (especially engineers and physicists) [Chan], [Geo85].

The variational principles for nonsymmetric operators involve not only the given equation but also its adjoint, *i.e.*

$$Au = 0, \quad A^*u^* = 0, \tag{3.2.42}$$

and with it, it is associated not a functional but a sesquilinear form $\mathcal{F}(u, u^*)$. This form is obtained from (Au, u^*) using by-parts integrations: the derivatives on u are passed on u^*. This process is continued up to the stage when the sum of the terms outside the integral vanishes. Sometimes several $\mathcal{F}$ can be constructed for the given equations (3.2.42). In this case we choose that one which is closer to the form (3.2.41) from the symmetric case, *i.e.* contains the derivatives of the lowest possible order, involving minimal computations effort. An example is carried out in Section 3.4.3.5.

More precisely, the constructed functional has the form

$$\mathcal{F}(u, u^*) = \int_a^b L_1 u L_2 u^* dx. \tag{3.2.43}$$

Under convenient boundary conditions we have

$$\mathcal{F}(u, u^*) = \int_a^b u L_2^* L_1 u u^* dx = (Au, u^*)$$

and

$$\mathcal{F}(u, u^*) = \int_a^b u L_1^* L_2 u^* dx = (u, A^*u^*),$$

where $L_2^* L_1 = A$ and $L_1^* L_2 = A^*$. Then, by the Euler method, defining $\delta\mathcal{F}(u, u^*)(v, v^*) = \lim_{\epsilon \to 0} \frac{d}{d\epsilon} \mathcal{F}(u + \epsilon v, u^* + \epsilon v^*)$ we are led to (3.2.42). The same result follows if the Lagrange method is applied. See the example in Section 3.4.3.5.

In linear stability theory, our main interest concerns differential operators with constant coefficients. They define the perturbation equations about steady states, which are affine functions or are approximated by such functions. The form of the resulting operators A is equal to that of their Lagrange-adjoint operators A^+ or can be reduced to it. Thus, their characteristics, *e.g.* symmetry, selfadjointness, square roots, strongly depend on the boundary conditions. This is reflected in the possibility to associate with them variational problems, hence some functionals $\mathcal{F}$ and to prove variational principles.

Remark 3.2.13. If in defining $\mathcal{F}$ we use the inner product $(A^{1/2}u, A^{1/2}u)$, then, in general, $\mathcal{D}(\mathcal{F}) \neq \mathcal{D}(A)$. The boundary conditions occurring in $\mathcal{D}(\mathcal{F})$ are called *principal boundary conditions* while the boundary conditions occurring in $\mathcal{D}(A) \setminus \mathcal{D}(\mathcal{F})$ are called the *natural boundary conditions* [Mikh3]. The existence of natural boundary conditions is the main difficulty in establishing variational principles. In this case the point of minimum of $\mathcal{F}$ belongs to a set smaller than that corresponding to the problem $Au = f$. In [Mikh5] the fact that $\mathcal{D}(\mathrm{grad}\mathcal{F}) \subset \mathcal{D}(\mathcal{F})$ is expressed in terms of the absolute continuity of some terms in $\mathcal{F}$, ensuring the boundedness of the first variation $\delta\mathcal{F}$ (see also Remark 3.2.5). For applications it is worth noting that this continuity is expressed by the presence of supplementary boundary conditions (3.14) in $\mathcal{D}(A)$.

3.2.5 *Direct methods. Minimizing sequences*

In the earlier stages only the first topic of calculus of variations, *i.e.* the derivation of the Euler equation was considered. It was only due to Riemann that the attention to the second was drawn [Mikh2]: he used the energy method in order to prove the existence of the solution of the Dirichlet problem. Later, the Ritz method, based on Fourier series, applied to solve a variational problem, increased the interest in the energy method. A next step was the truncated series method of Galerkin. Soon all these became basic tools in engineering applications. In both the Ritz and Galerkin methods the expansion functions satisfied all boundary conditions of the problem. In 1940, a variant of the Ritz method was used in elasticity theory by Budiansky and his co-workers. In this method only part of the boundary conditions were satisfied by the expansion functions. Later, DiPrima applied it to hydrodynamic stability theory. The B-D method was less successful and it is practically forgotten in spite of its advantages over the Ritz-Galerkin method, as shown in a series of papers of us and our collaborators. This method is extensively used in our book.

In this section we deal with the numerical approximation of solutions of variational problems. Assume that H is a real Hilbert space.

Definition 3.2.4 [Mikh2]. A function belonging to $\overline{H}_A$ is referred to as a *function with finite energy.*

It is understood that its energy norm is finite. Correspondingly, an operator A defined on $\overline{H}_A$ such that the images through A of elements of $\overline{H}_A$ have bounded energy norm is an operator from $\overline{H}_A$ to $\overline{H}_A$, hence, in older (improper) terms, A *can be defined on* $\overline{H}_A$.

Definition 3.2.5 [Mikh2]. Let A be a positive (and, so, also a positive definite) operator defining equation (3.2.28) and consider the associated functional $\mathcal{F}$ defined by (3.2.29″). Let u_g be the point of minimum for $\mathcal{F}$ and denote by m its minimum value. A sequence $\{u_n\}$ of elements of $u_n \in \mathcal{D}(A)$ converging in the energy norm $|\cdot|_A$ towards u_g is called a *minimizing sequence for* $\mathcal{F}$.

For every minimizing sequence $\{u_n\}$, $\{\mathcal{F}u_n\}$ converges to the minimum m [Mikh3].

Recall that, generally, u_g is a generalized solution of (3.2.28), defined by a positive definite or merely positive operator A.

The construction of minimizing sequences is in-depth treated in S. G. Mikhlin's books. It is related to the variational methods and direct methods [Vel2], [Dac1], [Ne]. These names were attributed at the beginning of the past century and it is not an unanimous consensus on their content. Anyhow, we quote them because they are related directly or through variational principles to the variational problems. Among variational methods we mention: energy method as involved into Theorem 3.2.2, Ritz method, method of least squares, of orthogonal projections, the Trefftz's method, Bubnov-Galerkin method. They reduce the problem of solving a differential equation to an equivalent variational problem. At the beginning of Section 3.2.4 we used the name of energy method instead of variational methods because its main ideas and results are basic for all others. Loosely speaking, by *direct methods* we understand [Mikh3] those methods used to approximate solutions of differential or integral equations and reduce them to finite systems of algebraic equations. Among them we quote: method of networks, method of straight lines, method of finite differences. Some of them are variational, some others, not.

In this section, we present briefly two of them because they concern numerical minimization of functionals [Mikh4], [Mikh3], [Mikh5].

Ritz method is a direct method applied to a variational problem. Let $\{\phi_n\}_{n\in\mathbf{N}}$ be a sequence of elements $\phi_n \in \mathcal{D}(A)$ and assume that it is total in $\overline{H}_A$. Ritz called ϕ_n *the coordinate functions*. Then an approximate extremal of $\mathcal{F}$ reads

$$u_n(x) = \sum_{i=1}^{n} a_j\phi_j(x), \qquad (3.2.44)$$

where a_j are arbitrary real constants. Introducing (3.2.44) in (3.2.29″) we obtain

$$\mathcal{F}(u_n) = \sum_{j,k=1}^{n} a_j a_k (A\phi_j, \phi_k) - 2\sum_{i=1}^{n} a_j(\phi_j, g). \qquad (3.2.45)$$

Thus $\mathcal{F}(u_n)$ becomes a function of n real variables $a_1,\ldots,a_n$. Denote it by $\mathcal{F}(a_1,\ldots,a_n)$. Its stationarity requires $\frac{\partial \mathcal{F}(a_1,\ldots,a_n)}{\partial a_i} = 0$, $i = 1,\ldots,n$, which leads to the Ritz algebraic system in $(a_1,\ldots,a_n)$

$$\sum_{k=1}^{n}(A\phi_i, \phi_k)a_k = (g, \phi_i), \quad i = 1,\ldots,n \qquad (3.2.46)$$

or, due to the symmetry of A, to

$$\left(A(\sum_{k=1}^{n} a_k\phi_k) - g, \phi_i\right) = 0, \quad i = 1,\ldots,n \qquad (3.2.47)$$

or, since $\mathcal{D}(A) \subset \overline{H}_A$, to

$$\sum_{k=1}^{n} a_k[\phi_i, \phi_k] = (g, \phi_i), \quad i = 1,\ldots,n. \qquad (3.2.48)$$

As $\phi_1, \ldots, \phi_n$ are linearly independent, the Gram determinant associated with (3.2.48) is nonzero and, so, the solution $(a_1, \ldots, a_n)$ of (3.2.48) exists and is unique. Substituting it in (3.2.44) we obtain an approximate solution u_0 for the point of minimum u_g.

Indeed, let $n \to \infty$ and assume that $\{u_n\}_{n \in \mathbf{R}}$ converges towards a function u_0. First suppose that $u_0 \in \mathcal{D}(A)$. Then, from (3.2.47), taking into account that $\phi_i \in \mathcal{D}(A)$, $\mathcal{D}(A)$ is dense in H, and that the inner product is a continuous sesquilinear form, it follows that u_0 is the solution of (3.2.28). As the unique solution of (3.2.28) was u_g, we have $\{u_n\}_{n \in \mathbf{N}} \to u_g$, *i.e.* this sequence is a minimizing sequence for $\mathcal{F}$.

Similarly, if $u_0 \in \overline{H}_A \setminus \mathcal{D}(A)$, then writing (3.2.48) in the form $[\phi_i, \sum_{i=1}^n a_k \phi_k] = (g, \phi_i)$, $i = 1, \ldots, n$, letting $n \to \infty$ and taking into account that $\mathcal{D}(A)$ is dense in $\overline{H}_A$, from this equality we get (3.2.33) defining u_g, *i.e.* $u_0 = u_g$.

Here we use the fact that $\overline{\mathcal{D}(A)}_H^{\|\cdot\|} = H$, $\overline{\mathcal{D}(A)}_A^{|\cdot|} = \overline{H}_A$, and Theorem 1.2.

In earlier papers, the theoretical frame of the Ritz method was incomplete. The definitions, (*e.g.* the symmetry of unbounded operators) were completed and the necessary hypotheses (*e.g.* the positive definiteness of operators) were formulated during the proofs. Moreover, the suitable spaces of functions (*e.g.* H_A or $\overline{H}_A$) were constructed following the convergence requirements. By now, all these are clearly settled, whence the brevity of proofs.

K. Friedrichs, Lord Rayleigh and R. Courant worked out the energy method [CouH]. They replaced the study of the convergence of the Ritz approximate solutions by the study of the more general problem of the convergence of the minimizing sequence [Mikh3].

In [Mikh3], S. G. Mikhlin gave the series representation of the solution of variational problems of mathematical physics and established the relationship between these series and the ones in the Ritz method.

In applying the Ritz method, in general, it is not necessary that the coordinate functions satisfy the natural boundary conditions, but only the principal ones [Gro].

Galerkin method. At a large extent, the problems in elastic and hydrodynamic stability motivated the development of the studies on the minimum of a functional and their solution by the Ritz method and other direct methods. It is in a boundary-value problem for a partial differential equation governing the linear stability of some elastica that another important direct method is originating. More precisely, in his paper, published in 1913, in order to solve such a problem I. G. Bubnov used orthogonal series expansions directly in the differential equation and not in the associated variational problem as in the Ritz method. His expansion functions formed a total orthonormal set and satisfied the boundary conditions of the problem. Imposing to the Cramer determinant of the linear algebraic system in the Fourier coefficients to vanish, he obtained the same secular equation as that obtained by other authors by the Ritz method.

Seventy years later, the same secular equation for a problem of linear hydrodynamic stability was obtained by substituting the series either directly in the equations or in the associated variational principle [Geo85]. This time, the expansion functions satisfied only part of the boundary conditions.

By now, in view of the variational principle 3.2.2, the equality of the secular equations, obtained by the two different methods, is natural, but in 1913 the Bubnov's idea to avoid the variational formulation and to use only the given differential problem was revolutionary. This liberation of variational formulation led to one of the most important direct methods in mathematical physics, known as *Galerkin method* or *Bubnov-Galerkin method* [Mikh3]. This method was initiated in B. G. Galerkin's paper of 1915.

This method is *the Bubnov method*, but where the orthogonality of the expansion functions was not required, no relationship with some variational problem is assumed, it applies to any kind of equations, not only differential and not only linear. This method quickly became a basic method for applications. The fundamentation of the Galerkin method came very late, beginning with 1940, and since then it was carried out for a lot of classes and types of equations (see [Mikh3], [Mikh2] and the papers quoted therein).

Consider the (in general nonlinear) abstract equation

$$Tu = g \tag{3.2.49}$$

where T is a mapping defined in a Hilbert space H, $g \in H$ is a given element and let $\phi_1, \ldots, \phi_n$ be a sequence of linearly independent coordinate elements of $\mathcal{D}(T)$. Consider the approximate solution of (3.2.49) of the form $u_n = \sum_{k=1}^{n} a_k \phi_k$, where the coefficients are unknown. Of course, in order to have the possibility of approximating every point $u \in H$ by this formula, it is necessary that H be a separable space. In order to determine a_k, *the Galerkin's idea was to impose to $Au - f$ to be orthogonal to all $\phi_1, \ldots, \phi_n$*, or, equivalently,

$$(T(u_n), \phi_m) = (g, \phi_m) \quad m = 1, \ldots, n \tag{3.2.50}$$

and this is generally a nonlinear algebraic system in $a_1, \ldots, a_n$.

If $T = A$, where A is a positive definite operator and $\overline{\mathcal{D}(A)} = H$, then (3.2.50) becomes (3.2.46) and the Galerkin method become the Ritz method. This was to be expected because taking the scalar product of (3.2.49) by ϕ_m is involved in the approximated definition of the associated functional $\mathcal{F}$.

If $g = 0$ and $T(u) = Au - \lambda Bu$ or $T(u) = Au - B(\lambda)u$, where A and B are (linear) operators and λ is a parameter, then (3.2.49) becomes an approximate eigenvalue problem. In this case the necessary condition that a nontrivial solution u_n exists is that the corresponding root $(a_1, \ldots, a_n)$ has at least one nonnull component, *i.e.* the Cramer determinant be null

$$det\|A(\phi_k, \phi_m) - B(\lambda)\phi_k, \phi_m)\| = 0. \tag{3.2.51}$$

The proof of the convergence of the sequence $\{u_n\}_{n \in \mathbf{R}}$ is a difficult problem solved for particular situations. It is in this problem that the concept of completely continuous operators (also referred to as compact operators) turns out to be important [Kre], [Mikh3].

Naturally, the methods involving an inner product are related to orthogonal projections (Appendix 2) and, thus, to the Hilbert space splittings into direct sums. In nonlinear functional analysis, the method of orthogonal projections reduces the solution of a problem to the solution of its projections on the subspaces forming the direct sum. Usually, one subspace is of a finite dimension as in the Galerkin method and its variants, *e.g.* the Galerkin-Faedo-Hopf method [Geo85]. Moreover, this subspace is spanned by the eigenvectors of the linearized operator (Section 2.7.3.3). The Ritz method coincides with the orthogonal projection method for determining the solution of the associated homogeneous equation. The energy methods for non-stationary equations (Section 3.2.7) use the orthogonal projection on infinite-dimensional subspaces of solenoidal vector fields in order to eliminate the pressure and the solenoidality restrictions (Chapter 4).

3.2.6   *General isoperimetric problems and inequalities and associate eigenvalue problems*

By a *general isoperimetric problem* in a Banach space we mean [Mikh5]

$$\min_{u \in \mathcal{D}(\mathcal{F})} \mathcal{F}(u), \quad \text{with constraints} \quad \mathcal{G}_j(u) = l_j, \quad j = 1, \dots, k \tag{3.2.52}$$

where $\mathcal{G}_j$ are functionals defined on $\mathcal{D}(\mathcal{G}_j) \subset X$, $\mathcal{D}(\mathcal{F})$, $\mathcal{D}(\mathcal{G}_j)$, $j = 1, \dots, k$ satisfy the conditions from Section 3.2.2, $l_{1,\dots,k}$ are given constants, $\mathcal{D}_0 = \mathcal{D}(\mathcal{F}) \cap_{j=1}^{k} \mathcal{D}(\mathcal{G}_j) \neq \emptyset$.

Theorem 3.2.4 (multiplier Euler's rule for the isoperimetric problem) [Mikh5]. *Let $u_0 \in \mathcal{D}_0$ be the minimum of the functional $\mathcal{F}$ with constraints (3.2.52). If elements $\eta_1, \dots, \eta_k$ of the linear manifold M_0, corresponding to $\mathcal{D}_0$ exist, such that $\det(\delta \mathcal{G}_j(u_0, \eta_i)) \neq 0$, then it is possible to find the constants μ_j, $j = 1, \dots, k$ such that*

$$\Big(\mathrm{grad}\big(\mathcal{F} + \sum_{j=1}^{k} \mu_j \mathcal{G}_j\big)\Big)(u_0) = 0. \tag{3.2.53}$$

Here μ_j are called the Lagrange multipliers. A rigorous treatment of isoperimetric problems is to be found in Mikhlin's books, while an elementary presentation, in [LavLy] and [El].

The determination of the solution of (3.2.52) proceeds as in Example 5.2.1: we determine a solution u_0 of (3.2.53). It depends on $\mu_1, \dots, \mu_k$. Introducing u_0 in the constraints, we obtain an algebraic system in $\mu_1, \dots, \mu_k$. Substituting in u_0 the root of this system, the solution u_0 of the isoperimetric problem (3.2.52) is obtained. Theorem 3.2.4 is a necessary condition for extremum and it gives a condition (the non-vanishing of the determinant) ensuring the existence of the unique root $\mu_1, \dots, \mu_k$. In fact, as it is remarked in Example 3.2.1', in general the equation in $\mu_1, \dots, \mu_k$ is a bifurcation equation because the determinant is not always non-vanishing and so, the implicit function theorem involved in determining $\mu_1, \ldots, \mu_k$ does not hold. In this case all possible roots $\mu_1, \ldots, \mu_k$ must be taken into account.

If $\mathcal{F}, \mathcal{G}_1, \ldots, \mathcal{G}_n$ are homogeneous quadratic functionals, then (3.2.53) is a linear equation which has a non-vanishing solution only for certain values of $\mu_1, \ldots, \mu_k$. This is why these values are called *eigenvalues* and, correspondingly, (3.2.53) is *an eigenvalue problem in a larger sense* [Leip], [Gou], [Mikh5]. If $n = 1$, writing (3.2.52) in the form $Au + \mu_1 Bu = 0$, where at least one of the linear operators A and B is invertible, we obtain the standard eigenvalue problem $L_1 u = \mu_1 u$ or $L_2 u = \mu_1^{-1} u$, where $L_1 = B^{-1}A$ and $L_2 = A^{-1}B$.

If in the isoperimetric problem (3.2.52) $\mathcal{F}$ is a homogeneous quadratic functional in $L^2(\Omega)$, $\Omega \subset \mathbf{R}$ defined by a symmetric operator A, namely $\mathcal{F}(u) = (Au, u)$, $n = 1$ and the unique restriction reads $\mathcal{G}_1(u) = \int_\Omega | u |^2 (x)dx = 1$, then the Euler equation (3.2.53) becomes

$$Au = \mu_1 u \tag{3.2.53$'$}$$

and, so,

$$\mu_1 = (Au, u)/\|u\|^2. \tag{3.2.54}$$

It follows that the smallest eigenvalue of the Euler equation is the minimum of the functional $\mathcal{F}_2(u) = (Au, u)/\|u\|^2$. If we denote $w = u/\|u\|$ then $\mathcal{F}_2(u) = (Aw, w) = \mathcal{F}(w)$, where $\mathcal{G}(w) = 1$. Consequently the minimum of $\mathcal{F}$ in the class $\mathcal{D}_1(\mathcal{F}) = \{w \in \mathcal{D}(\mathcal{F}) \mid \mathcal{G}(w) = 1\}$ is μ_1. In this way, the minimum of a quadratic functional *is associated with the smallest eigenvalue of the Euler problem* [Fai], [Fich], [Baz], [BazF], [Dia], [Col1], [We2], [We1]. Correspondingly, the minimum point is the eigenfunction of the operator A defining the functional.

The introduction of the new function w transforming the constraint into $\mathcal{G}(w) = 1$ is usual in isoperimetric problems and numerical methods used to solve them [GlLe], [Gl], [GlLT], [KanK], [Mikh4], [DautL], [Col2].

If in the previous case $A = -\Delta$, where Δ is the Laplacian, then

$$\mu_1 = \min_{u \in \mathcal{D}(-\Delta) \subset W^{\circ,1,2}(\Omega)} \{\|Du\|/\|u\|\}, \tag{3.2.55}$$

i.e. μ_1 is the best constant α in the Poincaré type inequality

$$\int_\Omega |Du|^2 dx \geq \alpha \int_\Omega |u|^2 dx. \tag{3.2.20$'$}$$

More exactly, μ_1 is the eigenvalue of the problem

$$-\Delta \mathbf{u} = \mu_1 \mathbf{u}, \tag{3.2.56}$$

$$\mathbf{u}_{|\partial\Omega} = 0, \tag{3.2.57}$$

where $\mathbf{u} \in W^{\circ,1,2}(\Omega)$. In fluid mechanics the Poincaré inequality (2.7.6) is taken for $\mathbf{u} \in N^1(\Omega)$ and $A = -P\Delta$, where $P : L^2(\Omega) \to N(\Omega)$ and $A : \mathcal{D}(A) = N^2(\Omega) \to$

$N(\Omega)$. Correspondingly, the Euler associated equation is the problem (3.2.57) for the equations

$$-\Delta \mathbf{u} + \mathrm{grad}\phi = \mu \mathbf{u} \qquad (3.2.58)$$
$$\mathrm{div}\mathbf{u} = 0. \qquad (3.2.59)$$

For other isoperimetric problems for functions defined in $\mathbf{R}^n$, see [Gi1], [Morr]. They are related to eigenvalue problems for more general elliptic operators [KapT], [GiMS], [PayW], [Ne], [Lav], [BraP1], [BraP2].

The inequalities relating integrals of functions and their derivatives in which the best constants are minima of some associated isoperimetric problems are called *isoperimetric inequalities* [BraP1], [BraP2], [Ban], [Pay1], [Pay2], [Pay3], [PayW], [PayW2], [Os], [Chav], [Moss]. Among them we quote those corresponding to *Sobolev embedding inequalities* (Appendix 1). For instance, the embedding $W^{\circ,1,2}(\Omega) \subset C^0(\Omega)$, for $\dim(\Omega) = 1$, from (1.8) corresponds to the one-dimensional Poincaré inequality.

For $\Omega = [a, b]$, (3.2.20′) becomes

$$\int_a^b \left(u'(y)\right)^2 dy \geq \frac{\pi^2}{(b-a)^2} \int_a^b u^2(y)dy, \qquad (3.2.20'')$$

which follows immediately from (3.2.20) by using the change of variables $y = x(b-a)+a$. This is why most of the isoperimetric inequalities are written on the simplest standard domains, *e.g.* $(0,1)$ or $(-0.5, 0.5)$. The last domain is used especially for odd and even functions.

There are also geometric isoperimetric inequalities, shortly also referred to as isoperimetric inequalities [Kryz]. They concern minimum length of curves, of area of surfaces etc. subject to constraints. Among them we quote geodesics as shortest paths on manifolds in Euclidean spaces [BomW], [AlAlm], [Aub1], [Fom], [Pot], [Mors1], [Mors2], [Mors3], [Mors4], [Cara], [Herm], [Gri], [LovR], [LySc], [Ly]. As simple applications we mention the shortest path between two points on the Earth and the path followed in a vertical plane by a material point moving between two given points in the shortest time [PolS].

For relationship between analytical and geometric theories of calculus of variations we recommend [Gar], [Chav], [GiMS].

In classical settings, the isoperimetric inequalities can be found among other integral inequalities [DuvL], [Pan], [DimPF], [HarLP], [Lak], [Mitr]. In proving the positive definiteness of an operator, even non-isoperimetric integral inequalities can be useful.

Example 3.2.4. Let $u : (0,1) \to \mathbf{R}$, $u = u(x)$, $u \in C^1(0,1) \cap C[0,1]$ be a function satisfying the boundary conditions $u(0) = u(1) = 0$. Then, from the identity (Leibniz-Newton formula) $u(x) = \int_0^x u'(y)dy$ we get

$$u^2(x) = \left(\int_0^x u'(y)dy\right)^2 = \left(\int_0^x 1 \cdot u'(y)dy\right)^2 \leq x \int_0^x (u'(y))^2 dy$$

$$\leq 1 \cdot \int_0^1 (u'(y))^2 dy = \int_0^1 (u'(y))^2 dy,$$

where the Schwarz-Buniakowsky inequality was used. Integrating over $[0,1]$ we obtain the inequality

$$\int_0^1 u^2(x)dx \leq \int_0^1 (u'(x))^2 dx,$$

which is weaker than (3.2.20). Consider the operator

$$A : \mathcal{D}(A) \to L^2(0,1), \quad Au = -u'',$$
$$\mathcal{D}(A) = \{u \in C^2(0,1) \cap C[0,1] \mid u(0) = u(1) = 0\}.$$

Then

$$(Au, u) = \int_0^1 -u''(x)u(x)dx = \int_0^1 (u'(x))^2 dx$$

and A is symmetric. Thus, by the above inequality, $(Au, u) \geq \|u\|$. As $\overline{\mathcal{D}(A)} = L^2(0,1)$, A is symmetric and this inequality holds, it follows that A is positive definite.

The best constant in integral inequalities can be deduced as in Example 3.2.1 by transforming an isoperimetric problem into a variational problem without constraints and then associating with it the Euler equation (and, so, also the boundary conditions contained in $\mathcal{D}(\mathcal{F})$). However, in applications, in order to compute the minimum of a functional and its corresponding extremal without using the Euler equation another way is used: if $\mathcal{D}(\mathcal{F})$ is a *closed subspace* of a Banach space X, the functional $\mathcal{F}$ is continuous on $\mathcal{D}(\mathcal{F})$ and $\lim_{n\to\infty} u_n = u_0$, where $\{u_n\}$ is a minimizing sequence, then u_0 is a point of minimum for $\mathcal{F}$ [Mikh5] and the corresponding minimum is the best constant looked for.

The following isoperimetric inequalities extensively used in hydrodynamic stability [J76] read

$$I_2^2 \geq (4.73)^2 I_0^2, \quad I_2^2 \geq 4\pi^2 I_1^2, \quad I_1^2 \geq \pi^2 I_0^2 \tag{3.2.60}$$

where $I_i^2 = \int_0^1 u^{(i)2}(x)dx$. Others are associated with the variational problems like

$$\max_{\mathbf{V} \in \mathbf{N}^1(\Omega)} \frac{|\,\theta\mathbf{v}\cdot\mathbf{k}\,|_1}{\|\mathbf{v}\|^2 + \|\theta\|^2}, \qquad \max_{\mathbf{V} \in \mathbf{N}^1(\Omega)} \frac{|\,\theta(\mathbf{v}\cdot\mathbf{k} + \mathbf{v}\cdot\mathbf{j})\,|_1}{\|\mathbf{v}\|^2 + \|\theta\|^2},$$

where

$$\mathbf{N}_1(\Omega) = \{\mathbf{V} \equiv (\mathbf{v}, \theta) \in N_1(\Omega), \ \theta \in L^2(\Omega), \ \mathbf{v}_{|\partial\Omega} = 0, \ \theta_{|\partial\Omega} = 0\},$$

$\mathbf{k}$ and $\mathbf{j}$ are two unit vectors and $\mathbf{v}$ are three-dimensional vectors. We also quote a weighted Poincaré inequality [GaldR]

$$\int_\Omega e^{-z} |\,\mathbf{v}\,|^2 (\mathbf{x})d\mathbf{x} \leq \int_\Omega |\,\nabla\mathbf{v}(\mathbf{x})\,|^2 \, d\mathbf{x}, \tag{3.2.20'''}$$

where $\Omega = \{\mathbf{x} \in \mathbf{R}^3 \mid \mathbf{x} = (x, y, z), z \geq 0\}$. They can be easily deduced by this alternative way where for $\{u_n\}$ we use the Fourier series expansions upon total set of $L^2(0,1)$.

In order to deduce integral (and, in particular, isoperimetric) inequalities, for functions defined on $\Omega \subset \mathbf{R}^n$, the so-called *representation* formulae are used [Mikh5], [Bal], [BesIN], [CarboD], [Sob1], [Sob2]. They are identities, like $u = \int_0^x u'(x)dx$, for $u(0) = 0$, studied by potential theory. Their existence and form depend on the form of $\partial\Omega$, values taken by the functions and their derivatives on $\partial\Omega$.

Assume that Ω is bounded at least in one direction. For instance, in $(3.2.20')$, μ_1 is the smallest eigenvalue of the operator $-\Delta$, in the class of functions $u \in W^{\circ 1,2}(\Omega)$, which vanish on $\partial\Omega$, *i.e.* $u_{|\partial\Omega} = 0$ (since $W^{\circ 1,2}(\Omega) = \overline{C_0^\infty}(\Omega)$ (Appendix 1)). Consider their following representation

$$u(x_1, x_2, \ldots, x_n) = u(a_1, x_2, \ldots, x_n) + \int_{a_1}^{x_1} \frac{\partial u}{\partial x_1} dx_1.$$

Then, by the Schwarz inequality and taking into account the boundary condition, as in Example 3.2.4, we have $1/\mu_1 \leq d^2$, where d is the width of an n-dimensional strip containing Ω [Lad69] and μ_1 is the minimum of the ratio $\int_\Omega |Du|^2 dx / \int_\Omega |u|^2 dx$. Therefore, $\mu_1 \to 0$, as $d \to \infty$. Thus, in general, for unbounded Ω, no isoperimetric inequalities exist and the corresponding theory of differential equations differs from and is much less developed than for bounded Ω.

The requirements concerning the form of $\partial\Omega$ is related to some kind of convexity properties. For instance, in the isoperimetric inequalities expressed by embeddings in Sobolev spaces, Ω must be *star like* [Sob1], [Sob2].

The representation formulae depend on the boundary conditions which must be satisfied by the solutions. For instance, in Example 3.2.4 and, in general, for Poincaré like isoperimetric inequalities, the solutions must vanish on $\partial\Omega$. This restricts the possibility of applying these inequalities to many important concrete situations.

The representation formulae are important also in proving the absolute continuity in $L^2(a, b)$, involved in the construction of the Euler equation. We recall [Smir], [Mikh5] that the function $u : [a, b] \to \mathbf{R}$ is *absolutely continuous* on $[a, b]$ if there exists a function v, Lebesgue integrable on $[a, b]$, such that $u(x) = \int_a^b v(x)dx + \text{const.}$, for $x \in [a, b]$.

The isoperimetric inequalities and their corresponding eigenvalue problems are related to the construction of total sets of eigensolutions [Bere], [Tit], [Col1], [Mikh5], [Kre].

Example 3.2.5. Let $A = -\frac{d^2}{dx^2}$ be the opposite of the one-dimensional Laplacian operator, $A : \mathcal{D}(A) \to L^2(0, b)$, $\mathcal{D}(A) = \{u \in L^2(0, b) \cap C^2[0, b] \mid u(0) = u(b) = 0\}$. It is positive definite and its eigenvalues $\lambda_n = n^2\pi^2/b^2$ and the eigenvectors $u_n(x) = \sin(n\pi x/b)$, $n = 1, \ldots$ satisfy the eigenvalue problem $-\frac{d^2 u}{dx^2} = \lambda u$ on $(0, b)$. The energy space of A is $W^{1,2}(a, b)$. The eigenvectors of A are mutually orthogonal and form a total set $\{\sin(n\pi x/b)\}$, $n = 1, 2, \ldots$ in $W^{1,2}(0, b)$ and, by $(3.2.31)$, also in $L^2(0, b)$. The eigenvalues λ_n are minima of the Rayleigh quotient functional in subspaces of $L^2(0, b)$ orthogonal to $\{u_1, \ldots, u_{n-1}\}$. In particular, the first eigenvalue

λ_1 is the minimum π^2/b^2 in the Poincaré inequality (3.2.20). In the $n > 1$ dimensional case, the operator $-\Delta$ is still densely defined in $L_2(\Omega)$, is positive definite (for corresponding Dirichlet or Neumann boundary conditions) and has a purely discrete spectrum [Mikh5]. Therefore its eigenvalues λ_i, $i = 1, 2, \ldots$ are real and simple, the sequence of the eigenvalues is ordered as $0 < \lambda_1 < \lambda_2 < \ldots$ and has the accumulation point at infinity. The corresponding eigenvectors are orthogonal and form a total set in H_A as well as in $L_2(\Omega)$.

Remark that the u_n belong to $\mathcal{D}(A)$, *i.e.* $\mathcal{D}(A)$ contains a set total in $L_2(\Omega)$. This holds also for Hilbert spaces more general than $L_2(\Omega)$.

3.2.7 *Energy method for non-stationary equations*

Variational energy setting. We saw that rigorous theoretical treatments by potential energy method in calculus of variations concern only a class of Euler equations and associated functionals with good properties. Among them the most restrictive is the linearity of the operator A in (3.2.28) and quadraticity of the associated functional (3.2.29) or (3.2.29′). This framework was quite satisfactory for the linear stationary theories of continua, *e.g.* linear elasticity, linear theory of electrodynamics, linear hydrodynamics, and corresponding to affine elliptic (hence stationary) equations of mathematical physics. But, with the advent of new nonlinear materials and new complex phenomena in continua, the linear theories of continua, which dominated the first part of the 20-th century, ceased to be appropriate. As a consequence, new energy methods were developed for the nonlinear and non-stationary models in mechanics of continua. They rest on the basic ideas of the by now classical energy method developed by S. L. Sobolev, V. I. Smirnov and S. G. Mikhlin.

Among these ideas we remark that of associating with a model another problem called the *variational setting* of the initial-value or boundary-value problems for nonlinear pde's. It is obtained by multiplying its equations by the unknown function, say $\mathbf{v}$, and then integrating the result over the domain Ω of $\mathbf{v}$. In fact, even the generalized setting is obtained by scalar multiplication of the classical equations by an arbitrary test function φ. Taking, formally, $\varphi = \mathbf{v}$, the variational formulation is obtained. As in the stationary case, it is also called the *energy relation*, but, unlike the stationary case, the energy is formed with the terms expressed as derivatives with respect to time. Moreover, in the non-stationary case, only the sum of terms containing the derivatives with respect to the space $\mathbf{x}$ corresponds to the associated functional, (3.2.29) or (3.2.29′) (called the potential energy) from the stationary case. In addition, this sum is a functional only for t fixed, otherwise it is a nonlinear transformation carrying vectors $\mathbf{v}(t, \mathbf{x})$ into vectors of t only, and, in general, it is no longer quadratic. The energy relation follows by taking the inner product of (3.2.28) by $\mathbf{v}$, *e.g.* (2.6.4), (2.7.10), (2.7.11), (2.7.35). In this way, some constraints, *e.g.* the solenoidality of the velocity of the incompressible fluid or the magnetic field, are taken into account, such that this inner product realizes a projection of the given equations on some space appropriate to the problem.

In the class of normal modes, *i.e.* $\mathbf{v}(t,\mathbf{x}) = e^{\sigma t}\mathbf{v}_0(\mathbf{x})$, the generalized setting becomes an eigenvalue problem if the given problem is linear, or a bifurcation problem if it is nonlinear. Similarly to the linear case, the integration involved in the variational settings makes easier the application and the justification of the numerical methods applied to the corresponding generalizes models. As a result, by now, their most accurate presentation can be found in treatises devoted to application of finite elements method, boundary element methods and the like [GirR], [Gl], [Te], [BirH].

The variational setting is the starting point in the stability theories. It involves generalized derivatives, the machinery of the geometry of Sobolev or other spaces. Its construction and study is much more complicated than for (3.2.28) mainly due to the presence of nonlinear terms. As a result, for a single model, *e.g.* the N-S incompressible model, several books [Lad69], [Te], [GirR], [ShiJ], [Gald94], and an impressive number of shortest papers were not sufficient to study it.

As already said (Section 2.1), the first generalized setting of the N-S incompressible model was introduced in 1933 by Jean Leray [Ler] as an attempt to prove existence and uniqueness results for its solutions. The problem of such global results, *i.e.* with respect to time and data (body forces, initial and boundary values, domain of motion, Reynolds number), is still an open problem. If this model is realistic, (as it proved to be), it must reflect the same properties as those observed in nature or in laboratory. These revealed that in laminar-turbulent transition regime the solution loses its uniqueness, at some Reynolds number R_e, then it becomes more and more irregular, so that even the generalized models seem inappropriate and, consequently, measure-theory settings [Foi73], [LadV] must be used. Naturally, in the laminar and early stages of transition, the only ones deal with in the book, the generalized setting and, correspondingly, the variational settings are the most suitable in qualitative studies such as ours.

Energy method. In Section 2.7, we started with the generalized incompressible N-S model for strong or Leray turbulent solutions. This model, perturbed around a basic solution $\bar{\mathbf{u}}$, was then orthogonally projected on the subspace of solenoidal vectors to provide the energy relations (2.7.10) and (2.7.11). We used only (2.7.11), since (2.7.10) leads to weaker results. By energy we meant $|\mathbf{v}|^2/2$, *i.e.* the kinetic energy up to a constant (density). In the generalized setting it was equal to half of the $L^2(\Omega)$ norm of the perturbation $\mathbf{v}$ and its rate of change occurring in the energy relation (2.7.11) was written as

$$\frac{1}{2}\frac{d}{dt}|\mathbf{v}|^2 = b(\mathbf{v},\mathbf{v},\bar{\mathbf{u}}) - \nu\|\mathbf{v}\|^2. \tag{3.2.61}$$

Then the best estimation leading to a stability criterion would be

$$\frac{1}{2}\frac{d}{dt}|\mathbf{v}|^2 \le \left(\max_{\mathbf{v}\in N^1(\Omega)} \frac{b(\mathbf{v},\mathbf{v},\bar{\mathbf{u}})}{\|\mathbf{v}\|^2} - \nu\right)\|\mathbf{v}\|^2. \tag{3.2.62}$$

Denote

$$\nu^* = \max_{\mathbf{v} \in N^1(\Omega)} \frac{b(\mathbf{v}, \mathbf{v}, \overline{\mathbf{u}})}{\|\mathbf{v}\|^2}.$$

If

$$\nu > \nu^* \tag{3.2.63}$$

then, by the Poincaré inequality (2.7.6), equation (3.2.62) implies

$$\frac{1}{2}\frac{d}{dt}|\mathbf{v}|^2 \leq -2 \min_{\mathbf{v} \in N^1(\Omega)} \frac{\|\mathbf{v}\|^2}{|\mathbf{v}|^2} \left[\nu - \max_{\mathbf{v} \in N^1(\Omega)} \frac{b(\mathbf{v}, \mathbf{v}, \overline{\mathbf{u}})}{\|\mathbf{v}\|^2}\right]\frac{|\mathbf{v}|^2}{2}, \tag{3.2.64}$$

or, equivalently, the energy inequality

$$\frac{d}{dt}E(t) \leq -2\alpha d^{-2}(\nu - \nu^*)E(t), \tag{3.2.64'}$$

which is of the form (2.7.2) and shows that $\overline{\mathbf{u}}$ is exponentially asymptotically nonlinearly stable. Correspondingly, (3.2.63) is the best stability criterion one can deduce from (3.2.61). The value ν^* is referred to as the *energy stability limit* and it is denoted also by ν_E. Its existence depends on the existence of the solutions of the two involved variational problems $\min_{\mathbf{v} \in N^1(\Omega)} \frac{\|\mathbf{v}\|^2}{|\mathbf{v}|^2}$ and

$$\max_{\mathbf{v} \in N^1(\Omega)} \frac{b(\mathbf{v}, \mathbf{v}, \overline{\mathbf{u}})}{\|\mathbf{v}\|^2}. \tag{3.2.65}$$

In addition, in order to determine the value of ν^* the variational method (3.2.65) must be solved effectively.

Since, for every t fixed, $b(\mathbf{v}, \mathbf{v}, \overline{\mathbf{u}})$ is a quadratic functional, the variational problem (3.2.65) is associate with an affine Euler equation with variable coefficients, depending on $\overline{\mathbf{u}}$. The study of the spectrum of the involved operator is very difficult and it was carried out only in a few simple cases [J76], [Geo85]. As a consequence, weaker criteria were searched.

Thus, the energy relation (2.7.11) contains the negative viscous dissipation energy and a variable-sign nonlinear advective term. Therefore, the better estimation of this term is the better estimation of the rate of change of the energy. In all generalized settings and dimensions of Ω the general estimation for b reads

$$|b(\mathbf{v}, \mathbf{v}, \overline{\mathbf{u}})| \leq C(\overline{\mathbf{u}})\|\mathbf{v}\|^{\frac{3}{4}}|\mathbf{v}|^{\frac{1}{4}}, \tag{3.2.66}$$

where $C(\overline{\mathbf{u}})$ is a simple positive function of the basic solution. Then from (3.2.61) one gets the inequality

$$\frac{1}{2}\frac{d}{dt}|\mathbf{v}|^2 \leq C(\overline{\mathbf{u}})\|\mathbf{v}\|^{\frac{3}{4}}|\mathbf{v}|^{\frac{1}{4}} - \nu\|\mathbf{v}\|^2. \tag{3.2.67}$$

As we intend to write (3.2.66) in the form of an energy inequality (2.7.2), the terms in $\|\mathbf{v}\|$ must no longer be present. An appropriate use of the Young inequality can eliminate these terms but then $\frac{1}{2}\frac{d}{dt}|\mathbf{v}|^2$ is bounded above by a positive quantity and, so, no criterion can be derived. In this situation the Poincaré inequality (2.7.6) was used.

Alternatively let us write (3.2.67) as

$$\frac{1}{2}\frac{d}{dt}|\mathbf{v}|^2 \leq C(\overline{\mathbf{u}})\|\mathbf{v}\|^2\left(\frac{|\mathbf{v}|^2}{\|\mathbf{v}\|^2}\right)^{\frac{1}{8}} - \nu\|\mathbf{v}\|^2 \leq C(\overline{\mathbf{u}})\max_{\mathbf{v}\in N^1(\Omega)}\left(\frac{|\mathbf{v}|^2}{\|\mathbf{v}\|^2}\right)^{\frac{1}{8}} - \nu\|\mathbf{v}\|^2$$

$$\leq -2\frac{\|\mathbf{v}\|^2}{|\mathbf{v}|^2}\left[\nu - C(\overline{\mathbf{u}})\max_{\mathbf{v}\in N^1(\Omega)}\left(\frac{|\mathbf{v}|^2}{\|\mathbf{v}\|^2}\right)^{\frac{1}{8}}\right]\frac{|\mathbf{v}|^2}{2}. \tag{3.2.68}$$

If

$$\nu - C(\overline{\mathbf{u}})\max_{\mathbf{v}\in N^1(\Omega)}\left(\frac{|\mathbf{v}|^2}{\|\mathbf{v}\|^2}\right)^{\frac{1}{8}} > 0, \tag{3.2.69}$$

then

$$\frac{1}{2}\frac{d}{dt}|\mathbf{v}|^2 \leq -C_1(\overline{\mathbf{u}}, \alpha, d)\frac{|\mathbf{v}|^2}{2}, \tag{3.2.70}$$

where

$$C_1(\overline{\mathbf{u}}, \alpha, d) = 2\max_{\mathbf{v}\in N^1(\Omega)}\left(\frac{\|\mathbf{v}\|^2}{|\mathbf{v}|^2}\right)\left[\nu - C(\overline{\mathbf{u}})\max_{\mathbf{v}\in N^1(\Omega)}\left(\frac{|\mathbf{v}|^2}{\|\mathbf{v}\|^2}\right)^{\frac{1}{8}}\right].$$

This is a constant, depending on the specific quantities. In this simple case the maximum is αd^{-2}, so (3.2.69) becomes the criteria (2.7.20) and (2.7.23). In addition, (3.2.70) is the energy inequality implying, by (2.4.23),

$$\frac{1}{2}|\mathbf{v}|^2(t) \leq e^{-C_1(\overline{\mathbf{u}},\alpha,d)t}\frac{|\mathbf{v}|^2(0)}{2}. \tag{3.2.71}$$

In other words, (3.2.69) is a sufficient condition for $\overline{\mathbf{u}}$ to be tubularly stable and a global attractor. (Here global means that the initial perturbations are arbitrary in the phase space $N^1(\Omega)$.)

Denote

$$C(\overline{\mathbf{u}}) = \max_{\mathbf{v}\in N^1(\Omega)}\left(\frac{|\mathbf{v}|^2}{\|\mathbf{v}\|^2}\right) = \nu_1. \tag{3.2.72}$$

In order to compare ν^* and ν_1 we must pass to nondimensional quantities, which in our case comes to put $d = 1$. Then, the Poincaré inequality (2.7.6) reads

$$\frac{|\mathbf{v}|}{\|\mathbf{v}\|} < \frac{1}{\sqrt{\alpha}} < 1.$$

From (3.2.66) we have

$$\nu^* = \max_{\mathbf{v}\in N^1(\Omega)}\frac{b(\mathbf{v},\mathbf{v},\overline{\mathbf{u}})}{\|\mathbf{v}\|^2} \leq C_1(\overline{u})\left(\frac{|\mathbf{v}|}{\|\mathbf{v}\|}\right)^{\frac{1}{4}} = C_1(\overline{\mathbf{u}})\left(\frac{|\mathbf{v}|}{\|\mathbf{v}\|}\right)^{\frac{1}{4}}$$

$$\leq C(\overline{\mathbf{u}})\max_{\mathbf{v}\in N^1(\Omega)}\left(\frac{|\mathbf{v}|^2}{\|\mathbf{v}\|^2}\right)^{\frac{1}{8}} = \nu_1.$$

This was natural, because every supplementary inequality weakens the results (Section 2.7.1). Therefore the criterion (3.2.69) is weaker than (3.2.63). As a consequence, ν_1 is called an *energy stability bound*.

For fluids presenting, apart from viscosity, several other physical properties, except for magnetic conductivity, the perturbation energy relation is much more complicated than (3.2.61) and contains parameters $R, R_1, \ldots, R_m$ (one for every physical property). The perturbation $\mathbf{v}$ is a vector the components of which are the state functions (except for the pressure) and/or their linear combinations or derivatives. Correspondingly, we define the space $N^1(\Omega)$, which, in this context, is a Cartesian product of the spaces of these functions, the boundary conditions they satisfy, and the solenoidality conditions in the generalized sense for the corresponding functions. Hence, in energy method for non-stationary equations, $N^1(\Omega)$ is the space of difference motions. In this case instead of (3.2.7) the energy relation reads

$$\frac{d}{dt}E(t) = \mathcal{I}_1(R, R_1, \ldots, R_m, t) - \frac{1}{R}\mathcal{I}_2(R, R_1, \ldots, R_m, t), \tag{3.2.73}$$

where $E(t) = \frac{|\mathbf{v}|^2}{2}$, $\mathcal{I}_2$ is a positive integral with respect to the space variable $\mathbf{x}$ and $\mathcal{I}_1$ is also such an integral but, in general, it is not positive. The energy is, in fact, a Lyapunov function. In the integrands of $\mathcal{I}_1$ there occur cross products of the state functions v_i and of their first derivatives Dv_i, while in $\mathcal{I}_2$ there exist only products of the form Dv_iDv_i coming, via Green formulae, from terms $v_i\Delta v_i$ fluxes through $\partial\Omega$. Therefore $\mathcal{I}_2$ is analogous to $\|\mathbf{v}\|^2$ and $\mathcal{I}_1$ to $b(\mathbf{v}, \mathbf{v}, \overline{\mathbf{u}})$ from (3.2.61). Similarly, the best stability criterion we can derive from the energy relation (3.2.73) follows from the inequality (deduced from (3.2.73))

$$\frac{d}{dt}E(t) \leq \mathcal{I}_2\left(\max_{\mathbf{v}\in N^1(\Omega)} \frac{\mathcal{I}_1}{\mathcal{I}_2} - \frac{1}{R}\right) \tag{3.2.74}$$

namely it reads

$$R < R^* \tag{3.2.75}$$

where

$$R^* = \max_{\mathbf{v}\in N^1(\Omega)} \frac{|\mathcal{I}_1|}{|\mathcal{I}_2|}. \tag{3.2.76}$$

Then, from (3.2.74), if (3.2.75) holds, we obtain the energy inequality

$$\frac{d}{dt}E(t) \leq \min_{\mathbf{v}\in N^1(\Omega)} \frac{\mathcal{I}_2}{\mathcal{I}_1}\left(\frac{1}{R} - \frac{1}{R^*}\right) \tag{3.2.77}$$

implying exponential tubular stability of the global attractor $\overline{\mathbf{u}}$. Sometimes, instead of (3.2.73), we obtain the inequality

$$\frac{d}{dt}(E + \Psi)(t) \leq \min_{\mathbf{v}\in N^1(\Omega)} \frac{\mathcal{I}_2}{\mathcal{I}_1}\left(\frac{1}{R} - \frac{1}{R^*}\right) \tag{3.2.78}$$

where Ψ is a positive function of t implying the boundedness of E in the sense $\int_0^T E(t)dt \leq [E(0) + \Psi(0)]/C$, where $C = \min_{\mathbf{v}\in N^1(\Omega)} \frac{\mathcal{I}_2}{\mathcal{I}_1}$. Hence, $\overline{\mathbf{u}}$ is locally tubularly stable (locally means for sufficiently small initial perturbation energy $E(0)$).

If the set of normal modes is total in $N^1(\Omega)$, then (3.2.73) becomes $\sigma E(0) = \mathcal{I}_1 - \frac{1}{R}\mathcal{I}_2$, showing that R_1^* is equal to R_{1G}. In the nonmagnetic case, for every fixed

t, $\mathcal{I}_1$ and $\mathcal{I}_2$ are quadratic functionals. Thus, with the variational problem (3.2.72), which can be reduced to an isoperimetric problem, a boundary-value problem for an affine pde in $N^1(\Omega)$ can be associated. In the magnetic case, additional nonlinear terms occur due to the fact that, apart from the nonlinear advective terms, the governing equations contain nonlinearities of the third order. The coefficients in this boundary-value problem depend on the derivatives of $\bar{\mathbf{u}}$. Taking into account the presence of several terms, state functions and parameters, it is improbable to find its solution. This is the mathematical reason why, as basic states $\bar{\mathbf{u}}$, often the vertical and horizontal convections are considered, for which the derivatives of $\bar{\mathbf{u}}$ are constant. The convections are not flows with pronounced local physical properties, hence their characterization by means of a global quantity, like the energy E of the fluid in the entire Ω, is suitable.

Definition 3.2.6. *The energy method for non-stationary equations* or, shortly, *the energy method* is that one permitting the reduction of the associated energy relation to an energy inequality of the form (3.2.77) or (3.2.78).

The limits of (3.2.76) type are determined by the energy method from the stationary case (Section 3.2.4.2) and are usually denoted by the index E, *e.g.* R_{1E}. From now on, if otherwise not specified, by energy method we understand that from the non-stationary case.

The most extensive presentation of this method and its applications is to be found in [J76]. In Chapter 4 we present some of its new variant and applications. Its reduction to the energy inequality is conditioned by the possibility to put the classical model into the form of the energy relation (3.2.61). First the generalized associated model must be written and the global existence and uniqueness theorem with respect to t and data of the model must be proved. Then the integral Green-type identities must be used to obtain the functionals $\mathcal{I}_1$ and $\mathcal{I}_2$. This can be done only if the regularity and boundary conditions satisfied by the state functions are appropriate. Finally, the operator in the associated Euler equation must be proved to satisfy the requirements in the variational principle 3.2.2 in the energy space H_A, which must be constructed. All these difficulties were overcome for operators not differing much from $-\Delta$, intensively studied about the last mid-century, in connection with the linear problems of mathematical physics [Mikh3], [Mikh2], [Lad69].

In many models governing viscous fluid flows subject to several effects (except for the magnetic one), the presence of $-\Delta$ guaranteed a treatment along the already existing lines. The supplementary terms in the generalized models except for the derivatives with respect to time and the viscous one containing $-\Delta$ were studied in the same way as the body forces and the nonlinear terms. Thus, although more difficult technically, the mathematical problem is in principle the same. After the global existence and uniqueness theorem was proved, the derivation of stability criteria by energy method is not very different either, *e.g.* in the Prodi's proof in

Section 2.5, the term $R(\mathbf{v})$ will contain additional nonlinear terms. In [J76] and [Strau] for several such models the results obtained by energy method are reported; possibly, some of them are only formal.

The application of the energy method to fluids in electromagnetic fields is much more difficult mainly due to: supplementary nonlinearities and difficulty in defining an energy. As a consequence, some of the governing equations must be differentiated and the energy will contain not only the state functions but also some of their derivatives. The proofs use special techniques (of fractional powers of operators) [SolM].

Given a mathematical model, several variants of the energy method can be applied to it. For the applications it is of interest that one yielding the largest stability domain. In Chapter 4 we analyze three of these variants.

In (3.2.77) and (3.2.78) two variational problems occur. They must be solved in the class $N^1(\Omega)$ and have a universal character, *i.e.* their solution does not depend on the given model. In the purely viscous case they are those corresponding to the Poincaré inequality. In hydrodynamic and hydromagnetic stability they are not solved and, so, the stability domain is studied in terms of R/R^*. In addition, the maximum problem for the coefficient outside the parentheses in (3.2.77) or (3.2.78), from the stability point of view is not of interest.

3.3 Symmetrization of some matricial ordinary differential operators

Undoubtedly, variational methods were ([Lag], [Lebe]) and still are ([Loi], [SiPM], [SmitDS], [Stru], [GiH]) basic in mechanics of material points and rigid bodies. Then they passed also in elasticity. Apart from the Mikhlin's books and other papers already quoted, in this field we mention a few: [Ped], [MosoM], [HlHNL], [Fun], [Cou2], [Chu], [BudDiP], [BudHC], [BudK], [Was], [Rek], [Sche], [SiPM]. Subsequently, these methods were applied not only to hydrodynamic stability but also to related problems of hydrodynamics [SelW], [Berd], [Bio], [DoPH], [FinS], [LamL], [LebL1], [LebL2], [Pr], [Ras], [Robe2], [Rud], [StriA]. Variational finite elements method was first developed by elasticians, then by hydrodynamicists and now is one of the most important numerical methods. From the huge literature on the topic we quote some which are either basic or lie close to our views: [AzB], [OdR], [RedV], [BreS], [AxB], [CheY].

Four simple Bénard convection eigenvalue two-point problems for the same system of ode's are introduced in Section 3.3.1 and regularity results up to the boundary are found for them. The question of derivation of extra boundary conditions from the very equations is analyzed. In Section 3.3.2 these problems are written by means of four nonsymmetric matricial differential operators. Then their adjoint, symmetric and skew-symmetric operators for the operators are deduced. In order to obtain the variational formulation of the four problems, in Section 3.3.3 the associated operators are symmetrized. To this aim four tricks are shown: the smoothness of the solution up to the boundary of their domain of definition is used and additional boundary conditions are successfully deduced from the very system by algebraic, differential and integral operations.

The undergoing ideas proved to be applicable in much more general cases (*e.g.* the generalized Joseph's method of parameters differentiation). On this basis, the variational settings of the four problems are presented in Section 3.4. In Section 3.5 they are used to derive the hydrodynamic instability criteria by means of the DiPrima variational method.

3.3.1 *Four eigenvalue problems of thermal convection*

The problem governing the linear instability of a fluid heated from below and subject or not to dielectrophoretic forces is an eigenvalue problem for a nonlinear non-selfadjoint differential matricial operator. For three types of boundary conditions, by means of the Chandrasekhar's method, it was reduced by R. J. Turnbull [Tur], to an equivalent variational formulation and approximate solutions have been found. For the most important case Turnbull found no variational principle, instead he presented some bounds. By suitable splitting of the equations, or by introducing some new unknown functions as linear combinations of the former unknowns and their derivative or, at least, by smoothness arguments, in [Geo77] for this case of physical interest three variational principles were established. In the other three cases the non-selfadjoint eigenvalue problem was reduced to a selfadjoint one. Then a further use of a method, employed by B. Budianski and R. C. DiPrima to reduce the stationarity problem for certain associated functionals to an isoperimetric problem, allowed the author to derive the exact solution in all the four cases. All these problems are discussed in Sections 3.3, 3.4 and 3.5 mainly following [Geo77].

Ideas employed here may be applied to more general systems.

Consider the stability against normal mode perturbations of the mechanical equilibrium of a fluid layer heated from below when a dielectrophoretic force is acting upon.

If the layer is bounded by the planes $x = \pm 0.5$, R is the Rayleigh number, El is a dimensionless number proportional to the electric field squared, k represents the wave number in the z-direction, v denotes the x component of the velocity of perturbation, θ stands for the temperature of perturbation and ϕ for the electrical potential of perturbation; v, θ, ϕ are functions of x only, $D = \frac{d}{dx}$, then the linear stability problem is governed by the following system of ode's

$$(D^2 - k^2)\theta + v = 0, \tag{3.3.1}$$

$$(D^2 - k^2)^2 v - k^2 R(1 + El)\theta - k^2 REl D\phi = 0, \qquad x \in (-0.5, 0.5) \tag{3.3.2}$$

$$(D^2 - k^2)\phi + D\theta = 0. \tag{3.3.3}$$

The functions must satisfy one among the following four sets of boundary conditions at $z = \pm 0.5$

$$v = D^2 v = \theta = \phi = 0, \tag{3.3.4}$$

$$v = Dv = \theta = D\phi = 0, \tag{3.3.5}$$

$$v = Dv = D\theta = \phi = 0, \tag{3.3.6}$$

$$v = Dv = \theta = \phi = 0. \tag{3.3.7}$$

The physical meaning of these conditions follows from the definition of v, θ, ϕ.

Thus $\theta = 0$ at $x = \pm 0.5$ means that the boundaries are kept at constant temperature; $v = 0$ at $x = \pm 0.5$ shows that the normal velocity vanishes at the boundaries; $Dv = 0$ at $x = \pm 0.5$ means that the tangential velocity vanishes at $x = \pm 0.5$; $D^2 v = 0$ at $x = \pm 0.5$ shows that the shear stress is zero at $x = \pm 0.5$; the boundary conditions $v = Dv = 0$ at $x = \pm 0.5$ indicate that the boundaries are rigid walls, while $v = D^2 v = 0$ at $x = \pm 0.5$ show that the boundaries are free; $D\theta = 0$ at $x = \pm 0.5$ - that the heat conduction away from the boundaries is kept constant; $\phi = 0$ - that the electric tangential field vanishes; $D\phi = 0$ - that the normal electric field is kept constant, or if $\theta = 0$ too, that the charge on the electrode is kept constant [Tur].

For k and El assigned we have four eigenvalue problems, where (v, θ, ϕ) is the eigenvector corresponding to the eigenvalue R.

The aim of a stability investigation is to determine the smallest value R as a function of k, where k is a parameter, the curve $R = R(k)$ representing the neutral stability curve. This aim can be accomplished in several ways. One of them associates with the mentioned problems, written as operator equations (in which the operator is selfadjoint), some variational (namely isoperimetric) problems.

Denote by $[C^k(-0.5, 0.5)]^n$ the class of the continuous n vector functions possessing derivatives up to the order k, continuous on $(-0.5, 0.5)$ and by $[C^k[-0.5, 0.5]]^n$ the set of the functions of $[C^k(-0.5, 0.5)]^n$ which, together with their derivatives up to the order k, admit continuous extensions to $[-0.5, 0.5]$. Sometimes, by $[C^k[-0.5, 0.5]]^n$ we mean even the class of these extensions.

The above eigenvalue problems are completely formulated if we specify the class to which the functions belong. First we consider the largest possible class such that these problems make a classical sense. Thus, for all these four problems we suppose $v \in C^4(-0.5, 0.5) \cap C^0[-0.5, 0.5]$, $\theta \in C^2(-0.5, 0.5) \cap C^0[-0.5, 0.5]$, $\phi \in C^2(-0.5, 0.5) \cap C^0[-0.5, 0.5]$ and, in addition, we assume $v \in C^2[-0.5, 0.5]$, for the problem $(3.3.1) - (3.3.4)$, $\phi \in C^1[-0.5, 0.5]$ for the problem $(3.3.1) - (3.3.3)$, $(3.3.5)$, $\theta \in C^1[-0.5, 0.5]$ for the problem $(3.3.1) - (3.3.3)$, $(3.3.6)$ and $v \in C^1[-0.5, 0.5]$ for the last one.

The following lemmas are proved in Appendix 4.

Lemma 3.3.1. *In the above conditions we have:* $v, \theta, \phi \in C^\infty(-0.5, 0.5)$.

Lemma 3.3.2. *The functions occurring in the above four problems belong to* $C^\infty[-0.5, 0.5]$.

Lemma 3.3.3. *The solutions of the above settled four problems are either θ, v even functions and ϕ odd function, or, conversely, θ, v odd functions and ϕ even function.*

In some other cases, *e.g.* for $El = 0$ and disregarding equation (3.3.3) and the boundary condition (3.3.4)$_4$, the given problem splits into the given problem for the even part of the unknown functions and the given problem for the odd part of the unknown functions.

As a consequence, the given problem is solved separately for the even part of the unknown functions and separately for the odd part. Then, experiments are invoked to consider only one among these two problems, usually that for the even part. This is the most frequent situation in hydrodynamic stability theory and especially in convection problems. There is also a plausible argument: for some limiting cases (*e.g.* $El \to 0$), the onset of instability occurs for even unknown functions. In general, we must be careful with these arguments, especially due to the presence of several other physical parameters. Cases of invalidity of these arguments are treated in [PalG04a], [PalG04b], [GeoPalR06].

Remark 3.3.1. Lemma 3.3.2 can be proved also in the following way: in view of Lemma 3.3.1, differentiating suitably equations (3.3.1) and (3.3.3) and adding the results (multiplied by some constants), to the (suitably differentiated) equation (3.3.3), we obtain that v, which satisfies (3.3.1), (3.3.3), must also satisfy the equation

$$A_5' \mathbf{U} \equiv (D^2 - k^2)^4 v + k^2 R D^2 v - k^4 R(1 + El)v = 0, \qquad (3.3.8)$$

which admits solutions of the form $P_m(x)e^{\lambda x}$ where $P_m(x)$ is a polynomial of the degree m and λ is the root of multiplicity $m + 1$ of the characteristic equation

$$f(\lambda) \equiv \lambda^8 - 4k^2\lambda^6 + 6k^4\lambda^4 + (-4k^6 + k^2 R)\lambda^2 + k^8 - k^4 R(1 + El) = 0. \quad (3.3.9)$$

These solutions belong to $C^\infty[-0.5, 0.5]$. In the same way we may write down the equations satisfied by θ and ϕ, whence Lemma 3.3.2.

The regularity results in Lemmas 3.3.1 and 3.3.2 are well-known and follow from the fact that the general solution of vector ode's with constants coefficients are exponentials multiplied by polynomials. However, the methods used in our proofs are less known by contemporary applied mathematicians. In addition, the regularity on $[a, b]$ and not only on (a, b) is crucial for deriving supplementary boundary conditions from the equations. As we saw, in general, this derivation is forbidden, though it is frequently used in applications.

Remark 3.3.2. The two alternative methods used to prove Lemma 3.3.2 show that if the system in v, θ, ϕ is reduced to an equation in v, at this equation we can attach only the conditions *a priori* existing for v. If the solution of this equation has no sufficiently many derivatives we cannot deduce some other boundary conditions from the given system and, so, the two-point problem for a system in v, θ, ϕ cannot be reduced to a two-point problem for an equation in v.

Remark 3.3.3. If we want to split an equation by defining a new function u on $[-0.5, 0.5]$, then u must be defined only as a linear combination of those functions and derivatives which occur in the boundary conditions; some other boundary conditions can be derived from the given system if its solution is sufficiently smooth.

The relation of definition of u represents a new equation valid in $[-0.5, 0.5]$, while the above-mentioned equation in v, *i.e.* (3.3.8), is valid initially on $(-0.5, 0, 5)$.

3.3.2 *Adjoint operators, their symmetric and skew-symmetric part for matricial ordinary differential operators defining problems (3.3.1)–(3.3.7)*

In Appendix 2 we defined $\mathbf{L}^2(a, b)$ as the real Hilbert space of vector functions $\mathbf{U} = (U_1, \ldots, U_n)$, where $U_i : (a, b) \to \mathbf{R}$, $i = 1, \ldots, n$ are square Lebesgue integrable functions. It is endowed with the scalar product $(\mathbf{U}, \mathbf{V}) = \int_a^b U_i(x) V_i(x) dx$, where i is a dummy index.

Many problems governing the linear stability of fluid flows consist of two-point problems for a system of ode's which can be written in the form

$$A\mathbf{U} = 0, \qquad \mathbf{U} \in \mathcal{D}(A), \tag{3.3.10}$$

where A is a matricial differential operator in $\mathbf{L}^2(a, b)$. Let us write the boundary conditions in the form

$$B\mathbf{U} = 0, \qquad \text{at} \qquad x = a, b. \tag{3.3.11}$$

Since in applications the problem (3.3.10), (3.3.11) is formulated in a classical sense, *i.e.* $\mathcal{D}(A) \subset C^l[a, b]$, it is understood that $\mathcal{D}(A)$ was embedded into $\mathbf{L}^2(a, b)$ and the subset of $\mathbf{L}^2(a, b)$ isometric to $\mathcal{D}(A)$ was also denoted by $\mathcal{D}(A)$ (Convention in Section 3.2.2). Therefore, $\mathbf{U}$ is a smooth representative of the corresponding element of $\mathcal{D}(A)$ for which the integrals are taken in the Riemann sense. Hence $\mathcal{D}(A) = \{\mathbf{U} \in \mathbf{L}^2(a, b) \mid \mathbf{U} \text{ satisfy } (3.3.11)\}$.

The adjoint operator A^* satisfies

$$\int_a^b (A\mathbf{U})\mathbf{V} dx = \int_a^b \mathbf{U}(A^*\mathbf{V}) dx, \qquad \forall \, \mathbf{U} \in \mathcal{D}(A), \, \mathbf{V} \in \mathbf{L}^2(a, b)$$

It is computed in practice by multiple integration by parts in the left term and by imposing to all coefficients of arbitrary values of $\mathbf{U}$ and its derivatives at $x = a$, and $x = b$ to vanish.

In the following we put the four problems in Section 3.3.1 in the form (3.3.10), (3.3.11) and then we deduce A^*, A_s and A_{ss} for the associated operators A. The symmetric A_s and skew-symmetric A_{ss} parts of A can be defined by means of A^* (Appendix 2).

Example 3.3.1. The operator defining problem $(3.3.1) - (3.3.4)$ in the form $(3.3.10)$ reads $A : \mathcal{D}(A) \to [L^2(-0.5, 0.5)]^3$,

$$A = \begin{pmatrix} I & D^2 - k^2 I & 0 \\ (D^2 - k^2 I)^2 & -k^2 R(1 + El)I & -k^2 RElD \\ 0 & D & D^2 - k^2 I \end{pmatrix} \tag{3.3.12}$$

where I is the identity operator on $L^2(a, b)$, 0 is the null operator on $L^2(a, b)$, D is the differential operator $\frac{d}{dx}$ on $L^2(a, b)$ and $\mathcal{D}(A)$, in the classical setting, is

$$\mathcal{D}(A) = \{ \mathbf{U} = (v, \theta, \phi) \in [C^\infty(-0.5, 0.5)]^3 \cap C^{2,0,0}[-0.5, 0.5] \mid (v, \theta, \phi) \text{ satisfy } (3.3.4)\},$$

where we used the notation $C^{\alpha,\beta,\gamma}[-0.5, 0.5] = C^\alpha[-0.5, 0.5] \times C^\beta[-0.5, 0.5] \times C^\gamma[-0.5, 0.5]$. Then, denoting $\mathbf{V} = (v^*, \theta^*, \phi^*)$, we have

$$(A\mathbf{U}, \mathbf{V}) = \int_{-0.5}^{0.5} \left[vv^* + (D^2 - k^2 I)\theta v^* + (D^2 - k^2 I)^2 v\theta^* - k^2 R(1 + El)\theta\theta^* \right.$$
$$\left. - k^2 RElD\phi\theta^* + D\theta\phi^* + (D^2 - k^2 I)\phi\phi^* \right] dx$$
$$= [D\theta v^* - \theta Dv^* + D^3 v\theta^* - D^2 v D\theta^* + DvD^2\theta^* - vD^3\theta^* - k^2 REl\phi\theta^*$$
$$+ \theta\phi^* + D\phi\phi^* - \phi D\phi^*]_{-0.5}^{0.5} + \int_{-0.5}^{0.5} \left[vv^* + \theta(D^2 - k^2 I)v^* \right.$$
$$+ v(D^2 - k^2 I)^2\theta^* - k^2 R(1 + El)\theta\theta^* + k^2 REl\phi D\theta^* - \theta D\phi^*$$
$$\left. + \phi(D^2 - k^2 I)\phi^* \right] dx = (\mathbf{U}, A^*\mathbf{V}).$$

Taking into account $(3.3.4)$ the boundary conditions which must be satisfied are $[DvD^2\theta^* + D^3 v\theta^* + D\theta v^* + D\phi\phi^*]_{-0.5}^{0.5}$.

Since Dv, $D^3 v$, $D\theta$ and $D\phi$ are arbitrary, it follows that we must have

$$D^2\theta^* = \theta^* = v^* = \phi^* = 0, \quad \text{at} \quad x = \pm 0.5. \tag{3.3.13}$$

In this way,

$$A^* = \begin{pmatrix} I & (D^2 - k^2 I)^2 & 0 \\ D^2 - k^2 I & -k^2 R(1 + El)I & -D \\ 0 & k^2 RElD & D^2 - k^2 I \end{pmatrix}, \tag{3.3.14}$$

and $\mathcal{D}(A^*)$ is the set

$$\{ \mathbf{V} = (v^*, \theta^*, \phi^*) \in [C^\infty(-0.5, 0.5)^3] \cap C^{0,2,0}[-0.5, 0.5] \mid (v^*, \theta^*, \phi^*) \text{ satisfy } (3.3.13)\}.$$

Taking into account the expression $(3.3.14)$ of A^* it follows that the symmetric part A_s of A is

$$A_s = \begin{pmatrix} I & \frac{(D^2 - k^2 I)^2 + (D^2 - k^2 I)}{2} & 0 \\ \frac{(D^2 - k^2 I)^2 + (D^2 - k^2 I)}{2} & -k^2 R(1 + El) & -\frac{D(1 + k^2 REl)}{2} \\ 0 & \frac{D(1 + k^2 REl)}{2} & D^2 - k^2 I \end{pmatrix},$$

while the skew-symmetric part A_{ss} of A is

$$A_{ss} = \begin{pmatrix} 0 & \frac{(D^2-k^2I)-(D^2-k^2I)^2}{2} & 0 \\ \frac{(D^2-k^2I)^2-(D^2-k^2I)}{2} & 0 & \frac{D(1-k^2REl)}{2} \\ 0 & \frac{D(1-k^2REl)}{2} & 0 \end{pmatrix},$$

$\mathcal{D}(A_s) = \mathcal{D}(A_{ss}) = \mathcal{D}(A)$. As the construction of A_s and A_{ss} does not involve the boundary conditions, it follows that the expression of A_s and A_{ss} for all four problems in Section 3.3.1 are the same.

Example 3.3.2. The operator defining problem $(3.3.1)-(3.3.3)$, $(3.3.5)$ is given by $(3.3.12)$, where $\mathcal{D}(A)$ is the set

$$\{\mathbf{U} = (v,\theta,\phi) \in [C^\infty(-0.5,0.5)^3] \cap C^{1,0,1}[-0.5,0.5] \mid (v,\theta,\phi) \text{ satisfy } (3.3.5)\},$$

A^* is given by $(3.3.14)$ and, after satisfying conditions $(3.3.5)$, we must satisfy the boundary conditions

$[-D^2vD\theta^* + D^3v\theta^* + D\theta v^* - \phi(k^2REl\theta^* + D\phi^*]^{0.5}_{-0.5} = 0$ for arbitrary D^2v, D^3v, $D\theta$ and ϕ. It follows that $D\theta^* = \theta^* = v^* = k^2REl\theta^* + D\phi^* = 0$, or, equivalently,

$$D\theta^* = \theta^* = v^* = D\phi^* = 0, \quad \text{at} \quad x = \pm 0.5. \tag{3.3.15}$$

Therefore $\mathcal{D}(A^*) =$

$$\{\mathbf{V} = (v^*,\theta^*,\phi^*) \in [C^\infty(-0.5,0.5)^3] \cap C^{0,1,1}[-0.5,0.5] \mid (v^*,\theta^*,\phi^*) \text{ satisfy } (3.3.15)\}.$$

Example 3.3.3. The operator defining problem $(3.3.1)-(3.3.3)$, $(3.3.6)$ is given by $(3.3.12)$, where $\mathcal{D}(A) =$

$$\{\mathbf{U} = (v,\theta,\phi) \in [C^\infty(-0.5,0.5)^3] \cap C^{1,1,0}[-0.5,0.5] \mid (v,\theta,\phi) \text{ satisfy } (3.3.6)\},$$

the boundary conditions for arbitrary D^2v, D^3v, θ and $D\phi$ are $Dv^* + \phi^* = \theta^* = D\theta^* = \phi^* = 0$, or, equivalently,

$$D\theta^* = \theta^* = Dv^* = \phi^* = 0, \quad \text{at} \quad x = \pm 0.5. \tag{3.3.16}$$

Therefore $\mathcal{D}(A^*) =$

$$\{\mathbf{V} = (v^*,\theta^*,\phi^*) \in [C^\infty(-0.5,0.5)]^3 \cap C^{1,1,0}[-0.5,0.5] \mid (v^*,\theta^*,\phi^*) \text{ satisfy } (3.3.16)\}$$

while A^* is given by $(3.3.14)$.

Example 3.3.4. The operator defining problem $(3.3.1)-(3.3.3)$, $(3.3.7)$ is given by $(3.3.12)$, where $\mathcal{D}(A) =$

$$\{\mathbf{U} = (v,\theta,\phi) \in [C^\infty(-0.5,0.5)^3] \cap C^{1,0,0}[-0.5,0.5] \mid (v,\theta,\phi) \text{ satisfy } (3.3.7)\},$$

the boundary conditions for arbitrary D^2v, D^3v, $D\theta$ and $D\phi$ are

$$D\theta^* = \theta^* = v^* = \phi^* = 0 \quad \text{at} \quad x = \pm 0.5. \tag{3.3.17}$$

Therefore $\mathcal{D}(A^*) =$

$$\{\mathbf{V} = (v^*,\theta^*,\phi^*) \in [C^\infty(-0.5,0.5)]^3 \cap C^{0,1,0}[-0.5,0.5] \mid (v^*,\theta^*,\phi^*) \text{ satisfy } (3.3.17)\}$$

and A^* is given by (3.3.14).

These examples show that, in spite of the fact that the equations are the same, the operators A defining the four different boundary-value problems for these equations are different in the smoothness properties and boundary conditions occurring in the definition of $\mathcal{D}(A)$. Therefore, these operators have the same expression (3.3.12) but different domains of definition. The same is true for A^*.

For the same problems, the operators A and A^* differ in three aspects: expression, smoothness and boundary conditions.

3.3.3 *Symmetrization of matricial ordinary differential operators defining problems (3.3.1)–(3.3.7)*

Due to the theoretical and numerical importance of symmetric densely defined operators A, it is highly desirable to transform a boundary-value problem involving a nonsymmetric operator A into an equivalent problem defined by a symmetric one. Of course, in $\mathbf{L}^2(a, b)$ this is always possible, since $A^* A$ is a symmetric operator. In addition, $A^* A$ is positive definite, therefore the solution of $Au = f$ is the point of minimum for the functional $\mathcal{F}(u) = (Au, Au) - (Au, f) - (f, Au) = \|Au - f\|^2 - \|f\|^2$. However, from a computational viewpoint it is unsatisfactory because $A^* A$ has twice the order of differentiation of A. Therefore, we are interested in transforming the given operator up to a symmetric one having the same order, *i.e.* to symmetrize it. In this sense, the symmetrization is not always possible, *i.e.* there exist nonsymmetrizable operators. To symmetrize a symmetrizable operator means to apply the tricks (some of them given in the following) permitting to obtain the form of A^* which is identical to that of A. For some problems governing the stability of fluids flows, this form is the same from the beginning. In this case we may wish to prove that A is selfadjoint. This reduces to investigate $\mathcal{D}(A^*)$, *i.e.* to find the boundary conditions occurring in $\mathcal{D}(A^*)$. If initially the form of A and A^* differ, then first we must try to do some changes leading to the fulfillment of the necessary form of the associated matrix (the so-called *rule*). In general, the necessary transformations (*e.g.* the definition of new unknown functions) are transparent. However, they are not always possible due to the lack of the boundary conditions, even if additional boundary conditions are derived from the equations themselves. This is why in most cases the boundary conditions are responsible for the biggest difficulties involved by the problem.

Remark 3.3.4. Sometimes it is sufficient to multiply one of the equations, *e.g.* of (3.3.12′), by -1 to lose or to get the symmetry of the associated operator.

For a more general setting of the symmetrization of operators in hydrodynamic stability we recommend [GaldS89].

In the following we apply four tricks to symmetrize the operators defining the problems $(3.3.1) - (3.3.7)$.

3.3.3.1 *Symmetrization by splitting and multiplying equations*

Example 3.3.5. The operator A in Example 3.3.1 is not symmetric. Nevertheless, by introducing, on $[-0.5, 0.5]$, a new unknown function u

$$u = -(D^2 - k^2)v, \qquad (3.3.18)$$

we can transform it into a symmetric operator $A_1 : \mathcal{D}_1 \to [C^0(-0.5, 0.5)]^4$, corresponding to the system

$$\begin{cases} (D^2 - k^2)\theta + v = 0, \\ (D^2 - k^2)v + u = 0, \\ (D^2 - k^2)u + k^2 R(1 + El)\theta + k^2 RElD\phi = 0, \\ - k^2 REl(D^2 - k^2)\phi - k^2 RElD\theta = 0, \end{cases} \qquad (3.3.12')$$

and to the boundary conditions

$$u = v = \theta = \phi = 0, \qquad \text{at} \qquad x = \pm 0.5 \qquad (3.3.4')$$

where

$\mathcal{D}_1 = \{\mathbf{U} = (u, v, \theta, \phi) \in [C^\infty(-0.5, 0.5)]^4 \cap C^{0,0,0,0}[-0.5, 0.5]^4 \mid \mathbf{U} = 0 \text{ at } x = \pm 0.5\}$.
We note that the condition $u = 0$ at $x = \pm 0.5$ follows from the fact that equation $(3.3.18)_1$ is, by definition, valid up to $x = \pm 0.5$ and $v = D^2 v = 0$ at $x = \pm 0.5$. In fact, for v we had only these last two boundary conditions. This restricted us to choose u as in $(3.3.18)$.

One can see that A_1 is symmetric. Indeed, its expression

$$A_1 = \begin{pmatrix} 0 & I & D^2 - k^2 I & 0 \\ I & D^2 - k^2 I & 0 & 0 \\ D^2 - k^2 I & 0 & k^2 R(1 + El)I & k^2 RElD \\ 0 & 0 & -k^2 RElD & -k^2 REl(D^2 - k^2 I) \end{pmatrix}$$

satisfies the rule from Remark 3.3.7. In addition, the boundary condition (B) reads

$$[D\theta u^* - \theta Du^* + Dvv^* - vDv^* + Du\theta^* - uD\theta^* + k^2 REl\phi\theta^* - k^2 REl\theta\phi^*$$
$$-k^2 REl(D\phi\phi^* - \phi D\phi^*)]_{-0.5}^{0.5} = 0.$$

After taking into account boundary condition $(3.3.4')$ this condition becomes

$$[Du\theta^* + Dvv^* + D\theta u^* - k^2 RElD\phi\phi^*]_{-0.5}^{0.5} = 0.$$

The fact that the values of Du, Dv, $D\theta$ and $D\phi$ are arbitrary at $x = \pm 0.5$ implies $u^* = v^* = \theta^* = \phi^* = 0$ at $x = \pm 0.5$. Therefore the boundary conditions for $\mathbf{U}^*$ are the same with those for $\mathbf{U}$. Therefore $D(A) = D(A^*)$, hence A_1 is not only symmetric but it is also selfadjoint.

Remark that in order to obtain the symmetric operator A_1 we modified the operator A not only by splitting equation $(3.3.2)$ (and therefore by introducing a new unknown function) but also by multiplying $(3.3.3)$ by $-k^2 REl$. Without this last modification the corresponding operator would have still been nonsymmetric.

3.3.3.2 *Symmetrization by introducing new unknown functions and by differentiating equations*

Example 3.3.6. Consider the problem $(3.3.1) - (3.3.3)$, $(3.3.5)$. In this case, defining a function as in $(3.3.18)$ is useless since from the boundary conditions $(3.3.5)$ no boundary condition for u can be obtained. That is why, this time we put

$$u = -k^2 R(1 + El)\theta - k^2 RElD\phi, \qquad \text{for} \qquad x \in [-0.5, 0.5]$$

so from $(3.3.5)$ it follows that $u = 0$ at $x = \pm 0.5$. Differentiating $(3.3.3)$ and expressing $D\phi$ in terms of u we get

$$(D^2 - k^2)\theta = -v, \tag{3.3.1$'$}$$

$$(D^2 - k^2)^2 v = -u, \tag{3.3.2$'$}$$

$$(D^2 - k^2)u + k^2 R(D^2 - k^2)\theta - k^4 REl\theta = 0, \tag{3.3.3$'$}$$

$$u = v = Dv = \theta = 0, \quad \text{at} \quad x = \pm 0.5. \tag{3.3.5$'$}$$

This problem can be written in the form $A_2 f = 0$, where $A_2 : \mathcal{D}_2 \to [C^0(-0.5, 0.5)]^3$, $\mathcal{D}_2 = \{\mathbf{U} = (u, v, \theta) \in [C^\infty(-0.5, 0.5)]^3 \cap [C^{0,1,0}[-0.5, 0.5]]^3 \mid u, v, \theta \text{ satisfy } (3.3.5')\}$ and

$$A_2 = \begin{pmatrix} 0 & I & D^2 - k^2 I \\ I & (D^2 - k^2 I)^2 & 0 \\ D^2 - k^2 I & 0 & k^2 R(D^2 - k^2) - k^4 RElI \end{pmatrix}.$$

Direct computation shows that A_2 and A_2^* have the same form. In addition, (B) reads

$$\left[D\theta u^* - \theta Du^* + D^3 vv^* - D^2 vDv^* + DvD^2 v^* - vD^3 v^* \right.$$
$$\left. + Du\theta^* - uD\theta^* + k^2 RD\theta\theta^* - k^2 R\theta D\theta^* \right]_{-0.5}^{0.5} = 0.$$

Taking into account $(3.3.5')$ and the fact that the values of $Du, D\theta, D^3 v$, and $D^2 v$ are arbitrary at $x = \pm 0.5$, it follows that their coefficients must be equal to zero implying

$$u^* = v^* = Dv^* = \theta^* = 0, \qquad \text{at} \qquad x = \pm 0.5,$$

i.e. $\mathcal{D}_2(A_2^*) = \mathcal{D}(A^*)$. Thus, the operator A_2 is not only symmetric but also a selfadjoint operator.

3.3.3.3 *Symmetrization by rescaling the unknown functions and differentiating equations*

In Example 3.3.6 the boundary conditions of the problem constrained us to introduce such a new function u for which the problem yields boundary conditions. In addition, the function u is a linear combination of one unknown function and the derivative of some other function. In Example 3.3.5 the new function was a linear combination of an unknown function and one of the derivatives of the same function. Its definition was also appropriate because of the given boundary conditions. Both definitions of the new function took into account the existing boundary conditions and were aimed to produce the validity of the rule in Remark 3.3.5. A related idea is to concentrate more on the symmetry properties occurring in this rule and deduce the missing boundary conditions from the very equations and not only to use the existing boundary conditions.

Example 3.3.7. In order to obtain a selfadjoint operator in the third case, when the boundary conditions (3.3.6) hold, we take into account the fact that $\theta, v, \phi \in [C^\infty\,(-0.5, 0.5)]^3$, which enables us to deduce additional boundary conditions using the equations. Thus, taking into account (3.3.1), from (3.3.3) we obtain

$$D^2\phi = 0, \qquad \text{at} \qquad x = \pm 0.5. \tag{3.3.19}$$

Applying the operator $D^2 - k^2 I$ to equation (3.3.3) and expressing θ in terms of v we get

$$(D^2 - k^2)^2\phi = Dv. \tag{3.3.20}$$

Let us put $\phi\sqrt{k^2 REl} = \Phi$, $\theta = \dfrac{\Theta}{\sqrt{k^2 R(1+El)}}$; then (3.3.1), (3.3.2) and (3.3.20) become

$$-(D^2 - k^2)\Theta = v\sqrt{k^2 R(1 + El)}, \tag{3.3.21}$$
$$(D^2 - k^2)^2 v = \Theta\sqrt{k^2 R(1 + El)} + \sqrt{k^2 REl}\,D\Phi, \tag{3.3.22}$$
$$-(D^2 - k^2)^2\Phi = -\sqrt{k^2 REl}\,Dv, \tag{3.3.23}$$

to which we add the boundary conditions (3.3.6) and (3.3.19). We are now in the position of defining the symmetric operator $A_3 : \mathcal{D}_3 \to [C^0(-0.5, 0.5)]^3$, where $\mathcal{D}_3$ is the set of all $\mathbf{U} = (\Theta, v, \Phi) \in [C^\infty(-0.5, 0.5)]^3$ such that Θ, v, Φ satisfy the boundary conditions (3.3.6) and (3.3.19).

The form of A_3 is

$$A_3 = \begin{pmatrix} -(D^2 - k^2 I) & -\sqrt{k^2 R(1 + El)}I & 0 \\ -\sqrt{k^2 R(1 + El)}I & (D^2 - k^2 I)^2 & -\sqrt{k^2 REl}\,D \\ 0 & \sqrt{k^2 REl}\,D & -(D^2 - k^2 I)^2 \end{pmatrix}$$

and it is equal to that of A_3^*. Condition (B) reads

$$[-D\Theta\Theta^* + \Theta D\Theta^* + D^3 v v * - D^2 v Dv^* + Dv D^2 v^* - v D^3 v^* - \sqrt{k^2 REl}\Phi\Phi^*$$
$$+\sqrt{k^2 REl}\Phi\Phi^* - D^3\Phi\Phi^* + D^2\Phi D\Phi^* - D\Phi D^2\Phi^* + \Phi D^3\Phi^*]_{0.5}^{0.5} = 0$$

and it is satisfied for $\mathbf{U}, \mathbf{U}^* \in \mathcal{D}_3(A_3)$. For $\mathbf{U} \in \mathcal{D}_3(A_3)$ condition (B) implies that

$$v^* = Dv^* = D\Theta^* = \Phi^* = D^2\Phi^* = 0, \qquad \text{at} \quad x = \pm 0.5,$$

i.e. $\mathcal{D}_3(A_3) = \mathcal{D}_3(A_3^*)$, consequently A_3 is selfadjoint.

3.3.3.4 *Symmetrization by algebraic operations*

Sometimes it is possible to transform a nonsymmetric matricial differential operator into a symmetric one by: changing the order of equations; changing the order in the components of $\mathbf{u}$; multiplying some of the equations by some constants; rescaling some of the unknown functions; introducing new unknown functions as linear combinations of the other ones and their derivatives; performing linear combinations of the equations. In spite of the simplicity of these tricks, by changing the position of equations and/or unknown functions, the contribution to the functional $(A\mathbf{U}, \mathbf{U})$ of the unknowns and equations can be dramatically changed and the efficiency of the trick may become important.

Usually a combination of these tricks is necessary. Some of them have been presented in the three previous sections. Here we exemplify the first two of them.

Example 3.3.8. In Section 6.1 we deal with the maximum ξ of the functional in a class $\mathcal{M}$ of vector functions $(w, h_3', \frac{\partial}{\partial z} j, \zeta, \theta)$. This class is embedded in the class $\mathcal{M}_1$ of vector functions $(w, w_z, h_3', \frac{\partial}{\partial z} j, \zeta, \theta)$, where w_z is taken as independent from w. In the class $\mathcal{M}_1$ the boundary-value problem for the associated Euler equations reads

$$- S\Delta_1\theta - 2\xi P_m\Delta_1\Delta_1 w = 0,$$

$$- 2\xi P_m(\Delta w_z + \Delta_1 w_z) - (P_m M^2 + \Delta)h_3' = 0,$$

$$\frac{\partial}{\partial z}\left(\frac{\partial^2}{\partial z^2} + P_m M^2\right)j + 2\xi P_m\Delta\zeta = 0,$$

$$\frac{\partial}{\partial z}(\Delta + P_m M^2)w - \beta_H\Delta_1\frac{\partial}{\partial z}j + 2\xi\Delta h_3' = 0,$$

$$- S\Delta_1 w + 2\xi\frac{P_m}{P_r}\Delta_1\Delta\theta = 0,$$

$$- \beta_H\Delta_1 h_3' + 2\xi\Delta\frac{\partial}{\partial z}j + \left(\frac{\partial^2}{\partial z^2} + P_m M^2\right)\zeta = 0,$$

$$w = w_z = \theta = \theta_{zz} = h_3' = j = \zeta = 0 \quad \text{at} \quad x = \pm 0.5,$$

where $S = (1 + \mathcal{R}\frac{P_m^2}{P_r})$. In the class of stationary normal mode perturbations (1.4.12), denoting by $W(z), V(z), K(z), X(z), Z(z), \Theta(z)$ the intensity of the quantities $w, w_z, h_3', \frac{\partial}{\partial z} j, \zeta, \theta$ and putting $a_x^2 + a_y^2 = a^2$ in (1.4.12), this problem becomes

$$Sa^2\Theta - 2\xi P_m a^2 W = 0, \tag{3.3.24}$$

$$2\xi P_m(D^2 - 2a^2)V + (P_m M^2 + D^2 - a^2)K = 0, \tag{3.3.25}$$

$$(P_m M^2 + D^2)X + 2\xi P_m(D^2 - a^2)Z = 0, \tag{3.3.26}$$

$$(P_m M^2 + D^2 - a^2)V + 2\xi(D^2 - a^2)K + \beta_H a^2 X = 0, \tag{3.3.27}$$

$$Sa^2W + 2\xi\frac{P_m}{P_r}a^2(D^2 - a^2)\Theta = 0, \tag{3.3.28}$$

$$\beta_H a^2 K + 2\xi(D^2 - a^2)X + (D^2 + P_m M^2)Z = 0, \tag{3.3.29}$$

$$W = V = K = X = Z = \Theta = 0, \qquad z = \pm 0.5. \tag{3.3.30}$$

The associated operator $B : \mathcal{D}(B) \to \left(L^2(-0.5, 0.5)\right)^6$, where

$$\mathcal{D}(B) = \{\mathbf{U} = (W(z), V(z), K(z), X(z), Z(z), \Theta(z))$$

$$\in \left(C^\infty(-0.5, 0.5)\right)^6 \cap C^{0,0,0,0,0,0}[-0.5, 0.5] \mid \mathbf{U} \text{ satisfy } (3.3.30)\}$$

is defined by

$$B = \begin{pmatrix} -\xi P_m a^2 & 0 & 0 & 0 & 0 & a^2 S \\ 0 & 2\xi P_m(\overline{D} - a^2) & P_m M^2 + \overline{D} & 0 & 0 & 0 \\ 0 & 0 & 0 & P_m M^2 + D^2 & 2\xi P_m \overline{D} & 0 \\ 0 & P_m M^2 + \overline{D} & 2\xi\overline{D} & \beta_H a^2 & 0 & 0 \\ a^2 S & 0 & 0 & 0 & 0 & 2\xi\frac{P_m}{P_r}a^2\overline{D} \\ 0 & 0 & \beta_H a^2 & 2\xi P_m \overline{D} & P_m M^2 + D^2 & 0 \end{pmatrix},$$

where $D^2 - a^2 = \overline{D}$, it is not symmetric but becomes a symmetric operator B_1 if the equations are taken in the following order: (3.3.24), (3.3.25), (3.3.27), (3.3.26), (3.3.29), (3.3.28) and the new vector function is $\mathbf{U}_1 = (W, V, K, Z, X, \Theta)$. In addition, if $4\xi^2 P_m \neq 1$, then B_1 is selfadjoint. In Section 6.1 we choose the class $\mathcal{M}$ and, so, the operator corresponding to the Euler equation is symmetric but, due to the boundary conditions, it is not selfadjoint.

Eliminating W between (3.3.24) and (3.3.28) and taking into account $(3.3.30)_6$ we obtain an eigenvalue problem

$$D^2\Theta + \left(\frac{S^2 P_r}{4P_m^2 a^2 \xi^2} - a^2\right)\Theta = 0, \quad \Theta(0) = \Theta(1) = 0,$$

which has the eigenvalue $\frac{S^2 P_r}{4P_m^2 a^2 \xi^2} - a^2 = \pi^2 n^2$, whence $\xi^2 = \frac{S^2 P_r}{4P_m^2(n^2\pi^2 + a^2)a^2}$ and, therefore, the criterion (corresponding to $n = 1$) $\xi^2 < 1$, *i.e.* $R < \left(\frac{2aP_m\sqrt{\pi^2 + a^2}}{\sqrt{P_r}} - 1\right)\frac{P_r}{P_m^2}$, implying another criterion, namely $\frac{P_m}{\sqrt{P_r}} > \frac{1}{2a\sqrt{\pi^2 + a^2}}$.

3.4 Variational principles for problems (3.3.1)–(3.3.7)

3.4.1 *Boundary-value problems versus associated variational functional*

Propositions asserting the equivalence of a boundary-value problem and a variational problem are referred to as *variational principles* (Section 3.2.1).

If we start with a variational problem for a functional written in the form of an integral of some differential operator, then with it we associate a boundary-value problem for the Euler equations (Section 3.2.2). Irrespective of the form of these equations, in dependence on the class of admissible functions, all these boundary-value problems are equivalent.

Given a boundary-value problem, then with it a variational problem can be associated by suitably multiplying the equations by the unknown vector function and then performing some by-parts integrations.

Assume that the operator associated with the equations of the boundary-value problem is linear, selfadjoint and positive. Then the theory of minimization of quadratic functionals (Section 3.2.5) ensures the equivalence between the boundary-value and the variational problems.

If the operator does not satisfy all these requirements, which is the general situation in applications, then the variational principle must be established separately for each particular problem. For some given boundary-value problems the classical variational problem has no solution. In this case a generalized variational problem must be associated: the solution of the classical boundary-value problem is generalized, *i.e.* it is a function of a Sobolev space and does not satisfy the classical equations but some others associated with the given equations. In spite of the fact that the governing boundary-value problems have a classical setting, in hydrodynamic and hydromagnetic stability theory we are forced to recourse to the generalized formulations because the Navier-Stokes and Navier-Stokes-Fourier equations do not always have classical solutions. More exactly, the existence results for them are lacking.

For the same reason, the generalized formulations for the nonlinear problems are referred to as the *variational formulations*.

When the generalized framework is necessary, the class of the boundary-value problem, given in applications in the classical formulation, differs from the class of the admissible functions of the associated variational problem.

These two classes can be different even when both boundary and variational problems admit a classical setting. In particular, *natural boundary conditions, i.e.* which occur in the boundary-value problem but not in the variational one, can occur (Section 3.2.4.3).

Given a boundary-value problem $A\mathbf{U} = 0$ (with the boundary conditions included in $\mathcal{D}(A)$) several variational principles $\delta j(\mathbf{U}) = 0$ for it may be obtained. Some of them correspond to various equivalent forms of this problem and they are as many as equivalent forms of the problem we have. Some other variational principles depend on the number of by-parts integration we perform. The best ones are for symmetric operators where the integrations are stopped at the half way between $(A\mathbf{U}, \mathbf{U}^*)$ and $(\mathbf{U}, A\mathbf{U}^*)$ and so, the functional corresponds in some sense to $(A^{\frac{1}{2}}\mathbf{U}, A^{\frac{1}{2}}\mathbf{U}^*)$: for selfadjoint and positive definite operators $A^{\frac{1}{2}}$ is the operator defining the energy space (Section 3.2.5).

The preferred form of the functional among several possible forms is that one nearest to $(A^{\frac{1}{2}}\mathbf{U}, A^{\frac{1}{2}}\mathbf{U}^*)$ because when solved by Fourier series techniques it involves the less computations.

Indeed, every boundary-value of the expanded function in lower order derivatives is preserved in a more and more amplified form in the series for the higher-order derivatives. Whence the interest in symmetric operators defining the problems and also in splitting equations leading to the increased number of equations but to a lower-order of them.

For $\mathbf{U} : [a, b] \to \mathbf{R}$, formally $A\mathbf{U} = 0$ implies (using by-parts integration) $(A\mathbf{U}, \mathbf{U}) = [B_i\mathbf{U}, B_i^*\mathbf{U}^*]_a^b + j(\mathbf{U}, \mathbf{U}^*) = (\mathbf{U}, A^*\mathbf{U}^*))$ and all functionals $j(\mathbf{U}, \mathbf{U}^*) : \mathcal{D}(j) \equiv \mathcal{D}(A) \times \mathcal{D}(A^*) \to \mathbf{R}$ can be associated with the given problem in a variational principle of the form: $A\mathbf{U} = 0$ *for* $\mathbf{U} \in \mathcal{D}(A)$, $A^*\mathbf{U}^* = 0$ *for* $\mathbf{U}^* \in \mathcal{D}(A^*)$, *if and only if* $\delta j = 0$ *in the class* $\mathcal{D}(j)$. If A is selfadjoint, denoting $j(\mathbf{U}, \mathbf{U}) = j(\mathbf{U})$, the variational principle reads: $A\mathbf{U} = 0$ *in* $\mathcal{D}(A)$ *if and only if* $\delta j = 0$ *in* $\mathcal{D}(j) \subset \mathcal{D}(A)$.

The set $\mathcal{D}(A) \setminus \mathcal{D}(j)$ consists in *natural boundary conditions*.

The methods of series expansions lead to the same results if applied directly either to the equation $A\mathbf{U} = 0$ or to the variational problem $\delta j = 0$. For instance, in [GeoS] this was proved for problem $(3.3.1) - (3.3.3)$, $(3.3.5)$. However, since j contains derivatives of orders lower than A, the variational formulations are preferable in applications. Whence the importance of variational principles. This can be immediately seen by writing $\delta J(\mathbf{U}, \mathbf{U}^*) = \delta \int_a^b A_1\mathbf{U} A_2\mathbf{U}^* dx = \int_a^b [\delta(A_1\mathbf{U})A_2\mathbf{U}^* + A_1\mathbf{U}\delta A_2\mathbf{U}^*]dx = \int_a^b A\mathbf{U}\delta\mathbf{U}^* dx + \int_a^b \delta\mathbf{U} A^*\mathbf{U}^* dx.$

3.4.2 Variational principles for the first three two-point problems for (3.3.1)–(3.3.3)

In Sections 3.3.2, 3.3.3 the encountered problems had two settings, namely corresponding to the nonsymmetric or symmetric associated operators. For each of them we give two corresponding variational principles. Three two-point problems equivalent to $(3.3.1) - (3.3.3)$, $(3.3.7)$ can also be found in [Geo77] together with three variational principles for them. All are of the same order as A^*A. In addition, two of them involved nonsymmetric operators and natural boundary conditions while the other involves a symmetric operator and no natural conditions. In the following we present briefly the variational principles for the first three quoted problems. The fourth problem is treated in Section 3.4.3.

Example 3.4.1. For the problem $(3.3.1) - (3.3.3)$, $(3.3.4)$ written in the form (Example 3.3.1)

$$AU = 0, \qquad U \in \mathcal{D}(A) \tag{3.4.1}$$

where $\mathcal{D}(A) = \{\mathbf{U} = (v, \theta, \phi) \in [C^\infty(-0.5, 0.5)]^3 \cap C^{2,0,0}[-0.5, 0.5] \mid \mathbf{U}$ satisfies (3.3.4)$\}$, we have $(A\mathbf{U}, \mathbf{U}^*) = (\mathbf{U}, \mathbf{V})$, where $\mathbf{V} \neq A\mathbf{U}$, because in the form (3.3.12) the operator A is not symmetric. Let us use the symmetric operator A_1 associated with the problem (3.3.12′), (3.3.4′), equivalent to (3.4.1), written in the form $A_1(\mathbf{U}) = 0$ and define the functional $j_1 : \mathcal{D}_1 \to \mathbf{R}$ by $j_1(\mathbf{U}) = (A_1\mathbf{U}, \mathbf{U})$.

We have

$$j_1(\mathbf{U}) = \int_{-0.5}^{0.5} [-2D\theta Du - 2k^2\theta u + 2uv - (Dv)^2 - k^2 v^2 + k^2 R(1 + El)\theta^2$$

$$+ 2k^2 REl\theta D\phi + k^2 REl(D\phi)^2 + k^4 REl\phi^2]dx.$$

If $\mathbf{U} \in \mathcal{D}_1$ is a solution of the equation $A_1\mathbf{U} = 0$, it follows that the variation of the functional j_1 at $\mathbf{U}$ in the class $\mathcal{D}_1$ vanishes $(\delta j_1(\mathbf{U}) = 0)$. In other words, $\mathbf{U}$ makes the functional j_1 stationary. Since

$$\delta j_1(\mathbf{U}) = 2 \int_{-0.5}^{0.5} \{(D^2\theta - k^2\theta + v)\delta u + (D^2 v - k^2 v + u)\delta v$$

$$+ [D^2 u - k^2 u + k^2 R(1 + El)\theta + k^2 RElD\phi]\delta\theta$$

$$+ [-k^2 RElD^2\phi + k^4 REl\phi - k^2 RElD\theta]\delta\phi\}dx,$$

the converse assertion is valid too: if $\mathbf{U}$ makes j_1 stationary, then $\mathbf{U}$ is a solution of the equation $A_1 f = 0$. Therefore, we have proved the following variational principle: *to solve the equation $A_1 f = 0$ it is equivalent to find the stationary points of the functional $(A_1 f, f)$.*

This can be expressed as

Theorem 3.4.1. $A_1\mathbf{U} = 0$ *if and only if* $\delta j_1(\mathbf{U}) = 0$.

The analysis of the functional j_1 points out that the splitting of (3.3.2) and the multiplication by a constant were chosen such that in the scalar product $(A_1\mathbf{U}, \mathbf{V})$ the terms (arising from two and only two equations) containing two different functions must have the same coefficients, while those containing a function and the first derivative of the another one, opposite coefficients (Remark 3.3.7). No attention has to be paid to terms containing the same function unless we introduce natural boundary conditions.

Although the smoothness of the unknown function in $(-0.5, 0.5)$ permitted us to differentiate equations $(3.3.1) - (3.3.3)$ and to obtain new equations, the avoidance of natural boundary conditions impeded us to use these new equations.

Remark 3.4.1. We also tried to obtain a variational principle by suitable multiplication of equation (3.4.1) by a vector function $\mathbf{H}$ whose components H_i are linear combinations of the derivatives of U_i. As a consequence, in the obtained functional $j' = (A\mathbf{U}, \mathbf{H})$ some products appeared more than two times (which is not possible if $\mathbf{U} = \mathbf{H}$). Since the definition of j' is connected with *quadratic* terms, in varying j' gives rise to the coefficient 2. We recall that a quadratic term is a function

$f(tx, ty) = t^2 f(x, y)$. The same coefficient multiplies also the non-quadratic terms containing different functions (or their derivatives) if the number of these products is *two*. As a result, $\delta j' = 2(A\mathbf{U}, \delta\mathbf{U})$, $\forall \delta\mathbf{U}$, whence (3.4.1), while if the same products appear in j' once or more than two times we can obtain only $\delta j' = (A'\mathbf{U}, \delta\mathbf{U})$, where $A' \neq A$, which implies $A'\mathbf{U} = 0$ and not $A\mathbf{U} = 0$.

Example 3.4.2. In the same way, for problem $(3.3.1) - (3.3.3)$, $(3.3.5)$ written in the symmetric form by means of A_2, define the functional $j_2 : \mathcal{D}_2 \to R$ such that $j_2(\mathbf{U}) = (A_2\mathbf{U}, \mathbf{U})$. We have

$$j_2(\mathbf{U}) = \int_{-0.5}^{0.5} [-2D\theta Du - 2k^2\theta u + 2uv + k^4 v^2 + (D^2 u)^2 + 2k^2 (Dv)^2 - k^2 R(D\theta)^2$$

$$- k^4 R(1 + El)\theta^2] dx.$$

The following variational principle can be immediately proved:

Theorem 3.4.2. $A_2\mathbf{U} = 0$ *if and only if* $\delta j_2(\mathbf{U}) = 0$.

Example 3.4.3. Consider the symmetric form of the problem $(3.3.1) - (3.3.3)$, $(3.3.6)$ defined by A_3 and define the functional
$j_3 : \mathcal{D}_3 \to \mathbf{R}$, $j_3(\mathbf{U}) = (A_3\mathbf{U}, \mathbf{U})$, to obtain

$$j_3(\mathbf{U}) = \int_{-0.5}^{0.5} \{(D\Theta)^2 + k^2\Theta^2 - 2v\Theta\sqrt{k^2 R(1 + El)} + (D^2 v)^2 + 2k^2 (Dv)^2 + k^4 v^2$$

$$+ 2\Phi Dv\sqrt{k^2 REl} - (D^2\Phi)^2 - 2k^2 (D\Phi)^2 - k^4\Phi^2\} dx,$$

whence

Theorem 3.4.3. $\delta j_3(\mathbf{U}) = 0$, *if* $A_3\mathbf{U}) = 0$.

Remark 3.4.2. As none of the symmetric operators in Examples $3.4.1 - 3.4.3$ were positively defined, we could not use the corresponding theory.

3.4.3 *Variational principles for problem (3.3.1)–(3.3.3), (3.3.7)*

For this problem we were not able to transform the associated operator up to a symmetric one the order of differentiation of which be not higher. This is why this problem require the use of more complicated variational principles involving higher order operators. These operators defined two-point problems equivalent to $(3.3.1) - (3.3.3)$, $(3.3.7)$. The proof of this equivalence is not immediate and is performed before associating the variational principles.

As for the other problems for $(3.3.1) - (3.3.3)$, in the following we first establish the equivalence of the boundary-value problems and then we prove the corresponding variational principle.

3.4.3.1 *First equivalent problem in one unknown function. Boundary conditions derived by inverse operator method*

By applying the direct method to the problem $(3.3.1) - (3.3.3)$, $(3.3.7)$, we derive the secular equations in two particular cases. However, in this section our aim is to show how to get equivalent forms for this problem, how to determine the closed-form solutions and their class rather than to study completely these secular equations (this study is carried out in Section 7.5). The system $(3.3.1) - (3.3.3)$ was written as a single eighth order equation in v, namely $(3.3.8)$. It is shown that the same equation is satisfied by θ and ϕ. Then the exact form of the solutions v, θ and ϕ are expressed in terms of the roots λ_i of the characteristic equation. In $(3.3.7)$ only four boundary conditions for v appear. Here the missing boundary conditions for v are derived directly from the boundary conditions for θ and ϕ by using the closed form solution $v = A_i \cosh(\lambda_i x)$, $\theta = B_i \cosh(\lambda_i x)$, $\phi = C_i \cosh(\lambda_i x)$, i being a dummy index. Namely, if A_i, B_i and C_i respectively stand for the coefficients in these forms, the derivation, directly from the equations, of B_i and C_i as functions of A_i enables us immediately to express the boundary conditions for θ and ϕ in terms of A_i.

The obtained conditions in A_i contain expression like $(\lambda_i^2 - k^2)^{-1}$ and $(\lambda_i^2 - k^2)^{-2}$ suggesting the involvement of the inverse operators $(D^2 - k^2)^{-1}$ and $(D^2 - k^2)^{-2}$. Indeed, by the variation of the coefficients method, the even solution of $(3.3.1)$, as a function of v, reads $\theta(x) = A_0 \cosh(kx) - \frac{1}{2k} \int^x v(x') \sinh(kx - kx')dx'$, where $\int^x f(x')dx'$ stands for the primitive of f. Imposing to θ an assigned value at $x = +0.5$, a well-determined value for A_0 follows. Therefore to every given $v(x)$ a unique θ corresponds, *i.e.* there exists an operator $-v(x) \to \left((D^2 - k^2)^{-1}(-v) \right)(x) \equiv \theta(x)$. If $v(x) = A_i \cosh(\lambda_i x)$ then $\theta(x) = A_0 \cosh(kx) - A_i \frac{\cosh(\lambda_i x)}{\lambda_i^2 - k^2}$. In the class of functions of the form $\cosh(\lambda_i x)$ we must have $A_0 = 0$, and, so, $\theta(x) = B_i \cosh \lambda_i x$, where $B_i = -A_i(\lambda_i^2 - k^2)^{-1}$.

This method of inverse operators, first more or less heuristically applied in [Geo77], is appropriate to difficult cases. By means of it, additional boundary conditions are deduced by integrating the equations. We shall make use of it every time no other method to derive boundary conditions is available.

Example 3.4.4. Suppose that v and θ are even functions while the function ϕ is odd. The case when v and θ are odd and ϕ is even can be treated in a similar way.

In Remark 3.3.3 from $(3.3.1) - (3.3.3)$ we deduced equation $(3.3.8)$, *i.e.* $A_5'v = 0$, the corresponding characteristic equation of which has the form $(3.3.9)$ $f(\lambda) = 0$. Writing $\lambda^2 - k^2 = \mu$, equation $(3.3.9)$ becomes

$$\mu^4 + k^2 R\mu - k^4 REl = 0. \tag{3.3.9'}$$

Then $\lambda_1 = \sqrt{\mu_1 + k^2}$, $\lambda_2 = \sqrt{\mu_2 + k^2}$, $\lambda_3 = \sqrt{\mu_3 + k^2}$, $\lambda_4 = \sqrt{\mu_4 + k^2}$, $\lambda_5 = -\lambda_1$, $\lambda_6 = -\lambda_2$, $\lambda_7 = -\lambda_3$, $\lambda_8 = -\lambda_4$, are the eight solutions of $(3.3.9)$. The closed form solutions depend on the multiplicity of the roots λ_i of the characteristic equation (3.3.9), therefore on the tree parameters k, R, El occurring in it. The multiplicity of some λ_i can be greater than 1 in two situations: a) the corresponding μ_i is multiple; b) $\mu_i = -k^2$. Case b) occurs only if $k^4 = R(1+El)$. In this section we are interested only in the case when λ_i, λ_j are mutually disjoint, hence we suppose that $k, R, El > 0$, $k^4 \neq R(1+El)$.

Then the closed-form solution of (3.3.8) is given by the following theorem. This theorem and Lemmas 3.4.2, 3.4.3, 3.4.4, 3.4.5, are immediate. We presented them because, by analogy, they make clear more complicated situations.

Theorem 3.4.4. *If $k, R, El > 0$, $k^4 \neq R(1 + El)$, Equation (3.3.8) has the general even solution of the form*

$$v = \sum_{i=1}^{4} A_i ch \lambda_i x, \tag{3.4.2}$$

where λ_i and $-\lambda_i$ $i = 1, \ldots, 4$ are all the roots of (3.3.9) and A_i are arbitrary constants.

Lemma 3.4.1. *Equation $(3.3.9')$ has no multiple solution.*

Proof. The Viète relations for $(3.3.9')$ read

$$\mu_1 + \mu_2 + \mu_3 + \mu_4 = 0, \quad \mu_1\mu_2 + \mu_1\mu_3 + \mu_1\mu_4 + \mu_2\mu_3 + \mu_2\mu_4 + \mu_3\mu_4 = 0,$$

$$\mu_1\mu_2\mu_3 + \mu_1\mu_2\mu_4 + \mu_1\mu_3\mu_4 + \mu_2\mu_3\mu_4 = -k^2 R, \quad \mu_1\mu_2\mu_3\mu_4 = -k^4 REl. \tag{3.3.9'}^*$$

1) Suppose that $\mu_1 = \mu_2 = \mu_3 = \mu_4 = \eta$. From $(3.3.9')_1^*$ it follows $\eta = 0$ while from $(3.3.9')_3^*$ we have $\eta = 0$ only for $k = 0$, or $R = 0$, which contradicts our assumption that $k, R \neq 0$. It follows that assumption 1) does not hold.

2) Assume that $\mu_1 = \mu_2 = \mu_3 = \eta$; from $(3.3.9')_1^*$ and $(3.3.9')_2^*$ we get $\eta = \mu_4 = 0$, which shows that assumption 2) does not hold either.

3) Let us now suppose that $\mu_1 = \mu_2 = \eta$; from $(3.3.9')_1^*$ we have $-2\eta = \mu_3 + \mu_4$ and from $(3.3.9')_2^*$ we get $\mu_3\mu_4 = 3\eta^2$, so $(3.3.9')_4^*$ reads $3\eta^4 = -k^2 REl < 0$ contradicting the assumption $R, El > 0$. It follows that case 3) cannot occur; this proves Lemma 3.4.1.

Lemma 3.4.2. *The functions $x \to e^{\lambda_i x}$ defined on $[-0.5, 0.5]$, $i = 1, \ldots, 8$, are linearly independent.*

Proof. Denote by $| a_{ij} |$ the determinant corresponding to some matrix (a_{ij}). Then the Wronsky determinant W corresponding to the set $e^{\lambda_i x}$, $i = 1, \ldots, 8$ reads

$$W = | e^{\lambda_j x} \lambda_j^{i-1} | = e^{(\lambda_1 + \ldots \lambda_8)x} | \lambda_j^{i-1} | = \Pi_{i>j}(\lambda_i - \lambda_j)$$

and it is non-vanishing since $\lambda_i \neq \lambda_j$ for $i \neq j$, whence Lemma 3.4.2.

Lemma 3.4.3. *The functions* $x \to \cosh(\lambda_i x)$ $i = 1, \ldots, 4$, *and* $x \to \sinh(\lambda_i x)$ *defined on* $[-0.5, 0.5]$ *where* $\cosh(\lambda_i x) = \frac{e^{\lambda_i x} + e^{-\lambda_i x}}{2}$ *and* $\sinh(\lambda_i x) = \frac{e^{\lambda_i x} - e^{-\lambda_i x}}{2}$, *are linearly independent.*

Proof. Let us add the first column from W to the second one and divide the resulting elements of the second column by 2. Then let us add the obtained second column to the first one. Next apply the same procedure to the third and fourth columns and so on. As a result we get that the Wronsky determinant corresponding to the set $e^{\lambda_i x}$, $i = 1, \ldots, 8$ is equal to the Wronsky determinant corresponding to the set $\{\sinh(\lambda_1 x), \cosh(\lambda_1 x), \ldots, \sinh(\lambda_4 x), \cosh(\lambda_4 x)\}$ multiplied by 2^4, whence Lemma 3.4.3.

Lemma 3.4.4. *The functions* $x \to sh\lambda_i x$, $i = 1, \ldots, 4$ *defined on* $[-0.5, 0.5]$ *are linearly independent.*

Indeed, if $\sinh(\lambda_i x)$, $i = 1, \ldots, 4$ would be a linearly dependent set, then would exist four constants α_i, $i = 1, \ldots, 4$ not all vanishing such that $\sum_{i=1}^{4} \alpha_i \sinh(\lambda_i x) = 0$, where 0 stands for the null function on $[-0.5, 0.5]$. But then, for $\beta_i = 0$, $i = 1, \ldots, 4$ it is also true that $\sum_{i=1}^{4} (\alpha_i \sinh(\lambda_i x) + \beta_i \cosh(\lambda_i x)) = 0$, which contradicts Lemma 3.4.3, whence Lemma 3.4.4.

Similarly we have,

Lemma 3.4.5. *The functions* $x \to \cosh(\lambda_i x)$, $i = 1, \ldots, 4$ *defined on* $[-0.5, 0.5]$ *are linearly independent.*

By Lemma 3.4.1 the general solution of (3.3.8) is of the form $v = \sum_{i=1}^{8} a_i e^{\lambda_i x}$ and, by Lemma 3.3.6, it can be written as $v = \sum_{i=1}^{4} (A_i \cosh(\lambda_i x) + C_i \sinh(\lambda_i x))$. But v is assumed to be an even function, therefore for every x we have $\sum_{i=1}^{4} (A_i \cosh(\lambda_i x) + C_i \sinh(\lambda_i x)) = \sum_{i=1}^{4} (A_i \cosh(\lambda_i x) - C_i \sinh(\lambda_i x))$, whence $2 \sum_{i=1}^{4} (C_i \sinh(\lambda_i x)) = 0$, and by Lemma 3.3.6'' we obtain $C_i = 0$, $i = 1, \ldots, 4$. Consequently, the general solution of the problem (3.3.8), (3.3.7) has the form asserted by Theorem 3.4.4. This concludes its proof.

Remark 3.4.3. Eliminating v and ϕ between $(3.3.1) - (3.3.3)$, we obtain for θ equation (3.3.8) too. Similarly, it follows that the same equation is satisfied by ϕ. Hence

$$v = \sum_{i=1}^{4} A_i \cosh(\lambda_i x), \quad \theta = \sum_{i=1}^{4} B_i \cosh(\lambda_i x), \quad \phi = \sum_{i=1}^{4} C_i \sinh(\lambda_i x). \qquad (3.4.2')$$

In addition, by (3.3.1) and (3.3.3), it follows that $B_i = A_i (k^2 - \lambda_i^2)^{-1}$ and $C_i = \lambda_i (k^2 - \lambda_i^2)^{-2}$.

Finally, we consider equation (3.3.8) and the boundary conditions (3.3.7), implying

$$\sum_{i=1}^{4} A_i \cosh(\frac{\lambda_i}{2}) = \sum_{i=1}^{4} B_i \cosh(\frac{\lambda_i}{2}) = \sum_{i=1}^{4} A_i \lambda_i \sinh(\frac{\lambda_i}{2}) = \sum_{i=1}^{4} C_i \sinh(\frac{\lambda_i}{2}) = 0,$$

which expressed in terms of v (namely A_i), become

$$\sum_{i=1}^{4} A_i \cosh(\frac{\lambda_i}{2}) = \sum_{i=1}^{4} \frac{A_i \cosh(\frac{\lambda_i}{2})}{k^2 - \lambda_i^2} = \sum_{i=1}^{4} A_i \lambda_i \sinh(\frac{\lambda_i}{2}) = \sum_{i=1}^{4} \frac{A_i \lambda_i \sinh(\frac{\lambda_i}{2})}{(k^2 - \lambda_i^2)^2} = 0.$$

$$(3.4.3)$$

Then, the requirement that A_i are not all vanishing leads to the following secular equation

$$\begin{vmatrix} \cosh \frac{\lambda_1}{2} & \cosh \frac{\lambda_2}{2} & \cosh \frac{\lambda_3}{2} & \cosh \frac{\lambda_4}{2} \\ \frac{1}{k^2 - \lambda_1^2} \cosh \frac{\lambda_1}{2} & \frac{1}{k^2 - \lambda_2^2} \cosh \frac{\lambda_2}{2} & \frac{1}{k^2 - \lambda_3^2} \cosh \frac{\lambda_3}{2} & \frac{1}{k^2 - \lambda_4^2} \cosh \frac{\lambda_4}{2} \\ \lambda_1 \sinh \frac{\lambda_1}{2} & \lambda_2 \sinh \frac{\lambda_2}{2} & \lambda_3 \sinh \frac{\lambda_3}{2} & \lambda_4 \sinh \frac{\lambda_4}{2} \\ \frac{\lambda_1}{(k^2 - \lambda_1^2)^2} \sinh \frac{\lambda_1}{2} & \frac{\lambda_2}{(k^2 - \lambda_2^2)^2} \sinh \frac{\lambda_2}{2} & \frac{\lambda_3}{(k^2 - \lambda_3^2)^2} \sinh \frac{\lambda_3}{2} & \frac{\lambda_4}{(k^2 - \lambda_4^2)^2} \sinh \frac{\lambda_4}{2} \end{vmatrix} = 0, \quad (3.4.3')$$

which can also be written as

$$[AB(a+b) + CD(c+d)](a-b)(c-d) + [BC(b+c) + AD(a+d)](b-c)(d-a)$$

$$+[BD(b+d) + AC(a+c)](d-b)(c-a) = 0, \qquad (3.4.3'')$$

where $A = \lambda_1 \tanh(\frac{\lambda_1}{2})$, $B = \lambda_2 \tanh(\frac{\lambda_2}{2})$, $C = \lambda_3 \tanh(\frac{\lambda_3}{2})$, $D = \lambda_4 \tanh(\frac{\lambda_4}{2})$, $a = \frac{1}{k^2 - \lambda_1^2}$, $b = \frac{1}{k^2 - \lambda_2^2}$, $c = \frac{1}{k^2 - \lambda_3^2}$, $d = \frac{1}{k^2 - \lambda_4^2}$. Equation (3.4.3'') must be solved together with equation (3.3.9'); for El fixed this yields the neutral curve $R = R(k)$ and the linear instability criterion

Theorem 3.4.5. [Geo77] *For $k, R, El > 0$, $k^4 \neq R(1 + El)$, the mechanical equilibrium is linearly unstable to normal mode perturbations satisfying (3.3.8), (3.4.3) if $R > R^*$, where $R^* = R^*(k, El)$ is the neutral curve given by (3.4.3'), (3.3.9').*

The complete treatment of our fourth eigenvalue problem assumes also that the odd solutions shall be analyzed. The calculations can be drastically simplified if there exists the possibility of solving (3.3.9') like in the case $El = 0$. This means that no electrophoretic force is acting, therefore in (3.3.1) − (3.3.3), (3.3.7) we must take $\phi = 0$ and equation (3.3.3) is not to be taken into account (it expresses an electrical law).

Example 3.4.5. Let $El = 0$. Then eliminating v between (3.3.1) and (3.3.2) we obtain the classical Bénard equation

$$(D^2 - k^2)^3 \theta = -k^2 R \theta, \quad x \in (-0.5, 0.5) \qquad (3.3.8')$$

while the conditions (3.3.7), where the expression of v from (3.3.1) is taken into account, read

$$\theta = (D^2 - k^2)\theta = D(D^2 - k^2)\theta = 0, \quad x = \pm 0.5. \qquad (3.3.7')$$

Remark 3.4.4. In fact, $(3.3.8')$ can be obtained from equation $(3.3.8)$ in θ, written for $El = 0$ in the form $(D^2 - k^2)[(D^2 - k^2)\theta + k^2 R\theta] = 0$ and taking into account that $(3.3.7')$ yields the boundary conditions $(D^2 - k^2)\theta + k^2 R\theta = 0$ at $x = \pm 0.5$. Denoting $\Theta = (D^2 - k^2)\theta + k^2 R\theta$, this last problem reads $(D^2 - k^2)\Theta = 0$, $\Theta = 0$, at $x = \pm 0.5$, which has only the null solution, which is just $(3.3.8')$. It follows that the operator in equation $(3.3.8)$ written in θ decomposes into $(D^2 - k^2) \cdot [D^2 - k^2 + k^2 R]$ where by the above, the operator $D^2 - k^2$ can be removed. Its presence was necessary in the case $El \neq 0$, when Φ was to be eliminated between $(3.3.2)$ and $(3.3.3)$. The problem $(3.3.8')$, $(3.3.7')$ has been treated in [DiP61] and in a more general case in [GeoPo]. Here we shall present it as a particular case of $(3.4.3'')$, $(3.3.9')$. In this case $(3.3.9')$ becomes

$$\mu^3 + k^2 R = 0 \qquad\qquad (3.3.9'')$$

whence $\mu_1 = -\sqrt[3]{k^2 R}$, $\mu_2 = -\epsilon\sqrt[3]{k^2 R}$, $\mu_3 = -\epsilon'\sqrt[3]{k^2 R}$, where $\epsilon = \frac{-1+i\sqrt{3}}{2}$ and $\epsilon' = \frac{-1-i\sqrt{3}}{2}$. Correspondingly, $\lambda_1 = \sqrt{k^2 + \mu_1}$, $\lambda_2 = \sqrt{k^2 + \mu_2}$, $\lambda_3 = \sqrt{k^2 + \mu_3}$, $\lambda_4 = -\lambda_1$, $\lambda_5 = -\lambda_2$ $\lambda_6 = -\lambda_3$.

Further on we consider only the case $R \neq k^4$, corresponding to mutually distinct λ_i. Then, by Theorem 3.4.4, the general solution of $(3.3.8')$ is $\theta = \sum_{i=1}^{3} E_i \cosh(\lambda_i x)$ and must satisfy $(3.3.7')$. Thus, instead of $(3.4.3')$, the determinant obtained from $(3.4.3')$ by neglecting the last row and column must vanish. Then the relation analogous to $(3.4.3'')$, providing the neutral curve and the linear instability criterion, (mainly following [Geo77]) reads

$$[\lambda_3(1 - \epsilon') + \lambda_2(\epsilon - 1) + \lambda_1(\epsilon' - \epsilon)]\sinh\frac{(\lambda_1 + \lambda_2 + \lambda_3)}{2}$$

$$+[\lambda_3(1 - \epsilon') + \lambda_2(\epsilon - 1) - \lambda_1(\epsilon' - \epsilon)]\sinh\frac{(-\lambda_1 + \lambda_2 + \lambda_3)}{2}$$

$$+[\lambda_3(1 - \epsilon') - \lambda_2(\epsilon - 1) + \lambda_1(\epsilon' - \epsilon)]\sinh\frac{(\lambda_1 - \lambda_2 + \lambda_3)}{2} \qquad (3.4.3''')$$

$$+[-\lambda_3(1 - \epsilon') + \lambda_2(\epsilon - 1) + \lambda_1(\epsilon' - \epsilon)]\sinh\frac{(\lambda_1 + \lambda_2 - \lambda_3)}{2} = 0.$$

By means of the same method, in Chapter 7 eleven other thermal convections were minutely investigated, pointing out the bifurcation sets for the characteristic manifolds.

Remark 3.4.5. Denote by $\mathcal{D}$ the characteristic determinant for the eigenvalue problem $(3.3.8') - (3.3.7')$, whose general solution is $\theta = \sum_{i=1}^{6} \gamma_i e^{\lambda_i x}$, by $\mathcal{D}_e$ the characteristic determinant corresponding to the general even solution $\theta = \sum_{i=1}^{3} E_i \cosh(\lambda_i x)$, and by $\mathcal{D}_o$ the characteristic determinant corresponding to the general odd solution $\theta = \sum_{i=1}^{3} F_i \sinh(\lambda_i x)$. It can be shown that $\mathcal{D} = \mathcal{D}_e \mathcal{D}_o$. It follows that Theorem 3.4.4 reduces by half the computations. On the other hand, for the even solution the characteristic equation reads $\mathcal{D}_e = 0$ (which is expressed as $(3.4.3''')$) while for odd solutions the characteristic equation is $\mathcal{D}_o = 0$. This last assertion will be particularly useful in dealing with the odd solutions of our fourth eigenvalue problem.

3.4.3.2 *First variational principle*

By Remark 3.4.3, the function θ satisfies the same equation (3.3.8) as v, *i.e.*

$$(D^2 - k^2)^4\theta + k^2 R(D^2 - k^2)\theta - k^4 REl\theta = 0, \quad x \in (-0.5, 0.5); \qquad (3.4.4)$$

from (3.3.1) we deduce that $D^2\theta(\pm 0.5) = 0$. Differentiating (3.3.1) and taking into account that $Dv(\pm 0.5) = 0$ we get $[D(D^2 - k^2)\theta](\pm 0.5) = 0$. Finally, differentiating (3.3.2) and replacing v by its expression given by (3.3.1) and $D^2\phi$ by the expression $k^2\phi - D\theta$ given by (3.3.3), we obtain, taking into account that $\phi(\pm 0.5) = 0$, $\{[D(D^2 - k^2)^3 + k^2 RD]\theta\}(\pm 0.5) = 0$. Consequently, condition (3.3.7), written in terms of θ are

$$\theta = D^2\theta = D(D^2 - k^2)\theta = [D(D^2 - k^2)^3 + k^2 RD]\theta = 0, \quad \text{at} \quad x = \pm 0.5. \quad (3.4.5)$$

If (3.4.4) reads $A_5\theta = 0$, where $\mathcal{D}(A_5) = \{\theta \in C^\infty(-0.5, 0.5) \cap C^7[-0.5, 0.5] \mid \theta \text{ satisfies } (3.4.5)\}$, then the form of A_5^* is equal to that of A_5, *i.e.* A_5 is symmetric, but not selfadjoint, because the boundary conditions defining A_5^* are

$$D\theta^* = D^3\theta^* = (D^2 + 3k^2)\theta^* = (D^6 + k^2 D^4 - 2k^2)\theta^* = 0, \quad \text{at} \quad x = \pm 0.5.$$

In this way, the eigenvalue problem $(3.3.1) - (3.3.3), (3.3.7)$ is equivalent to the eigenvalue problem (3.4.4), (3.4.5) which can be written as

$$A_5\theta = 0, \qquad \theta \in \mathcal{D}_5, \qquad (3.4.6)$$

where $A_5 : \mathcal{D}_5 \to C^\infty[-0.5, 0.5]$, $\mathcal{D}_5 = \{\theta \in C^\infty[-0.5, 0.5] \mid \theta \text{ satisfies } (3.4.5)\}$.

Theorem 3.4.6. *The problem* (3.4.6) *is equivalent to the problem*

$$A_5^2\theta = 0, \qquad x \in (-0.5, 0.5) \qquad (3.4.7)$$

$$\begin{aligned}
\theta &= D^2\theta = D(D^2 - k^2)\theta = [D(D^2 - k^2)^3 + k^2 RD]\theta = A_5\theta \\
&= D(A_5\theta) = D^2(A_5\theta) = D^3(A_5\theta) = 0,
\end{aligned} \qquad (3.4.8)$$

at $x = \pm 0.5$. Indeed, denoting $A_5\theta = z$, from (3.4.7) and (3.4.8) we have

$$A_5 z = 0, \qquad (3.4.7')$$

$$z = Dz = D^2 z = D^3 z = 0, \quad \text{at} \quad x = \pm 0.5. \qquad (3.4.8')$$

By Theorem 3.4.4, we can write $z = \sum_{i=1}^{4} C_i \cosh(\lambda_i x)$. Imposing to it condition $(3.4.8')$ we obtain that C_i must satisfy a system whose determinant is just the Wronsky determinant corresponding to $\cosh(\lambda_i x)$, $i = 1, \ldots, 4$, at $x = 0.5$. By Lemma 3.4.5, this determinant is non-vanishing at any x. Whence $C_i = 0$, $i = 1, \ldots, 4$ consequently $z \equiv 0$, *i.e.* $A_5\theta = 0$. In other words, all solutions of (3.4.7), (3.4.8) satisfy (3.4.6). Hence Theorem 3.4.6 is proved.

By Lemma 3.3.2, it follows that the solution of (3.4.4), (3.4.5) satisfies the conditions $D^l z = 0$, at $x = \pm 0.5$ for every integer l. Hence, Theorem 3.4.6 can be written as

Theorem 3.4.6′. *The problem* (3.4.6) *is equivalent to the problem* (3.4.7), (3.4.8), (3.4.9), *where*

$$D^4(A_5\theta) = D^6(A_5\theta) = 0, \quad \text{at} \quad x = \pm 0.5. \tag{3.4.9}$$

Define the operator $A_4 : \mathcal{D}_4 \to C^\infty[-0.5, 0.5]$, where $A_4\theta = A_5^2\theta$ and

$$\mathcal{D}_4 = \{\theta \in C^\infty[-0.5, 0.5] \mid \theta \text{ satisfies } (3.4.8) \text{ and } (3.4.9)\}.$$

Then, by Theorem 3.4.6′, $A_5\theta = 0$ if and only if $A_4\theta = 0$. By simple computations we have

$$(A_4\theta, \theta^*) = \int_{-0.5}^{0.5} (A_4\theta)\theta^* dx = \int_{-0.5}^{0.5} (A_5^2\theta)\theta^* dx = \int_{-0.5}^{0.5} (A_5\theta)A_5\theta^* dx,$$

which means that the operator A_4 is selfadjoint and positive definite. Let

$$\tilde{\mathcal{D}}_4 = \{\theta \in C^\infty[-0.5, 0.5] \mid \theta \text{ satisfies } (3.4.5) \text{ and } A_5\theta = D^4 A_5\theta = 0 \text{ at } x = \pm 0.5\}$$

and denote by $\tilde{A}_4$ an extension of A_4 to $\tilde{\mathcal{D}}_4$. We can define the functional $j_4 : \tilde{\mathcal{D}}_4 \to R$, $j_4\theta = (\tilde{A}_4\theta, \theta) = \int_{-0.5}^{0.5}(A_5\theta)^2 dx$.

Obviously, if θ is a solution of $A_4\theta = 0$, it makes j_4 stationary. Conversely, $\delta j_4(\theta) = 0$ implies that $A_5^2\theta = 0$ while $D^2 A_5\theta = D^6 A_5\theta = 0$ at $x = \pm 0.5$ appear as natural conditions. Summing up we got (see also Theorem 3.2.2)

Theorem 3.4.7. $A_4\theta = 0$ *if and only if* $\delta j_4(\theta) = 0$.

3.4.3.3   *Second variational principle*

Consider now the system (3.3.2), (3.3.20), (3.3.1), (3.4.4) and the boundary conditions (3.3.7). Multiplying (3.3.2) by v, (3.3.20) by $-k^2 REl\phi$, (3.3.1) by $-k^2 R(1 + El)\theta$, and (3.4.4) by θ, summing and integrating over $[-0.5, 0.5]$ we get

$$\begin{aligned}
j_4' =\,& I_2^2(v) + 2k^2 I_1^2(v) + k^4 I_0^2(v) - k^2 R(1 + El)\int_{-0.5}^{0.5} \theta v dx \\
& - k^2 REl\int_{-0.5}^{0.5} vD\phi dx + k^2 R(1 + El)I_1^2(\theta) + k^4 R(1 + El)I_0^2(\theta) \\
& - k^2 R(1 + El)\int_{-0.5}^{0.5} \theta v dx - k^2 REl[D\theta D\phi]_{-0.5}^{0.5} - k^2 RElI_2^2(\phi) - 2k^4 RElI_1^2(\phi) \\
& - k^6 RElI_0^2(\phi) + k^2 REl\int_{-0.5}^{0.5} \phi Dv dx + k^2 REl[D\theta D\phi]_{-0.5}^{0.5} + I_4^2(\theta) \\
& + 4k^2 I_3^2(\theta) + 6k^4 I_2^2(\theta) + (4k^6 - k^2 R)I_1^2(\theta) + [k^8 - k^4 R(1 + El)]I_0^2(\theta) = 0,
\end{aligned}$$

where $I_j^2(f) = \int_{-0.5}^{0.5}(D^j f)^2 dx$. The terms in $[D\theta D\phi]_{-0.5}^{0.5}$ are canceled out, nevertheless if we vary the functional j_4' we shall see that they introduce natural conditions. $\delta j_4' = 0$ implies that θ, v, ϕ satisfy (3.3.2), (3.3.20) and the equation obtained adding to (3.4.4) equation (3.3.1) multiplied by $-k^2R(1+El)\theta$. The natural conditions can arise only from the terms

$$-k^2REl\delta I_2^2 = -k^2REl[D^2\phi\delta D\phi]_{-0.5}^{0.5} - k^2REl\int_{-0.5}^{0.5}D^4\phi\delta\phi dx$$

and

$$\delta I_4^2(\theta) + 4k^2\delta I_3^2(\theta) = [(D^6\theta - 3k^2D^4\theta)\delta D\theta]_{-0.5}^{0.5} + \int_{-0.5}^{0.5}(D^8\theta - 4k^2D^6\theta)\delta\theta dx.$$

But from (3.3.2) we have $D^6\theta - 3k^2D^4\theta = -k^2RElD\phi$ at $x = \pm0.5$ and (3.3.3) implies $D^2\phi = -D\theta$ at $x = \pm0.5$, whence

$$\delta(-k^2RElI_2^2(\phi) + I_4^2(\theta) + 4k^2I_3^2(\theta)) = 2k^2REl[D\theta\delta D\phi - D\phi\delta D\theta]_{-0.5}^{0.5}$$
$$-2k^2REl\int_{-0.5}^{0.5}D^4\phi\delta\phi dx + 2\int_{-0.5}^{0.5}(D^8\theta - 4k^2D^6\theta)\delta\theta dx.$$

On the other hand, from $-k^2REl\int_{-0.5}^{0.5}\phi(3.3.20)dx$ we obtain $-k^2REl[D\theta D\phi]_{-0.5}^{0.5}$ and from $\int_{-0.5}^{0.5}\theta(3.4.4)dx$ we have $k^2REl[D\theta D\phi]_{-0.5}^{0.5}$. This suggests that we can avoid natural conditions by doing the above operations, but instead of adding (3.4.4) multiplied by θ we must subtract (3.4.4) multiplied by θ. In other words, consider the equations for $x \in (-0.5, 0.5)$

$$(D^2 - k^2)^2v - k^2R(1 + El)\theta - k^2RElD\phi = 0, \qquad (3.3.2)$$
$$-k^2REl(D^2 - k^2)^2\phi + k^2RElDv = 0, \qquad (3.3.20')$$
$$-(D^2 - k^2)^4\theta + k^4REl\theta - k^2R(2 + El)(D^2 - k^2)\theta - k^2R(1 + El)v = 0, \quad (3.4.10)$$

and the boundary conditions

$$v = Dv = \phi = \theta = D^2\theta = D(D^2 - k^2)\theta = D^2\phi + D\theta$$
$$= D^6\theta - 3k^2D^4\theta + k^2RElD\phi = 0, \quad \text{at} \quad x = \pm0.5. \qquad (3.4.11)$$

Equation (3.4.10) is obtained by multiplying (3.4.4) by -1 and adding with (3.3.1) multiplied by $-k^2R(1+El)$. Defining now the operator $A_4'' : \mathcal{D}_4'' \to [C^\infty[-0.5, 0.5]]^3$, where

$$\mathcal{D}_4'' = \{\mathbf{U} \equiv (v, \phi, \theta) \in [C^\infty[-0.5, 0.5]]^3 \mid v, \phi, \theta \text{ satisfy } (3.4.11)\},$$

the problem (3.3.2), (3.3.20'), (3.4.10), (3.4.11) reads

$$A_4''\mathbf{U} = 0. \qquad (3.4.12)$$

Putting $\tilde{\mathcal{D}}_4'' = \{\mathbf{U} \equiv (v, \phi, \theta) \in [C^\infty[-0.5, 0.5]]^3 \mid v, \phi, \theta \text{ satisfy the first six of}$ (3.4.11) at $x = \pm0.5\}$ and denoting by $\tilde{A}_4''$ an extension of $\mathcal{A}_4''$ to $\tilde{\mathcal{D}}_4''$ we can define the functional $j_4'' : \tilde{\mathcal{D}}_4'' \to \mathbf{R}$, $j_4''(\mathbf{U}) = (\tilde{A}_4''\mathbf{U}, \mathbf{U})$, such that

$$j_4''(\mathbf{U}) = I_2^2(v) + 2k^2I_1^2(v) + k^4I_0^2(v) - 2k^2R(1+El)\int_{-0.5}^{0.5}\theta v dx - 2k^2REl\int_{-0.5}^{0.5}vD\phi dx$$
$$+k^2R(1 + El)I_1^2(\theta) + k^4R(1 + El)I_0^2(\theta) - 4k^2REl(D\theta D\phi)(0.5)$$
$$-k^2RElI_2^2(\phi) - 2k^4RElI_1^2(\phi) - k^6RElI_0^2(\phi)$$
$$-I_4^2(\theta) - 4k^2I_3^2(\theta) - 6k^4I_2^2(\theta) - (4k^6 - k^2R)I_1^2(\theta) - [k^8 - k^4R(1 + El)]I_0^2(\theta) = 0.$$

The stationarity of j_4'' in $\tilde{\mathcal{D}}_4''$ implies the system (3.3.2), (3.3.20'), (3.4.10) and the last two boundary conditions (3.4.11) being satisfied. Therefore, these last conditions are not natural ones. It follows

Theorem 3.4.8. $\delta j_4''(\mathbf{U}) = 0$ *if and only if* $\mathcal{A}_4''\mathbf{U} = 0$.

3.4.3.4   *Third variational principle*

Consider now the operator $\mathcal{A}_4'''$, which is the restriction of $\mathcal{A}_4''$ to the space of functions (v, ϕ, θ) of $\mathcal{D}_4''$ which, in addition, satisfy the boundary conditions

$$D^2[(D^2 - k^2)\theta + v] = D^3[(D^2 - k^2)\theta + v] = 0, \quad \text{at} \quad x = \pm 0.5. \tag{3.4.13}$$

Thus, the problem (3.3.2), (3.3.20'), (3.4.10), (3.4.11), (3.4.13), reads

$$\mathcal{A}_4'''\mathbf{U} = 0. \tag{3.4.14}$$

The following result holds:

Theorem 3.4.9. $\mathcal{A}_4'''\mathbf{U} = 0$ *if and only if* $\mathbf{U}$ *satisfies the problem* (3.3.1) − (3.3.3), (3.3.7).

Proof. Define the operator $\mathcal{A}_5' : \left[[C^\infty[-0.5, 0.5]]\right]^3 \to \left[[C^\infty[-0.5, 0.5]]\right]^3$ by

$$A_5'(\theta) = \{(D^2 - k^2)^4 + k^2 R(D^2 - k^2) - k^4 REl\}\theta.$$

Then equation (3.4.10) reads

$$-A_5'\theta - k^2 R(1 + El)[(D^2 - k^2)\theta + v] = 0. \tag{3.4.15}$$

Eliminating ϕ between (3.3.2) and (3.3.20') we get

$$-A_5'v - k^2 R(1 + El)(D^2 - k^2)[(D^2 - k^2)\theta + v] = 0. \tag{3.4.16}$$

Applying now to (3.4.15) the operator $-(D^2 - k^2) : C^\infty[-0.5, 0.5] \to C^\infty[-0.5, 0.5]$ and adding the result to (3.4.16) we obtain

$$A_5'[(D^2 - k^2)\theta + v] = 0. \tag{3.4.17}$$

Finally, applying to (3.4.15) the operator A_5' we have, taking into account (3.4.17),

$$A_5'^2\theta = 0. \tag{3.4.18}$$

Now, introducing in the relation (3.4.15) (valid on $[-0.5, 0.5]$) the conditions (3.4.11) and (3.4.13), we get $A_5'\theta = D(A_5'\theta) = D^2(A_5'\theta) = D^3(A_5'\theta) = 0$ at $x = \pm 0.5$. A reasoning similar to that from the proof of Theorem 3.4.6 implies that $A_5'\theta = 0$. Then from (3.4.15) it follows that equation (3.3.1) holds and from (3.4.16) that $A_5'v = 0$. Consequently, by Theorem 3.4.4,

$$v = \sum_{i=1}^{4} A_i \cosh(\lambda_i x), \quad \theta = \sum_{i=1}^{4} B_i \cosh(\lambda_i x). \tag{3.4.2''}$$

Eliminating v between (3.3.2) and (3.3.20') we get

$$A_5'\phi - k^2 R(1 + El)[(D^2 - k^2)\phi + D\theta] = 0. \tag{3.4.19}$$

Applying $(D^2 - k^2) : C^\infty[-0.5, 0.5] \to C^\infty[-0.5, 0.5]$ to (3.4.19) and subtracting the derivative of (3.4.15) from the result we have

$$A_5'[(D^2 - k^2)\phi + D\theta] = 0. \tag{3.4.20}$$

Finally, (3.4.19) and (3.4.20) imply $\mathcal{A}_5'^2(\phi) = 0$. The characteristic equation corresponding to this equation is $f^2(\lambda) = 0$; in other words, λ_i shall have now the multiplicity 2. Then, by a theorem analogous to Theorem 3.4.4, odd solutions of $\mathcal{A}_5'^2(\phi) = 0$ can be written as

$$\phi = \sum_{i=1}^{4} C_i \sinh(\lambda_i x) + x \sum_{i=1}^{4} D_i \cosh(\lambda_i x). \tag{3.4.21}$$

On the other hand, taking into account Theorem 3.4.4, from (3.4.20), for the odd solution, we have

$$(D^2 - k^2)\phi + D\theta = \sum_{i=1}^{4} E_i \sinh(\lambda_i x). \tag{3.4.22}$$

Introducing (3.4.2″) and (3.4.21) into (3.4.22) and taking into account the linear independence of $\sinh(\lambda_i x)$ and $x \cosh(\lambda_i x)$ we deduce $D_i = 0$. Hence $\mathcal{A}_5'\phi = 0$ and (3.4.19) implies (3.3.3). In this way the solution of (3.4.14) satisfies the problem (3.3.1) $-$ (3.3.3), (3.3.7). The converse implication is immediate, hence Theorem 3.4.9 is proved.

Obviously, $A_4'''\mathbf{U} = 0$ if and only if $\delta j_4'''(\mathbf{U}) = 0$, where j_4''' is the restriction of j_4'' to the subspace of the functions of $\mathcal{D}_4''$ satisfying (3.4.13), whence the second variational principle

Theorem 3.4.9′. $\delta j_4'''(\mathbf{U}) = 0$ *if and only if* $\mathbf{U}$ *is a solution of* (3.3.1) $-$ (3.3.3), (3.3.7).

We emphasize again that the stationary points of j_4''' are among the vector functions $(v, \phi, \theta) \in [C^\infty[-0.5, 0.5]]^3$ satisfying the first six boundary conditions (3.4.11) and (3.4.13) (the last two conditions (3.4.11) being natural ones).

The fine results in this section are due, among others, to the use of the boundary conditions derived by the inverse operators method, involved into (3.4.21) and Theorem 3.4.4.

3.4.3.5 *The best variational principle*

Consider the problem (3.3.1) $-$ (3.3.3), (3.3.7), this time taking into account its adjoint problem

$$(D^2 - k^2)\Theta = D\Phi + k^2 R(1 + El)V, \tag{3.4.23}$$

$$(D^2 - k^2)^2 V = -\Theta, \tag{3.4.24}$$

$$(D^2 - k^2)\Phi = -k^2 RElDV, \tag{3.4.25}$$

where Θ, V, Φ satisfy (3.3.7). It is interesting to note that in this case the operator and its adjoint have the same boundary conditions but not the same form. Writing the problem $(3.3.1) - (3.3.3)$, $(3.3.7)$ in the form $A\mathbf{U} = 0$ where $\mathbf{U} = (v, \theta, \phi)$, $A : \mathcal{D}_{11} \to [C^\infty[-0.5, 0.5]]^3$, $\mathcal{D}_{11} = \{\mathbf{U} \in [C^\infty[-0.5, 0.5]]^3 \mid \mathbf{U}\ \text{satisfies}\ (3.3.7)\}$ and the adjoint problem $(3.4.23) - (3.4.25)$, $(3.3.7)$ in the form $A^*\mathbf{U}^* = 0$, where $\mathbf{U}^* = (\Theta, V, \Phi)$, $A^* : \mathcal{D}_{11} \to [C^\infty[-0.5, 0.5]]^3$, define the functional $j_4^{IV} : \mathcal{D}_{11}^2 \to \mathbf{R}$, $j_4^{IV}(\mathbf{U}, \mathbf{U}^*) = (A\mathbf{U}, \mathbf{U}^*)$, *i.e.*

$$j_4^{IV}(\mathbf{U}, \mathbf{U}^*) = \int_{-0.5}^{0.5} [-D\theta D\Theta - k^2\theta\Theta + v\Theta + D^2vD^2V + 2k^2DvDV + k^4vV$$
$$- k^2R(1 + El)\theta V - k^2REl D\phi V - D\phi D\Phi - k^2\phi\Phi + \Phi D\theta] dx.$$

Obviously, if $A\mathbf{U} = 0$ and $A^*\mathbf{U}^* = 0$, then $j_4^{IV}(\mathbf{U}) = 0$. The first Fréchet variation of j_4^{IV} in $\mathcal{D}_{11}^2$ is $\delta j_4^{IV} = (A\mathbf{U}, \delta\mathbf{U}^*) + (\delta\mathbf{U}, A^*\mathbf{U}^*)$ whence

Theorem 3.4.10. $\delta j_4^{IV}(\mathbf{U}, \mathbf{U}^*) = 0$ *if and only if* $\mathbf{U}$ *and* $\mathbf{U}^*$ *are solutions of* $(3.3.1) - (3.3.3)$, $(3.3.7)$ *and* $(3.4.23) - (3.4.25)$, $(3.3.7)$ *respectively.*

Remark 3.4.6. The existence of $D\phi$ into $(3.3.2)$ and the lack of more than two boundary conditions for lower order derivatives of the solution of $(3.3.1) - (3.3.3)$, $(3.3.7)$, conferred to this problem a higher degree of difficulty; all the usual trials of establishing a variational principle led to the appearance of natural conditions.

The smoothness analysis from Section 3.3.1 allowed us to write for θ and v boundary conditions equivalent to $(3.3.7)$

$$v = Dv = (D^2 - k^2)^3v = [D(D^2 - k^2)^3 + k^2El D(D^2 - k^2)^2]v = 0;$$

$$\theta = D^2\theta = D(D^2 - k^2)\theta = [D(D^2 - k^2)^3 + k^2RD]\theta = 0, \quad \text{at} \quad x = \pm 0.5.$$

No more information about possible boundary conditions of lower-order for a single function (among θ, v, ϕ) was obtained from the very encouraging form of the general solution (θ, v, ϕ) of $(3.3.1) - (3.3.3)$, $(3.3.7)$ mainly due to the complicated connection between the coefficients of $\cosh(\lambda_i x)$ and $\sinh(\lambda_i x)$ occurring in this solution. This is the reason why we were led to a variational principle corresponding to a selfadjoint differential operator of order 16 and not of order 8.

Nevertheless, the existence of some boundary conditions connecting lower-order derivatives of θ and ϕ has been exploited to write in Section 3.4.3.4 the variational principle for $(3.3.1) - (3.3.3)$, $(3.3.7)$. In the principle from this section, we have had as few as possible products of different functions.

To conclude with, we presented a few tricks, procedures and methods to derive variational principles, more exactly, to associate with a boundary-value problem an equivalent variational problem. The interest in functionals containing as lower an order of differentiation as possible will be understood in the following sections devoted to the use of Fourier series in variational as well as in boundary-value problems.

3.5 Fourier series solutions for variational problems of Sections 3.3 and 3.4

For a wide class of fluid flows (*e.g.* fluids heated from below, Couette flow between rotating cylinders) the mathematical problem governing their linear stability is an eigenvalue problem for a non-selfadjoint linear differential equation or system with constant coefficients. As their order is usually higher than six, a standard direct or variational solution of this eigenvalue problem based on Fourier series is often very laborious. Indeed, there is a significant increase in the complexity of the Fourier coefficients as the order of differentiation is increased. This is why some additional unknown functions are introduced and, consequently, some additional equations occur, but the order of the initial equations is lowered. Then, in some cases, to this new eigenvalue problem, an equivalent variational formulation is attached. Known as the Chandrasekhar method, this approach was first applied to a hydrodynamic stability problem, namely to the classical Bénard problem by Pellew and Southwell [PeS]. Then, in the 50's, Chandrasekhar [Chan] solved by this method several problems in hydrodynamic and hydromagnetic stability theory. Subsequently his method was extensively used by other authors. In particular, it was applied to the four problems from Section 3.3.1 by Turnbull [Tur].

In this method one has to integrate some differential equations as a part of solution; besides this, in order to know an exact solution of the variational problem (solution represented by an expansion in a complete set of orthogonal functions, each of them satisfying the boundary conditions of the problem) one must solve an equation $det(A_{ij}) = 0$, where $det(A_{ij})$ is a determinant of an infinite order. So, in practice only approximate solutions are available.

DiPrima [DiP61] used an alternative method: his expansion functions do not satisfied all boundary conditions but only part of them. Imposing to the solution (*i.e.* to the series) to satisfy the rest of the boundary conditions some constraints are obtained, so the DiPrima's variational problem is an isoperimetric problem. In his approach the least eigenvalue λ is determined from an equation of the form $\sum_{n=1}^{\infty} A_n(B_n + \lambda C_n)^{-1} = 0$ where A_n, B_n, C_n are known constants. DiPrima's method gives an exact solution and reduced substantially the amount of computations. Nevertheless, no algorithm indicating how to split the equations exists.

Such a splitting in a stability problem of a non-Newtonian fluid flow was realized in [Geo73], [GeoPo]. The splitting depends in an essential way on the form of the equations and boundary conditions: in order to establish the variational formulation, a special care has to be taken to avoid the introduction of natural conditions.

The splitting of equations leads to the symmetrization of the operator defining the problem and to a square root operator in the associated variational problem. As a result, the order of differentiation is halved. This means that in order to simplify the numerical determination of the solution, first we must symmetrize the associated operator (not necessary by splitting) or to put the associated functional in a form as close to the square root as possible.

In this section, by closely following [Geo77], the functions occurring in the functionals from Sections 3.3, 3.4 shall be expressed as sums of converging Fourier series on total sets of expansion functions satisfying part of the boundary conditions imposed to the corresponding function itself. Requiring the series to satisfy the rest of the boundary conditions (which must be satisfied by the corresponding functions) we introduce some constraints. In this way, all our four eigenvalue problems are turned into isoperimetric problems. The results coincide with those obtained in [Geo77] by applying methods based on Fourier series directly to those eigenvalue problems.

We consider the case where θ and v are even functions and ϕ is an odd function. The treatment of the case where θ and v are odd functions and ϕ is an even function is similar.

The problem $(3.3.1) - (3.3.3)$, $(3.3.4)$

Case El $\neq$ 0. Consider the form $(3.3.12') - (3.3.4')$ (Section 3.3.3.1) of this problem. By the analysis of Appendix 5, the unknown functions u, v, θ admit on $[-0.5, 0.5]$ the following expansions:

$$u = \sum_{n=1}^{\infty} u_{2n-1} E_{2n-1}, \quad v = \sum_{n=1}^{\infty} v_{2n-1} E_{2n-1}, \quad \theta = \sum_{n=1}^{\infty} \theta_{2n-1} E_{2n-1},$$

$$Du = \sum_{n=1}^{\infty} [-(2n-1)\pi] u_{2n-1} F_{2n-1}, \quad Dv = \sum_{n=1}^{\infty} [-(2n-1)\pi] v_{2n-1} F_{2n-1},$$

$$D\theta = \sum_{n=1}^{\infty} [-(2n-1)\pi] \theta_{2n-1} F_{2n-1},$$

while on $(-0.5, 0.5)$ for the other unknown function we have

$$\phi = \sum_{n=1}^{\infty} \phi_{2n-1} F_{2n-1}, \quad D\phi = \sum_{n=1}^{\infty} [-(2n-1)\pi] \phi_{2n-1} E_{2n-1}.$$

Since ϕ must satisfy $(3.3.12')$ we obtain the constraint

$$\Gamma = \sum_{n=1}^{\infty} (-1)^{n+1} \phi_{2n-1} = 0. \tag{3.5.1}$$

Then the associated functional j_1 becomes

$$j_1(\mathbf{U}) = \sum_{n=1}^{\infty} \{ -2(2n-1)^2 \pi^2 u_{2n-1}\theta_{2n-1} - 2k^2 u_{2n-1}\theta_{2n-1} + 2u_{2n-1}v_{2n-1}$$

$$- (2n-1)^2 \pi^2 v_{2n-1}^2 - k^2 R(1+El)\theta_{2n-1}^2 + 2k^2 REl(2n-1)\pi\theta_{2n-1}\phi_{2n-1}$$

$$+ k^2 REl(2n-1)^2 \pi^2 \phi_{2n-1}^2 + k^4 REl\phi_{2n-1}^2 \}.$$

Now impose on the variation of j_1 subject to the constraint (3.5.1) to vanish. This comes to impose

$$\frac{\partial(j_1 - \mu\Gamma)}{\partial u_{2n-1}} = \frac{\partial(j_1 - \mu\Gamma)}{\partial v_{2n-1}} = \frac{\partial(j_1 - \mu\Gamma)}{\partial\theta_{2n-1}} = \frac{\partial(j_1 - \mu\Gamma)}{\partial\phi_{2n-1}} = 0$$

where μ is a Lagrange multiplier, whence

$$2v_{2n-1} - 2A_n\theta_{2n-1} = 0, \quad 2u_{2n-1} - 2A_n v_{2n-1} = 0,$$

$$2k^2 R(1 + El)\theta_{2n-1} + 2k^2 REl(2n-1)\pi\phi_{2n-1} - 2A_n u_{2n-1} = 0,$$

$$2k^2 REl(2n-1)\pi\theta_{2n-1} + 2k^2 REl(2n-1)^2\pi^2\phi_{2n-1} + 2RElk^4\phi_{2n-1} - \mu(-1)^{n+1} = 0.$$

We denoted $A_n = (2n-1)^2\pi^2 + k^2$. Solving this system for u_{2n-1}, v_{2n-1}, θ_{2n-1} and ϕ_{2n-1} and introducing in (3.5.1) we have

$$\sum_{n=1}^{\infty} \frac{A_n^3 - k^2 R(1 + El)}{A_n^4 - R(k^2 A_n + k^4 El)} = 0, \tag{3.5.2}$$

which is a transcendental equation in $\mathbf{R}$ as a function of k^2 for given El. The curve $R_1 = R_1(k^2)$, where R_1 is the least solution corresponding to k^2, is the neutral curve in the (R, k) plane. The series in (3.5.2) is convergent like $(2n-1)^{-2}$. The convergence can be made like $(2n-1)^{-4}$ by using the formula $\frac{\pi^2}{8} = \sum_{n=1}^{\infty} \frac{1}{(2n-1)^2}$ (that is $\frac{1}{8} = \sum_{n=1}^{\infty} \frac{1}{A_n - k^2}$) so instead of (3.5.2) we have

$$\frac{1}{8} - \sum_{n=1}^{\infty} \frac{k^2 A_n^3 + k^2 RA_n + k^4 R}{[A_n^4 - R(k^2 A_n + k^4 El)](A_n - k^2)} = 0. \tag{3.5.2'}$$

The linear instability criterion reads $R > R_1$.

Case El=0. Instead of the functional j_1 we have now
$$j_1'(\mathbf{U}) = \int_{-0.5}^{0.5}\{-2D\theta Du - 2k^2\theta u + 2uv - (Dv)^2 - k^2 v^2 + k^2 R\theta^2\}dx,$$
$\mathbf{U} = (u, \theta, v)$ and $\delta j_1(\mathbf{U}) = 0$ if and only if $A_1'(\mathbf{U}) = 0$, where A_1' corresponds to the first three equations, $A_1' : \mathcal{D}_1' \to \left[C^\infty[-0.5, 0.5]\right]^3$ and $\mathcal{A}_1' = \{\mathbf{U} = (u, v, \theta) \in \left[C^\infty[-0.5, 0.5]\right]^3 \mid u, v, \theta \text{ satisfy } (3.3.4')_{1,2,3}\}$. Choose u, v, θ in the form (5.11), (5.12), *i.e.*

$$\theta = \frac{\theta_0}{2} + \sum_{n=1}^{\infty}\theta_{2n}E_{2n}(x), \quad v = \frac{v_0}{2} + \sum_{n=1}^{\infty}v_{2n}E_{2n}(x), \quad u = \frac{u_0}{2} + \sum_{n=1}^{\infty}u_{2n}E_{2n}(x),$$

$$D\theta = \sum_{n=1}^{\infty}(-2n\pi)\theta_{2n}F_{2n}(x), \quad Dv = \sum_{n=1}^{\infty}(-2n\pi)v_{2n}F_{2n}(x), \quad Du = \sum_{n=1}^{\infty}(-2n\pi)u_{2n}F_{2n}(x).$$

This choice introduces the constraints

$$\Gamma_1 = \frac{u_0}{2} + \sum_{n=1}^{\infty}u_{2n}(-1)^n\sqrt{2} = 0, \qquad \Gamma_2 = \frac{v_0}{2} + \sum_{n=1}^{\infty}v_{2n}(-1)^n\sqrt{2} = 0,$$

$$\Gamma_3 = \frac{\theta_0}{2} + \sum_{n=1}^{\infty} \theta_{2n}(-1)^n \sqrt{2} = 0.$$

Then the functional j_1' becomes

$$j_1'(\mathbf{U}) = \sum_{n=1}^{\infty} \{-2(2n\pi)^2 u_{2n}\theta_{2n} + 2u_{2n}v_{2n} - (2n\pi)^2 v_{2n-1}^2 - 2k^2 u_{2n}\theta_{2n} - k^2 v_{2n}^2 + k^2 R\theta_{2n}^2\}$$

$$+ \frac{u_o v_0}{2} - \frac{k^2 v_0^2}{2} + \frac{k^2 R\theta_0^2}{2} - \frac{k^2\theta_0 u_0}{2}.$$

Denoting the Lagrange multipliers by μ_1, μ_2 and μ_3, the stationarity of j_1' is equivalent to the requirement that the equalities

$$\frac{\partial(j_1' - \mu_1\Gamma_1 - \mu_2\Gamma_2 - \mu_3\Gamma_3)}{\partial u_{2n}} = \frac{\partial(j_1' - \mu_1\Gamma_1 - \mu_2\Gamma_2 - \mu_3\Gamma_3)}{\partial v_{2n}}$$

$$= \frac{\partial(j_1' - \mu_1\Gamma_1 - \mu_2\Gamma_2 - \mu_3\Gamma_3)}{\partial\theta_{2n}} = 0 \quad n = 0, 1, \ldots.$$

hold subject to the above-mentioned constraints. Solving this system with respect to the unknowns θ_{2n}, u_{2n}, v_{2n} and introducing their values into the constraints we obtain a system of three homogeneous linear equations in the unknowns μ_1, μ_2 and μ_3. Imposing to the determinant of this system to vanish (in order to have non-vanishing solutions (μ_1, μ_2 and μ_3)) we get, finally, the equation which yields R in terms of k^2

$$\begin{aligned}
&\left[\frac{k^4}{\Delta_0} + 2\sum_{n=1}^{\infty}\frac{B_n^2}{\Delta_n}\right]^3 + k^4 R^2\left[\frac{1}{\Delta_0} + 2\sum_{n=1}^{\infty}\frac{1}{\Delta_n}\right]^3 + k^2 R\left[\frac{k^2}{\Delta_0} + 2\sum_{n=1}^{\infty}\frac{B_n}{\Delta_n}\right]^3 \\
&- 3k^2 R\left[\frac{k^4}{\Delta_0} + 2\sum_{n=1}^{\infty}\frac{B_n^2}{\Delta_n}\right]\left[\frac{1}{\Delta_0} + 2\sum_{n=1}^{\infty}\frac{1}{\Delta_n}\right]\left[\frac{k^2}{\Delta_0} + 2\sum_{n=1}^{\infty}\frac{B_n}{\Delta_n}\right] = 0,
\end{aligned} \tag{3.5.2$''$}$$

where $\Delta_0 = k^2 R - k^6$, $\Delta_n = k^2 R - B_n^3$, $B_n = (2n\pi)^2 + k^2$. This secular equation provides the neutral curve and the linear instability criterion.

The problem $(3.3.1) - (3.3.3)$, $(3.3.5)$

By the above analysis, for the even functions vanishing at $x = \pm 0.5$, on $[-0.5, 0.5]$ we have

$$\theta = \sum_{n=1}^{\infty}\theta_{2n-1}E_{2n-1}(x), \quad v = \sum_{n=1}^{\infty}v_{2n-1}E_{2n-1}(x), \quad u = \sum_{n=1}^{\infty}u_{2n-1}E_{2n-1}(x),$$

$$D\theta = -\sum_{n=1}^{\infty}(2n-1)\pi)\theta_{2n-1}F_{2n-1}(x), \quad Dv = -\sum_{n=1}^{\infty}(2n-1)\pi v_{2n-1}F_{2n-1}(x),$$

$$Du = -\sum_{n=1}^{\infty}(2n-1)\pi u_{2n-1}F_{2n-1}(x).$$

The only boundary condition which remains to be satisfied is $Dv = 0$ at $x = \pm 0.5$. It is expressed as the constraint $\Gamma_1 = \sum_{n=1}^{\infty}(2n-1)(-1)^{n+1}v_{2n-1} = 0$. Introducing the above series into j_2 we get

$$j_2(\mathbf{U}) = \sum_{n=1}^{\infty}\{-2(2n-1)^2\pi^2 u_{2n-1}\theta_{2n-1} - 2k^2 u_{2n-1}\theta_{2n-1} + 2u_{2n-1}v_{2n-1} + k^4 v_{2n-1}^2$$

$$+2k^2(2n-1)^2\pi^2 v_{2n-1}^2 + (2n-1)^4\pi^4 v_{2n-1}^2 - k^2 R(2n-1)^2\pi^2\theta_{2n-1}^2 - k^4 R(1+El)\theta_{2n-1}^2\}.$$

If μ stands for the Lagrange multiplier, the stationarity of j_2 requires that

$$\frac{\partial(j_2 - \mu\Gamma_1)}{\partial u_{2n-1}} = \frac{\partial(j_2 - \mu\Gamma_1)}{\partial v_{2n-1}} = \frac{\partial(j_2 - \mu\Gamma_1)}{\partial\theta_{2n-1}} = 0.$$

These equations have θ_{2n-1}, u_{2n-1}, v_{2n-1} as unknowns. Introducing the expression of v_{2n-1} from them into $\Gamma_1 = 0$ we obtain the equation for R in terms of k^2

$$\sum_{n=1}^{\infty}\frac{A_n^2(A_n - k^2)}{A_n^4 - R(k^2 A_n + k^4 El)} = 0, \tag{3.5.3}$$

or, improving the convergence of this series by adding $\frac{1}{8} = \sum_{n=1}^{\infty}\frac{1}{A_n - k^2}$,

$$\frac{1}{8} + \sum_{n=1}^{\infty}\frac{-2k^2 A_n^3 + R(k^2 A_n + k^4 El) + k^4 A_n^2}{[A_n^4 - R(k^2 A_n + k^4 El)](A_n - k^2)} = 0. \tag{3.5.3'}$$

The problem $(3.3.1) - (3.3.3)$, $(3.3.6)$

In this case at $x = \pm 0.5$ we must have $v = Dv = D\Theta = \Phi = D\Phi = 0$. Accordingly, we take

$$\Theta = \frac{\theta_0}{2} + \sum_{n=1}^{\infty}\theta_{2n}E_{2n}(x), \qquad v = \frac{v_0}{2} + \sum_{n=1}^{\infty}v_{2n}E_{2n}(x),$$

$$\Phi = \sum_{n=1}^{\infty}\phi_{2n}F_{2n}(x), \qquad D\Theta = -\pi\sum_{n=1}^{\infty}2n\theta_{2n}F_{2n}(x),$$

$$Dv = -\pi\sum_{n=1}^{\infty}2nv_{2n}F_{2n}(x), \qquad D^2 v = -\pi^2\sum_{n=1}^{\infty}(2n)^2 v_{2n}E_{2n}(x),$$

$$D\Phi = \frac{\phi_{01}}{2} + \pi\sum_{n=1}^{\infty}2n\phi_{2n}E_{2n}(x), \qquad D^2\Phi = -\pi^2\sum_{n=1}^{\infty}(2n)^2\phi_{2n}F_{2n}(x),$$

$$D^3\Phi = \frac{\phi_{03}}{2} - \pi^3\sum_{n=1}^{\infty}(2n)^3\phi_{2n}E_{2n}(x), \qquad D^4\Phi = \pi^4\sum_{n=1}^{\infty}(2n)^4\phi_{2n}F_{2n}(x).$$

The conditions $Dv = D\Theta = \Phi = D^2\Phi = 0$ are automatically satisfied, so the only constraint generated by $v = 0$ at $x = \pm 0.5$ impose $\Gamma_1 = \frac{v_0}{2} + \sum_{n=1}^{\infty}v_{2n}(-1)^n\sqrt{2} = 0$. The functional j_3 becomes

$$j_3(\mathbf{U}) = \sum_{n=1}^{\infty}\{B_n\Theta_{2n}^2 - 2\sqrt{k^2 R(1+El)}\,v_{2n}\theta_{2n} + B_n^2 v_{2n}^2$$

$$-2\sqrt{k^2 R E l} 2n\pi v_{2n}\phi_{2n} - B_n^2 \Phi_{2n}^2\} + \frac{k^2\theta_0^2}{4} - 2\sqrt{k^2 R(1 + El)}\frac{v_0\theta_0}{4} + \frac{k^4 v_0^2}{4} - 2\frac{k^2\phi_{01}^2}{4},$$

where $(2n\pi)^2 + k^2 = B_n$. Denoting by μ the Lagrange multiplier, the stationarity conditions become

$$\frac{\partial(j_3 - \mu\Gamma_1)}{\partial v_{2n}} = \frac{\partial(j_3 - \mu\Gamma_1)}{\partial\phi_{2n}} = \frac{\partial(j_3 - \mu\Gamma_1)}{\partial\theta_{2n}} = 0.$$

In the usual manner these lead to the secular equation (which yields the equation of the neutral curve $R = R_1$).

$$\frac{1}{k^4 - R(1 + El)} + \sum_{n=1}^{\infty}\frac{2B_n^2}{B_n^4 - R(k^2 B_n + k^4 El)} = 0. \tag{3.5.4}$$

The corresponding linear instability criterion reads $R > R_1$.

The problem $(3.3.1) - (3.3.3)$, $(3.3.7)$

We take into account Theorem 4.12, where $\mathbf{U} = (\theta, v, \phi)$ and $\mathbf{U}^* = (\Theta, V, \Phi)$. On $[-0.5, 0.5]$ we can put

$$\theta = \sum_{n=1}^{\infty}\theta_{2n-1}E_{2n-1}(x), \ v = \sum_{n=1}^{\infty}v_{2n-1}E_{2n-1}(x), \ \phi = \sum_{n=1}^{\infty}\phi_{2n-1}F_{2n-1}(x),$$

$$D\theta = -\sum_{n=1}^{\infty}(2n - 1)\pi\theta_{2n-1}F_{2n-1}(x), \ Dv = -\sum_{n=1}^{\infty}(2n - 1)\pi v_{2n-1}F_{2n-1}(x),$$

and on $[-0.5, 0.5]$ we take

$$D^2 v = -\pi^2\sum_{n=1}^{\infty}(2n - 1)^2 v_{2n}E_{2n-1}(x), \ D\phi = \pi\sum_{n=1}^{\infty}(2n - 1)\phi_{2n-1}E_{2n-1}(x).$$

The conditions $\theta = v = 0$ are automatically satisfied. The conditions $Dv = \phi = 0$ at $x = \pm 0.5$ imply the constraints

$$\Gamma_1 = \sum_{n=1}^{\infty}(2n - 1)\pi v_{2n-1}(-1)^{n+1} = 0, \quad \Gamma_2 = \sum_{n=1}^{\infty}\phi_{2n-1}(-1)^{n+1} = 0.$$

For V, Θ, Φ we have the same relations, so j_4^{IV} becomes

$$j_4^{IV}(\mathbf{U}, \mathbf{U}^*) = \sum_{n=1}^{\infty}\bigg[-A_n\theta_{2n-1}\Theta_{2n-1} + v_{2n-1}\Theta_{2n-1} + A_n^2 v_{2n-1}V_{2n-1}$$

$$- k^2 R(1 + El)\theta_{2n-1}V_{2n-1} - k^2 REl(2n - 1)\pi\phi_{2n-1}V_{2n-1}$$

$$- A_n\phi_{2n-1}\Phi_{2n-1} - (2n - 1)\pi\theta_{2n-1}\Phi_{2n-1}\bigg],$$

where $A_n = (2n - 1)^2\pi^2 + k^2$.

Finally, for the case $El = 0$, we get

$$\sum_{n=1}^{\infty}\frac{A_n^2(A_n - k^2)}{A_n^4 - Rk^2 A_n} = 0. \tag{3.5.5}$$

Chapter 4

Variants of the energy method for non-stationary equations

Three such variants are presented through examples. Section 4.1 is devoted to an extension of Joseph's parameter differentiation method to the case when additional thermal parameters are present in a concrete vertical thermal convection problem. Through the same problem, in Section 4.2 the G-P-R method is described in terms of the symmetrization of the involved operators. It is shown that this method generalizes the approach from Section 4.2 in several aspects, and, so, it applies to a wider class of problems. Sections 4.2.2 and 4.2.3 present two other variants of ours, preserving from Joseph's variant only the optimality condition. For a horizontal convection problem in Section 4.2.4 a variant slightly different from the standard energy method is applied and an energy stability criterion is obtained. In Section 4.2.5 the G-P-R method is extended to a case when the linear combinations of the unknown functions are vector functions. In Section 4.3.1 the main ideas in the Rionero and Mulone variants are shown and in Section 4.3.2, an energy criterion is derived for a very complicated magnetohydrodynamics convection problem.

4.1 Variant based on differentiation of parameters

Joseph's parameter differentiation method is extended, allowing us the treatment of a more complicated thermal convection problem containing two additional thermodynamic parameters N and λ. The coincidence of the linear and energy stability limits is proved. For $N = \lambda = 0$ Joseph's results are regained.

4.1.1 *Classical mathematical model governing the conduction and convection in a binary mixture*

Let $(O, \mathbf{i}, \mathbf{j}, \mathbf{k})$ be the Cartesian system of coordinates with $\mathbf{k}$ pointing upwards along the vertical direction, let $\Omega = \{(x_1, x_2, x_3) \in \mathbf{R}^3 \mid 0 < x_1 < \frac{2\pi}{m_1}d, \, 0 < x_2 <$

151

$\frac{2\pi}{m_2}d,\ 0 < x_3 < d\}$ be the domain of motion, let $\partial\Omega$ be its boundary, denote by $\partial\Omega_1 = \{(x_1, x_2, x_3) \in \mathbf{R}^3 \mid 0{\leq}x_1{\leq}\frac{2\pi}{m_1}d,\ 0{\leq}x_2{\leq}\frac{2\pi}{m_2}d,\ x_3 = 0\}$ $(\partial\Omega_2 = \{(x_1, x_2, x_3) \in \mathbf{R}^3 \mid 0{\leq}x_1{\leq}\frac{2\pi}{m_1}d,\ 0{\leq}x_2{\leq}\frac{2\pi}{m_2}d,\ x_3 = d\})$ the rigid (free) lower (upper) plane part of $\partial\Omega$ and let $\mathbf{n}$ be the outer normal to $\partial\Omega$.

In the dimensional form, the mathematical model describing the conduction or convection in Ω of a binary fluid mixture, with two competing effects, temperature and concentration, in the presence of thermodiffusive Soret-Dufour currents, in the O-B approximation, reads [Shap], [Vert], [SuS], [Shin], [Pal97]

$$\begin{cases} \dfrac{\partial \mathbf{u}}{\partial t} + \mathbf{u}\nabla\mathbf{u} = -\dfrac{1}{\rho}\nabla p + [1 - \alpha(T - T_0) + s\beta(C - C_0)] + \nu\Delta\mathbf{u}, \\[2mm] \dfrac{\partial T}{\partial t} + \mathbf{u}\cdot\nabla T = (k_T + N\lambda^2 k_C)\Delta T + N\lambda k_C\Delta\gamma, \\[2mm] \dfrac{\partial C}{\partial t} + \mathbf{u}\cdot\nabla C = k_C\Delta C + \lambda k_C\Delta T, \\[2mm] \qquad \nabla\cdot\mathbf{u} = 0 \end{cases} \tag{4.1.1}$$

$$\mathbf{u} = \mathbf{u}^0,\ T = T^0,\ C = C^0 \quad \text{at} \quad t = 0 \tag{4.1.2}$$

$$\begin{cases} \mathbf{u} = 0 \quad \text{on} \quad \partial\Omega_1 \qquad \mathbf{u}\cdot\mathbf{n} = (\mathbf{n}\cdot\mathbf{D})\times\mathbf{n} = 0 \quad \text{on} \quad \partial\Omega_2, \\[2mm] T = T_0, \quad C = C_0 \quad \text{on} \quad \partial\Omega_1, \quad T = T_{\mathrm{d}}, \quad C = C_{\mathrm{d}} \quad \text{on} \quad \partial\Omega_2, \end{cases} \tag{4.1.3}$$

where t is the time, $\mathbf{x} \equiv (x_1, x_2, x_3)$ is the current point in Ω, $\mathbf{u}$, T, C are the kinetic, thermal and concentration field respectively, $\mathbf{g}$ is the acceleration due to the gravity, p is the pressure. The positive constants ν, ρ, are the kinematic viscosity, the constant density, T_0, C_0 represent a reference temperature and a reference concentration, respectively; $s = 1(-1)$ if the solute density is greater (less) than the solvent density; k_T and k_C are the thermal and solute diffusivity respectively, N and λ are the thermodynamic coefficients, which govern the interaction between thermal diffusivity and diffusive thermal conductivity. Moreover, $(4.1.1)_{1,2,3,4}$ are the balance equations for momentum, energy, solute mass and mixture mass respectively.

The domain Ω is a periodicity cell in a horizontal layer of width d. This choice agrees with the observations and experiments [Béna], [KoS], [GerJ] that a fluid layer heated from below, *i.e.* $T_{|x_3=0} > T_{|x_3=d}$, for $T_{|x_3=0} - T_{|x_3=d}$ not so large, is in a mechanical equilibrium, called the *conduction state*. For larger such differences the fluid has a stationary motion periodic in x_1 and x_2 directions, called the *thermal horizontal convection*. For still larger differences the convection becomes non-stationary and more and more irregular motions emerge before the fluid become turbulent [Chol]. Each change in type of the flow takes place at the values of $T_{|x_3=0} - T_{|x_3=d}$ at which the solution set of the initial (4.1.2) and boundary-value problem (4.1.3) for the set of partial differential equations (4.1.1) undergoes a bifurcation. At a bifurcation value of the parameter $T_{|x_3=0} - T_{|x_3=d}$ the previous observed solution exchanges its stability with the emerged solution. In Section 4.1

we are dealing with the first bifurcation only, namely we determine the value at which the conduction becomes unstable, by using the Joseph variant of the energy method.

Various types of vertical convections exist, according to the type of the upper and lower surface of the layer (rigid or free), or type of fluid, and the effects acting in Ω. In the case $(4.1.1) - (4.1.3)$, the setting in of the convection is influenced by the existence of a constant vertical concentration gradient.

The model $(4.1.1) - (4.1.3)$ and all other models involving the concentration are not presented in Section 1.2 due to the fact that the dependence of ρ on T and C is not consistent with the modelization principles in Section 1.1. On the other hand, no rigorous proof is known for the O-B approximation.

Existence, uniqueness and regularity of the solutions of $(4.1.1) - (4.1.3)$ can be found in [J65], [J66], [J70b], [RioM84], [MuloS85], [RioM88], [MuloR84], [J70a], [J76].

The basic solution is the conduction state

$$m_0 \equiv \{\mathbf{u} = 0, \quad \overline{T} = \frac{T_d - T_0}{d} x_3 + T_0, \quad \overline{C} = \frac{C_d - C_0}{d} x_3 + C_0, \quad \overline{p} = p_0(x_3)\},$$

Let us use the characteristic quantities: d for distance, ν/d for velocity, d^2/ν for time, $g\nu^2/d^3$ for pressure, T_0 for temperature, C_0 for concentration. The nondimensional form for $(4.1.1)$ is

$$\begin{cases} \dfrac{\partial \mathbf{u}}{\partial t} + \mathbf{u}\nabla\mathbf{u} = -\nabla p + (\mathcal{R}T - s\mathcal{C}C)\mathbf{k} + \Delta\mathbf{u}, \\[2mm] P_r\left(\dfrac{\partial T}{\partial t} + \mathbf{u}\cdot\nabla T\right) = (1 + N\lambda^2\tau^{-1})\Delta T + N\lambda\sigma\Delta C, \\[2mm] S_c\left(\dfrac{\partial C}{\partial t} + \mathbf{u}\cdot\nabla C\right) = \Delta C + \lambda\sigma^{-1}\tau^{-1}\Delta T, \end{cases} \qquad (4.1.4)$$

where, apart from x_1, x_2, x_3 whose nondimensional correspondents are x, y, z, the nondimensional quantities were denoted by the same letters as their corresponding dimensional ones. Here $\tau = S_c/P_r$, $\sigma = \beta/\alpha$ and $s = 1(-1)$ if the solute density is greater (less) than the solvent density. The other nondimensional parameters were defined in Section 1.1. The nondimensional basic state, namely the mechanical equilibrium, becomes

$$m_0 = \{\mathbf{u} = 0, T = -\beta_1 z + T_0, \quad C = -\beta_2 z + C_0, \quad p = p_0(z)\}. \qquad (4.1.5)$$

The initial and boundary conditions $(4.1.2) - (4.1.3)$ keep their form.

4.1.2 *Perturbation model*

Due to its importance in astrophysics, geophysics, oceanography and meteorology, the study of the stability of the mechanical equilibrium $(4.1.5)$ received much attention [RioM84], [ShiJ], [Gald85], [GaldS85], [GeoM99], [GeoO90], [GeoOP], [J65], [J66], [J70a], [J70b], [J76], [MuloR89], [MuloR94], [MuloR97], [MuloR03].

We dealt with this problem since 1994 [GeoPal96a], [GeoPR96b], using Joseph's variant of the energy method (called of parameters differentiation [J70a]) and then [GeoPalR00], [GeoPal04] creating our variant based on symmetrization of some associated operators.

Consider the model (4.1.4), (4.1.2), (4.1.3) and assume that $\mathbf{u}$, T and C are the perturbed quantities around $\overline{\mathbf{u}}$, $\overline{T}$, and $\overline{C}$ such that $\mathbf{u} = \overline{\mathbf{u}} + \mathbf{v}$, $T = \overline{T} + \theta$, $C = \overline{C} + \gamma$, where the basic quantities $\overline{\mathbf{u}}$, $\overline{T}$, and $\overline{C}$ are given by (4.1.5). Then we have

$$\frac{\partial \mathbf{v}}{\partial t} + \mathbf{v}\nabla \mathbf{v} = -\nabla p + (\mathcal{R}\theta - s\mathcal{C}\gamma)\mathbf{k} + \Delta \mathbf{v}, \tag{4.1.6}$$

$$P_r\left(\frac{\partial \theta}{\partial t} + \mathbf{v}\cdot\nabla\theta\right) = (1+N\lambda^2\tau^{-1})\Delta\theta + \mathcal{R}\mathbf{v}\cdot\mathbf{k} + N\lambda\sigma\Delta\gamma, \quad (t,\mathbf{x}) \in (o,\infty)\times\Omega, \tag{4.1.7}$$

$$S_c\left(\frac{\partial \gamma}{\partial t} + \mathbf{v}\cdot\nabla\gamma\right) = \Delta\gamma + \lambda\sigma^{-1}\tau^{-1}\Delta\theta + \mathcal{C}\mathbf{v}\cdot\mathbf{k}, \tag{4.1.8}$$

in the following subspace of $L_2(\Omega)$

$$\mathbf{N}^1 = \Big\{ (\mathbf{v},\theta,\gamma) \ |\mathbf{v},\theta,\gamma \in W^{2,2}(\Omega), \ \mathbf{v} = 0 \text{ on } \partial\Omega_1, \ \mathbf{v}\cdot\mathbf{n} = (\mathbf{n}\cdot\mathbf{D})\times\mathbf{n}$$
$$= 0 \text{ on } \partial\Omega_2, \ \theta = \gamma = 0 \text{ on } \partial\Omega_1 \cup \partial\Omega_2, \ \nabla\cdot\mathbf{v} = 0 \Big\}. \tag{4.1.9}$$

The initial boundary-value problem (4.1.6) − (4.1.9), (4.1.2) governs the perturbation evolution of the mechanical equilibrium m_0 of a horizontal layer of a thermally-conducting binary mixture, in the presence of the Soret and Dufour effects, characterized by N and λ, in a horizontal layer. The fluid is heated from below and has a greater concentration at the bottom. In (4.1.6) − (4.1.9), $\Omega \subset \mathbf{R}^3$ is the bounded periodicity cell of boundary $\partial\Omega$. The intersection of $\partial\Omega$ with the rigid bottom boundary is denoted by $\partial\Omega_1$. The unknown functions $\mathbf{v} \equiv (u,v,w)$, θ and γ represent the perturbation velocity, temperature and concentration field, respectively, $\mathbf{x}(x,y,z)$ is the position vector of a material point, $\mathbf{k}$ is the unit vector of the vertical axis in the direction opposite to gravity, t stands for the time and $\mathbf{g}$ is the gravitational acceleration. Except for s, all other parameters occurring in the perturbation model, are positive. Among them, 7 are independent.

Denote $\mathbf{V} \equiv (\mathbf{v},\theta,\gamma)$. Then, apart for the pressure term, the right-hand side of (4.1.6) − (4.1.8), can be written as $A\mathbf{V}$, where A is the nonsymmetric matricial differential operator associated with the matrix

$$\mathbf{A} = \begin{pmatrix} \Delta & \mathcal{R}\mathbf{k} & -s\mathcal{C} \\ \mathcal{R}\mathbf{k} & N_1\Delta & N\lambda\sigma\Delta \\ \mathcal{C} & \lambda\sigma^{-1}\tau^{-1}\Delta & \Delta \end{pmatrix}, \tag{4.1.10}$$

where

$$N_1 = 1 + N\lambda^2\tau^{-1}. \tag{4.1.11}$$

In standard energy method, in order to determine the energy relation, equations (4.1.6) − (4.1.8) are projected on $\mathbf{N}_1$ by simply multiplying them scalarly in $L^2(\Omega)$

by $\mathbf{V}$. Then, for $s = 1$, one of the two most important physical parameters $\mathcal{R}$ and $\mathcal{C}$, namely $\mathcal{C}$, no longer occurs in the resulted energy relation. Indeed, $(A\mathbf{V}, \mathbf{V}) = (A_s\mathbf{V}, \mathbf{V}) + (A_{ss}\mathbf{V}, \mathbf{V})$, where A_s is the symmetric part of A and A_{ss} the skew-symmetric part of A, where (Appendix 2) $A_s = \frac{A+A^+}{2}$ and $A_{ss} = \frac{A-A^+}{2}$, A^+ is the formal adjoint of A (and its form is equal to that of the adjoint A^*). Since A contains only even order derivatives, A^+ is defined by the transpose matrix of $\mathbf{A}$ in (4.1.10) (Remark 4.5), *i.e.*

$$\mathbf{A}_s = \begin{pmatrix} \Delta & \mathcal{R}\mathbf{k} & \frac{1-s}{2}\mathcal{C} \\ \mathcal{R}\mathbf{k} & N_1\Delta & \frac{(N\lambda\sigma+\lambda\sigma^{-1}\tau^{-1})}{2}\Delta \\ \frac{1-s}{2}\mathcal{C} & \frac{(N\lambda\sigma+\lambda\sigma^{-1}\tau^{-1})}{2}\Delta & \Delta \end{pmatrix},$$

$$\mathbf{A}_{ss} = \begin{pmatrix} 0 & 0 & -\frac{1+s}{2}\mathcal{C} \\ 0 & 0 & \frac{(N\lambda\sigma-\lambda\sigma^{-1}\tau^{-1})}{2}\Delta \\ \frac{1+s}{2}\mathcal{C} & -\frac{(N\lambda\sigma-\lambda\sigma^{-1}\tau^{-1})}{2}\Delta & 0 \end{pmatrix},$$

where $\mathbf{A}_s$ and $\mathbf{A}_{ss}$ are the matrices defining the operators A_s and A_{ss} respectively. Remind that $(A_{ss}\mathbf{V}, \mathbf{V}) = 0$, whence the announced disappearance of $\mathcal{C}$ for $s = 1$. In this way, we cannot expect that the stability bound deduced from the corresponding energy relation can be a nonlinear stability limit. Therefore the standard energy method must be changed.

4.1.3 *Energy relation*

In [J70a] Joseph presented his variant of the energy method for the case $N = \lambda = 0$, $s = 1$. The modifications operated by Joseph on the standard energy method are slight but efficient. In order to present them, let us remind that the projection of $(4.1.6) - (4.1.8)$ on $\mathbf{N}^1$ means to multiply (4.1.6) by $\mathbf{v}$, (4.1.7) by θ, (4.1.8) by γ, integrate the resulted equations over Ω, and take into account the boundary conditions from (4.1.9), obtaining

$$\frac{1}{2}\frac{d|\mathbf{v}|^2}{dt} = -\|\nabla\mathbf{v}\|^2 + \mathcal{R}|\theta\mathbf{v}\cdot\mathbf{k}|_1 - s\mathcal{C}|\gamma\mathbf{v}\cdot\mathbf{k}|_1, \tag{4.1.12}$$

$$\frac{1}{2}P_r\frac{d|\theta|^2}{dt} = \mathcal{R}|\theta\mathbf{v}\cdot\mathbf{k}|_1 - N_1\|\theta\|^2 - N\lambda\sigma|\nabla\theta\cdot\nabla\gamma|_1, \tag{4.1.13}$$

$$\frac{1}{2}S_c\frac{d|\gamma|^2}{dt} = \mathcal{C}|\gamma\mathbf{v}\cdot\mathbf{k}|_1 - \lambda\sigma^{-1}\tau^{-1}|\nabla\theta\cdot\nabla\gamma|_1 - \|\gamma\|^2 \tag{4.1.14}$$

and then to sum up (4.1.12), (4.1.13) and (4.1.14) to obtain the energy relation, where $|\cdot|_1, |\cdot|$ and $\|\cdot\|$ stand for the norm in $L^1(\Omega)$, $L^2(\Omega)$ and $N^1(\Omega)$ respectively. As this sum leads to weak results, before performing the sum, Joseph introduced two new functions ϕ_1 and ϕ_2 as known linear combination of θ and γ and defined a new energy E in terms of $|\mathbf{v}|^2$ and $|\phi_1|^2$. In order to succeed in obtaining the rate of change for E, from (4.1.7) and (4.1.8) he deduced an expression for the rate of change of $|\theta\gamma|_1$. Then he multiplied (4.1.13), (4.1.14) and the equality for $|\theta\gamma|_1$ by arbitrary constants a, b and c and summed up these relations to obtain the energy relation for E. As c was a known function of the parameters a and b, he determined first a relationship between a and b from the requirement that the energy relation should have the simplest form. Thus, one parameter, say b, was a function of the other parameter a. Based on this remark, he obtained a new relation between a and b by imposing on the energy limit to be optimal. More exactly, this limit was a function of a and b. By this remark, he imposed on the derivative of $\mathcal{R}_e$ with respect to a to vanish, which yielded the second relationship between a and b. Whence the name of *parameter differentiation method*.

Here are the main steps we used to extend the Joseph stability criterion from the case $N = \lambda = 0$, $s = 1$, to the case $N, \lambda \neq 0$, $s = \pm 1$ [GeoPal96a].

Integrating over Ω the sum of equation (4.1.7) multiplied by $P_r^{-1}\gamma$ and equation (4.1.8) multiplied by $S_c^{-1}\theta$, we obtained the following expression of the rate of change of $|\theta\gamma|_1$

$$
\begin{aligned}
\frac{d|\theta\gamma|_1}{dt} =\ &\tau\mathcal{R}S_c^{-1}|\gamma w|_1 + \mathcal{C}S_c^{-1}|\theta w|_1 - (1 + \tau + N\lambda^2)S_c^{-1}|\nabla\theta\cdot\nabla\gamma|_1 \\
&- N\lambda\sigma\tau S_c^{-1}\|\gamma\|^2 - \lambda\sigma^{-1}\tau^{-1}S_c^{-1}\|\theta\|^2.
\end{aligned}
\tag{4.1.15}
$$

Then, adding (4.1.12) to (4.1.13) multiplied by $a > 0$, (4.1.14) multiplied by $b > 0$, and (4.1.15) multiplied by $c < 0$, we have

$$
\begin{aligned}
\frac{1}{2}\frac{d}{dt}\big(|\mathbf{v}|^2 + aP_r|\theta|^2 + bS_c|\gamma|^2 + 2c|\theta\gamma|_1\big) = &-\|\mathbf{v}\|^2 - d_4\|\theta\|^2 - d_5\|\gamma\|^2 \\
&- d_6|\nabla\theta\cdot\nabla\gamma|_1 + d_7\mathcal{R}|\theta\mathbf{v}\cdot\mathbf{k}|_1 + d_8\mathcal{C}\alpha^{-1}\gamma|\mathbf{v}\cdot\mathbf{k}|_1,
\end{aligned}
\tag{4.1.16}
$$

where d_4, d_5, d_6, d_7, and d_8 have simple expressions in terms of the seven independent physical parameters and the multiplication constants a, b and c.

Recall that in the standard energy method we assumed $c = 0$, hence equation (4.1.15) is ignored.

In (4.1.14) the term $\mathcal{C}|\gamma\mathbf{v}\cdot\mathbf{k}|_1$ comes from the term $-\mathcal{C}\mathbf{v}\nabla\overline{C}/\overline{\beta}$, where $\overline{C}$ is the basic concentration and $\overline{\beta} = sup|\overline{C}|$ in Ω [RioM84]. In (4.1.16) it corresponds to $\mathcal{C}\mathbf{v}\cdot\mathbf{k}$, that is, our model $(4.1.6) - (4.1.8)$ corresponds to the realistic physical situation in which $\nabla\overline{C}/\overline{\beta} = -\mathbf{k}$: since $\overline{\beta} > 0$, it follows that the concentration gradient is opposite to $\mathbf{k}$. The treatment from [RioM84] was based on the hypothesis $\nabla\overline{C}/\overline{\beta} = skN^{-1}\sigma^{-2}$ and on the choice $a = \tau^{-1}$, $b = N\sigma^2$, and $c = 0$. Correspondingly, in (4.1.16), instead of $\mathcal{C}\mathbf{v}\cdot\mathbf{k}$, Rionero and Mulone found $-\mathcal{C}sN^{-1}\sigma^{-2}\mathbf{v}\cdot\mathbf{k}$. In this way, in (4.1.16) instead of $d_8\mathcal{C}\alpha^{-1}|\gamma\mathbf{v}\cdot\mathbf{k}|_1$ they obtained $-2s\mathcal{C}|\gamma\mathbf{v}\cdot\mathbf{k}|_1$ and the corresponding linear operator associated with the right-hand side of $(4.1.6) - (4.1.8)$ was symmetric. If, in their hypothesis, we had changed the sign, to obtain the real case, which corresponded to a skew-symmetric operator, then instead of $-\mathcal{C}sN^{-1}\sigma^{-2}\mathbf{v}\cdot\mathbf{k}$ we would have obtained zero and, since $c = 0$, $\mathcal{C}$ would no longer occur in the energy relation (4.1.16). Consequently, the resulting criterion would not have been influenced by the basic concentration field, as in the standard energy method.

As it stands, (4.1.16) is not yet an energy relation because the energy is still undefined. We define it such that in the resulting energy inequality we keep the contribution of all terms in the right-hand side of (4.1.16).

Hence, introduce the functions

$$E(t) = (|\mathbf{v}|^2 + d_1|\phi_1|^2 + d_2|\phi_1\phi_2|_1)/2, \qquad \Psi(t) = d_3|\phi_2|^2/2, \tag{4.1.17}$$

where ϕ_1 and ϕ_2 are linear combinations of θ and γ

$$\phi_1 = a_1'\theta + a_2'\gamma, \quad \phi_2 = b_1'\theta + b_2'\gamma \tag{4.1.18}$$

and thus, conversely, θ and γ are linear combinations of ϕ_1 and ϕ_2

$$\theta = a_1\phi_1 + a_2\phi_2, \quad \gamma = b_1\phi_1 + b_2\phi_2, \tag{4.1.19}$$

where

$$a_1' = b_2/M, \quad a_2' = -a_2/M, \quad b_1' = -b_1/M, \quad b_2' = a_1/M,$$

$$a_1 = b_2'/M', \quad a_2 = -a_2'/M', \quad b_1 = -b_1'/M', \quad b_2 = a_1'/M',$$

$M = a_1b_2 - a_2b_1$ and $M' = a_1'b_2' - a_2'b_1'$.

Then, (4.1.17) has the form

$$\begin{aligned}
\frac{dE}{dt} + \frac{d\Psi}{dt} = &- \left(\|\mathbf{v}\|^2 + (a_1^2 d_4 + b_1^2 d_5 + a_1 b_1 d_6)|\nabla\phi_1|^2\right) \\
&+ (a_2^2 d_4 + b_2^2 d_5 + a_2 b_2 d_6)|\nabla\phi_2|^2 \\
&+ \left[2a_1 a_2 d_4 + 2b_1 b_2 d_5 + (a_1 b_2 + a_2 b_1)d_6\right]|\nabla\phi_1 \cdot \nabla\phi_2|_1 \\
&+ \mathcal{R}(a_1 d_7 + b_1 d_8)|\phi_1 \mathbf{v} \cdot \mathbf{k}|_1 + \mathcal{R}(a_2 d_7 + b_2 d_8)|\phi_2 \mathbf{v} \cdot \mathbf{k}|_1.
\end{aligned} \tag{4.1.20}$$

Here, a_1, a_2, b_1 and b_2 are unknown parameters while d_1, d_2, d_3, d_4 are known functions of a, b, c and the physical parameters, namely

$$\begin{cases}
d_1 = aP_r a_1^2 + bS_c b_1^2 + 2ca_1 b_1; \quad d_2 = aP_r a_1 a_2 + bS_c b_1 b_2 + c(a_1 b_2 + a_2 b_1); \\
d_3 = aP_r a_2^2 + bS_c b_2^2 + 2ca_2 b_2; \quad d_4 = a(1 + N\lambda^2\tau^{-1}) + cS_c^{-1}\lambda\sigma^{-1}\tau^{-1}; \\
d_5 = b + cN\lambda\sigma\tau S_c^{-1}; \qquad\qquad d_6 = aN\lambda\sigma + b\lambda\sigma^{-1}\tau^{-1} + cS_c^{-1}(1 + \tau + N\lambda^2); \\
d_7 = 1 + a + c\alpha S_c^{-1}; \qquad\qquad d_8 = b\alpha - s\alpha + c\tau S_c^{-1},
\end{cases} \tag{4.1.21}$$

where $\alpha = \mathcal{C}/\mathcal{R}$.

Five relationships between the seven constants a, b, c, a_1, b_1, a_2 and b_2 are determined by imposing on (4.1.20) to assume the form

$$\frac{dE}{dt} + \frac{d\Psi}{dt} = -(\|\mathbf{v}\|^2 + \|\phi_1\|^2) + \mathcal{R}(a_1 d_7 + b_1 d_8)|\phi_1 \mathbf{v} \cdot \mathbf{k}|_1, \tag{4.1.22}$$

and the energy E, the form

$$E(t) = (|\mathbf{v}|^2 + d_1|\phi_1|^2)/2, \tag{4.1.23}$$

where $d_1, d_3 > 0$. This leads to the system of five nonlinear algebraic equations

$$\begin{cases} d_2 = 0, \\ a_1^2 d_4 + b_1^2 d_5 + a_1 b_1 d_6 = 1 \\ a_2^2 d_4 + b_2^2 d_5 + a_2 b_2 d_6 = 0, \\ 2a_1 a_2 d_4 + 2b_1 b_2 d_5 + (a_1 b_2 + a_2 b_1) d_6 = 0, \\ a_2 d_7 + b_2 d_8 = 0. \end{cases} \qquad (4.1.24)$$

The constants a_2 and b_2 are determined up to a multiplicative factor. Indeed, the product $d_3 |\varphi|^2$ is unchanged if d_3 is divided by some constant, say δ, and a_2 and b_2 are multiplied by δ. However, the ratio a_2/b_2 must have a well-determined value.

The system (4.1.24) can be considered as yielding $a_1, b_1, a_2/b_2, b, c$ as functions of a and, hence, another relationship between these parameters is necessary.

4.1.4 *Energy inequality and the stability criterion*

Denote

$$\xi^2 = min_{\mathbf{v}, \theta \in \mathbf{N}^1(\Omega)} \frac{2\left(\|\mathbf{v}\|^2 + \|\theta\|^2\right)}{|\mathbf{v}|^2 + |\theta|^2}, \quad \frac{1}{\sqrt{R_{a*}}} = max_{\mathbf{v}, \theta \in \mathbf{N}^1(\Omega)} \frac{2|\theta w|}{\|\mathbf{v}\|^2 + \|\theta\|^2}, \quad (4.1.25)$$

$$2A = \mathcal{R}|a_1 d_7 + b_1 d_8|.$$

Then the energy relation (4.1.22) implies

$$\frac{dE}{dt} + \frac{d\Psi}{dt} \leq - \left(\|\mathbf{v}\|^2 + \|\theta\|^2\right)\left(1 - A/\sqrt{R_{a*}}\right). \qquad (4.1.26)$$

If the nonlinear stability criterion

$$\mathcal{R} < 2\sqrt{R_{a*}}/|a_1 d_7 + b_1 d_8| \qquad (4.1.27)$$

holds, then (4.1.26) implies the energy inequality

$$\frac{dE}{dt} + \frac{d\Psi}{dt} \leq - \xi^2 \left(1 - A/\sqrt{R_{a*}}\right) E(t). \qquad (4.1.28)$$

This inequality is of the form (3.2.78). Thus, the basic state of conduction is stable.

The energy stability limit

$$\mathcal{R}_E = 2\sqrt{R_{a*}}/|a_1 d_7 + b_1 d_8| \qquad (4.1.29)$$

is maximal if $|a_1 d_7 + b_1 d_8|$ is minimal. Since $|a_1 d_7 + b_1 d_8|$ is a function of the parameter a, by Joseph's idea of parameter differentiation, this requirement is fulfilled if and only if

$$\frac{d(a_1 d_7 + b_1 d_8)}{da} = 0. \qquad (4.1.30)$$

Thus we obtained a closed algebraic system (4.1.24), (4.1.30) for the determination of the multiplication constants a, b, c and the coefficients a_1, b_1, a_2 and b_2/a_2.

Further, in [GeoPal96a] it was shown that

$$a_2/b_2 = -d_8/d_7, \quad a_1 = (cd_8 - bS_c d_7)/Q, \quad b_1 = (cd_7 - aP_r d_8)/Q, \qquad (4.1.24')$$

(where Q is a quadratic form in a, b, c with coefficients depending on $d_4, \ldots, d_8$),

$$d_6^2 = 4d_4 d_5, \quad d_8^2/d_7^2 = d_5/d_4 \quad a_1 d_7 + b_1 d_8 = d_7/\sqrt{d_4}, \qquad (4.1.24'')$$

so the (optimality) requirement (4.1.30) has the form

$$\left(d_7/d_4\right)^2 c_1 - \left(d_7/d_4\right)\left(c_2 + c_3 d_6/d_4\right) + \left(d_6/d_4\right)^2 c_4 + \left(d_6/d_4\right)c_5 + c_6 = 0, \qquad (4.1.31)$$

where the expressions of $c_1, \ldots, c_6$ are given in [GeoPal96a], and has the solutions

$$d_7/d_4 = \left[\alpha(1 + N\lambda^2\tau^{-1}) - \lambda\sigma^{-1}\tau^{-1}\right]d_6/d_4 + 2(1 - \alpha N\lambda\sigma), \qquad (4.1.32)$$

$$d_7/d_4 = \left(\alpha d_6/d_4 + 2\tau\right)/(1 + \tau + N\lambda^2). \qquad (4.1.33)$$

Using these formulae, (4.1.24) and (4.1.30) were reduced to a system of three equations in a, b, c: one affine in a, b, c, one of the form $a^2 C_1 + b^2 C_2 + c^2 C_3 + abC_4 + acC_5 + bcC_6 = 0$ and the last of the form $a^2 C_7 + b^2 C_8 + c^2 C_9 + abC_{10} + acC_{11} + bcC_{12} + aC_{13} + bC_{14} + cC_{15} = 0$, where $C_1, \ldots, C_{15}$ were complicated expressions of all the physical parameters of the problem. Thus, the validity of the stability criterion (4.1.27) was conditioned by the solution of this system.

However, when particularized to the case $N = \lambda = 0, s = 1$, we regained Joseph's criterion. We show this in Sections 4.1.5 and 4.1.6. Hence, our extension of Joseph's method was good but a further investigation of the system (4.1.24), (4.1.30) was necessary.

The difficulty to solve the mentioned system in a, b, c forced us to look for another system equivalent to (4.1.24), (4.1.30). Thus, in [GeoPalR00] we reduced it to the solution of a biquadratic equation in d_5. The closed-form expression of d_5 was found for (4.1.32). The corresponding nonlinear stability limit $\mathcal{R}_E$ was determined, showing the equality between the linear $\mathcal{R}_L$ and (nonlinear) energy $\mathcal{R}_E$ stability limits.

In addition, the treatment in [GeoPalR00] and relations (4.1.24'), (4.1.24'') revealed that if the constants d_4, d_5 and d_6 were used, then the computations simplified significantly. However, the derivation of d_5 still implied cumbersome calculations. As a consequence, in [GeoPalR96c], another way of deriving the equation d_5 was searched for, bearing in mind the quoted importance of using d_4, d_5 and d_6 if simpler computations were in view. Thus, a direct solution of (4.1.24)$_{2,3,4}$ yielded

$$d_4 = a_1'^2, \qquad d_5 = a_2'^2, \qquad d_6 = 2a_1' a_2', \qquad (4.1.34)$$

explaining why the use of d_4, d_5 and d_6 was preferred and showing that instead of the coefficients a_1, a_2, b_1 and b_2 we must use the coefficients a_1', a_2', b_1' and b_2'. The definitive form of the $\mathcal{R}_E$ and the proof of the equality $\mathcal{R}_E = \mathcal{R}_L$, given in [GeoPalR96c], are presented in Section 4.1.7.

4.1.5   *Case $N = \lambda = 0$, $s = 1$*

In this case, which is the only one considered by Joseph in [J70a], letting $a = \lambda_T^2$, $b = \lambda_C^2$, we have $d_1 = \lambda_T^2$, $d_2 = \lambda_C^2$, $d_3 = cS_c^{-1}(1+\tau)$; relation $d_6 = -\sqrt{d_4 d_5}$ implies $cS_c^{-1} = -2\lambda_T\lambda_C/(1+\tau)$; relation $d_8/d_7 = -\sqrt{d_5/d_4}$ becomes relation (14) from [J70a]

$$\frac{1 + \lambda_T^2 - 2\alpha\lambda_T\lambda_C(1+\tau)^{-1}}{\lambda_T} = -\frac{\lambda_C^2 - s\alpha - 2\tau\lambda_T\lambda_C(1+\tau)^{-1}}{\lambda_C}, \qquad (4.1.35)$$

which, for $s = 1$ becomes

$$\frac{1 + \lambda_T^2 - 2\alpha\lambda_T\lambda_C(1+\tau)^{-1}}{\lambda_T} = -\frac{\lambda_C^2 - \alpha - 2\tau\lambda_T\lambda_C(1+\tau)^{-1}}{\lambda_C}. \qquad (4.1.35')$$

Therefore Joseph's calculations correspond to the case $d_6 = -\sqrt{d_4 d_5}$, $d_8/d_7 = -\sqrt{d_5/d_4}$.

Relations for a_1 and b_1 in terms of

$$Q = -\left(\frac{d_4}{\sqrt{d_1}}\right)\lambda_T^2\lambda_C^2 S_C(\tau - 1)^2[\tau(1+\tau)]^{-1}$$

imply $a_1 = \tau/[\lambda_T(\tau - 1)]$, $b_1 = 1/[\lambda_C(\tau - 1)]$, $d_6 = S_C/(1 + \tau)$; $a_2/b_2 = -d_8/d_7$ shows that $a_2/b_2 = +\lambda_C/\lambda_T$. In particular, if $b_2 = -1/[\lambda_C(\tau - 1)]$, we obtain $a_2 = -1/[\lambda_T(\tau - 1)]$, yielding Joseph's linear combinations $\phi_1 = \lambda_T\theta - \lambda_C\gamma$, $\phi_2 = \lambda_T\theta - \tau\lambda_C\gamma$. (We can also have $a_2 = 1/[\lambda_T(\tau - 1)]$, and $b_2 = 1/[\lambda_C(\tau - 1)]$, because their quotient is the same and d_5 depends only on b_2^2. We took the former values in order to compare with Joseph's constants.)

One equation relating λ_T and λ_C is (4.1.35$'$), another is the following, deduced from (4.1.31)

$$\left(\frac{d_4}{d_1}\right)^2(1+\tau) - \left(\frac{d_4}{d_1}\right)\left[4\tau + 2 - 2\alpha(\tau + 2)\sqrt{\frac{b}{a}}\right] + \frac{4\alpha^2 b}{a} - 4\alpha(1+\tau)\sqrt{\frac{b}{a}} + 4\tau = 0 \quad (4.1.36)$$

which gives the solutions $d_4/a = 2 - 2\alpha\sqrt{\frac{b}{a}}$ and $d_4/a = (2\tau - 2\alpha\sqrt{\frac{b}{a}})(1 + \tau)$. The first solution implies

$$\alpha = -\frac{(1+\tau)(1 - \lambda_T)^2}{2\tau\lambda_T\lambda_C}. \qquad (4.1.37)$$

The system (4.1.35$'$), (4.1.37) yields λ_C and λ_T as functions of α and τ

$$\lambda_T = \frac{\sqrt{1 - \alpha^2\tau^2} - \tau\sqrt{1 - \alpha^2}}{1 - \tau}, \qquad \lambda_C = \frac{\sqrt{1 - \alpha^2\tau^2} - \sqrt{1 - \alpha^2}}{\alpha(1 - \tau)}. \qquad (4.1.38)$$

In order to obtain these solutions, Joseph proceeded as follows: he replaced (4.1.37) in (4.1.35$'$) and deduced the relation

$$(\lambda_T^2 - 1)(1 + \lambda_C^2)(1 + \tau)^2 - 4\tau\lambda_T^2\lambda_C^2 = 0, \qquad (4.1.39)$$

which is relation (21) from [J70a]. The solutions (4.1.38) correspond to the first branch of Joseph's curve and are valid for $\alpha < 1 \leq 1/\tau$. The case $\tau = 1$ was also treated in [Drag05a].

The second solution of (4.1.36) implies $\lambda_T^2 = (\tau+1)/(\tau-1)$, which, introduced into (4.1.35′) yields $\lambda_C^2 = (\tau+1)/(\tau-1)$. It corresponds to the second branch and holds for $\tau > 1$. The two solutions of (4.1.36) coincide if $\lambda_T/\lambda_C = \alpha\tau$. In this case (4.1.35′) and the relation $d_8/d_7 = -\sqrt{d_5/d_4}$ imply $\alpha\tau = \pm 1$. Since $\alpha, \tau > 0$ we must have $\alpha\tau = 1$, therefore, $\lambda_T = \lambda_C = [(\tau+1)/(\tau-1)]^{1/2}$. For $\tau = 1$ Joseph's formulae for $a_1, b_1, a_2, b_2, \lambda_T$, and λ_C cease to hold and a special treatment is necessary.

Formally, the case $d_6 = 2\sqrt{d_4 d_5}$, $d_8/d_7 = \sqrt{d_5/d_4}$ can be obtained by replacing λ_C by $-\lambda_C$ in (4.1.35) and (4.1.37) and the constant $\sqrt{\frac{b}{a}}$ by $-\sqrt{\frac{b}{a}}$ in (4.1.36). As these lead to the same a and b we obtain the same criterion, concluding the study of this case.

4.1.6   *Case $N = \lambda = 0$, $s = -1$*

In this case, which extends Joseph's analysis to the case $s = -1$, all relations from the case $N = \lambda = 0$, $s = 1$ hold, except for (4.1.35), which, in the case $d_6 = -2\sqrt{d_4 d_5}$, $d_8/d_7 = -\sqrt{d_5/d_4}$, must be replaced by

$$\frac{1 + \lambda_T^2 - 2\alpha\lambda_T\lambda_C(1+\tau)^{-1}}{\lambda_T} = -\frac{\lambda_C^2\alpha + \alpha - 2\tau\lambda_T\lambda_C(1+\tau)^{-1}}{\lambda_C}. \tag{4.1.40}$$

The second solution of (4.1.36) is unacceptable because it yields $\lambda_T^2 = (\tau+1)/(\tau-1)$ and then (4.1.40) implies $\lambda_C^2 = (\tau+1)/(1-\tau)$ and, therefore, λ_T^2 and λ_C^2 have opposite signs, which contradicts the fact that they are both positive.

Hence we consider the first solution of (4.1.36) which yields (4.1.37). Replacing it in (4.1.40) we obtain

$$(\lambda_T^2 - 1)(\lambda_C^2 - 1)(1+\tau)^2 - 4\tau\lambda_T^2\lambda_C^2 = 0. \tag{4.1.41}$$

Therefore, in order to determine λ_T and λ_C, instead of considering the system (4.1.40), (4.1.37), we solve (4.1.40), (4.1.41). From (4.1.41) we have

$$\lambda_T^2 = \frac{(\lambda_C^2 - 1)(1+\tau)^2}{\lambda_C^2(\tau-1)^2 - (1+\tau)^2}. \tag{4.1.41′}$$

Since in (4.1.40) we must have $\alpha > 0$, it follows that $\lambda_T^2 > 1$, if λ_T, λ_C have the same sign, and $\lambda_T^2 < 1$ if λ_T, λ_C have opposite signs. Due to the fact that if (λ_T, λ_C) is a solution of (4.1.40), (4.1.41) then $(-\lambda_T, -\lambda_C)$ is also a solution, we consider only one of the two possible cases. Then from (4.1.41′) it follows that for positive λ_T and λ_C we must have $\lambda_C^2 < (1+\tau)^2/(2\tau)$ for $2-\sqrt{3}\leq\tau\leq 2+\sqrt{3}$ and $\lambda_C^2 > (1+\tau)^2/(2\tau)$ for $\tau > 2+\sqrt{3}$ and $0 < \tau < 2-\sqrt{3}$; for λ_T and λ_C of opposite signs we must have $\lambda_C^2 > (1+\tau)^2/(\tau-1)^2$ for $2-\sqrt{3}\leq\tau\leq 2+\sqrt{3}$ and $\lambda_C^2 < (1+\tau)^2/(2\tau)$ for $\tau > 2+\sqrt{3}$ and $0 < \tau < 2-\sqrt{3}$.

Writing (4.1.40) in the form $\lambda_C\left[1 + \lambda_T^2 - \frac{2\tau\lambda_T^2}{(1+\tau)}\right] = \alpha\lambda_T\left[-\lambda_C^2 + \frac{2\lambda_C^2}{(1+\tau)} - 1\right]$ which, if we take into account (4.1.41′), becomes

$$\lambda_C\frac{2[\lambda_C^2(1-\tau) - (1+\tau)]}{\lambda_C^2(1-\tau)^2 - (1+\tau)^2} = \alpha\lambda_T\frac{\lambda_C^2(1-\tau) - (1+\tau)}{(1+\tau)},$$

and taking into account that $\lambda_C^2 \neq (1+\tau)(1-\tau)$, we have $2(1+\tau)\lambda_C = \alpha\lambda_T[\lambda_C^2(1-\tau)^2 - (1+\tau)^2]$ and, so,

$$\lambda_C^4(1-\tau)^2 - 2\lambda_C^2(1+\tau^2+2\alpha^{-2}) + (1+\tau)^2 = 0. \qquad (4.1.42)$$

Writing (4.1.41) as $\lambda_C^2 = \frac{(\lambda_T^2-1)(1+\tau)^2}{\lambda_T^2(1-\tau)^2-(1+\tau)^2}$, (4.1.42) implies $\lambda_T^4(1-\tau)^2 - 2\lambda_T^2(1+\tau^2+2\alpha^2\tau^2) + (1+\tau)^2 = 0$. The solutions of these two equations read

$$\lambda_C = \pm\frac{\sqrt{\tau^2+\alpha^{-2}} \pm \sqrt{1+\alpha^{-2}}}{\tau-1}, \quad \lambda_T = \pm\frac{\tau\sqrt{1+\alpha^{-2}} \pm \sqrt{1+\alpha^2\tau^2}}{\tau-1}. \qquad (4.1.43)$$

For $\alpha^2\tau = 1$ they coincide and become $\lambda_C = \lambda_T = \pm\frac{\sqrt{\tau^2+\tau}\pm\sqrt{1+\tau}}{\tau-1}$.

4.1.7   *Coincidence of nonlinear (energy) and linear stability limits*

The application of Joseph's energy variant to problem (4.1.6), (4.1.3) led to the criterion (4.1.27), where the energy stability limit $\mathcal{R}_E$ was given by (4.1.29). In order to express $\mathcal{R}_E$ in terms of the physical parameters, the system (4.1.24), (4.1.31) must be solved. Its solution (a, b, c, a_1, a_2, b_1) was given in [GeoPal96a], [GeoPalR00], [GeoPalR96c], after successive investigations and refinements. Correspondingly, the desired expression for $\mathcal{R}_E$ and the equality of stability limits $\mathcal{R}_E = \mathcal{R}_L$ was obtained. Since the implied algebra is complicated, we derive the same results but considering $(a, b, c, a_1', a_2', b_1')$ as a solution of (4.1.24), (4.1.31). Thus, in [GeoPalR96c], from (4.1.24), (4.1.32) it followed

$$\begin{aligned}
a_1'^2 &= aN_1 + cS_c^{-1}\lambda\sigma^{-1}\tau^{-1}, \\
a_2'^2 &= b + cS_c^{-1}N\lambda\sigma\tau, \\
2a_1'a_2' &= aN\lambda\sigma + b\lambda\sigma^{-1}\tau^{-1} + cS_c^{-1}(1+\tau N_1), \\
aa_2'b_2' &+ b\tau a_1'b_1' - cS_c^{-1}\tau(a_1'b_2' + b_1'a_2') = 0, \\
\frac{a_2'}{a_1'} &= \frac{\alpha(b-s)+\tau cS_c^{-1}}{1+a+\alpha cS_c^{-1}} = \frac{d_8}{d_7}, \\
1 + a &+ \alpha cS_c^{-1} = (\alpha N_1 - \lambda\sigma^{-1}\tau^{-1})2a_1'a_2' + 2a_1'^2(1-\alpha N\lambda\sigma),
\end{aligned} \qquad (4.1.44)$$

while in the case when (4.1.32) was replaced by (4.1.33), equation $(4.1.44)_6$ was replaced by

$$1 + a + \alpha cS_c^{-1} = (2a_1'^2\tau + 2a_1'a_2'\alpha)/(1+\tau N_1). \qquad (4.1.44)_7$$

In this section we present two other approaches from [GeoPalR01]: one is used to derive much easier the two systems (4.1.44) and $(4.1.44)_{1...5} - (4.1.44)_7$, and the other to solve them even easier. Then we give their closed-form solutions, which completes the derivation of our criterion for $(4.1.6) - (4.1.9)$. The first part of this criterion, namely that corresponding to the system (4.1.44), was deduced in [GeoPalR00].

The two approaches can also be applied to governing equations more general than $(4.1.6) - (4.1.9)$.

First approach. Take into account (4.1.19) and $(4.1.24)_1$ in (4.1.22) to obtain

$$\frac{d}{dt}\Big(|\mathbf{v}|^2 + d_1(a_1'^2|\theta|^2 + a_2'^2|\gamma|^2 + 2a_1'a_2'|\theta\gamma|_1) + d_3(b_1'^2|\theta|^2 + b_2'^2|\gamma|^2$$
$$+ 2b_1'b_2'|\theta\gamma|_1)\Big)/2 = \Big(-\|\mathbf{v}\|^2 + a_1'^2\|\theta\|^2 + a_2'^2\|\gamma\|^2 \tag{4.1.45}$$
$$+ 2a_1'a_2'|\nabla\theta\cdot\nabla\gamma|_1\Big)+(a_1d_7 + b_1d_8)\mathcal{R}\Big(a_1'|\theta\mathbf{v}\cdot\mathbf{k}|_1 + a_2'|\gamma\mathbf{v}\cdot\mathbf{k}|_1\Big)$$

and identify (4.1.45) and (4.1.16) to get (4.1.34) (and, therefore $(4.1.44)_{1,2,3}$) and

$$a_1'^2 d_1 + b_1'^2 d_3 = aP_r, \qquad a_2'^2 d_1 + b_2'^2 d_3 = bS_c, \qquad a_1'a_2'd_1 + b_1'b_2'd_3 = 2c, \tag{4.1.46}$$

$$a_1'(a_1d_7 + b_1d_8) = d_7, \qquad a_2'(a_1d_7 + b_1d_8) = d_8. \tag{4.1.47}$$

From $(4.1.47)_1$ we get

$$a_1d_7 + b_1d_8 = d_7(a_1 + b_1d_8/d_7) = d_7(b_2' - b_1'a_2'/a_1')/M' = d_7/a_1' \tag{4.1.48}$$

and the quotient of $(4.1.47)_2$ and $(4.1.47)_1$ is just $(4.1.44)_5$. Finally, elimination of d_1 and d_3 between the three equations (4.1.46) implies $(4.1.44)_4$.

Second approach. The central idea is to introduce two new unknown constants

$$Y = \frac{d_7^{(*1)}}{a_1'}, \qquad\qquad Z = \frac{d_7^{*(2)}}{a_1'}.$$

This choice is imposed by simplicity reasons. Thus, in the case of the system (4.1.44) the energy bound simply reads $\mathcal{R}_E = 2\sqrt{R*_a}/Y$, while in the case of the second system $(4.1.44)_{1-5,7}$, we have $\mathcal{R}_E = 2\sqrt{R*_a}/Z$, hence, in order to find $\mathcal{R}_E$ it is sufficient to deduce Y and Z.

Then, for the first system, $(4.1.44)_{5,6}$ read

$$a_2'Y = -\alpha s + \alpha b + \tau cS_c^{-1} \tag{4.1.49}$$

and

$$a_1'Y = 1 + a + \alpha cS_c^{-1} \tag{4.1.50}$$

respectively. For the second system, $(4.1.44)_5$ and $(4.1.44)_7$ read

$$a_2'Z = -\alpha s + \alpha b + \tau cS_c^{-1}$$

and

$$a_1'Z = 1 + a + \alpha cS_c^{-1} \tag{4.1.51}$$

respectively.

Using the notation

$$1 + \tau N_1 = f, \quad 1 - \alpha N\lambda\sigma = r, \quad \lambda\sigma^{-1}\tau^{-1} - \alpha N_1 = m,$$
$$\lambda\sigma^{-1}\tau^{-1} = d, \quad N\lambda\sigma = e \tag{4.1.52}$$

from $(4.1.44)_{1,2,3}$ we have immediately

$$\begin{cases} a = [a_1'^2(1+\tau) + a_2'^2 d^2 - 2a_1'a_2'd]/f, \\ b = [a_1'^2 e^2\tau + a_2'^2(1+\tau N_1^2) - 2a_1'a_2'N_1 e\tau]/f, \\ cS_c^{-1} = [-a_1'^2 e - a_2'^2 N_1 d + 2a_1'a_2'N_1]/f, \end{cases} \tag{4.1.53}$$

and, thus, $(4.1.21)_1$, $(4.1.24)_2$ give

$$d_7 = [a_1'^2(\tau+r) + a_2'^2 md - 2a_1'a_2'm]/f + 1,$$

$$d_8 = -\alpha s + \{-a_1'^2\tau er + a_2'^2[\alpha - (f-1)m] + 2a_1'a_2'(f-1)r\}/f.$$

Introduce also the notation

$$\delta = 4(r\alpha + m\tau)f^{-1} \tag{4.1.54}$$

and take into account the following identities

$$m - dr = -\alpha, \qquad d\tau = \lambda\sigma^{-1}, \qquad me - rN_1 = -1$$

$$\alpha e + \tau N_1 = f - r, \qquad f\delta = 4[\alpha + \lambda\sigma^{-1} - \alpha(f-r)]$$

to obtain

$$\begin{aligned}
Y &= 2ra_1' - 2ma_2', & Z &= 2\tau f^{-1}a_1' + 2\alpha f^{-1}a_2', \\
a_1' &= (2\alpha/f\delta)Y + 2m\delta^{-1}Z, & a_2' &= (-2\tau/f\delta)Y + 2r\delta^{-1}Z, \\
d_7 &= 4(\alpha^2\tau + \alpha^2 r + \tau^2 md + 2m\alpha\tau)f^{-3}\delta^{-2}Y^2 + m\delta^{-1}Z^2 + 1, \\
d_8 &= 4\tau(\alpha\tau - \alpha^2 er - \tau^2 mN_1 - 2r\alpha\tau N_1)f^{-3}\delta^{-2}Y^2 + r\delta^{-1}Z^2 - \alpha s.
\end{aligned}$$

Then (4.1.50), (4.1.49), read

$$d_7 = 2\alpha f^{-1}\delta^{-1}Y^2 + 2m\delta^{-1}YZ,$$

and

$$d_8 = -2\tau f^{-1}\delta^{-1}Y^2 + 2r\delta^{-1}YZ,$$

respectively, or, equivalently,

$$\begin{aligned}
&[4(\alpha^2\tau + \alpha^2 r + \tau^2 md + 2m\alpha\tau)f^{-3}\delta^{-2} - 2\alpha f^{-1}\delta^{-1}]Y^2 \\
&+ m\delta^{-1}Z^2 - 2m\delta^{-1}YZ + 1 = 0,
\end{aligned} \tag{4.1.55}$$

$$\begin{aligned}
&[4\tau(\alpha\tau - \alpha^2 er - \tau^2 mN_1 - 2r\alpha\tau N_1)f^{-3}\delta^{-2} + 2\tau f^{-1}\delta^{-1}]Y^2 \\
&+ r\delta^{-1}Z^2 - 2r\delta^{-1}YZ - \alpha s = 0.
\end{aligned} \tag{4.1.56}$$

Multiplying (4.1.55) by $-r$ and adding the obtained equation to (4.1.56) multiplied by m it follows

$$Y^2 = 4(\alpha ms + r) = 4\mathcal{R}_E^{(1)2}, \tag{4.1.57}$$

whence the first part of the extended criterion [GeoPalR00]

$$\mathcal{R} < \mathcal{R}_E^{(1)} \equiv 2\sqrt{R_{a*}}/2\sqrt{\alpha ms + r}$$

$$= \sqrt{R_{a*}}/\sqrt{-\alpha^2(1 + N\lambda^2\tau^{-1})s + \alpha(s\lambda\sigma^{-1}\tau^{-1} - N\lambda\sigma) + 1}. \qquad (4.1.58)$$

Theorem 4.1.1. *The conduction state m_0 (4.1.5) is nonlinearly stable in energy in the domain (4.1.27), where $\mathcal{R}_E$ (defined by (4.1.29)) has the expression (4.1.58).*

Similarly, (4.1.51) and (4.1.53) read

$$d_7 = 2m\delta^{-1}Z^2 + 2\alpha f^{-1}\delta^{-1}YZ, \quad d_8 = 2r\delta^{-1}Z^2 - 2\tau f^{-1}\delta^{-1}YZ,$$

or, equivalently,

$$[4(\alpha^2\tau + \alpha^2 r + \tau^2 md + 2m\alpha\tau)f^{-3}\delta^{-2}]Y^2 - m\delta^{-1}Z^2 - 2\alpha f^{-1}\delta^{-1}YZ + 1 = 0, \quad (4.1.59)$$

$$[4\tau(\alpha\tau - \alpha^2 er - \tau^2 mN_1 - 2r\alpha\tau N_1)f^{-3}\delta^{-2}]Y^2 - r\delta^{-1}Z^2 + 2\tau f^{-1}\delta^{-1}YZ - \alpha s = 0. \qquad (4.1.60)$$

Multiplying (4.1.59) by $-r$ and (4.1.60) by m and adding the obtained equations we have

$$-Y^2 + 2YZ = 4(m\alpha s + r). \qquad (4.1.61)$$

Multiplying (4.1.59) by τ and (4.1.60) by α and adding the obtained equations we have

$$\tau Y^2 - f^2 Z^2 = 4f(\alpha^2 s - \tau). \qquad (4.1.62)$$

Then (4.1.61) and (4.1.62) imply the equation

$$Y^4(4\tau - f^2) - 8fY^2[f(m\alpha s + r) - 2(\tau - \alpha^2 s)] - 16f^2(m\alpha s + r)^2 = 0,$$

whose solutions read

$$Y^2 = \frac{4f[f(m\alpha s + r) - 2(\tau - \alpha^2 s)]}{4\tau - f^2}$$
$$\pm \frac{8f\sqrt{(\tau - \alpha^2 s)^2 + \tau(m\alpha s + r)^2 - f(m\alpha s + r)(\tau - \alpha^2 s)}}{4\tau - f^2}. \qquad (4.1.63)$$

Taking into account (4.1.62) we get the expression $Y^2 = \frac{f^2 Z^2 + 4f(\alpha^2 s - \tau)}{\tau}$, which introduced into (4.1.63) leads to

$$Z^2 = \frac{4[f\tau(m\alpha s + r) - (f^2 - 2\tau)(\tau - \alpha^2 s)]}{f(4\tau - f^2)}$$

$$\mp \frac{8\tau\sqrt{(\tau - \alpha^2 s)^2 + \tau(m\alpha s + r)^2 - f(m\alpha s + r)(\tau - \alpha^2 s)}}{f(4\tau - f^2)}. \qquad (4.1.64)$$

In this way the second (last) part of our criterion for $(4.1.6) - (4.1.9)$ becomes

$$\mathcal{R} < \mathcal{R}_E^{(2)} \equiv \sqrt{R_{a*}} \cdot \left\{ f(4\tau - f^2)/[f\tau(m\alpha s + r) - (f^2 - 2\tau)(\tau - \alpha^2 s)] \right.$$

$$\left. \mp 2\tau\sqrt{(\tau - \alpha^2 s)^2 + \tau(m\alpha s + r)^2 - f(m\alpha s + r)(\tau - \alpha^2 s)]} \right\}^{1/2}. \qquad (4.1.65)$$

This formula reduces to the corresponding one from [J70a] for $N = \lambda = 0$ and $s = 1$.

The expression of $\mathcal{R}_E^{(1)}$ from (4.1.57) represents the energy stability limit while in the domain of validity of (4.1.65) we expect to have subcritical instability.

Equality $\mathcal{R}_E = \mathcal{R}_L$. Consider now the steady problem obtained by linearizing the stationary equations $(4.1.6) - (4.1.8)$ about the trivial solution

$$\Delta\mathbf{v} + (\mathcal{R}\theta - s\mathcal{C}\gamma)\mathbf{k} = \nabla p, \tag{4.1.66}$$

$$\mathcal{R}\mathbf{v}\cdot\mathbf{k} + N_1\Delta\theta + N\lambda\sigma\Delta\gamma = 0, \tag{4.1.67}$$

$$\mathcal{C}\mathbf{v}\cdot\mathbf{k} + \lambda\sigma^{-1}\tau^{-1}\Delta\theta + \Delta\gamma = 0. \tag{4.1.68}$$

Elimination of $\mathbf{v}\cdot\mathbf{k}$ between (4.1.67) and (4.1.68) implies

$$\Delta\Big[(N_1\mathcal{C} - \mathcal{R}\lambda\sigma^{-1}\tau^{-1})\theta - (\mathcal{R} - \mathcal{C}N\lambda\sigma)\gamma\Big] = 0, \tag{4.1.69}$$

In addition, (4.1.9) implies

$$\theta = \gamma = 0 \quad \text{on} \quad \partial\Omega. \tag{4.1.70}$$

But $-\Delta : W^2(\Omega) \to L^2(\Omega)$ is a positively defined operator and, by the Green formula, $(-\Delta h, h) = \|h\|^2$. Therefore, by the variational principle 3.2.2 the boundary-value problem $(4.1.69) - (4.1.70)$, written as $-\Delta h = 0$ in Ω, $h = 0$ on $\partial\Omega$, is equivalent to the associate variational problem $\min_{h\in W^{2,2}(\Omega)} \mathcal{F}(h)$, defined by the functional $\mathcal{F}(h) = \|h\|^2$. As $\mathcal{F}\geq 0$ and $\mathcal{F}(0) = 0$, it follows $h = 0$. Hence, according to the minimum Dirichlet principle (Section 3.2.4.2) for the Laplace equation, we have

$$\gamma = (N_1\mathcal{C} - \mathcal{R}\lambda\sigma^{-1}\tau^{-1})\theta/(\mathcal{R} - \mathcal{C}N\lambda\sigma) = -\frac{\lambda\sigma^{-1}\tau^{-1} - \alpha N_1}{1 - \alpha N\lambda\sigma}\theta. \tag{4.1.71}$$

Taking into account (4.1.71), (4.1.66) and (4.1.67) become

$$\Delta\mathbf{v} + \mu_1\theta\mathbf{k} = \nabla p, \tag{4.1.72}$$

$$\Delta\theta + \mu_2\mathbf{v}\cdot\mathbf{k} = 0. \tag{4.1.73}$$

The operator associated with (4.1.72), (4.1.73) is not symmetric but it can be symmetrized, so (4.1.72), (4.1.73) read equivalently

$$\Delta\mathbf{v}_1 + \sqrt{\mu_1\mu_2}\theta_1\mathbf{k} = \sqrt{\mu_2}\nabla p, \tag{4.1.74}$$

$$\Delta\theta_1 + \sqrt{\mu_1\mu_2}\mathbf{v}_1\cdot\mathbf{k} = 0, \tag{4.1.75}$$

where $\mathbf{v}_1 = \mathbf{v}\sqrt{\mu_2}$, $\theta_1 = \theta\sqrt{\mu_1}$, $\mu_2 = \mathcal{R}(1 - \alpha N\lambda\sigma)$ and

$$\mu_1 = \mathcal{R}\Big[1 - \alpha N\lambda\sigma + s\alpha(\lambda\sigma^{-1}\tau^{-1} - \alpha N_1)\Big]/(1 - \alpha N\lambda\sigma).$$

The system (4.1.74), (4.1.75) also reads

$$\begin{pmatrix} -\Delta & -\sqrt{\mu_1\mu_2}\mathbf{k} \\ -\sqrt{\mu_1\mu_2}\mathbf{k} & -\Delta \end{pmatrix}\mathbf{V} = -\mathbf{f},$$

where $\mathbf{V} = (\mathbf{v}_1, \theta)^T$ and $\mathbf{f} = (\sqrt{\mu_2}\nabla p, 0)^T$. Let A be the matricial partial differential operator associated with this equation and let $\mathbf{A} = \begin{pmatrix} -\Delta & -\sqrt{\mu_1\mu_2}\mathbf{k} \\ -\sqrt{\mu_1\mu_2}\mathbf{k} & -\Delta \end{pmatrix}$ be the matrix defining it. The system (4.1.74), (4.1.75) written as $(A\mathbf{V}, \mathbf{V}) = 0$ represents the Euler-Lagrange equations for the functional $\mathcal{F}(\mathbf{V}) = (A\mathbf{V}, \mathbf{V})$, namely

$$\mathcal{F}(\mathbf{V}) = \|\mathbf{v}_1\|^2 + \|\theta\|^2 - 2\sqrt{\mu_1\mu_2}\mid \theta\mathbf{v}_1 \cdot \mathbf{k} \mid_1 = 2 \mid \theta\mathbf{v}_1 \cdot \mathbf{k} \mid_1 \left[\frac{\|\mathbf{v}_1\|^2 + \|\theta\|^2}{2 \mid \theta\mathbf{v}_1 \cdot \mathbf{k} \mid_1} - \sqrt{\mu_1\mu_2}\right].$$

Since

$$(A\mathbf{V}, \mathbf{V}) = \|\mathbf{v}_1\|^2 + \|\theta\|^2 - 2\sqrt{\mu_1\mu_2}\int \theta\mathbf{v}_1 \cdot \mathbf{k}d\mathbf{x}$$

$$\geq \|\mathbf{v}_1\|^2 + \|\theta\|^2 - 2\sqrt{\mu_1\mu_2}\int (|\theta|^2 + |\mathbf{v}_1 \cdot \mathbf{k}|^2)d\mathbf{x}$$

$$\geq (\|\mathbf{v}_1\|^2 + \|\theta\|^2)(1 - \frac{\sqrt{\mu_1\mu_2}}{\alpha_p^2}),$$

A is a positive definite operator for $\sqrt{\mu_1\mu_2} < \alpha_p^2$, where α_p^2 is the constant in the Poincaré inequality $\|\mathbf{V}\|^2 \geq \alpha_p^2|\mathbf{V}|^2$, therefore we have $\min_{\mathbf{V}\in\mathbf{N}^1(\Omega)}\mathcal{F}(\mathbf{V}) = 0$, implying that the minimum of the functional $[\|\mathbf{v}\|^2 + \|\theta\|^2]/(2|\theta\mathbf{v} \cdot \mathbf{k}|_1)$, is $\sqrt{\mu_1\mu_2}$.

As a consequence, the mechanical equilibrium has the linear stability bound $\mathcal{R}_L$ which satisfies the relation

$$\mu_1\mu_2 \equiv \mathcal{R}_L^2\left[1 - \alpha N\lambda\sigma + s\alpha(\lambda\sigma^{-1}\tau^{-1} - \alpha N_1)\right] = R_{a*}.$$

This implies that

$$\mathcal{R}_L = \mathcal{R}_E, \tag{4.1.76}$$

if

$$1 - \alpha N\lambda\sigma > 0, \quad 1 - \alpha N\lambda\sigma + s\alpha(\lambda\sigma^{-1}\tau^{-1} - \alpha N_1) > 0, \tag{4.1.77}$$

whence

Theorem 4.1.2. *For physical parameters $N, \lambda, \sigma, \tau, s, \alpha$ in the domain defined by the inequality (4.1.77), the null solution of (4.1.6)−(4.1.9), corresponding to the basic conduction state m_0, is nonlinearly stable in the sense $\lim_{t\to\infty}\int_0^t E(t')\,dt' < \infty$ if $\mathcal{R} < \mathcal{R}_E$, where $\mathcal{R}_E$ is given by (4.1.58). Moreover, (4.1.76) holds, i.e. the linear and nonlinear energy stability limits coincide.*

Relation (4.1.58) shows that in the $(\mathcal{R}^2, \mathcal{C}^2)$-plane, keeping all other parameters fixed, the stability domain is defined by

$$\mathcal{R}^2 - \mathcal{R}\mathcal{C}(N\lambda\sigma - s\lambda\sigma^{-1}\tau^{-1}) - \mathcal{C}^2 s(1 + N\lambda^2\tau^{-1}) < R_{a*}. \tag{4.1.78}$$

Hence, for $s = 1$ it is situated under the hyperbola and for $s = -1$ under the curve defined by (4.1.78).

4.2   Variant based on simplest symmetric part of operators

Our variant of the energy method [GeoPalR96c], [GeoPalR00], [GeoPalR01], based on symmetry properties of the two operators involved into the energy relation is presented and applied to problem (4.1.6) − (4.1.10). Other variants of ours preserving only Joseph's optimality condition are subsequently described. Then a problem of horizontal convection is treated and, finally, the G-P-R method is extended to the case of two vector and one scalar unknown functions in a complicated hydromagnetic convection problem.

4.2.1   *Symmetry and optimality condition*

The energy method variant used in Section 4.1 generalizes the standard energy method in three aspects: it uses the additional equation (4.1.15), introduces the two unknown functions Φ_1 and Φ_2 as linear combination of θ and γ, and maximizes the coefficients in the energy relation (4.1.22). This leads to the optimality condition (4.1.30).

In addition we demonstrate the necessity of introducing as the new unknown constants the solutions d_7^*/a_1' of (4.1.30) if a closed form of the bound $\mathcal{R}_E$ is preferred.

Here we reformulate them all in more general functional analytic terms, suitable to a larger class of governing evolution equations. However, we keep as close to Joseph's variant as possible, since it yields optimal numerical results. Since in the energy relation only the symmetric part of the reaction-diffusion operator occurs, the central idea of the present variant is to change just these differential equations, and not those integral deduced from them. Moreover this modification must be done in such a way that the energy relation assumes the simplest form. This implies that the symmetric part of the operator must be the simplest possible. At the same time it must contain the main features of the fluid flow contained in all physical parameters. This idea emerged from [Geo77] where the symmetrization operated on the given problem enabled the derivation of variational principles (Sections 3.3, 3.4).

In the study of fluid flow stability, the symmetrization of and handling with the symmetric part of operators arises as early as 1961 in [DiP61]. The previous variant used algebraic operations in the energy relation (4.1.22) deduced from the averaged equations (4.1.12) − (4.1.15). Therefore from the very beginning an integral relation obtained by suitable multiplications of (4.1.6) − (4.1.8) by multiples of $\mathbf{v}$, θ and γ was used. On the other hand, relation (4.1.15) contributed heavily in hiding the fact that (4.1.22) was nothing else but a projection on N^1 of the following system, equivalent to (4.1.6) − (4.1.8)

$$\frac{\partial \mathbf{v}}{\partial t} + \mathbf{v}\nabla\mathbf{v} = -\nabla p + \Delta\mathbf{v} + \mathcal{R}\theta\mathbf{k} - s\alpha\mathcal{R}\gamma\mathbf{k}, \qquad (4.2.1)$$

$$aP_r\left(\frac{\partial\theta}{\partial t}+\mathbf{v}\cdot\nabla\theta\right)+ag_3S_c\left(\frac{\partial\gamma}{\partial t}+\mathbf{v}\cdot\nabla\gamma\right)=a\mathcal{R}(1+\alpha g_3)\mathbf{v}\cdot\mathbf{k} \tag{4.2.2}$$
$$+\,a(N_1+g_3\lambda\sigma^{-1}\tau^{-1})\Delta\theta+a(N\lambda\sigma+g_3)\Delta\gamma,$$

$$bS_c\left(\frac{\partial\gamma}{\partial t}+\mathbf{v}\cdot\nabla\gamma\right)+bg_2P_r\left(\frac{\partial\theta}{\partial t}+\mathbf{v}\cdot\nabla\theta\right)=b\mathcal{R}(\alpha+g_2)\mathbf{v}\cdot\mathbf{k} \tag{4.2.3}$$
$$+\,(\lambda\sigma^{-1}\tau^{-1}+g_2N_1)\Delta\theta+(1+g_2N\lambda\sigma)\Delta\gamma.$$

Equation (4.2.1) is identical to (4.1.6). Equation (4.2.2) was derived by adding equation (4.1.8) multiplied by some constant g_3 to (4.1.7) and then multiplying the resulted equation by some constant $a>0$. Similarly, multiplying (4.1.7) by some constant g_2, adding the obtained equation to (4.1.8) and then multiplying this sum by some positive constant b we obtained (4.2.3).

In this way, into (4.2.2) and (4.2.3) we included the equations which led to (4.1.15). Furthermore we required that, in the criterion for $(4.1.6)-(4.1.9)$ the effect of the changes operated in (4.1.7) and (4.1.8) and leading to (4.2.2) and (4.2.3) would be the same as those in (4.1.15). Thus, we first introduced the condition for $(4.2.1)-(4.2.3)$ to possess a symmetrizable linear part for a suitable choice of the constants a,b,g_3 and g_2. Indeed, due to the simplicity of the basic state, many linear nonsymmetric problems in the vertical thermal convection can be reduced to symmetrical form by suitable multiplications, additions, splittings (decomposings) and recomposings of the given equations. By splitting equations additional unknown functions are introduced while in recomposing equations, fewer unknown functions are used. In energy variant of parameters differentiation, apart from multiplications by constants and addition of equations, the number of the unknown functions was reduced (instead of $\mathbf{u},\theta$ and γ the variational problem defining $\mathcal{R}_E$ had $\mathbf{u}$ and ϕ_1 only). The system $(4.2.1)-(4.2.3)$ can be written in the form

$$L_2\mathbf{V}+(\mathbf{v}\nabla)\mathbf{V}_1=L_1\mathbf{V}+(-\nabla p,0,0)^T \tag{4.2.4}$$

or in the form

$$L_2\mathbf{V}=L_1\mathbf{V}-(\mathbf{v}\nabla)L_2\mathbf{V}+(-\nabla p,0,0)^T \tag{4.2.4'}$$

or, equivalently,

$$L_2\mathbf{V}=L_1\mathbf{V}+N(\mathbf{V}) \tag{4.2.4}$$

where $\mathbf{V}=(\mathbf{v},\theta,\gamma)^T$, $\mathbf{V}_1=(\mathbf{v},aP_r\theta+ag_3S_c\gamma,bS_c\gamma+bg_2P_r\theta)^T$ and L_1 and L_2 are the (linear) operators defined by

$$L_1=\begin{pmatrix}\Delta & \mathcal{R}\mathbf{k} & -s\alpha\mathcal{R}\mathbf{k}\\ a\mathcal{R}(1+\alpha g_3)\mathbf{k} & a(N_1+g_3\lambda\sigma^{-1}\tau^{-1})\Delta & a(N\lambda\sigma+g_3)\Delta\\ b\mathcal{R}(\alpha+g_2)\mathbf{k} & (\lambda\sigma^{-1}\tau^{-1}+g_2N_1)\Delta & b(1+g_2N\lambda\sigma)\Delta\end{pmatrix}. \tag{4.2.5}$$

$$L_2=\begin{pmatrix}\frac{\partial}{\partial t} & 0 & 0\\ 0 & aP_r\frac{\partial}{\partial t} & ag_3S_c\frac{\partial}{\partial t}\\ 0 & bg_2P_r\frac{\partial}{\partial t} & bS_c\frac{\partial}{\partial t}\end{pmatrix}. \tag{4.2.5'}$$

The nonlinear part of $(4.2.1) - (4.2.3)$ was denoted by $N(\mathbf{V})$. We recall that the rate of the change term in $(4.2.4)$ is $L_2\mathbf{V}+(\mathbf{v}\nabla)\mathbf{V}_1$. It contains a linear part $L_2\mathbf{V}$ and a nonlinear advective part $(\mathbf{v}\nabla)\mathbf{V}_1$. Therefore the nonlinear mapping $N(\mathbf{V})$ consists of the advective and pressure terms, L_2 is the linear rate of the change operator, L_1 is the linear operator corresponding to the perturbation diffusion $(\Delta\mathbf{V})$ and reaction terms. The reaction terms coming from $(4.2.2)$ and $(4.2.3)$ and not depending on g_2 and g_3 are advective terms for the basic thermal equilibrium. These basic advective terms are different from those in [RioM84] and other papers, due to their form, which is such that L_1 assumes different symmetry properties. Thus we can use them at our best convenience. Project the new system $(4.2.1) - (4.2.3)$ from L^2 to N^1 to obtain the energy relation for this system

$$(L_2\mathbf{V}_1, \mathbf{V}) = (L_1\mathbf{V}, \mathbf{V}) + (N(\mathbf{V}), \mathbf{V}). \qquad (4.2.6)$$

It must be equal to $(4.1.22)$. Since we have in view to obtain a rate of change term as in $(4.1.16)$, this can be obtained if the coefficients of $\theta\dfrac{d\gamma}{dt}$ and $\gamma\dfrac{d\theta}{dt}$ in $(4.2.6)$ are equal

$$ag_3 S_c = bg_2 P_r. \qquad (4.2.7)$$

Therefore we have only three parameters, a, b, and g_3, multiplying $(4.1.7)$ and $(4.1.8)$ as in Joseph's variant. The relationship between g_3 and c from Section 4.1.3 is $c = ag_3 S_c$. The choice $(4.2.7)$ implies the null contribution in $(4.2.6)$ of the nonlinear terms from the left-hand side of $(4.2.2)$ and $(4.2.3)$.

Indeed, the symmetric part of L_2 became (Remark 4.5)

$$L_{2s} = \begin{pmatrix} \frac{\partial}{\partial t} & 0 & 0 \\ 0 & aP_r\frac{\partial}{\partial t} & 0 \\ 0 & 0 & bS_c\frac{\partial}{\partial t} \end{pmatrix}.$$

In addition, in $(4.2.6)$ the term in ∇p is no longer present being orthogonal to solenoidal vectors $\mathbf{v}$. Consequently,

$$(N(\mathbf{V}), \mathbf{V}) = 0. \qquad (4.2.8)$$

The symmetric part of L_1 reads

$$L_{1s} = \begin{pmatrix} \Delta & \frac{1}{2}d_7\mathcal{R}\mathbf{k} & \frac{1}{2}d_8\mathcal{R}\mathbf{k} \\ \frac{1}{2}d_7\mathcal{R}\mathbf{k} & d_4\Delta & \frac{1}{2}d_6\Delta \\ \frac{1}{2}d_8\mathcal{R}\mathbf{k} & \frac{1}{2}d_6\Delta & d_5\Delta \end{pmatrix}. \qquad (4.2.9)$$

Since $(L_1\mathbf{V}, \mathbf{V}) = (L_{1s}\mathbf{V}, \mathbf{V})$, we have

$$(L_1\mathbf{V}, \mathbf{V}) = - \|\mathbf{u}\|^2 + \mathcal{R}\Big(d_7|\theta\mathbf{v}\cdot\mathbf{k}|_1 + d_8|\gamma\mathbf{v}\cdot\mathbf{k}|_1\Big)$$
$$- \Big(d_4\|\theta\|^2 + d_6|D\theta\cdot D\gamma|_1 + d_5\|\gamma\|^2\Big), \qquad (4.2.10)$$

hence $(4.2.10)$ becomes $(4.1.22)$ if the two identities hold: the terms in the accolades form the perfect square $\|\phi_1\|^2$; the coefficient of $\mathcal{R}$ represents a factor of $|\phi_1\mathbf{v}\cdot\mathbf{k}|_1$.

In order for (4.2.6) to coincide with (4.1.22) it remains to make equal also their rate of change terms. From (4.2.6) we have $(L_{2s}\mathbf{V}, \mathbf{V}) = \frac{1}{2}\frac{d}{dt}|aP_r\theta^2 + 2c\theta\gamma + bS_c\gamma^2|$, while in (4.2.22) the corresponding terms are $\frac{1}{2}\frac{d}{dt}(|\mathbf{v}|^2 + d_1|\phi_1|^2 + d_3|\phi_2|^2)$, hence we must have

$$d_1 a_1'^2 + d_3 b_1'^2 = aP_r, \qquad d_1 a_2'^2 + d_3 b_2'^2 = bS_c, \qquad d_1 a_1' a_2' + d_3 b_1' b_2' = c,$$

where d_3, b_1' and b_2' are determined up to some factor. Eliminating d_1 and d_3 between these relations we obtain (4.1.44). Therefore, the same results were obtained by the two variants of the energy method. Joseph's idea of using (4.1.15) was generalized by including it into the modified governing equations, in which the equations generating (4.1.15) were present by means of the terms in g_2 and g_3. These terms drastically changed the linear part of the governing equations leading to a much more advantageous symmetric part of the diffusion operator. On the contrary, the symmetric operator in $(4.1.6) - (4.1.8)$ does not contain the effect of terms in $\mathbf{v}$ from (4.1.8) and those of terms in θ from (4.1.6) because they are opposite.

Consequently, our variant of energy method proceeds as follows: 1) *the balance equations but the momentum equation (in $\mathbf{v}$) are replaced by their linear combinations.* Thus, a new system of governing equations is obtained; 2) *the symmetric part of the corresponding new linear rate of change operator must be the simplest possible.* As a consequence, the contribution of the (nonlinear) advective terms to the energy relation vanishes. In exchange, all physical parameters are taken into account: 3) *the symmetric part of the linear diffusion-reaction operator L_1 must be the simplest possible.* This enables the definition of a new function Φ_1 as a linear combination of the unknown functions (except for $\mathbf{v}$); 4) *the standard energy method is applied to the new system of governing equations* to produce the energy relation; 5) *the energy relation must have only five terms as in the standard energy method,* i.e. to have the form $\frac{dE}{dt} + \frac{d\Psi}{dt} = -(\|\mathbf{v}\|^2 + \|\Phi_1\|^2) + \mathcal{R}k|\Phi_1\mathbf{v}\cdot\mathbf{k}|_1$, where k is a certain constant; 6) *it is required that the undetermined coefficient k in the energy relation must be optimal;* 7) *the system of the multiplication constants and the parameters defining the new unknown functions must be reduced to solving the system in which the new constant unknowns are the solutions of the optimality relation.*

4.2.2 *More general setting of our variant*

The derivation of $(4.2.1) - (4.2.3)$ still reminds the peculiarities involved into the variant in Section 4.1. Here we present our variant independently of the Joseph's one, but preserving its optimality condition.

In order to obtain the energy relation (4.1.22), written, equivalently, as

$$\frac{1}{2}\frac{d}{dt}(|\mathbf{v}|^2 + d_1|\phi_1|^2 + d_2|\phi_2|^2) = -(\|\mathbf{v}\|^2 + \|\phi_1\|^2) + \mathcal{R}c_1|\phi_1\mathbf{v}\cdot\mathbf{k}|_1, \qquad (4.2.11)$$

we perform $\frac{a_1'}{P_r}(4.1.7) + \frac{a_2'}{S_c}(4.1.8)$ and $\frac{b_1'}{P_r}(4.1.7) + \frac{b_2'}{S_c}(4.1.8)$ yielding

$$\frac{\partial\phi_1}{\partial t} + \mathbf{v}\cdot\nabla\phi_1 = \mathcal{R}c_2\mathbf{v}\cdot\mathbf{k} + c_3\Delta\phi_1 + c_4\Delta\phi_2, \qquad (4.2.12)$$

$$\frac{\partial \phi_2}{\partial t} + \mathbf{v} \cdot \nabla \phi_2 = \mathcal{R} c_5 \mathbf{v} \cdot \mathbf{k} + c_6 \Delta \phi_1 + c_7 \Delta \phi_2, \tag{4.2.13}$$

where

$$
\begin{aligned}
c_2 &= \frac{a'_1}{P_r} + \frac{a'_2}{S_c}\frac{\mathcal{C}}{\mathcal{R}} = \frac{1}{S_c}(a'_1 \tau + \alpha a'_2), \\[4pt]
c_3 &= \frac{1}{M'S_c}[N_1 \tau a'_1 b'_2 + \lambda \sigma^{-1}\tau^{-1} a'_2 b'_2 - a'_2 b'_1 - N\lambda\sigma\tau a'_1 b'_1], \\[4pt]
c_4 &= \frac{1}{M'S_c}[-N_1 \tau a'_1 a'_2 - \lambda \sigma^{-1}\tau^{-1} a'^2_2 + a'_1 a'_2 + N\lambda\sigma\tau a'^2_1], \\[4pt]
c_5 &= \frac{1}{S_c}(\alpha b'_2 + \tau b'_1), \\[4pt]
c_6 &= \frac{1}{M'S_c}[N_1 \tau b'_1 b'_2 + \lambda \sigma^{-1}\tau^{-1} b'^2_2 - b'_1 b'_2 - N\lambda\sigma\tau b'^2_1], \\[4pt]
c_7 &= \frac{1}{M'S_c}[-N_1 \tau b'_1 a'_2 - \lambda \sigma^{-1}\tau^{-1} a'_2 b'_2 + a'_1 b'_2 + N\lambda\sigma\tau a'_1 b'_1].
\end{aligned}
\tag{4.2.14}
$$

Writing θ and γ in terms of ϕ_1 and ϕ_2, and multiplying (4.2.12) by a parameter p_1^2 and (4.2.13) by p_2^2, where $p_1, p_2 \in \mathbf{R}$, we have the system

$$\frac{\partial \mathbf{v}}{\partial t} + \mathbf{v} \cdot \nabla \mathbf{v} = -\nabla p + \Delta \mathbf{v} + \mathcal{R} c_8 \phi_1 \cdot \mathbf{k} + \mathcal{R} c_9 \phi_2 \cdot \mathbf{k} \tag{4.2.15}$$

$$p_1^2 \frac{\partial \phi_1}{\partial t} + p_1^2 \mathbf{v} \cdot \nabla \phi_1 = p_1^2 \mathcal{R} c_2 \mathbf{v} \cdot \mathbf{k} + p_1^2 c_3 \Delta \phi_1 + p_1^2 c_4 \Delta \phi_2, \tag{4.2.16}$$

$$p_2^2 \frac{\partial \phi_2}{\partial t} + p_2^2 \mathbf{v} \cdot \nabla \phi_2 = p_2^2 \mathcal{R} c_5 \mathbf{v} \cdot \mathbf{k} + p_2^2 c_6 \Delta \phi_1 + p_2^2 c_7 \Delta \phi_2, \tag{4.2.17}$$

where $c_8 = (b'_2 + s\alpha b'_1)/M'$, $c_9 = -(a'_2 + s\alpha a'_1)/M'$, whose associated diffusion-reaction operator B, and its symmetric B_s read $B\mathbf{W} = \mathbf{B}\mathbf{W}$, and $B_s\mathbf{W} = \mathbf{B}_s\mathbf{W}$, where $\mathbf{W} = (\mathbf{v}, \phi_1, \phi_2)^T$, and are associated with the matrices

$$
\mathbf{B} = \begin{pmatrix} \Delta & \mathcal{R} c_8 \mathbf{k} & \mathcal{R} c_9 \mathbf{k} \\ \mathcal{R} p_1^2 c_2 \mathbf{k} & p_1^2 c_3 \Delta & p_1^2 c_4 \Delta \\ \mathcal{R} p_2^2 c_5 \mathbf{k} & p_2^2 c_6 \Delta & p_2^2 c_7 \Delta \end{pmatrix}, \quad
\mathbf{B}_s = \begin{pmatrix} \Delta & \mathcal{R}\frac{c_8 + p_1^2 c_2}{2}\mathbf{k} & \mathcal{R}\frac{c_9 + p_2^2 c_5}{2}\mathbf{k} \\ \mathcal{R}\frac{c_8 + p_1^2 c_2}{2}\mathbf{k} & p_1^2 c_3 \Delta & \frac{p_2^2 c_6 + p_1^2 c_4}{2}\Delta \\ \mathcal{R}\frac{c_9 + p_2^2 c_5}{2}\mathbf{k} & \frac{p_2^2 c_6 + p_1^2 c_4}{2}\Delta & p_2^2 c_7 \Delta \end{pmatrix}.
$$

The energy relation, obtained by taking the inner product of $(4.2.15) - (4.2.17)$ in $\mathbf{N}^1$ by $\mathbf{W}$, reads

$$\frac{1}{2}\frac{d}{dt}(|\mathbf{v}|^2 + p_1^2|\phi_1|^2 + p_2^2|\phi_2|^2) = -(\|\mathbf{v}\|^2 + p_1^2 c_3\|\phi_1\|^2 + p_2^2 c_7\|\phi_2\|^2)$$
$$+ \mathcal{R}(c_8 + p_1^2 c_2)|\phi_1 \mathbf{v} \cdot \mathbf{k}|_1 + \mathcal{R}(c_9 + p_2^2 c_5)|\phi_2 \mathbf{v} \cdot \mathbf{k}|_1 - (p_2^2 c_6 + p_1^2 c_4)|\nabla \phi_1 \cdot \nabla \phi_2|_1. \tag{4.2.18}$$

From (4.2.18) we obtain (4.2.11) if

$$p_1^2 c_3 = 1, \tag{4.2.19}$$

$$p_2^2 c_7 = 0, \tag{4.2.20}$$

$$\mathcal{R}(c_9 + p_2^2 c_5) = 0, \tag{4.2.21}$$

$$p_2^2 c_6 + p_1^2 c_4 = 0, \tag{4.2.22}$$

$$\mathcal{R}\frac{\partial(c_8 + p_1^2 c_2)}{\partial p} = 0, \tag{4.2.23}$$

when $\mathbf{B}$ and $\mathbf{B}_s$ become

$$\mathbf{B} = \begin{pmatrix} \Delta & \mathcal{R}c_8\mathbf{k} & \mathcal{R}c_9\mathbf{k} \\ \mathcal{R}p_1^2 c_2\mathbf{k} & \Delta & p_1^2 c_4\Delta \\ -\mathcal{R}c_9\mathbf{k} & -p_1^2 c_4\Delta & 0 \end{pmatrix}, \quad \mathbf{B}_s = \begin{pmatrix} \Delta & \mathcal{R}\max\frac{c_8+p_1^2 c_2}{2}\mathbf{k} & 0 \\ \mathcal{R}\max\frac{c_8+p_1^2 c_2}{2}\mathbf{k} & \Delta & 0 \\ 0 & 0 & 0 \end{pmatrix}. \tag{4.2.24}$$

In (4.2.23) the parameter p is a convenient function of the unknown parameters $p_1, p_2, a_1', a_2', b_1'/b_2'$.

It is interesting to note that the simplicity of $\mathbf{B}_s$ (*i.e.* null entries) implies that the corresponding entries in the skew-symmetric part of $\mathbf{B}$ are not null. Thus, we leave aside much information of the given problem and we expect to have weak results. That it is not the case; this follows from the equality of R_E and R_L. Therefore, we must be very cautious in handling $\mathbf{B}_s$ and $\mathbf{B}_{ss}$.

For the choice (4.2.19) $-$ (4.2.23), the energy relation (4.2.18) becomes (4.2.11), where $d_1 = p_1^2$, $d_2 = p_2^2$ and $c_1 = \max(c_8 + p_1^2 c_2)$, the maximum being taken with respect to parameter p. It is understood that from the system (4.2.19) $-$ (4.2.22) all unknown parameters are expressed in terms of p.

It can be proved that using this general setting the same results as in Section 4.1.3 (and, so, Section 4.2.1) are obtained. In particular, the energy stability limit is (4.2.29), where $|a_1 d_7 + b_1 d_8|$ is replaced by c_1, but the numerical value is the same. Moreover, this setting is easier to apply as compared with the variant from Section 4.2.1.

Example 4.2.1. In the particular case $N = \lambda = 0$, $s = 1$, we have $c_2 = (a_1'\tau + \alpha a_2')/S_c$, $c_3 = (\tau a_1' b_2' - a_2' b_1')/(M'S_c)$, $c_4 = a_1' a_2'(1-\tau)/(M'S_c)$, $c_5 = (\alpha b_2' + \tau b_1')/S_c$, $c_6 = b_1' b_2'(\tau - 1)/(M'S_c)$, $c_7 = (b_2' a_1' - \tau b_1' a_2')/(M'S_c)$, $c_8 = (b_2' + \alpha b_1')/M'$, $c_9 = -(a_2' + \alpha a_1')/M'$, while (4.1.19) $-$ (4.1.22) imply

$$p_1^2/S_c = M'/(\tau a_1' b_2' - a_2' b_1'), \tag{4.2.25}$$

$$b_2' a_1' = \tau b_1' a_2', \tag{4.2.26}$$

$$-(a_2' + \alpha a_1')/M' + p_2^2/[S_c(\alpha b_2' + \tau b_1')] = 0, \tag{4.2.27}$$

$$p_2^2 b_1' b_2'(\tau - 1)/(M'S_c) + p_1^2 a_1' a_2'(1 - \tau)/(M'S_c) = 0. \tag{4.2.28}$$

Taking into account (4.2.26), it follows that $M' = b_1' a_2'(\tau - 1)$, so (4.2.25) implies $p_1^2/S_c = 1/(1 + \tau)$. Then (4.2.28) provides $p_2^2/S_c = \frac{a_1' a_2'}{b_1' b_2'(1+\tau)}$, while (4.2.27) gives $p_2^2/S_c = \frac{(a_2'+\alpha a_1')}{(\alpha b_2'+\tau b_1')M'}$. Equating these two expressions of p_2, introducing the parameter $p = \tau b_1'/b_2'$ and taking into account that $a_1'/a_2' = p$, we obtain

$$a_1'^2 = \frac{\tau+1}{\tau-1}\frac{(1+\alpha p)p}{\alpha+p}, \quad a'^2_2 = \frac{\tau+1}{\tau-1}\frac{(1+\alpha p)}{p(\alpha+p)}, \quad 2a_2'\frac{da_2'}{dp} = -\frac{(\tau+1)(\alpha p^2+2p+\alpha)}{(\tau-1)p^2(\alpha+p)^2}.$$
$$(4.2.29)$$

Then (4.2.23) reads successively

$$\frac{d}{dp}\left[\frac{b_2'+\alpha b_1'}{(\tau-1)a_2'b_1'} + \frac{a_1'\tau+\alpha a_2'}{\tau+1}\right] = \frac{d}{dp}\left[\frac{(\tau+\alpha p)}{(\tau-1)a_2'p} + \frac{a_2'(\alpha+\tau p)}{\tau+1}\right]$$

$$= \frac{d}{dp}\left\{\frac{1}{(\tau^2-1)pa_2'}[(\tau+\alpha p)(\tau+1) + p(\alpha+\tau p)(\tau-1)a'^2_2\right\}$$

$$= \frac{d}{dp}\left\{\frac{1}{(\tau^2-1)pa_2'}[(\tau+\alpha p)(\tau+1) + p(\alpha+\tau p)(\tau-1)\frac{\tau+1)}{(\tau-1)}\frac{1+\alpha p}{p(\alpha+p)}]\right\}$$

$$= \frac{1}{\tau-1}\frac{d}{dp}\left[\frac{p^2\alpha(1+\tau)+2(\alpha^2+\tau)p+\alpha+\alpha\tau}{pa_2'}\right] = 0,$$

implying

$$\alpha^2(\tau+1)p^4 + 2\alpha(1+\alpha^2\tau)p^3 - 2\alpha(1+\alpha^2\tau)p - \alpha^2(\tau+1) = 0. \qquad (4.2.30)$$

The solutions of (4.2.30) are ± 1, $\frac{-(1+\alpha^2\tau)\pm\sqrt{(1-\alpha^2)(1-\alpha^2\tau^2)}}{\alpha(\tau+1)}$. Joseph's case (Section 4.1.5) corresponds to $p = -1$ and $p = \frac{-(1+\alpha^2\tau)-\sqrt{(1-\alpha^2)(1-\alpha^2\tau^2)}}{\alpha(\tau+1)}$.

For the choice $(4.2.19) - (4.2.23)$, the energy relation (4.2.18) becomes (4.2.11), where $d_1 = p_1^2$, $d_2 = p_2^2$ and $c_1 = \max(c_8 + p_1^2 c_2)$, the maximum being taken with respect to parameter p. It is understood that from the system $(4.2.19) - (4.2.22)$ all unknown parameters $p_1, p_2, a_1', a_2', b_1'/b_2'$ are expressed in terms of p.

It can be proved that using the general setting the same results as in Section 4.1.3 and, so, Section 4.1.2, are obtained. In particular the energy stability limit in (4.1.29), where $| a_1d_7 + b_1d_8 |$ is replaced by c_1, but the numerical value is the same.

This setting is easier to apply compared to the variant from Section 4.2.1.

4.2.3   *Symmetry instead of optimality condition (4.2.23)*

The operator B and, since all the involved derivatives are of an even order, the matrix $\mathbf{B}$ are symmetric if and only if

$$c_8 = p_1^2 c_2, \qquad (4.2.31)$$

$$c_9 = p_2^2 c_5, \qquad (4.2.32)$$

$$p_2^2 c_6 = p_1^2 c_4. \qquad (4.2.33)$$

Then, taking into account $(4.2.20) - (4.2.23)$, it follows $c_2 = c_4 = c_5 = c_6 = c_7 = c_8 = c_9 = c_{10}$, *i.e.* an over-determined system for the five unknown parameters.

Consequently, if it has a root, then the corresponding stability criterion is valid on a submanifold of the parameter space. This criterion will be valid for every initial perturbation energy, hence it asserts the exponential asymptotic stability of m_0.

Therefore we look either for a partial symmetry of B, or we renounce to the simplicity of B_s leading to criteria valid for larger domains of the parameter space.

Case 1. Suppose that all reasonings from Section 4.2.2 are valid except for (4.2.23), which is replaced by a requirement of partial symmetry for B, namely

$$c_8 = p_1^2 c_2. \tag{4.2.34}$$

i.e. $b_{12} = b_{21}$, where b_{ij} are the entries of $\mathbf{B}$. Correspondingly, $c_1 = c_8(= p_1^2 c_2)$, implying, for small initial perturbation energy, the energy stability bound $\mathcal{R}_E = 2\sqrt{R_{a*}}/c_8$, where the unknown constants occurring in c_8, namely $p_1, p_2, a_1', a_2', b_1'/b_2'$, are the roots of $(4.2.19) - (4.2.22)$, (4.2.34). As $\mathbf{B}$ is not (completely) symmetric and the optimality condition (4.2.23) no longer holds, this bound is possibly not equal to the nonlinear stability limit $\mathcal{R}_G$.

Case 2. Let us renounce to the very simple form (4.2.11), by keeping in (4.2.18) the term in $\|\phi_2\|^2$. Therefore, replace (4.2.20) by

$$p_2^2 c_7 = 1. \tag{4.2.35}$$

Note that (4.2.19) and (4.2.35) are necessary for the variational problem

$$\max_{\mathbf{v}\in N^1} \frac{|\,\phi_1 \mathbf{v} \cdot \mathbf{k}\,|_1}{\|\mathbf{v}\|^2 + \|\phi_1\|^2 + \|\phi_2\|^2} \tag{4.2.36}$$

be universal, *i.e.* independent of physical parameters. In this case $\mathbf{B}_s$ becomes

$$\mathbf{B}_s = \begin{pmatrix} \Delta & \mathcal{R}\frac{c_8+p_1^2 c_2}{2}\mathbf{k} & 0 \\ \mathcal{R}\frac{c_8+p_1^2 c_2}{2}\mathbf{k} & \Delta & 0 \\ 0 & 0 & \Delta \end{pmatrix}. \tag{4.2.37}$$

Therefore, in order to determine the five unknown parameters, besides (4.2.19), (4.2.35), (4.2.21) and (4.2.22) we need another relation, which we choose to be the equality $b_{12} = b_{21}$, *i.e.* (4.2.34). In these conditions m_0 is exponentially asymptotically stable.

General case. Conditions like (4.2.21) and (4.2.22) induced to the operator B and to the matrix $\mathbf{B}$, given by (4.2.24), a partial skew-symmetry. However, if the optimality condition (4.2.23) is not to be considered, we are interested to have $\mathbf{B}$ symmetric, to be sure that the variational problem defining $\mathcal{R}_E$ contains all information of the model $(4.1.6) - (4.1.9)$. In addition, the requirements (4.2.19) and (4.2.36) can be relaxed. More exactly, the most general variational problem defining $\mathcal{R}_E$ is

$$\frac{1}{\mathcal{R}_E} = \frac{1}{2}\max_{\mathbf{v}\in N^1} \frac{\frac{(c_8+p_1^2 c_2)}{\sqrt{p_1^2 c_3}}\,|\,\sqrt{p_1^2 c_3}\,\phi_1 \mathbf{v} \cdot \mathbf{k}\,|_1 + \frac{(c_9+p_2^2 c_5)}{\sqrt{p_2^2 c_7}}\,|\,\sqrt{p_2^2 c_7}\,\phi_1 \mathbf{v} \cdot \mathbf{k}\,|_1}{\|\mathbf{v}\|^2 + \|\sqrt{p_1^2 c_3}\,\phi_1\|^2 + \|\sqrt{p_2^2 c_7}\,\phi_2\|^2} \tag{4.2.38}$$

and it is universal if

$$\frac{(c_8 + p_1^2 c_2)}{\sqrt{p_1^2 c_3}} \, \frac{\sqrt{p_2^2 c_7}}{(c_9 + p_2^2 c_5)} = p_3, \tag{4.2.39}$$

where p_3 is a constant which does not depend on physical parameters. Hence, the conditions (4.2.19) and (4.2.35) are sufficient but not necessary and we no longer impose them. In exchange, we impose (4.2.22), (4.2.34) and (4.2.39) and the additional symmetry conditions

$$c_9 = p_2^2 c_5. \tag{4.2.40}$$

$$p_1^2 c_4 = p_2^2 c_6. \tag{4.2.41}$$

Note that (4.2.22) must always be imposed. Otherwise we must estimate its absolute value introducing positive terms in the derivatives of ϕ_1 and ϕ_2, hence taking an important part of the negative terms on the right-hand side of (4.2.18). As a result we obtain a stability bound (not limit) and it is expected to be very small.

From (4.2.22) and (4.2.41) it follows

$$c_4 = c_6 = 0 \tag{4.2.42}$$

restricting the validity of the resulting $\mathcal{R}_E$ to a manifold of the parameter space. Further, several other supplementary restrictions follow taking into account the other two conditions for symmetry of B, *i.e.* (4.2.34) and (4.2.35).

Thus, if in (4.2.39) we take into account (4.2.34) and (4.2.40), we obtain $\frac{p_1 c_2 \sqrt{c_7}}{p_2 c_5 \sqrt{c_3}} = p_3$, implying

$$\frac{\partial}{\partial p} \left(\frac{p_1 c_2 \sqrt{c_7}}{p_2 c_5 \sqrt{c_3}} \right) = 0. \tag{4.2.43}$$

Consequently, the over-determinacy of the system determining the parameters increases, being necessary to impose (4.2.43). It implies $\frac{1}{\mathcal{R}_E} = \mid \frac{2(c_9 + p_2^2 c_5)}{(\sqrt{p_2^2 c_7}} \mid \frac{1}{\sqrt{\mathcal{R}^{**}}}$, where

$$\frac{1}{\sqrt{\mathcal{R}^{**}}} = \max_{\mathbf{W} \in \mathbf{N}^1} \frac{p_3 \mid \psi_1 \mathbf{v} \cdot \mathbf{k} \mid_1 + \mid \psi_2 \mathbf{v} \cdot \mathbf{k} \mid_1}{\|\mathbf{v}\|^2 + \|\psi_1\|^2 + \|\psi_2\|^2},$$

where $\mathbf{W} = (\mathbf{v}, \psi_1, \psi_2)$ is an arbitrary difference motion of $\mathbf{N}^1$.

Since, in addition, (4.2.39) implies that the derivatives with respect to all physical parameters of the expression of p are equal to zero, it is possible to have a stability limit valid for a few values of the physical parameters or to have no such limit at all. Each such case is to be studied separately.

4.2.4 *Case of nonsymmetric operators in a horizontal convection problem*

In spite of its importance for phase transitions phenomena (*e.g.* artificial crystal growth), large scale atmospheric motions, astrophysics and geophysics, the horizontal convection was investigated much less than the vertical one. Among the causes we quote: the basic state cannot have a closed-form; the perturbation equations contain two reaction terms, namely in the vertical and horizontal directions; the O-B approximation used is not yet rigorously proved to hold; the reaction-diffusion operator cannot be conveniently symmetrized.

The first attempt to study rigorously the onset of the stationary horizontal convection from the conduction state belongs to one of the authors (A.G.) and Daniela Mansutti [GeoM99]. Partially we follow their paper.

The domain of motion Ω is the parallelepipedic box with rigid faces parallel to the coordinate planes, the origin O of the Cartesian system of coordinates is located at a corner of the bottom horizontal (x, z) plane, Oy-axis is pointing vertically upwards. Let $\mathbf{i}, \mathbf{j}, \mathbf{k}$ stand for the unit vectors in the x, y and z directions respectively. The phase functions are the velocity $\mathbf{u}$ and the temperature T.

Define the characteristic quantities: $T_1 - T_0$, temperature; d, (box height), length; $\sqrt{G_r}\nu/d$, velocity; d^2/ν, time; $G_r\rho_0\nu^2/d^2$, pressure; ρ_0, density. Here $G_r = g\beta(T_1 - T_0)d^3/\nu^2$ is the Grashof number, so $\mathcal{R} = P_rG_r$ is the Rayleigh number. As usual, ν is the coefficient of kinematic viscosity, β is the coefficient of volumetric expansion, k is the thermal diffusivity, g is the gravitational acceleration and $P_r = \nu/k$ is the Prandtl number. Then the mathematical model governing the conduction state and the various types of horizontal convection consists of conservation equations for mass, momentum and energy (in the O-B approximation)

$$\frac{1}{\sqrt{G_r}}\frac{\partial\mathbf{u}}{\partial t} + \mathbf{u}\cdot\nabla\mathbf{u} = -\nabla p - T\mathbf{j} + \frac{1}{\sqrt{G_r}}\Delta\mathbf{u}, \tag{4.2.44}$$

$$P_r\left(\frac{1}{\sqrt{G_r}}\frac{\partial T}{\partial t} + \mathbf{u}\cdot\nabla T\right) = \frac{1}{\sqrt{G_r}}\Delta T, \tag{4.2.45}$$

$$\nabla\cdot\mathbf{u} = 0, \tag{4.2.46}$$

the boundary conditions on $\partial\Omega$

$$\mathbf{u} = 0 \text{ on } \partial\Omega, \quad T = T_0/(T_1 - T_0) \text{ at } x = 0, \quad T = T_1/(T_1 - T_0) \text{ at } x = a_1,$$

$$T = T_0/(T_1 - T_0) + x/a_1 \text{ at } y = 0, 1 \tag{4.2.47}$$

and initial conditions for $\mathbf{u}$ and T.

By a detailed asymptotic analysis, using [Geo95], and based on experimentally confirmed assumptions concerning the order of magnitude of all quantities in $(4.2.44) - (4.2.46)$, the conduction basic state was *approximated* by

$$m_0 = \{(\overline{\mathbf{u}}, \overline{T}, \overline{P}), \overline{\mathbf{u}} = 0, \overline{T} = T_0/(T_1 - T_0) + x/a_1,$$
$$\overline{P} = p_u - [T_0/(T_1 - T_0) + x/a_1]y\}, \tag{4.2.48}$$

where p_u is a reference pressure. It was for the first time that such a basic state was advanced. This opened the possibility of theoretically studying its stability in many situations consistent with the assumptions leading to (4.2.48).

Remark that (4.2.48) is not an exact solution of problem (4.2.44) $-$ (4.2.47), which is expected to influence the remarks on stability investigation. So far we are not able to estimate this influence. Likewise no one is able to say the same thing in similar cases, *e.g.* for boundary layer flows.

At a first glance, in the horizontal convection there must be a basic motion and not a basic state. In fact, this is the case, because our basic state (4.2.48) is only the first approximation, while the higher order approximation terms depend on time.

Finally, remark that the key point in our approximation is the order of the pressure.

Let $\mathbf{u} = \mathbf{v}$, $T = \overline{T} + \theta$, $p = \overline{P} + p'$ be the perturbed fields around m_0. Therefore, $\mathbf{v}$, θ, p' stand for the perturbation velocity, temperature and pressure respectively and satisfy the following perturbation equations

$$\frac{1}{\sqrt{G_r}}\frac{\partial \mathbf{v}}{\partial t} + (\mathbf{v} \cdot \nabla)\mathbf{v} = -\nabla p' + \frac{1}{\sqrt{G_r}}\Delta \mathbf{v} - \mathbf{j}\theta, \tag{4.2.49}$$

$$P_r\left(\frac{1}{\sqrt{G_r}}\frac{\partial \theta}{\partial t} + \mathbf{v} \cdot \nabla\theta\right) = \frac{1}{\sqrt{G_r}}\Delta\theta - P_r\mathbf{v} \cdot \mathbf{i}, \tag{4.2.50}$$

in

$$\mathbf{N}_2 = \left\{(\mathbf{v}, \theta) \in \left(L^2(\Omega)\right)^3 \times L^2(\Omega) \mid \nabla \cdot \mathbf{v} = 0, \ \mathbf{u} = 0 \quad \theta = 0 \quad \text{on} \quad \partial\Omega\right\}.$$

In [GeoM99] it is treated the particular case where two characteristic lengths are equal, while in [HS] are different. In the definition of $\mathbf{N}_2$ it is understood that at least $\mathbf{v} \in (W^{1,2}(\Omega))^3$ and $\nabla p' \in L^2(\Omega)$. Then (4.2.49) implies that $\mathbf{v} \in (W^{2,2}(\Omega))^3$. Assuming that $\nabla\theta \in L^2(\Omega)$, (4.2.50) implies that $\theta \in W^{2,2}(\Omega)$ and, since $\mathbf{v} \in (W^{2,2}(\Omega))^3$ from (4.2.50) it follows that $\theta \in W^{3,2}(\Omega)$. Coming back to (4.2.49) and assuming that $\nabla p'$ has a better regularity, it can be proved in a similar way a better and better regularity for $\mathbf{v}$ and θ.

The associated reaction-diffusion operator, say, $\mathbf{C}$, defined by the matrix

$$\mathbf{C} \equiv \begin{pmatrix} \frac{1}{\sqrt{G_r}}\Delta & -\mathbf{j} \\ -P_r\mathbf{i} & \frac{1}{\sqrt{G_r}}\Delta \end{pmatrix} \tag{4.2.51}$$

is not symmetric due to the fact that in the entries c_{12} and c_{21} different unit vectors occur. Writing $\mathbf{C}$ as a 4×4 matrix we do not get a symmetric matrix either. Indeed, we cannot act on $\mathbf{v}$ due to the solenoidality constraint.

In the following we use the notation $\mathbf{w} = P_r\mathbf{v}$, $\gamma = P_r\theta$, to write equations (4.2.49), (4.2.50) in the form

$$\frac{\partial \mathbf{w}}{\partial t} + \frac{\sqrt{\mathcal{R}}}{P_r\sqrt{P_r}}\mathbf{w} \cdot \nabla\mathbf{w} = -\sqrt{\mathcal{R}P_r}\nabla p' + \Delta\mathbf{w} - \sqrt{\mathcal{R}}\gamma\mathbf{j}, \tag{4.2.52}$$

$$P_r \frac{\partial \gamma}{\partial t} + \frac{\sqrt{G_r}}{P_r} \mathbf{w} \cdot \nabla \gamma = \Delta \gamma - \sqrt{\mathcal{R}} \mathbf{w} \cdot \mathbf{i}. \tag{4.2.53}$$

Furthermore we consider solutions $W = (\mathbf{w}, \gamma)$ which are null on $\partial \Omega$. Project the system (4.2.52), (4.2.53) on $\mathbf{N}_2$ to obtain the energy relation

$$\frac{1}{2}\frac{d}{dt}\left(|\mathbf{w}|^2 + P_r|\gamma|^2\right) = -\left(\|\mathbf{w}\|^2 + \|\gamma\|^2\right)\left[1 - \sqrt{\mathcal{R}}\frac{|\,\gamma(\mathbf{w}\cdot\mathbf{j} + \mathbf{w}\cdot\mathbf{i})\,|_1}{\|\mathbf{w}\|^2 + \|\gamma\|^2}\right]. \tag{4.2.54}$$

Remark that it contains only the contribution of the linear terms in (4.2.51), (4.2.52).

Let $a = \min\{1, P_r\}$. Then the last relation implies

$$\frac{a}{2}\frac{d}{dt}\left(|\mathbf{w}|^2 + |\gamma|^2\right) \leq - P_r^{-1}\left(\|\mathbf{w}\|^2 + \|\gamma\|^2\right)\left[1 - \sqrt{\mathcal{R}}\frac{|\,\gamma(\mathbf{w}\cdot\mathbf{j} + \mathbf{w}\cdot\mathbf{i})\,|_1}{\|\mathbf{w}\|^2 + \|\gamma\|^2}\right],$$

leading to the exponential asymptotic stability of m_0 if the criterion

$$\mathcal{R} < \mathcal{R}_E \tag{4.2.55}$$

holds, where

$$\sqrt{\mathcal{R}_E^{-1}} = \max_{(\mathbf{w},\theta)\in\mathbf{N}_2} \frac{|\,\theta(\mathbf{w}\cdot\mathbf{j} + \mathbf{w}\cdot\mathbf{i})\,|_1}{\|\mathbf{w}\|^2 + \|\theta\|^2}. \tag{4.2.56}$$

Here $\mathcal{R}_L$ is the smallest eigenvalue of the linear stationary problem

$$\Delta \mathbf{w} - \sqrt{\mathcal{R}}\gamma\mathbf{j} = \sqrt{\mathcal{R}P_r}\nabla p', \tag{4.2.57}$$

$$\Delta \gamma - \sqrt{\mathcal{R}}\mathbf{w}\cdot\mathbf{i} = 0, \tag{4.2.58}$$

$$\mathbf{w} = \mathbf{0}, \quad \gamma = 0 \quad \text{on} \quad \partial\Omega, \tag{4.2.59}$$

deduced from (4.2.52), (4.2.53).

In problems in which the energy equation (in our case (4.2.53)) contains no contribution from the nonlinear terms, all the eigenvalues of the problem are values of the associated functional. In particular, among them is the smallest eigenvalue $\mathcal{R}_L$ [Ser1].

Furthermore we must calculate $\mathcal{R}_E$ as the smallest eigenvalue of the Euler-Lagrange equations

$$\theta(\mathbf{i} + \mathbf{j}) + \frac{1}{\mathcal{R}_E}\Delta\mathbf{w} = -\nabla\lambda, \tag{4.2.60}$$

$$\mathbf{w}\cdot\mathbf{i} + \mathbf{w}\cdot\mathbf{j} + \frac{1}{\mathcal{R}_E}\Delta\theta = 0 \tag{4.2.61}$$

associated with the functional (4.2.56). As in [GeoM] one term was lost in these equations, and the case in [GeoM] is a particular case of that treated in [HS], further on we follow [HS].

In [HS] using numerical methods are determined $\mathcal{R}_E$ and $\mathcal{R}_L$ and it is found that

$$\mathcal{R}_E < \mathcal{R}_L, \tag{4.2.62}$$

showing that there are some subcritical instabilities.

4.2.5   *Case of new vector unknown functions in a vertical hydromagnetic convection*

In Sections 4.1, 4.2.1, 4.2.2 and 4.2.3 we treat vertical convections characterized by velocity, temperature and concentration fields. In Section 4.2.4 we deal with a horizontal convection characterized by velocity and temperature fields. Herein by following [BicGP] and by applying an adapted version of our variant from Section 4.2.1, we study a vertical hydromagnetic convection where velocity, temperature and magnetic fields occur.

Since in the corresponding perturbation governing model one scalar and two vector unknown functions are involved, in order to reduce the number of these functions we define two new linear combinations ϕ_1 and ϕ_2 of the vector unknown functions and, at the end, in the energy relation only ϕ_1 occurs. Unlike this situation, in Section 4.2.1 the new linear combinations were scalar functions.

Let us investigate the vertical convection of a viscous incompressible homogeneous thermoelectrically conducting fluid, bounded by two horizontal unbounded rigid thermally perfectly conducting and electrically nonconducting walls. It is governed by problem (1.4.3), (1.4.5), (1.4.6). Using a slightly different non-dimensionalization, the model reads as an initial and boundary-value problem for the system

$$
\begin{cases}
\dfrac{\partial}{\partial t}\mathbf{v} = -\mathbf{v}\cdot\nabla\mathbf{v} - \nabla p' + \Delta\mathbf{v} + P_m\mathbf{h}\cdot\nabla\mathbf{h} + \mathcal{R}\theta\mathbf{k} + Q\dfrac{\partial}{\partial z}\mathbf{h}, \\[2mm]
\dfrac{\partial}{\partial t}\mathbf{h} = -\mathbf{v}\cdot\nabla\mathbf{h} - \mathbf{h}\cdot\nabla\mathbf{u} + \dfrac{1}{P_m}\Delta\mathbf{h} + \dfrac{Q}{P_m}\dfrac{\partial}{\partial z}\mathbf{h}, \\[2mm]
\dfrac{\partial}{\partial t}\theta = -\mathbf{v}\cdot\nabla\theta + \dfrac{\mathcal{R}}{P_r}\mathbf{v}\cdot\mathbf{k} + \dfrac{1}{P_r}\Delta\theta, \\[2mm]
\nabla\cdot\mathbf{v} = 0, \\[2mm]
\nabla\cdot\mathbf{h} = 0,
\end{cases}
\tag{4.2.63}
$$

in the periodicity cell Ω defined in Section 4.2.1. The space of the problem is

$$
\mathcal{N}_1 = \Big\{(\theta,\mathbf{v},\mathbf{h}) \in (W^{2,2}(\Omega))^7 \mid \nabla\cdot\mathbf{v}=0,\ \nabla\cdot\mathbf{h}=0,\ \mathbf{v}=\mathbf{h}=\mathbf{0},\ \theta=0 \text{ on } \partial\Omega_1\cup\partial\Omega_2\Big\}
\tag{4.2.64}
$$

A system equivalent to (4.2.63) reads

$$
\frac{\partial\theta}{\partial t} + \mathbf{v}\cdot\nabla\theta = \frac{1}{P_r}\Delta\theta + \frac{\mathcal{R}}{P_r}\mathbf{v}\cdot\mathbf{k},
\tag{4.2.65}
$$

$$
a\Big(\frac{\partial}{\partial t}\mathbf{v} + \mathbf{v}\cdot\nabla\mathbf{v}\Big) + ag_3 P_m\Big(\frac{\partial}{\partial t}\mathbf{h} + \mathbf{v}\cdot\nabla\mathbf{h}\Big) = -a\nabla p' + a\mathcal{R}\theta\mathbf{k} + a\Delta\mathbf{v}
$$
$$
+ aQ\frac{\partial}{\partial z}\mathbf{h} + aP_m\mathbf{h}\cdot\nabla\mathbf{h} + ag_3 Q\frac{\partial}{\partial z}\mathbf{v} + ag_3\Delta\mathbf{h} + ag_3 P_m\mathbf{h}\cdot\nabla\mathbf{v},
\tag{4.2.66}
$$

$$
bP_m\Big(\frac{\partial}{\partial t}\mathbf{h} + \mathbf{v}\cdot\nabla\mathbf{h}\Big) + bg_2\Big(\frac{\partial}{\partial t}\mathbf{v} + \mathbf{v}\cdot\nabla\mathbf{v}\Big) = bQ\frac{\partial}{\partial z}\mathbf{v} + b\Delta\mathbf{h} + bP_m(\mathbf{h}\cdot\nabla\mathbf{v}
$$
$$
- bg_2\nabla p') + bg_2\mathcal{R}\theta\mathbf{k} + bg_2\Delta\mathbf{v} + bg_2 Q\frac{\partial}{\partial z}\mathbf{h} + bg_2 P_m\mathbf{h}\cdot\nabla\mathbf{h}.
\tag{4.2.67}
$$

It is obtained by performing the algebraic operations: $(4.2.66) = a(4.2.63)_1 + ag_3P_m$ $(4.2.63)_2$, $(4.2.67) = bP_m(4.2.63)_2 + bg_2(4.2.63)_1$ where a, b, g_2 and g_3 are, so far, undetermined nonzero constants. In addition, $(4.2.65)$ is identical to $(4.2.63)_3$.

Consider on $\mathcal{N}_1$ the scalar product $(\cdot, \cdot)$ of $(\mathbf{L}^2(\Omega)(\equiv L^2(\Omega)^7)$. Introduce two linear operators $L_1 \in L(\mathcal{N}_1, \mathbf{L}_2(\Omega))$, $L_2 \in L(\mathcal{N}_1, \mathcal{N}_1)$ and use the notation $\mathbf{V} = (\theta, \mathbf{v}, \mathbf{h})^T$, $\mathbf{V}_1 = (\theta, a\mathbf{v} + ag_3P_m\mathbf{h}, bg_2\mathbf{v} + bP_m\mathbf{h})^T$, where L_1 and L_2 are defined by the matrices

$$\mathbf{L}_1 = \begin{pmatrix} P_r^{-1}\Delta & P_r^{-1}\mathcal{R}\mathbf{k} & 0 \\ a\mathcal{R}\mathbf{k} & a\Delta + ag_3Q\partial/\partial z & ag_3\Delta + aQ\partial/\partial z \\ bg_2\mathcal{R}\mathbf{k} & bg_2\Delta + bQ\partial/\partial z & b\Delta + bg_2Q\partial/\partial z \end{pmatrix}, \quad \mathbf{L}_2 = \begin{pmatrix} 1 & 0 & 0 \\ 0 & a & ag_3P_m \\ 0 & bg_2 & bP_m \end{pmatrix}.$$

In addition, define the nonlinear mapping $\mathbf{T}$ described by the matrix

$$\mathbf{T} = \begin{pmatrix} 0 & 0 & 0 \\ 0 & ag_3P_m\mathbf{h}\cdot\nabla\mathbf{h} & aP_m\mathbf{h}\cdot\nabla\mathbf{h} \\ 0 & bP_m\mathbf{h}\cdot\nabla\mathbf{h} & bg_2P_m\mathbf{h}\cdot\nabla\mathbf{h} \end{pmatrix}.$$

It follows that the system $(4.2.65) - (4.2.67)$ in $\mathbf{V} \in \mathcal{N}_1$ reads

$$(\frac{\partial}{\partial t} + \mathbf{v}\cdot\nabla)\mathbf{V}_1 = L_1\mathbf{V} + (0, -a\nabla p', -bg_2\nabla p')^T + T(\mathbf{V})$$

or, equivalently, $\frac{\partial \mathbf{V}_1}{\partial t} = L_1\mathbf{V} + N(\mathbf{V}) + T(\mathbf{V})$, where the mapping $N(\mathbf{V})$ corresponds to the advective and pressure terms, *i.e.* $N(\mathbf{V}) = -\mathbf{v}\cdot\nabla\mathbf{V}_1 + (0, -a\nabla p', -bg_2\nabla p')^T$. By applying to the last relation the projector of $L^2(\Omega)$ on $\mathcal{N}_1$ we obtain the energy relation

$$(\frac{\partial \mathbf{V}_1}{\partial t}, \mathbf{V}) = (L_1\mathbf{V}, \mathbf{V}) + (N(\mathbf{V}), \mathbf{V}) + (T(\mathbf{V}), \mathbf{V}). \tag{4.2.68}$$

If $a = b$, then $(T(\mathbf{V}), \mathbf{V}) = 0$ because $\int_\Omega \mathbf{h}\cdot\nabla\mathbf{v}\mathbf{v}d\Omega = \int_\Omega \mathbf{h}\cdot\nabla\mathbf{h}\mathbf{h}d\Omega = 0$ and $\int_\Omega \mathbf{h}\cdot\nabla\mathbf{h}\mathbf{v}d\Omega = -\int_\Omega \mathbf{h}\cdot\nabla\mathbf{v}\mathbf{h}d\Omega$. Moreover, in order for the coefficients of $\frac{\partial}{\partial t}\mathbf{v}\cdot\mathbf{h}$ and $\frac{\partial}{\partial t}\mathbf{h}\cdot\mathbf{v}$ in the left-hand side of $(4.2.68)$ be equal, we must have $g_3P_m = g_2$. Then $(4.2.68)$ becomes

$$\frac{1}{2}\frac{d}{dt}(\mathbf{V}_1, \mathbf{V}) = (L_1\mathbf{V}, \mathbf{V}) + (N(\mathbf{V}), \mathbf{V}).$$

Using Green identities, the relation $g_3P_m = g_2$ and the fact that $\nabla p'$ is orthogonal to the solenoidal vectors $\mathbf{v}$ and $\mathbf{h}$, it follows that $N(\mathbf{V}) + T(\mathbf{V}) = 0$. Consequently, the energy relation becomes

$$\frac{1}{2}\frac{d}{dt}(\mathbf{V}_1, \mathbf{V}) = (L_1\mathbf{V}, \mathbf{V}). \tag{4.2.69}$$

The symmetric part of L_1 reads

$$\mathbf{L}_{1s} = \begin{pmatrix} P_r^{-1}\Delta & \delta_1\mathbf{k} & \delta_2\mathbf{k} \\ \delta_1\mathbf{k} & a\Delta & \delta_3\Delta \\ \delta_2\mathbf{k} & \delta_3\Delta & b\Delta \end{pmatrix},$$

where $\delta_1 = 0.5(a + P_r^{-1})$, $\delta_2 = 0.5ag_2 R$, $\delta_3 = 0.5a(g_3 + g_2)$. Since $(L_1 \mathbf{V}, \mathbf{V}) = (L_{1s}\mathbf{V}, \mathbf{V})$, (4.2.69) becomes

$$\frac{1}{2}\frac{d}{dt}(\mathbf{V}_1, \mathbf{V}) = -P_r^{-1} \mid \nabla\theta \mid^2 -P_r a \mid \nabla\mathbf{v} \mid^2 -P_r a \mid \nabla\mathbf{h} \mid^2$$
$$- P_r a(g_3 + g_2)(\nabla\mathbf{u} \cdot \nabla\mathbf{h}) + P_r R(a + P_r^{-1})(\theta, w) + P_r Rag_2(\theta, h_3). \qquad (4.2.70)$$

In the following, we adapt the method in Section 4.2.1. Thus, introduce the new vector functions $\boldsymbol{\Phi_1} = \mathbf{a_1 v} + \mathbf{a_2 h}$, $\quad \boldsymbol{\Phi_2} = \mathbf{b_1 v} + \mathbf{b_2 h}$, where the constants $a_1, a_2, b_1, b_2 \in \mathbf{R}$ are to be determined and $\boldsymbol{\Phi_1} = \boldsymbol{\Phi_1}(\mathbf{t}, \mathbf{x})$, $\boldsymbol{\Phi_2} = \boldsymbol{\Phi_2}(\mathbf{t}, \mathbf{x})$. Remark that this choice represents an extension of the G-P-R method, because here $\boldsymbol{\Phi_1}$ and $\boldsymbol{\Phi_2}$ are vector functions. Thus the expression $(\mathbf{V}_1, \mathbf{V}) = \mid \theta \mid^2 + a < \mid \mathbf{v} \mid^2 + 2g_2 \mathbf{v} \cdot \mathbf{h} + P_m \mid \mathbf{h} \mid^2 >$ must be read, equivalently, as $(\mathbf{V}_1, \mathbf{V}) = \mid \theta \mid^2 + d_1 \mid \boldsymbol{\Phi_1} \mid^2 + \mathbf{d_2} \mid \boldsymbol{\Phi_2} \mid^2$, where $d_1, d_2 \in \mathbf{R}_+$, implying

$$d_1 a_1^2 + d_2 b_1^2 = a, \quad d_1 a_1 a_2 + d_2 b_1 b_2 = ag_2, \quad d_1 a_2^2 + d_2 b_2^2 = aP_m, \qquad (4.2.71)$$

where d_1, d_2, b_1, b_2 are determined up to some factor. Eliminating d_1 and d_2 between these equalities, we obtain the relationship between b_1 and b_2

$$(a_2 b_2 + P_m a_1 b_1) - g_2(a_2 b_1 + a_1 b_2) = 0, \quad a_2^2 b_1^2 - a_1^2 b_2^2 \neq 0, \qquad (4.2.72)$$

defining $\boldsymbol{\Phi_2}$ up to a factor.

Let us find a_1 and a_2 such that (4.2.70) has the simplest form depending only on $\boldsymbol{\Phi_1}$

$$\frac{1}{2}\frac{d}{dt}(\mathbf{V}_1, \mathbf{V}) = -P_r^{-1}[\mid \nabla\theta \mid^2 + \mid \nabla\boldsymbol{\Phi_1} \mid^2 + P_r Rk'(\theta\boldsymbol{\Phi_1} \cdot \mathbf{k})], \qquad (4.2.73)$$

where k' is an undetermined factor. By identifying (4.2.70) and (4.2.73), it follows

$$-P_r a \mid \nabla\mathbf{v} \mid^2 -P_r a \mid \nabla\mathbf{h} \mid^2 -P_r a(g_3 + g_2)(\nabla\mathbf{u} \cdot \nabla\mathbf{h}) = - \mid \nabla\boldsymbol{\Phi_1} \mid^2,$$

$$P_r R(a + P_r^{-1})(\theta, w) + P_r Rag_2(\theta, h_3) = P_r Rk'(\theta, \boldsymbol{\Phi_1} \cdot \mathbf{k}).$$

If $P_m > 1$, we obtain

$$a = (P_m + 1)[P_r(P_m - 1)]^{-1}, \quad a_1 = \pm a_2, \quad g_2 = \pm 2P_m(P_m + 1)^{-1}, \quad g_3 = g_2 P_m^{-1},$$

where the signs $+$ and $-$ correspond, and

$$a_1 = \pm\sqrt{(P_m + 1)(P_m - 1)^{-1}}, \quad k' = \pm 2P_m(P_r\sqrt{(P_m^2 - 1)^{-1}}),$$

where the signs $+$ and $-$ correspond.

From $(4.2.71)_{1,3}$ it follows that $d_1 = (b_2^2 - P_m b_1^2)/[P_r(b_2^2 - b_1^2)]$, $d_2 = a(P_m - 1)/(b_2^2 - b_1^2)$. Then, for $a_1 = a_2$, $(4.2.71)_2$ implies $b_2/b_1 = P_m$, while for $a_1 = -a_2$, $(4.2.71)_2$ implies $b_2/b_1 = -P_m$. In both these cases we have

$$d_1 = P_m/[P_r(P_m + 1)], \qquad d_2 = a/[b_2^2(P_m + 1)].$$

Therefore, all these four solutions $(a, b, g_2, g_3, a_1, a_2, b_2/b_1, k')$ are convenient. Furthermore we show that they lead to the same stability criterion.

Introduce the functions

$$E(t) = (|\,\theta\,|^2 + d_1\,|\,\mathbf{\Phi_1}\,|^2)/2, \qquad \mathbf{\Psi}(t) = |\,\mathbf{\Phi_1}\,|^2/2$$

and the notation

$$\xi^2 = min_{\mathbf{\Phi}_1,\theta\in\mathcal{N}_1(\Omega)}\frac{2\big(|\nabla\mathbf{\Phi_1}|^2 + |\nabla\theta|^2\big)}{|\mathbf{\Phi_1}|^2 + |\theta|^2}, \quad \frac{1}{\sqrt{R_{a*}}} = max_{\mathbf{\Phi}_1,\theta\in\mathcal{N}_1(\Omega)}\frac{2(\theta,\mathbf{\Phi_1}\cdot\mathbf{k})}{|\nabla\mathbf{\Phi_1}|^2 + |\nabla\theta|^2}.$$

Then, due to the fact that $\mathbf{\Phi_1} = \mathbf{0}$ on $\partial\Omega$, for $k' > 0$ the energy relation (4.2.69) becomes successively

$$\frac{dE}{dt} + \frac{d\Psi}{dt} = -P_r^{-1}[|\nabla\mathbf{\Phi_1}|^2 + |\nabla\theta|^2 - \mathbf{P_r}\mathbf{R}k'(\theta,\mathbf{\Phi_1}\cdot\mathbf{k})]$$

$$= -P_r^{-1}[|\nabla\mathbf{\Phi_1}|^2 + |\nabla\theta|^2]\Big[1 - \frac{\mathbf{P_r}\mathbf{R}k'(\theta,\mathbf{\Phi_1}\cdot\mathbf{k})}{|\nabla\mathbf{\Phi_1}|^2 + |\nabla\theta|^2}\Big],$$

implying, if the expression in the square brackets is positive,

$$\frac{dE}{dt} + \frac{d\Psi}{dt} \leq -P_r^{-1}\xi^2\frac{1}{max\{1,d_1\}}[1 - P_r Rk'\frac{1}{2\sqrt{R_a*}}]E, \qquad (4.2.74)$$

whence the stability criterion

Theorem 4.2.1. *Suppose that $P_m > 1$. If $R < \sqrt{(P_m^2 - 1)}\sqrt{R_a}*/P_m$, then the basic state m_0 is nonlinearly stable.*

Let $k' < 0$ and remark that from definition of $\mathcal{N}_1$ it follows that if $(\theta,\mathbf{v},\mathbf{h}) \in \mathcal{N}_1$, then $(\theta,\mathbf{\Phi_1},\mathbf{\Phi_2}) \in \mathcal{N}_1$ and $(-\theta,\mathbf{\Phi_1},\mathbf{\Phi_2}) \in \mathcal{N}_1$. Introduce the space $\tilde{\mathcal{N}}_1 = \{(\theta,\mathbf{\Phi_1}) \in (\mathbf{H}^2(\Omega))^4 \,|\, \nabla\cdot\mathbf{\Phi_1} = 0,\ \mathbf{\Phi_1} = \mathbf{0}, \theta = 0,$ on $\partial\Omega_1 \cup \partial\Omega_2\}$. Obviously $\tilde{\mathcal{N}}_1$ is embedded in $\mathcal{N}_1$. In addition, if $(\theta,\mathbf{\Phi_1})$ runs over $\tilde{\mathcal{N}}_1$, then $(-\theta,\mathbf{\Phi_1})$ runs over $\tilde{\mathcal{N}}_1$ too.

We have $k'(\theta,\mathbf{\Phi_1}\cdot\mathbf{k}) = |\,k'\,|\,(-\theta,\mathbf{\Phi_1}\cdot\mathbf{k})$ and $|\nabla\theta|^2 = |\nabla(-\theta)|^2$. Therefore ξ^2 and $R*$ are the same if θ is replaced by $-\theta$. Consequently, (4.2.74) holds for k' replaced by $|\,k'\,|$. In this way, for the case $k' < 0$, Theorem 4.2.1 holds too.

Stability criterion 4.2.1 is valid for all solutions $(a, b, a_1, a_2, b_2/b_1, g_2, g_3, k')$.

4.3 Variants based on energy splitting

The brief description of these variants is given in Section 4.3.1. The most important ideas are exposed and their connection with those involved in other variants are pointed out. In Section 4.3.2 a nonlinear magnetohydrodynamic convection problem is treated by using some of these ideas and, basically, the results for the corresponding linear case presented in Section 6.2.

4.3.1 *Splitting of Lyapunov functional*

The weakness of the energy method has two main sources. One is that the embedding inequalities are one-sided. The other resides in the impossibility to evaluate sharply the varying-sign terms from the energy relation. It is the Joseph's idea to introduce suitable constants and additional equations [J70b], *e.g.* (4.1.15), in order to take into account the correct effect of those terms. In the Joseph variant or in our generalization of it (Section 4.1) the additional equations are deduced from the governing problem while in our variant from Section 4.2 they are included from the very beginning by starting with modified governing equations. The corresponding energy is, in fact, a Lyapunov functional expressed only in terms of physical quantities.

As early as 1986 [Rio88a], Salvatore Rionero initiated another variant based on the splitting of the Lyapunov functional of energy into two parts. The first part is derived from the governing equations reformulated in terms of the so-called essential variables (toroidal and poloidal fields) for representing the solenoidal vectors [J76], where the energy of the corresponding linear problem is evidenced. The second part is derived from the additional equations.

In [Rio05] Rionero studies the coincidence of linear and nonlinear stability bounds using functionals depending on the eigenvalues of the involved linear operator.

Unlike the variants in Sections 4.1 and 4.2, the supplementary equations are obtained from the governing equations not only by algebraic multiplications by constants and unknown functions but also by applying differential operators and multiplying the equations by derivatives of the physical quantities. In this way, the resulted energy contains derivatives of those physical quantities and is more appropriate to be called the Lyapunov functional of energy (Section 2.2).

As this approach heavily rests on the results of the linear theory, it can be applied only to those problems where linear studies are available.

In the Lyapunov functional of energy, the second part, namely that derived from the additional equations, is multiplied by a constant which is to be determined from the requirement that the difference from the linear and energy bounds be minimum.

The variant of the energy method based on the splitting of the Lyapunov functional of energy was applied by Salvatore Rionero and Giuseppe Mulone to many stability problems, especially of hydromagnetic convection. A good description of them can be found in [Bic]. As an example in the following we present a Bénard magnetic problem by following [RioM88c]. To this aim first we must describe the Rionero guideline, synthesizing the algorithmic procedure (steps) behind this variant. This guideline is intended to enable finding the functional for studying the stability of the null solution of a nonlinear evolution pde, by decomposing the corresponding vector field into the linear L and nonlinear N parts. The first preliminary step concerns the linear stability and essential variables: the linear stability problem is solved using, instead of the given state variables, some convenient functions of them, called the essential variables $X_1, X_2, \ldots$. Here the intuition of the researcher is important. In the second preliminary step the so-called admissible balances $f_{ij} = X_i - c_{ij} X_j$, $c_{ij} \in \mathbf{R}$, are defined, where X_i (resp X_j) are the variables inhibiting (resp. favoring) the stability. Then the Lyapunov functional is splitted as $V = V_0 + bV_1$, embedding $b > 0$ and V_0 and V_1 are chosen according to the guide: 1) V_0 is supposed to depend on the essential variables, their balances and their space derivatives; 2) V_0 is the Lyapunov function for the linear stability problem; 3) if this problem is written in a variational form, then V_0 must be chosen to be as close to the corresponding Lagrangian as possible; 4) V_0 must be equivalent to a fixed norm (this condition is not imposed in [RioM88c]); 5) V_1 must depend on the space derivatives of the essential variables and state functions; 6) V_1 must be larger than the nonlinear terms N; 7) V_1 must be chosen such that the radius of the attraction domain of the null solution be as large as possible.

In the Bénard magnetic convection governed by perturbation equations (1.4.3), the boundary conditions (1.4.5′), (1.4.6)$_1$ and (1.4.6)$_3$ and the initial conditions, the state functions (also referred to as variables) are $u, v, w, h_1, h_2, h_3 \equiv h, \theta$ and p', while the essential variables, used in [Chan], are $u, \xi \equiv \mathbf{k}\nabla \times \mathbf{u}, h, \xi^{(m)} \equiv \mathbf{k}\nabla \times \mathbf{h}$ and θ. The inhibiting quantity of interest is $\mathbf{h}$, while the favorable field is $\frac{\partial\theta}{\partial z}$, so the balance reads $f = h - c\frac{\partial\theta}{\partial z}$, where c is a positive constant. By applying the operators *curl* to (1.4.3)$_1$, *curl curl* to (1.4.3)$_1$, and *curl* to (1.4.3)$_2$, the evolution equations for ξ, Δw and $\xi^{(m)}$ follow. Then, from (1.4.3)$_2$, (1.4.3)$_3$ and differentiating f with respect to time, the evolution equations for $\frac{\partial\theta}{\partial z}, \frac{\partial\theta}{\partial x}, \frac{\partial\theta}{\partial y}$, and f are obtained. Further, assuming that the physical quantities are periodic in x and y, the corresponding linearized evolution equations for ξ, Δw, $\xi^{(m)}$, $\frac{\partial\theta}{\partial z}$, and f are written and their Lyapunov functional V_0 is deduced. Along the trajectories of the linearized system the rate of change of V_0 is written in terms of three functional, J_0, D_0 and B_0 as $\frac{dV_0}{dt} = RJ_0 - D_0 + B_0$ (R is the Rayleigh number) and its simpler form for periodic fields is shown. Denoting $1/R_\epsilon = \max_G J_0/D_0$, where G is the space of the periodic functions $w, \xi, \xi^{(m)}, \theta, f$ satisfying the boundary conditions, and applying various forms of the Poincaré inequality, we get the energy inequality

$$\frac{dV_0}{dt} \leq -2\lambda_0\pi^2 P_r^{-1} P_m^{-1}(1 - R/R_\epsilon)V_0,$$

yielding the energy limit for conditional exponential asymptotic stability R_ϵ. It is shown that J_0/D_0 is a bounded ratio and the Euler-Lagrange equations for the variational problem for R_ϵ is written. It involves two constants which, in order for the requirement 3) be fulfilled, particular values for them are taken. The obtained equation in θ is solved by the normal mode method. Then it is proved that $V_0(t)$ is positively defined, all these implying the stability criterion $R^2 < R_\epsilon^2$, ensuring the linear asymptotic exponential stability, namely,

$$V_0(t) \leq V_0(0)exp[-2\lambda_0\pi P_r^{-1} P_m^{-1}(1 - R/R_\epsilon)t,$$

where $\lambda_0 = \min(P_r P_m, P_r, P_m)$. In addition, it is found that

$$R_\epsilon^2 = R_L^2, \quad \text{for} \quad P_m < P_r, \ 0 < Q^2 < \pi^2(4\sqrt{2} + 5), \ R_\epsilon^2 \in [27/4\pi^4, 2\pi^4(5\sqrt{2} + 7)].$$

Here R_L is the linear stability limit. The case $P_m > P_r$ is also studied. Further on, as nonlinear evolution equation for V_0 one takes $\dot{V}_0 = RJ_0 - D_0 + N_0$, where N_0 has a given (intuited) expression. Taking into account the requirement 6), V_1 is chosen and V is written as $V = V_0 + bV_1$ and the evolution equation for V_1 is derived by multiplying $(1.4.3)_1$, $(1.4.3)_2$ and the evolution equations for $\frac{\partial\theta}{\partial x}$ and $\frac{\partial\theta}{\partial y}$ by $-\Delta u$, $-\Delta h$, $-\Delta\frac{\partial\theta}{\partial x}$, $-\Delta\frac{\partial\theta}{\partial y}$ respectively and integrating over the periodicity cell. Several inequalities lead, in certain particular cases, to the energy inequality for V in the form $\dot{V}\leq-D_2(1-AV^{1/2})$, implying the conditional exponential asymptotic stability criterion $R^2 < R_\epsilon^2$, for $V_0 < A^{-2}$, so $V_0(t)\leq V_0(0)exp[-k^*(1 - AV^{1/2}(0))]t$ for $t\geq 0$, where k^* is a positive constant depending on P_r, P_m, Q^2 and R^2. Coincidence of linear and nonlinear stability Rayleigh numbers $R_\epsilon^2 = R_L^2$ is proved for $P_m < P_r$, $0\leq Q^2\leq\pi^2(4\sqrt{2}+5)$, and $V_0 < A^{-2}$.

Remark that the cases $P_m > P_r$ and $P_m < P_r$ require different treatments. This conclusion follows also from the treatment in Section 4.2.5 where the same equations as in [RioM88c] are governing the convection. This sketch shows that the treatment is difficult. The interested reader can find in [RioM88c] the complete proofs and the expression of the involved quantities.

In [Mulo06], if the symmetric part of L_1 in (4.2.4) and (4.2.6) is a uniformly elliptic operator, N is a more general nonlinear application and $(N(V),V) \neq 0$, an optimal (generalized-energy) Lyapunov functional is constructed. More exactly, Mulone presents a method in eight steps leading to this functional. The key point of his approach is to define a new state function of the dynamical system associated with a physical problem. This is realized by means of the modal matrix of the eigenvalues of the matrix associated with L_1 via the principal eigenvalues of uniformly elliptic operators of the particular problems. Several interesting applications to concrete problems (*e.g.* convections) are done.

4.3.2   *Nonlinear stability of the MHD anisotropic Bénard problem*

In this section, by using the Lyapunov direct method and following [Pal05], we study the nonlinear Lyapunov stability of the conduction-diffusion solution of the anisotropic magnetic Bénard problem, for a fully ionized fluid. We show that, if the conduction-diffusion solution is linearly stable, it is asymptotically nonlinearly stable too. The linear corresponding case is treated in Section 6.2 and it is basic in the method used in this section. In addition, we use the Rionero idea [Rio88a] of the energy splitting.

For a partially ionized fluid the same nonlinear study was carried out in [Pal06].

4.3.2.1   *Perturbation problem reformulated*

Consider a homogeneous thermoelectrically fully ionized conducting fluid, in a horizontal layer S, in the presence of a uniform imposed magnetic field $\mathbf{H}_0$ normal to the layer, in which a constant vertical adverse temperature gradient β is maintained. In an orthonormal reference frame $\{O, \mathbf{i}, \mathbf{j}, \mathbf{k}\}$, with $\mathbf{k}$ pointing upwards positive, the layer is bounded by the planes $\pi_0 : z = 0$ and $\pi_1 : z = 1$, both stress-free, thermally conducting but electrically nonconducting, satisfying, at $z = 0, 1$, the boundary conditions (1.4.4), (1.4.6). In the framework of physics of continua and in the O-B approximation the dimensionless equations governing the evolution of perturbation (convection) $\mathbf{v}, \mathbf{h}, \theta, p$ of the thermodiffusive equilibrium m_0, given by (1.4.2), are (1.4.16).

As a domain of motion we take the periodicity cell $\Omega = [0, \dfrac{2\pi}{a_x}] \times [0, \dfrac{2\pi}{a_y}] \times [0, 1]$. Like in Sections 6.1 and 6.2, we use the variables $\mathbf{k} \cdot \mathbf{v}$, $\mathbf{k} \cdot \nabla \times \mathbf{v}$, $\mathbf{k} \cdot \mathbf{h}$, $\mathbf{k} \cdot \nabla \times \mathbf{h}$ suitable to representation of solenoidal fields in a plane layer [J76]. Applying the operators $\mathbf{k} \cdot \nabla \times \nabla \times$, $\mathbf{k} \cdot \nabla \times \partial_z$ to equation $(1.4.16)_1$, and $\mathbf{k} \cdot \partial_z$, $\mathbf{k} \cdot \nabla \times$ to equation $(1.4.16)_2$, (1.4.16) becomes

$$
\left\{
\begin{aligned}
-\frac{\partial}{\partial t}\Delta w =\ & - \Delta\Delta w - M^2 \Delta h_3' - \frac{\mathcal{R}}{P_r}\Delta_1\theta \\
& - \mathbf{k} \cdot \nabla \times \nabla \times (\mathbf{v}\nabla\mathbf{v}) + M^2 \mathbf{k} \cdot \nabla \times \nabla \times (\mathbf{h}\nabla\mathbf{h}), \\[4pt]
\frac{\partial}{\partial t}\zeta' =\ & + \Delta\zeta' + M^2 \partial_{zz} j \\
& - \mathbf{k} \cdot \nabla \times \partial_z (\mathbf{v}\nabla\mathbf{v}) + M^2 \mathbf{k} \cdot \nabla \times \partial_z (\mathbf{h}\nabla\mathbf{h}), \\[4pt]
\frac{\partial}{\partial t}h_3' =\ & \partial_{zz} w + \frac{P_m}{P_r}\Delta h_3' - \beta_H \frac{P_m}{P_r}\partial_{zz} j \\
& + \beta_H \frac{P_m}{P_r}\mathbf{k} \cdot \nabla \times \partial_z (\mathbf{h} \times \nabla \times \mathbf{h}) + \mathbf{k} \cdot \nabla \times \partial_z (\mathbf{v} \times \mathbf{h}), \\[4pt]
\frac{\partial}{\partial t}j =\ & \zeta' + \frac{P_m}{P_r}\Delta j + \beta_H \frac{P_m}{P_r}\Delta h_3' \\
& + \mathbf{k} \cdot \nabla \times \nabla \times (\mathbf{v} \times \mathbf{h}) + \beta_H \frac{P_m}{P_r}\mathbf{k} \cdot \nabla \times \nabla \times (\mathbf{h} \times \nabla \times \mathbf{h}),
\end{aligned}
\right.
\tag{4.3.1}
$$

where

$$
w = \mathbf{k} \cdot \mathbf{v}, \quad h_3' = \partial_z \mathbf{k} \cdot \mathbf{h}, \quad \zeta' = \partial_z \mathbf{k} \cdot \nabla \times \mathbf{v}, \quad j = \mathbf{k} \cdot \nabla \times \mathbf{h}, \quad \Delta_1 = \frac{\partial^2}{\partial x^2} + \frac{\partial^2}{\partial y^2}.
$$

Remark that the meaning of h_3' differs from that in Section 6.1. The boundary conditions (1.4.4), expressed in the new unknown functions $w, \zeta', j, h_3', \theta$ are

$$
w = \partial_{zz} w = \Delta w = h_3' = \Delta h_3' = j = \Delta j = \zeta' = \theta = \theta_{zz} = \Delta\theta_{zz} = 0
\tag{4.3.2}
$$

at $z = 0$ and $z = 1$. They follow, as in [Chan], from $(1.4.16)_4$ and $(1.4.16)_5$, (1.4.4) and (4.3.1) if we assume that the perturbation equations (4.3.1) hold on the boundaries of the layer. Consequently, for any t fixed, the space of problem (4.3.1), (4.3.2) is

$$
\mathcal{M} = \{(w, \zeta', j, h_3', \theta) \in C^\infty(\Omega)^5 \mid (w, \zeta', j, h_3', \theta) \text{ satisfy } (4.3.2)\}.
\tag{4.3.3}
$$

4.3.2.2 *Nonlinear conditional Lyapunov stability*

In Section 6.2, by following [PalG04a], by the Lyapunov direct method, we study the linear stability of the conduction solution (1.4.2) with respect to normal modes perturbations belonging to the space $\mathcal{M}$, and we proved Theorem 6.2.10. In [GeoPal-PasB], [PalGPashB], numerical results concerning critical stability bounds for a thermal equilibrium state in the presence of Hall and ion-slip currents are provided, writing the equation $\frac{dR_H}{dx} = 0$ as an equation of second degree in β_H^2, showing the dependence of the stability results on the given physical parameters of the problem. For a non-conducting or isotropic electrically conducting fluid, in the condition $R < R_H$, where R_H is defined in (6.2.16), we recover the conditions [Chan] (6.2.17), (6.2.18) of the linear instability of the thermodiffusive equilibrium for the hydrodynamic and isotropic magnetohydrodynamic Bénard problem, respectively.

In order to evaluate the effect of the nonlinear terms on the stability of the conduction diffusion solution m_0 we consider the following Lyapunov function [Rio88a] [MuloR94] [RioM88c]

$$V(t) = E_l(t) + bE_n(t), \qquad b > 0, \tag{4.3.4}$$

where E_l is the Lyapunov function of the linear case [PalG04a] given by

$$E_l(t) = \frac{1}{2}\{(\nabla w, \nabla w) + d_1(\zeta', \zeta') + d_2(j, j) + d_3(h_3', h_3') + d_4(\nabla_1\theta, \nabla_1\theta)\}, \tag{4.3.5}$$

b is a suitable positive parameter to be determined further, and

$$\begin{aligned}
E_n(t) =\frac{1}{2}\{&(\nabla\partial_3\mathbf{v}, \nabla\partial_3\mathbf{v}) + d_5(\nabla\zeta', \nabla\zeta') + d_7(\Delta h_3', \Delta h_3') + d_7(\nabla j', \nabla j') \\
&+ d_5(\Delta\partial_3 w, \Delta\partial_3 w) + d_{10}(\Delta\theta, \Delta\theta) + d_{11}(\nabla\Delta_1\partial_3\mathbf{u}, \nabla\Delta_1\partial_3\mathbf{u}) \\
&+ d_{12}(\nabla\Delta_1\partial_3\mathbf{h}, \nabla\Delta_1\partial_3\mathbf{h}) + d_{13}(\nabla\nabla \times \partial_3\mathbf{h}, \nabla\nabla \times \partial_3\mathbf{h})\}.
\end{aligned} \tag{4.3.6}$$

In (11) and (12) $(f, f) = \int_\Omega f^2 d\Omega$. In order to determine the largest stability domain, the coefficients d_i, $i\leq 4$ are derived in [PalG04a] by applying Joseph's idea of differentiation of parameters. The remained parameters d_i, $i\geq 5$, so far are arbitrary.

Consider the following additional equations derived from (4.3.1) and suitable to our study

$$\nabla\partial_3\mathbf{v}_{,t} = -\nabla\partial_3(\mathbf{v}\nabla\mathbf{v}) - \nabla\partial_3\nabla p + \nabla\partial_3\Delta\mathbf{v} + M^2\nabla\partial_3[(\mathbf{k} + \mathbf{h})\nabla\mathbf{h}] + \frac{R}{P_r}\nabla\partial_3\theta\mathbf{k},$$

$$\Delta\zeta'_{,t} = \mathbf{k}\cdot\partial_3\nabla \times \Delta(\mathbf{v}\nabla\mathbf{v}) + \Delta\Delta\zeta' + M^2\Delta\partial_{33}j + M^2\mathbf{k}\cdot\partial_3\nabla \times (\mathbf{h}\Delta\mathbf{h}),$$

$$\begin{aligned}
\Delta h'_{3,t} =&\Delta\partial_{33}w + \frac{P_m}{P_r}\Delta\Delta h_3' + \beta_H\frac{P_m}{P_r}\mathbf{k}\cdot\partial_3\nabla \times \Delta(\mathbf{h} \times \nabla \times \mathbf{h}) \\
&- \beta_H\frac{P_m}{P_r}\Delta\partial_{33}j + \mathbf{k}\cdot\partial_3\nabla \times \Delta(\mathbf{v} \times \mathbf{h}),
\end{aligned}$$

$$\tag{4.3.7}$$

$$\Delta j^{'}{}_{,t} = \Delta \zeta^{''} + \frac{P_m}{P_r}\Delta\Delta j^{'} + \beta_H \frac{P_m}{P_r}\Delta\Delta\partial_3 h^{'}_3$$

$$+\, \mathbf{k}\cdot\partial_3\nabla\times\nabla\times\Delta(\mathbf{v}\times\mathbf{h}) + \beta_H\frac{P_m}{P_r}\mathbf{k}\cdot\partial_3\nabla\times\nabla\times\Delta(\mathbf{h}\times\nabla\times\mathbf{h}),$$

$$-\Delta\partial_3 w_{,t} = -\,\mathbf{k}\cdot\partial_3\nabla\times\nabla\times(\mathbf{v}\nabla\mathbf{u}) - \Delta\Delta\partial_3 w - M^2\Delta\partial_{33}h_3$$

$$-\,\frac{R}{P_r}\Delta_1\partial_3\theta + M^2\mathbf{k}\cdot\partial_3\nabla\times\nabla\times(\mathbf{h}\nabla\mathbf{h}),$$

$$\Delta\theta_{,t} = \Delta w + \frac{P_m}{P_r}\Delta\Delta\theta - \Delta(\mathbf{v}\nabla\theta),$$

$$\Delta_1\partial_3\mathbf{v}_{,t} = -\,\Delta_1\partial_3(\mathbf{v}\nabla\mathbf{v}) - \Delta_1\partial_3 p + M^2\Delta_1\partial_3[(\mathbf{k}+\mathbf{h})\nabla\mathbf{h}] + \frac{R}{P_r}\Delta_1\partial_3\theta\mathbf{k},$$

$$\Delta_1\partial_3\mathbf{h}_{,t} = \Delta_1\partial_3\nabla\times[\mathbf{v}\times(\mathbf{k}+\mathbf{h})] + \frac{P_m}{P_r}\Delta_1\Delta\partial_3\mathbf{h}$$

$$+\,\beta_H\frac{P_m}{P_r}\Delta_1\partial_3\nabla\times[(\mathbf{k}+\mathbf{h})\times\nabla\times\mathbf{h}],$$

$$\nabla\nabla\times\partial_3\mathbf{h}_{,t} = \nabla\nabla\times\nabla\times\partial_3[\mathbf{v}\times(\mathbf{k}+\mathbf{h})] + \frac{P_m}{P_r}\nabla\nabla\times\Delta\partial_3\mathbf{h}$$

$$+\,\beta_H\frac{P_m}{P_r}\nabla\nabla\times\nabla\times\partial_3[(\mathbf{k}+\mathbf{h})\times\nabla\times\mathbf{h}].$$

The role of these equations is similar to that of Joseph's equation (4.1.15) enabling us to take into account the variable-sign terms in the energy relation. From the previous equations we obtain

$$\frac{d}{dt}V(t) = \mathcal{I}_l - \mathcal{D}_l + \mathcal{N}_l + b(\mathcal{I}_n - \mathcal{D}_n + \mathcal{N}_n), \tag{4.3.8}$$

where $\mathcal{I}_l$ and $\mathcal{D}_l$, deduced in [PalG04a], are given in Section 6.2 by formulae (6.2.3) and (6.2.4) and

$$\mathcal{N}_l = -\,(w, \mathbf{k}\cdot\nabla\times\nabla\times(\mathbf{v}\nabla\mathbf{v})) + M^2(w, \mathbf{k}\cdot\nabla\times\nabla\times(\mathbf{h}\nabla\mathbf{h}))$$

$$-\,d_1(\zeta^{'}, \mathbf{k}\cdot\partial_3\nabla\times(\mathbf{v}\nabla\mathbf{v}))$$

$$+\,d_1 M^2(\zeta^{'}, \mathbf{k}\cdot\partial_3\nabla\times(\mathbf{h}\nabla\mathbf{h})) + d_2(j, \mathbf{k}\cdot\nabla\times\nabla\times(\mathbf{v}\times\mathbf{h}))$$

$$+\,\beta_H\frac{P_m}{P_r}d_2(j, \mathbf{k}\cdot\nabla\times\nabla\times(\mathbf{h}\times\nabla\times\mathbf{h})) + d_3(h^{'}_3, \mathbf{k}\cdot\partial_3\nabla\times(\mathbf{v}\times\mathbf{h}))$$

$$+\,\beta_H\frac{P_m}{P_r}d_3(h^{'}_3, \mathbf{k}\cdot\partial_3\nabla\times(\mathbf{h}\times\nabla\times\mathbf{h})) - d_4(\Delta_1\theta, \mathbf{v}\nabla\theta), \tag{4.3.9}$$

$$\begin{aligned}
\mathcal{I}_n = &- M^2(\mathbf{k}\cdot\nabla\partial_3\mathbf{h},\partial_3\Delta\mathbf{v}) - \frac{R}{P_r}(\partial_3\theta,\mathbf{k}\cdot\partial_3\Delta\mathbf{v}) + (d_7 - M^2 d_5)(\Delta\zeta',\partial_{33}j) \\
&+ (d_7 - d_5 M^2)(\Delta h_3',\partial_{33}\Delta w) - \frac{R}{P_r}d_5(\Delta_1\theta,\partial_{33}\Delta w) + d_{10}(\Delta\theta,\Delta w) \\
&- (M^2 d_{11} - d_{12})(\partial_3\Delta_1\mathbf{v},\Delta\Delta_1\partial_3(\mathbf{k}\cdot\nabla\mathbf{h})) - \frac{R}{P_r}d_{11}(\Delta\Delta_1\partial_3\theta\mathbf{k},\Delta_1\partial_3\mathbf{v}) \\
&- d_{12}\beta_H\frac{P_m}{P_r}(\Delta_1\partial_3\mathbf{h},\Delta\Delta_1\nabla\times\partial_3(\mathbf{k}\times\nabla\times\mathbf{h})) \\
&+ d_{13}(\nabla\nabla\times\partial_3\mathbf{h},\nabla\nabla\times\nabla\times\partial_3(\mathbf{v}\times\mathbf{k}) \\
&+ d_{13}\beta_H\frac{P_m}{P_r}(\nabla\nabla\times\partial_3\mathbf{h},\nabla\nabla\times\nabla\times\partial_3(\mathbf{k}\times\nabla\times\mathbf{h})),
\end{aligned}$$

$$(4.3.10)$$

$$\begin{aligned}
\mathcal{D}_n = &(\Delta\partial_3\mathbf{v},\Delta\partial_3\mathbf{v}) + d_5(\Delta\zeta',\Delta\zeta') + d_7\frac{P_m}{P_r}(\nabla\Delta h_3',\nabla\Delta h_3') + d_7\frac{P_m}{P_r}(\Delta j',\Delta j') \\
&+ d_5(\nabla\Delta\partial_3 w,\nabla\Delta\partial_3 w) + d_{10}\frac{1}{P_r}(\nabla\Delta\theta,\nabla\Delta\theta) + d_{11}(\Delta\Delta_1\partial_3\mathbf{v},\Delta\Delta_1\partial_3\mathbf{v}) \\
&+ d_{12}\frac{P_m}{P_r}(\Delta\Delta_1\partial_3\mathbf{h},\Delta\Delta_1\partial_3\mathbf{h}) + d_{13}\frac{P_m}{P_r}(\Delta\nabla\times\partial_3\mathbf{h},\Delta\nabla\times\partial_3\mathbf{h}),
\end{aligned}$$

$$(4.3.11)$$

$$\begin{aligned}
\mathcal{N}_n = &(\partial_3(\mathbf{v}\nabla\mathbf{v}),\partial_3\Delta\mathbf{v}) - M^2(\partial_3(\mathbf{h}\nabla\mathbf{h}),\partial_3\Delta\mathbf{v}) + d_5(\Delta\zeta',\mathbf{k}\cdot\nabla\times\partial_3(\mathbf{v}\nabla\mathbf{v})) \\
&- M^2 d_5(\Delta\zeta',\mathbf{k}\cdot\nabla\times\partial_3(\mathbf{h}\nabla\mathbf{h})) + d_7(\Delta h_3',\mathbf{k}\cdot\Delta\nabla\times\partial_3(\mathbf{v}\times\mathbf{h})) \\
&+ d_7\beta_H\frac{P_m}{P_r}(\Delta h_3',\mathbf{k}\cdot\Delta\nabla\times\partial_3(\mathbf{h}\times\nabla\times\mathbf{h})) \\
&- d_7(\Delta j',\mathbf{k}\cdot\nabla\times\nabla\times\partial_3(\mathbf{v}\times\mathbf{h})) \\
&- d_7\beta_H\frac{P_m}{P_r}(\Delta j',\mathbf{k}\cdot\nabla\times\nabla\times\partial_3(\mathbf{h}\times\nabla\times\mathbf{h})) \\
&- d_5(\Delta\partial_{33}w,\mathbf{k}\cdot\nabla\times\nabla\times(\mathbf{v}\nabla\mathbf{v})) \\
&+ M^2 d_5(\Delta\partial_{33}w,\mathbf{k}\cdot\nabla\times\nabla\times(\mathbf{h}\nabla\mathbf{h})) - d_{10}(\Delta\theta,\Delta(\mathbf{v}\nabla\theta)) \\
&+ d_{11}(\Delta_1\partial_3\mathbf{v},\Delta\Delta_1\partial_3(\mathbf{v}\nabla\mathbf{v})) - M^2 d_{11}(\Delta_1\partial_3\mathbf{v},\Delta\Delta_1\partial_3(\mathbf{h}\nabla\mathbf{h})) \\
&- d_{12}((\Delta_1\partial_3\mathbf{h},\Delta\Delta_1\nabla\times\partial_3(\mathbf{v}\times\mathbf{h}) \\
&- \beta_H\frac{P_m}{P_r}d_{12}((\Delta_1\partial_3\mathbf{h},\Delta\Delta_1\nabla\times\partial_3(\mathbf{h}\times\nabla\times\mathbf{h})) \\
&+ d_{13}(\nabla\nabla\times\partial_3\mathbf{h},\nabla\nabla\times\nabla\times\partial_3(\mathbf{v}\times\mathbf{h})) \\
&+ \beta_H\frac{P_m}{P_r}d_{13}(\nabla\nabla\times\partial_3\mathbf{h},\nabla\nabla\times\nabla\times\partial_3(\mathbf{h}\times\nabla\times\mathbf{h})).
\end{aligned}$$

$$(4.3.12)$$

Remark that, in the class of normal mode perturbations, in order to have the positivity of E_l and E_n, we are forced to assume [PalG04a] that d_1, d_3, d_5, d_7, < 0.

Taking into account the fact that the stability of m_0 is meaningful only in a class in which m_0 is unique, to exclude any other rigid solution we require the average velocity condition [HarLP], [RioM88c]

$$\int_\Omega u d\Omega = \int_\Omega v d\Omega = 0. \qquad (4.3.13)$$

By using Schwarz, Poincaré and Wirtinger inequalities [HarLP], [J76], [Strau], and some inequalities derived in [Lad69] as particular cases of some theorems for elliptic operators, it follows

$$|\nabla \mathbf{h}| \le \frac{2}{\pi}|\nabla \partial_3 \mathbf{h}| \le \frac{4}{\pi^2}|\Delta \partial_3 \mathbf{h}| \le \frac{8}{\pi^3}|\nabla \Delta \partial_3 \mathbf{h}| = \frac{8}{\pi^3}|\nabla \times \Delta \partial_3 \mathbf{h}|,$$

$$|\nabla \mathbf{v}| \le \frac{2}{\pi}|\nabla \partial_3 \mathbf{v}| \le \frac{4}{\pi^2}|\Delta \partial_3 \mathbf{v}|, \quad |\Delta \mathbf{f}| \le \frac{2}{\pi}|\Delta \partial_3 \mathbf{f}|, \quad \text{where} \quad \mathbf{f} \in \{\mathbf{h}, \mathbf{u}\},$$

$$|\mathbf{h}| \le \frac{2}{\pi}|\partial_3 \mathbf{h}| \le \frac{4}{\pi^2}|\nabla \partial_3 \mathbf{h}|, \quad |\Delta j| \le \frac{2}{\pi}|\Delta j'| \le \frac{2}{\pi^2}|\nabla \Delta j'|,$$

$$|\Delta \nabla \times \mathbf{h}| \le \frac{2}{\pi}|\Delta \nabla \times \partial_3 \mathbf{h}|, |\Delta \Delta_1 \mathbf{v}| \le \frac{2}{\pi}|\Delta \Delta_1 \partial_3 \mathbf{v}|, |\nabla \Delta_1 \partial_3 \mathbf{h}| \le \frac{2}{\pi}|\Delta \Delta_1 \partial_3 \mathbf{h}|,$$

$$|\nabla \Delta \mathbf{v}| \le \frac{4\sqrt{2}}{\pi^2}|\Delta \Delta_1 \partial_3 \mathbf{v}| + \sqrt{2}|\Delta \partial_3 \mathbf{v}|,$$

$$|\Delta \mathbf{v}| \le \frac{4\sqrt{2}}{\pi^2}|\nabla \Delta_1 \partial_3 \mathbf{v}| + \sqrt{2}|\nabla \partial_3 \mathbf{v}|, |\Delta \Delta \mathbf{h}| \le \sqrt{2}(\frac{2}{\pi}|\Delta \Delta_1 \partial_3 \mathbf{h}| + |\nabla \Delta \partial_3 \mathbf{h}|).$$
$$(4.3.14)$$

From (4.3.10), (4.3.13), (4.3.14), the boundary conditions (4.3.2) and

$$\Delta h_1 = \Delta h_2 = \Delta \Delta h_1 = \Delta \Delta h_2 = \Delta \Delta w = \Delta j = \Delta \zeta' = 0, \qquad z = 0,1 \qquad (4.3.15)$$

which follow from the assumption that the perturbation equations hold on the boundary, and choosing $d_7 = d_5 M^2$, $d_{12} = M^2 d_{11}$, we have

$$\mathcal{I}_n \le \frac{4}{\pi^2} M^2 |\nabla \times \Delta \partial_3 \mathbf{h}||\Delta \partial_3 \mathbf{v}| + \frac{1}{\pi}(\frac{R}{P_r} + d_{10})|\nabla \Delta \theta||\Delta w|$$

$$+ \frac{R}{P_r}\frac{|d_5|}{\pi}\sqrt{2}|\nabla \Delta \theta||\nabla \Delta \partial_3 w| + \frac{R}{P_r}d_{11}\sqrt{2}|\Delta \Delta_1 \partial_3 \mathbf{v}||\nabla \Delta \theta| + \beta_H \frac{P_m}{P_r} M^2 d_{11}$$

$$|\Delta \Delta_1 \partial_3 \mathbf{h}|^2 + d_{13}\sqrt{2}|\Delta \nabla \times \partial_3 \mathbf{h}||\Delta \partial_3 \mathbf{v}| + d_{13}\beta_H \frac{P_m}{P_r}|\Delta \nabla \times \partial_3 \mathbf{h}|^2. \qquad (4.3.16)$$

Assuming that $d_{11} \le \frac{\beta_H^2 P_r}{4R^2} d_{10}$, it follows

$$-\frac{\beta_H}{2}d_{11}|\Delta \Delta_1 \partial_3 \mathbf{v}|^2 + \frac{R}{P_r}d_{11}\sqrt{2}|\Delta \Delta_1 \partial_3 \mathbf{v}||\nabla \Delta \theta| - \frac{\beta_H}{4}\frac{d_{10}}{P_r}|\nabla \Delta \theta|^2 \le 0. \qquad (4.3.17)$$

Applying the Young's inequality, from (4.3.10), (4.3.16), (4.3.17), we obtain

$$b\mathcal{I}_n \le A_0 b \mathcal{D}_n + A_1 b \mathcal{D}_l^{\frac{1}{2}} \mathcal{D}_n^{\frac{1}{2}} \le (A_0 + \mu^2 \frac{A_1}{2})b\mathcal{D}_n + \frac{A_1}{2\mu^2}b\mathcal{D}_l, \qquad (4.3.18)$$

where μ is a positive parameter and

$$A_0 = \sqrt{\frac{P_r}{P_m}}\left\{\frac{4M^2}{\pi^2\sqrt{d_{13}}} + \sqrt{2}\sqrt{d_{13}}\right\} + R\frac{\sqrt{2}}{\pi\sqrt{P_r}}\sqrt{\frac{|d_5|}{d_{10}}} + \beta_H, \; A_1 = (\frac{R}{P_r} + d_{10})\frac{1}{\pi}\sqrt{\frac{P_r}{d_{10}}}.$$

Let us estimate the $\mathcal{N}_l$ terms by using the inequalities (4.3.14) and the inequality for the supremum of a function [Lad69]

$$sup_\Omega \mid \mathbf{f} \mid \leq C|\Delta \mathbf{f}|, \tag{4.3.19}$$

where the value for the constant C, depending on the periodicity cell Ω, is calculated in [Strau].

For each term of $\mathcal{N}_l$ we have

$$(w, \mathbf{k} \cdot \nabla \times \nabla \times (\mathbf{v}\nabla\mathbf{v})) = (\nabla \times \nabla \times (\mathbf{k}w), \mathbf{v}\nabla\mathbf{v}) \leq C\sqrt{2}|\Delta\mathbf{v}||\nabla\mathbf{v}||\Delta w|$$

$$\leq C\frac{4\sqrt{2}}{\pi^2}|\Delta\partial_3\mathbf{v}||\nabla\partial_3\mathbf{v}||\Delta w|, \tag{4.3.20}$$

$$(w, \mathbf{k} \cdot \nabla \times \nabla \times (\mathbf{h}\nabla\mathbf{h})) = (\nabla \times \nabla \times (\mathbf{k}w), \mathbf{h}\nabla\mathbf{h}) \leq C\sqrt{2}|\Delta\mathbf{h}||\nabla\mathbf{h}||\Delta w|$$

$$\leq C\frac{16\sqrt{2}}{\pi^4}|\Delta\partial_3\mathbf{h}||\Delta\nabla \times \partial_3\mathbf{h}||\Delta w|, \tag{4.3.21}$$

$$(\zeta', \mathbf{k} \cdot \nabla \times \partial_3(\mathbf{v}\nabla\mathbf{v})) \leq C|\Delta\mathbf{v}||\nabla\mathbf{v}||\Delta\zeta'| \leq C\frac{4}{\pi^2}|\Delta\partial_3\mathbf{v}||\nabla\partial_3\mathbf{v}||\Delta\zeta'|, \tag{4.3.22}$$

$$(\zeta', \mathbf{k} \cdot \nabla \times \partial_3(\mathbf{h}\nabla\mathbf{h})) \leq C|\Delta\mathbf{h}||\nabla\mathbf{h}||\Delta\zeta'| \leq \frac{16}{\pi^4}C \parallel \Delta\partial_3\mathbf{h}||\Delta\nabla \times \partial_3\mathbf{h}||\Delta\zeta'|, \tag{4.3.23}$$

$$(j, \mathbf{k}\cdot\nabla\times\nabla\times(\mathbf{v}\times\mathbf{h})) \leq 2\sqrt{2}C|\Delta\mathbf{v}||\mathbf{h}||\Delta j| \leq C\frac{64\sqrt{2}}{\pi^5}|\Delta\partial_3\mathbf{v}||\nabla\nabla\times\partial_3\mathbf{h}||\Delta j'|, \tag{4.3.24}$$

$$(j, \mathbf{k} \cdot \nabla \times \nabla \times (\mathbf{h} \times \nabla \times \mathbf{h})) \leq 2\sqrt{2}C|\Delta\mathbf{h}||\nabla\mathbf{h}||\Delta j|$$

$$\leq C\frac{64\sqrt{2}}{\pi^5}|\Delta\partial_3\mathbf{h}||\Delta\nabla \times \partial_3\mathbf{h}||\Delta j'|, \tag{4.3.25}$$

$$(h_3', \mathbf{k} \cdot \nabla \times \partial_3(\mathbf{v} \times \mathbf{h})) = -(\nabla \times \partial_3 h_3' \cdot \mathbf{k}, (\mathbf{v} \times \mathbf{h})) \leq 2C|\Delta\mathbf{v}||\mathbf{h}||\Delta h_3'|$$

$$\leq \frac{32}{\pi^4}C|\Delta\partial_3\mathbf{v}||\Delta\partial_3\mathbf{h}||\Delta h_3'|, \tag{4.3.26}$$

$$(h_3', \mathbf{k} \cdot \nabla\times\partial_3(\mathbf{h} \times \nabla \times \mathbf{h})) = (\nabla \times \partial_3 h_3' \cdot \mathbf{k}, (\mathbf{h} \times \nabla \times \mathbf{h}))$$

$$\leq 2C|\Delta\mathbf{h}||\nabla\mathbf{h}||\Delta h_3'| \leq \frac{32}{\pi^5}C|\Delta\partial_3\mathbf{h}|\Delta\nabla \times \partial_3\mathbf{h}||\nabla\Delta h_3'|, \tag{4.3.27}$$

$$(\Delta_1\theta, \mathbf{v}\nabla\theta) \leq C|\nabla_1\theta|(|\nabla_1\mathbf{v}||\nabla\Delta\theta| + |\Delta\mathbf{v}||\nabla\nabla_1\theta|)$$

$$\leq \frac{2}{\pi}C|\nabla_1\theta|(|\nabla\partial_3\mathbf{v}||\nabla\Delta\theta| + |\Delta\partial_3\mathbf{v}||\nabla\nabla_1\theta|). \tag{4.3.28}$$

Applying the Young's inequality, from $(4.3.20) - (4.3.28)$, it follows

$$\mathcal{N}_l \leq A_2 \mathcal{D}_n V^{\frac{1}{2}} + A_3 \mathcal{D}_l^{\frac{1}{2}} \mathcal{D}_n^{\frac{1}{2}} V^{\frac{1}{2}} \leq \mathcal{D}_n V^{\frac{1}{2}}(A_2 + \frac{A_3\epsilon^2}{2}) + \frac{A_3}{2\epsilon^2}\mathcal{D}_l V^{\frac{1}{2}}, \tag{4.3.29}$$

where ϵ is a positive parameter and

$$A_2 = \frac{1}{\sqrt{bd_5}}\left\{\frac{4C}{\pi^4}\sqrt{2} + \frac{16CM^2}{\pi^6}\frac{\sqrt{2}}{d_{13}}\sqrt{\frac{P_r}{P_m}} + \frac{M128C}{\pi^5\sqrt{d_{13}}}\sqrt{\frac{P_r}{P_m}} + M\beta_H\frac{64C\sqrt{2}}{\pi^5 d_{13}}\right.$$

$$\left. + \frac{32C\sqrt{2}}{\pi^5\sqrt{d_{13}}}\sqrt{\frac{P_r}{P_m}}M\frac{(\pi^2+a^2)}{\pi^2} + \beta_H M\frac{(\pi^2+a^2)}{\pi^2}\frac{32C\sqrt{2}}{\pi^5 d_{13}}\right\} \equiv \frac{A_2'}{\sqrt{b}},$$

$$A_3 = \frac{2C}{\pi}\sqrt{2P_r}(1+\frac{1}{\pi}) + \frac{8C}{\pi^2\sqrt{b}} + \frac{32CM^2}{\pi^4 d_{13}\sqrt{b}}\sqrt{\frac{P_r}{P_m}} \equiv \frac{A_3'}{\sqrt{b}} + A_3''.$$

In deriving the expressions for A_2 and A_3 we took into account (Section 6.2) that $d_2 = M^2$, $|d_3| = \frac{M^2(\pi^2+a^2)}{\pi^2}$, and we chose $d_4 = d_{10}$. Therefore we obtain

$$\frac{d}{dt}V(t) \leq \mathcal{D}_l(\lambda - 1 + \frac{A_1 b}{2\mu^2}) + b\mathcal{D}_n(A_0 + \frac{A_1\mu^2}{2} - 1) + (A_2 + \frac{A_3\epsilon^2}{2})\mathcal{D}_n V^{\frac{1}{2}}$$

$$+ \frac{A_3}{2\epsilon^2}\mathcal{D}_l V^{\frac{1}{2}} + b\mathcal{N}_n, \tag{4.3.30}$$

where [PalG04a] $\lambda = \max_{\mathcal{M}}\frac{\mathcal{J}_l}{\mathcal{D}_l}$. Choosing $\epsilon^2 = \frac{A_3\mu^2}{A_1 b}$, $\mu^2 = \frac{2A_1 b(b-A_2)}{A_1^2 b^2 + A_3^2}$, where $b > \sqrt[3]{A_2'^2}$, and taking into account that $\lambda < 1$, the inequality (4.3.30) becomes

$$\frac{d}{dt}V(t) \leq \left[A_0 b - (A_2 + \frac{A_3\epsilon^2}{2})\right]\mathcal{D}_n + \frac{A_3}{2\epsilon^2}\mathcal{D}_l + (A_2 + \frac{A_3\epsilon^2}{2})\mathcal{D}_n V^{\frac{1}{2}}$$

$$+ \frac{A_3}{2\epsilon^2}\mathcal{D}_l V^{\frac{1}{2}} + b\mathcal{N}_n. \tag{4.3.31}$$

From (6.2.4) and (4.3.11), assuming $|d_5| \geq \frac{\pi^2+a^2}{\pi^4}$, it follows $\mathcal{D}_l < \mathcal{D}_n$, so (4.3.31) implies

$$\frac{d}{dt}V(t) \leq -C\mathcal{D}_n + B\mathcal{D}_n V^{\frac{1}{2}} + b\mathcal{N}_n, \tag{4.3.32}$$

where

$$C = -\left[A_0 b - (A_2 + \frac{A_3\epsilon^2}{2}) + \frac{A_3}{2\epsilon^2}\right], \qquad B = \left[(A_2 + \frac{A_3\epsilon^2}{2}) + \frac{A_3}{2\epsilon^2}\right].$$

Assuming that $\epsilon^2 > \bar{\epsilon}^2 \equiv max\{1, \epsilon_1^2\}$, where ϵ_1^2 is the positive root of the equation

$$-\epsilon^4(A_3''\sqrt[3]{A_2'} + A_3') + 2\epsilon^2 A_2'(A_0 - 1) + (A_3''\sqrt[3]{A_2'} + A_3') = 0$$

and, consequently, $\mu^2 > \frac{A_1 b}{A_3}\bar{\epsilon}^2$, we obtain that the condition $C > 0$, equivalent to

$$f(\sqrt{b}) \equiv 2\epsilon^2 A_0(\sqrt{b})^3 + \sqrt{b}A_3''(1 - \epsilon^4) + [A_3'(1 - \epsilon^4) - 2\epsilon^2 A_2'] < 0,$$

is satisfied if we choose $\sqrt[3]{A_2'} < \sqrt{b} < (\sqrt{b})_1$, where $(\sqrt{b})_1$ is the smallest positive root of the equation $f(\sqrt{b}) = 0$.

In order to estimate the $\mathcal{N}_n$ terms we remark that, from the representation theorem of solenoidal doubly periodic vectors [J76] as the sum of poloidal and toroidal fields, for a plane layer the identity $-\Delta_1\mathbf{v} = \nabla \times \nabla \times (\mathbf{e}_3 w) + \nabla \times (\zeta\mathbf{e}_3)$ holds. In addition, in [MuloR89], it is showed that, for a regular function $\mathbf{u}$ defined on a periodicity cell Ω and periodic in x and y direction, there exists a positive constant k such that $\|\nabla_1\Delta\mathbf{u}\| \leq k \|\Delta_1\Delta\mathbf{u}\|, \quad \|\nabla_1\nabla\mathbf{u}\| \leq k \|\Delta_1\nabla\mathbf{u}\|$. Using the Hölder inequality in [Strau], for functions doubly periodic in x and y directions, with periodicity a_x and a_y respectively, the following general inequality

$$\left(\int_\Omega u^4 d\Omega\right)^{\frac{1}{2}} \leq 32(a_x a_y)^{\frac{1}{6}}\left(\int_\Omega u^2 d\Omega + \int_\Omega \partial_i u \partial_i u d\Omega\right)$$

follows irrespective of the boundary conditions at $z = 0, 1$.

For each term of $\mathcal{N}_n$ we have

$$(\partial_3(\mathbf{v}\nabla\mathbf{v}), \partial_3\Delta\mathbf{v}) \leq C|\Delta\partial_3\mathbf{v}|(|\Delta\partial_3\mathbf{v}||\nabla\mathbf{v}| + |\Delta\mathbf{v}||\nabla\partial_3\mathbf{v}|)$$

$$\leq \frac{4}{\pi}C|\Delta\partial_3\mathbf{v}|^2 \|\nabla\partial_3\mathbf{v}| \leq \frac{4}{\pi}C\frac{\sqrt{2}}{\sqrt{b}}\mathcal{D}_n V^{\frac{1}{2}} \equiv A_4\mathcal{D}_n V^{\frac{1}{2}}, \tag{4.3.33}$$

$$(\partial_3(\mathbf{h}\nabla\mathbf{h}), \partial_3\Delta\mathbf{v}) \leq C|\Delta\partial_3\mathbf{v}|(|\Delta\partial_3\mathbf{h}||\nabla\mathbf{h}| + |\Delta\mathbf{h}||\nabla\partial_3\mathbf{h}|)$$

$$\leq \frac{4}{\pi}C|\Delta\partial_3\mathbf{v}||\Delta\partial_3\mathbf{h}||\nabla\partial_3\mathbf{h}| \leq \frac{16}{\pi^3}C\frac{\sqrt{2}}{\sqrt{b}}\frac{1}{d_{13}}\sqrt{\frac{P_r}{P_m}}\mathcal{D}_n V^{\frac{1}{2}} \equiv A_5\mathcal{D}_n V^{\frac{1}{2}}, \tag{4.3.34}$$

$$(\Delta\zeta', \mathbf{k}\cdot\nabla\times\partial_3(\mathbf{v}\nabla\mathbf{v})) - (\Delta\partial_{33}w, \mathbf{k}\cdot\nabla\times\nabla\times(\mathbf{v}\nabla\mathbf{v}))$$

$$=(\Delta\Delta_1\partial_3\mathbf{v}, \partial_3(\mathbf{v}\nabla\mathbf{v}))$$

$$\leq |\Delta\Delta_1\partial_3\mathbf{v}|C(|\Delta\partial_3\mathbf{v}||\nabla\mathbf{v}| + |\Delta\mathbf{v}||\nabla\partial_3\mathbf{v}|)$$

$$\leq \frac{4}{\pi}|\Delta\Delta_1\partial_3\mathbf{v}|C|\Delta\partial_3\mathbf{v}||\nabla\partial_3\mathbf{v}| \tag{4.3.35}$$

$$\leq \frac{4}{\pi}C\frac{\sqrt{2}}{\sqrt{b}}\frac{1}{\sqrt{d_{11}}}\mathcal{D}_n V^{\frac{1}{2}} \equiv A_6\mathcal{D}_n V^{\frac{1}{2}},$$

$$(\Delta\zeta', \mathbf{k}\cdot\nabla\times\partial_3(\mathbf{h}\nabla\mathbf{h})) - (\Delta\partial_{33}w, \mathbf{k}\cdot\nabla\times\nabla\times(\mathbf{h}\nabla\mathbf{h}))$$

$$=(\Delta\Delta_1\partial_3\mathbf{v}, \partial_3(\mathbf{h}\nabla\mathbf{h}))$$

$$\leq |\Delta\Delta_1\partial_3\mathbf{v}|C(|\Delta\partial_3\mathbf{h}||\nabla\mathbf{h}| + |\Delta\mathbf{h}||\nabla\partial_3\mathbf{h}|)$$

$$\leq \frac{4}{\pi}|\Delta\Delta_1\partial_3\mathbf{v}|C|\Delta\partial_3\mathbf{h}||\nabla\partial_3\mathbf{h}| \tag{4.3.36}$$

$$\leq \frac{16}{\pi^3}C\frac{\sqrt{2}}{\sqrt{b}}\frac{1}{d_{13}\sqrt{d_{11}}}\sqrt{\frac{P_r}{P_m}}\mathcal{D}_n V^{\frac{1}{2}} \equiv A_7\mathcal{D}_n V^{\frac{1}{2}},$$

$$(\Delta h_3', \mathbf{k}\cdot\Delta\nabla\times\partial_3(\mathbf{v}\times\mathbf{h})) - (\Delta j', \nabla\times\nabla\times\partial_3(\mathbf{u}\times\mathbf{h}))$$

$$=(\Delta\Delta_1\partial_3\mathbf{h}, \nabla\times\partial_3(\mathbf{v}\times\mathbf{h})$$

$$\leq |\Delta\Delta_1\partial_3\mathbf{h}|C(|\Delta\partial_3\mathbf{h}||\nabla\mathbf{v}| + |\Delta\mathbf{h}||\nabla\partial_3\mathbf{v}|$$

$$+ |\Delta\partial_3\mathbf{v}||\nabla\mathbf{h}| + |\Delta\mathbf{v}||\nabla\partial_3\mathbf{h}|) \tag{4.3.37}$$

$$\leq |\Delta\Delta_1\partial_3\mathbf{h}|\frac{4}{\pi}C\Big(|\Delta\partial_3\mathbf{h}||\nabla\partial_3\mathbf{v}| + |\Delta\partial_3\mathbf{v}||\nabla\partial_3\mathbf{h}|\Big)$$

$$\leq \frac{16}{\pi^2}C\frac{\sqrt{2}}{\sqrt{b}}\frac{1}{\sqrt{d_{12}d_{13}}}\sqrt{\frac{P_r}{P_m}}\mathcal{D}_n V^{\frac{1}{2}} \equiv A_8\mathcal{D}_n V^{\frac{1}{2}},$$

$$
\begin{aligned}
(\Delta h_3', \mathbf{k} \cdot \Delta\nabla\times\partial_3(\mathbf{h}\times\nabla\times\mathbf{h})) &- (\Delta j', \nabla\times\nabla\times\partial_3(\mathbf{h}\times\nabla\times\mathbf{h})) \\
&= (\Delta\Delta_1\partial_3\mathbf{h}, \nabla\times\partial_3(\mathbf{h}\times\nabla\times\mathbf{h}) \\
&\leq |\Delta\Delta_1\partial_3\mathbf{h}| C(|\Delta\nabla\mathbf{h}||\nabla\times\partial_3\mathbf{h}| + |\Delta\nabla\times\mathbf{h}||\nabla\partial_3\mathbf{h}| \\
&\quad + |\Delta\partial_3\mathbf{h}||\nabla\nabla\times\mathbf{h}| + |\Delta\mathbf{h}||\nabla\nabla\times\partial_3\mathbf{h}|) \\
&\leq |\Delta\Delta_1\partial_3\mathbf{h}| C\frac{2}{\pi}(|\Delta\nabla\times\partial_3\mathbf{h}||\nabla\times\partial_3\mathbf{h}| \\
&\quad + |\Delta\partial_3\mathbf{h}||\nabla\nabla\times\partial_3\mathbf{h}|) \\
&\leq \frac{8}{\pi^2}C\frac{\sqrt{2}}{\sqrt{b}}\frac{1}{\sqrt{d_{12}d_{13}}}\frac{P_r}{P_m}\mathcal{D}_n V^{\frac{1}{2}} \equiv A_9\mathcal{D}_n V^{\frac{1}{2}},
\end{aligned}
\tag{4.3.38}
$$

$$
\begin{aligned}
(\Delta\theta, \Delta(\mathbf{v}\nabla\theta)) &= (\nabla\Delta\theta, \nabla(\mathbf{v}\nabla\theta)) \leq \|\nabla\Delta\theta| C(|\nabla\Delta\theta||\nabla\mathbf{v}| + |\Delta\mathbf{v}||\nabla\nabla\theta|) \\
&\leq \frac{2}{\pi}|\nabla\Delta\theta| C(|\nabla\Delta\theta||\nabla\partial_3\mathbf{v}| + |\Delta\partial_3\mathbf{v}||\nabla\nabla\theta|) \\
&\leq \frac{2}{\pi}C\frac{\sqrt{2}}{\sqrt{b}}\frac{\sqrt{P_r}}{d_{10}}(1 + \sqrt{P_r})\mathcal{D}_n V^{\frac{1}{2}} \equiv A_{10}\mathcal{D}_n V^{\frac{1}{2}},
\end{aligned}
\tag{4.3.39}
$$

$$
\begin{aligned}
(\Delta_1\partial_3\mathbf{v}, \Delta\Delta_1\partial_3(\mathbf{v}\nabla\mathbf{v})) &= (\Delta\Delta_1\partial_3\mathbf{u}, \Delta_1\partial_3(\mathbf{v}\nabla\mathbf{v})) \\
&\leq |\Delta\Delta_1\partial_3\mathbf{v}| C(|\Delta\Delta_1\partial_3\mathbf{v}||\nabla\mathbf{v}| + |\Delta\Delta_1\mathbf{v}||\nabla\partial_3\mathbf{v}| + |\Delta\partial_3\mathbf{v}||\Delta_1\nabla\mathbf{v}| \\
&\quad + |\Delta\mathbf{v}||\Delta_1\nabla\partial_3\mathbf{v}| + 2|\Delta\nabla_1\partial_3\mathbf{v}||\nabla\nabla_1\mathbf{v}| + 2|\Delta\nabla_1\mathbf{v}||\nabla\nabla_1\partial_3\mathbf{v}|) \\
&\leq |\Delta\Delta_1\partial_3\mathbf{v}|\frac{4}{\pi}C\big(|\Delta\Delta_1\partial_3\mathbf{v}||\nabla\partial_3\mathbf{v}| + |\Delta\partial_3\mathbf{v}||\Delta_1\nabla\partial_3\mathbf{v}| \\
&\quad + 2|\Delta\nabla_1\partial_3\mathbf{v}||\nabla\nabla_1\partial_3\mathbf{v}|\big) \\
&\leq \frac{4}{\pi}C\frac{\sqrt{2}}{\sqrt{b}}\frac{1}{d_{11}}2(1 + \frac{k^2}{\sqrt{d_{11}}})\mathcal{D}_n V^{\frac{1}{2}} \equiv A_{11}\mathcal{D}_n V^{\frac{1}{2}},
\end{aligned}
\tag{4.3.40}
$$

$$
\begin{aligned}
(\Delta_1\partial_3\mathbf{v}, \Delta\Delta_1\partial_3(\mathbf{h}\nabla\mathbf{h})) &= (\Delta\Delta_1\partial_3\mathbf{v}, \Delta_1\partial_3(\mathbf{h}\nabla\mathbf{h})) \\
&\leq |\Delta\Delta_1\partial_3\mathbf{v}| C(|\Delta\Delta_1\partial_3\mathbf{h}||\nabla\mathbf{h}| + |\Delta\Delta_1\mathbf{h}||\nabla\partial_3\mathbf{h}| + |\Delta\partial_3\mathbf{h}||\Delta_1\nabla\mathbf{h}| \\
&\quad + |\Delta\mathbf{h}||\Delta_1\nabla\partial_3\mathbf{h}| + 2|\Delta\nabla_1\partial_3\mathbf{h}||\nabla\nabla_1\mathbf{h}| + 2|\Delta\nabla_1\mathbf{h}||\nabla\nabla_1\partial_3\mathbf{h}|) \\
&\leq |\Delta\Delta_1\partial_3\mathbf{u}| C\frac{4}{\pi}\big(|\Delta\Delta_1\partial_3\mathbf{h}||\nabla\partial_3\mathbf{h}| + |\Delta\partial_3\mathbf{h}||\Delta_1\nabla\partial_3\mathbf{h}| \\
&\quad + 2|\Delta\nabla_1\partial_3\mathbf{h}||\nabla\nabla_1\partial_3\mathbf{h}|)\big) \\
&\leq \frac{8}{\pi}C\frac{\sqrt{2}}{\sqrt{b}}\frac{1}{\sqrt{d_{11}d_{12}}}\sqrt{\frac{P_r}{P_m}}\Big(\frac{2}{\pi\sqrt{d_{13}}} + \frac{k^2}{\sqrt{d_{12}}}\Big)\mathcal{D}_n V^{\frac{1}{2}} \equiv A_{12}\mathcal{D}_n V^{\frac{1}{2}},
\end{aligned}
\tag{4.3.41}
$$

$$(\Delta_1\partial_3\mathbf{h},\Delta\Delta_1\nabla\times\partial_3(\mathbf{v}\times\mathbf{h})) = (\Delta\Delta_1\partial_3\mathbf{h}, \Delta_1\nabla\times\partial_3(\mathbf{v}\times\mathbf{h}))$$

$$\leq|\Delta\Delta_1\partial_3\mathbf{h}|C(|\Delta\Delta_1\mathbf{h}||\nabla\partial_3\mathbf{v}| + |\Delta\mathbf{h}||\nabla\Delta_1\partial_3\mathbf{v}|$$

$$+ 2|\Delta\nabla_1\mathbf{h}||\nabla\nabla_1\partial_3\mathbf{v}| + |\Delta\Delta_1\partial_3\mathbf{h}||\nabla\mathbf{v}| + |\Delta\partial_3\mathbf{h}||\Delta_1\nabla\mathbf{v}|$$

$$+ 2|\Delta\nabla_1\partial_3\mathbf{h}||\nabla_1\nabla\mathbf{v}| + |\Delta\Delta_1\partial_3\mathbf{v}||\nabla\mathbf{h}| + 2|\Delta\nabla_1\partial_3\mathbf{v}||\nabla_1\nabla\mathbf{h}|$$

$$+ |\Delta\partial_3\mathbf{v}||\Delta_1\nabla\mathbf{h}| + |\Delta\Delta_1\mathbf{v}||\nabla\partial_3\mathbf{h}| + 2|\Delta\nabla_1\mathbf{v}||\nabla\nabla_1\partial_3\mathbf{h}|$$

$$+ |\Delta\mathbf{v}||\Delta_1\nabla\partial_3\mathbf{h}|)$$

$$\leq|\Delta\Delta_1\partial_3\mathbf{h}|C\frac{4}{\pi}\Big(|\Delta\Delta_1\partial_3\mathbf{h}||\nabla\partial_3\mathbf{v}| + |\Delta\partial_3\mathbf{h}||\nabla\Delta_1\partial_3\mathbf{v}| \tag{4.3.42}$$

$$+ 2|\Delta\nabla_1\partial_3\mathbf{h}||\nabla\nabla_1\partial_3\mathbf{v}|\Big)+|\Delta\Delta_1\partial_3\mathbf{h}|C\frac{4}{\pi}\Big(|\Delta\Delta_1\partial_3\mathbf{v}||\nabla\partial_3\mathbf{h}|$$

$$+ 2|\Delta\nabla_1\partial_3\mathbf{v}||\nabla_1\nabla\partial_3\mathbf{h}| + |\Delta\partial_3\mathbf{v}||\Delta_1\nabla\partial_3\mathbf{h}|\Big)$$

$$\leq\frac{4}{\pi}C\frac{\sqrt{2}}{\sqrt{b}}\frac{1}{\sqrt{d_{12}}}\sqrt{\frac{P_r}{P_m}}\Big[(1 + \frac{2k^2}{\sqrt{d_{11}}})\frac{1}{\sqrt{d_{12}}}(1 + \sqrt{\frac{P_r}{P_m}})$$

$$+ \frac{4}{\pi\sqrt{d_{11}d_{13}}}\Big]\mathcal{D}_n V^{\frac{1}{2}} \equiv A_{13}\mathcal{D}_n V^{\frac{1}{2}},$$

$$(\Delta_1\partial_3\mathbf{h}, \Delta\Delta_1\nabla\times\partial_3(\mathbf{h}\times\nabla\times\mathbf{h})) = (\Delta\Delta_1\partial_3\mathbf{h}, \Delta_1\nabla\times\partial_3(\mathbf{h}\times\nabla\times\mathbf{h}))$$

$$=(\Delta\Delta_1\partial_3\mathbf{h}, \Delta_1\nabla\times\partial_3\mathbf{h}\nabla\mathbf{h} + \nabla\times\partial_3\mathbf{h}\Delta_1\nabla\mathbf{h} + 2\nabla_1\nabla\times\partial_3\mathbf{h}\nabla_1\nabla\mathbf{h}$$

$$+ \Delta_1\nabla\times\mathbf{h}\cdot\nabla\partial_3\mathbf{h} + \nabla\times\mathbf{h}, \Delta_1\nabla\partial_3\mathbf{h} + 2\nabla_1\nabla\times\mathbf{h}\nabla_1\nabla\partial_3\mathbf{h}$$

$$- \Delta_1\partial_3\mathbf{h}\nabla\nabla\times\mathbf{h} - \partial_3\mathbf{h}\Delta_1\nabla\nabla\times\mathbf{h} - 2\nabla_1\partial_3\mathbf{h}\nabla_1\nabla\nabla\times\mathbf{h}$$

$$- \Delta_1\mathbf{h}\nabla\nabla\times\partial_3\mathbf{h} - \mathbf{h}\Delta_1\nabla\nabla\times\partial_3\mathbf{h} - 2\nabla_1\mathbf{h}\nabla_1\nabla\nabla\times\partial_3\mathbf{h})$$

$$\leq|\Delta\Delta_1\partial_3\mathbf{h}|C\Big(|\Delta_1\nabla\times\partial_3\mathbf{h}||\Delta\nabla\times\partial_3\mathbf{h}|\frac{4}{\pi} + |\Delta\Delta_1\partial_3\mathbf{h}||\Delta\partial_3\mathbf{h}|\frac{6\sqrt{2}}{\pi} \tag{4.3.43}$$

$$+ |\Delta\nabla\times\partial_3\mathbf{h}||\Delta_1\nabla\partial_3\mathbf{h}|\frac{4}{\pi} + 2|\Delta\nabla_1\partial_3\mathbf{h}||\nabla_1\nabla\nabla\times\mathbf{h}|$$

$$+ |\Delta\mathbf{h}||\Delta_1\nabla\nabla\times\partial_3\mathbf{h}| + 2|\Delta\nabla_1\mathbf{h}||\nabla_1\nabla\nabla\times\partial_3\mathbf{h}|\Big)$$

$$+ 2|\Delta\Delta_1\partial_3\mathbf{h}|(|\nabla_1\nabla\times\partial_3\mathbf{h}\cdot\nabla_1\nabla\mathbf{h}| + |\nabla_1\nabla\times\mathbf{h}\nabla_1\nabla\partial_3\mathbf{h}|).$$

In order to estimate the last three terms we note that

$$|\nabla_1\nabla\nabla\times\mathbf{h}|\leq\sqrt{2(k^2 + \frac{4}{\pi^2})}|\nabla\Delta_1\partial_3\mathbf{h}|, \tag{4.3.44}$$

$$|\nabla_1\nabla\times\mathbf{h}\nabla_1\nabla\partial_3\mathbf{h}|)\leq4\overline{\Omega}\sqrt{(1 + \frac{1}{\pi^2})(1 + \frac{4}{\pi^2})}|\nabla\Delta_1\partial_3\mathbf{h}|$$

$$\Big(|\Delta_1\Delta\partial_3\mathbf{h}|\frac{2}{\pi}\sqrt{\frac{4}{\pi^2} + k^2}) + |\Delta\nabla\times\partial_3\mathbf{h}|\frac{2}{\pi}\Big), \tag{4.3.45}$$

$$|\nabla_1\nabla\times\partial_3\mathbf{h}\nabla_1\nabla\mathbf{h}|)\leq4\overline{\Omega}(1 + \frac{1}{\pi^2})\Big(\frac{2}{\pi}|\nabla\Delta_1\partial_3\mathbf{h}|^2$$

$$+ |\Delta\nabla\times\partial_3\mathbf{h}||\nabla\Delta_1\partial_3\mathbf{h}|\sqrt{1 + \frac{4}{\pi^2}}\Big), \tag{4.3.46}$$

$$|\Delta\nabla_1\mathbf{h}|\leq\sqrt{2(k^2+\frac{4}{\pi^2})}|\nabla\Delta_1\partial_3\mathbf{h}|, \tag{4.3.47}$$

where $\overline{\Omega}=32\sqrt[6]{\frac{4\pi^2}{a_x a_y}}$. From $(58)-(62)$ we get

$$\begin{aligned}
(\Delta_1\partial_3\mathbf{h},&\Delta\Delta_1\nabla\times\partial_3(\mathbf{h}\times\nabla\times\mathbf{h}))\leq\Big\{C\frac{P_r}{P_m}\frac{1}{d_{12}\sqrt{b}}(1+\sqrt{2})\Big[\frac{8\sqrt{2}}{\pi\sqrt{d_{13}}}\\
&+\frac{4k}{\sqrt{d_{12}}}\sqrt{k^2+\frac{4}{\pi^2}}\Big]+\frac{P_r}{P_m}8\sqrt{2}\frac{1}{d_{12}\sqrt{b}}\overline{\Omega}\sqrt{1+\frac{1}{\pi^2}}\\
&\Big[\sqrt{1+\frac{1}{\pi^2}}\Big(\frac{4}{\pi^2\sqrt{d_{12}}}+\frac{1}{\sqrt{d_{13}}}\sqrt{1+\frac{4}{\pi^2}}\Big)\\
&+\sqrt{1+\frac{4}{\pi^2}}\Big(\frac{2}{\pi\sqrt{d_{12}}}\sqrt{k^2+\frac{4}{\pi^2}}+\frac{2}{\pi\sqrt{d_{13}}}\Big)\Big]\Big\}\mathcal{D}_n V^{\frac{1}{2}}\\
&\equiv A_{14}\mathcal{D}_n V^{\frac{1}{2}},
\end{aligned} \tag{4.3.48}$$

$$\begin{aligned}
(\nabla\nabla\times\partial_3\mathbf{h},&\nabla\nabla\times\nabla\times\partial_3(\mathbf{v}\times\mathbf{h}))=-(\Delta\nabla\times\partial_3\mathbf{h},\nabla\times\nabla\times\partial_3(\mathbf{v}\times\mathbf{h}))\\
&=-(\Delta\nabla\times\nabla\times\partial_3\mathbf{h},\nabla\times\partial_3(\mathbf{v}\times\mathbf{h}))\\
&=-(\Delta\partial_{33}\mathbf{h},\Delta(\mathbf{h}\nabla\mathbf{v}-\mathbf{v}\nabla\mathbf{h}))\leq 2C|\Delta\nabla\partial_3\mathbf{h}|(|\Delta\mathbf{h}||\Delta\nabla\mathbf{v}|\\
&+|\Delta\nabla\mathbf{h}||\nabla\nabla\mathbf{v}|+|\Delta\nabla\mathbf{h}||\Delta\mathbf{v}|+|\Delta\nabla\mathbf{v}||\nabla\nabla\mathbf{h}|)\\
&\leq\frac{8}{\pi}C|\Delta\nabla\partial_3\mathbf{h}|\Big(|\Delta\nabla\mathbf{v}||\Delta\partial_3\mathbf{h}|+|\Delta\nabla\partial_3\mathbf{h}||\Delta\mathbf{v}|\Big)\\
&\leq\frac{16C}{\pi}\sqrt{\frac{P_r}{P_m}}\frac{1}{\sqrt{b}}\frac{1}{d_{13}}(1+\frac{4}{\pi^2\sqrt{d_{11}}})(1+\sqrt{\frac{P_r}{P_m}})\mathcal{D}_n V^{\frac{1}{2}}\equiv A_{15}\mathcal{D}_n V^{\frac{1}{2}},
\end{aligned} \tag{4.3.49}$$

$$\begin{aligned}
(\nabla\nabla\times\partial_3\mathbf{h},&\nabla\nabla\times\nabla\times\partial_3(\mathbf{h}\times\nabla\times\mathbf{h})\\
&=-(\Delta\nabla\times\partial_3\mathbf{h},\nabla\times\nabla\times\partial_3(\mathbf{h}\times\nabla\times\mathbf{h}))\\
&=-(\Delta\nabla\times\nabla\times\partial_3\mathbf{h},\partial_3(\nabla\times\mathbf{h}\nabla\mathbf{h}-\mathbf{h}\nabla\nabla\times\mathbf{h}))\\
&=-(\Delta\partial_{33}\mathbf{h},\Delta(\nabla\times\mathbf{h}\nabla\mathbf{h}-\mathbf{h}\nabla\nabla\times\mathbf{h}))\\
&\leq|\nabla\Delta\partial_3\mathbf{h}|\Big(2C(|\nabla\nabla\times\mathbf{h}||\Delta\nabla\nabla\mathbf{h}|+|\Delta\mathbf{h}||\Delta\Delta\mathbf{h}|)\\
&+2|\nabla\mathbf{h}\nabla\nabla\times\mathbf{h}|+|\Delta\nabla\times\mathbf{h}\nabla\mathbf{h}|+|\nabla\times\mathbf{h}\Delta\nabla\mathbf{h}|\Big).
\end{aligned} \tag{4.3.50}$$

The inequalities

$$|\nabla\mathbf{h}\nabla\nabla\times\mathbf{h}|\leq\overline{\Omega}\frac{8\sqrt{2}}{\pi}(1+\frac{1}{\pi^2})|\nabla\nabla\times\partial_3\mathbf{h}|\Big(\frac{2}{\pi}|\Delta\Delta_1\partial_3\mathbf{h}|+|\Delta\nabla\times\partial_3\mathbf{h}|\Big), \tag{4.3.51}$$

$$|\Delta\nabla\times\mathbf{h}\nabla\mathbf{h}|\leq\overline{\Omega}\frac{2\sqrt{2}}{\pi}(1+\frac{1}{\pi^2})|\nabla\nabla\times\partial_3\mathbf{h}|(\frac{2}{\pi}|\Delta\Delta_1\partial_3\mathbf{h}|+|\Delta\nabla\times\partial_3\mathbf{h}|), \tag{4.3.52}$$

$$|\nabla\times\mathbf{h}\Delta\nabla\mathbf{h}|\leq\overline{\Omega}\frac{2\sqrt{2}}{\pi}(1+\frac{1}{\pi^2})|\nabla\nabla\times\partial_3\mathbf{h}|(\frac{2}{\pi}|\Delta\Delta_1\partial_3\mathbf{h}|+|\Delta\nabla\times\partial_3\mathbf{h}|), \tag{4.3.53}$$

imply

$$(\nabla\nabla\times\partial_3\mathbf{h}, \nabla\nabla \times \nabla \times \partial_3(\mathbf{h} \times \nabla \times \mathbf{h})$$

$$\leq \frac{P_r}{P_m} \frac{\sqrt{2}}{\sqrt{b}} \frac{1}{d_{13}} \Big(\frac{2}{\pi\sqrt{d_{12}}} + \frac{1}{\sqrt{d_{13}}} \Big) \Big[\frac{8C\sqrt{2}}{\pi} + \frac{24\sqrt{2}}{\pi}\overline{\Omega}(1 + \frac{1}{\pi^2}) \Big] \mathcal{D}_n V^{\frac{1}{2}} \quad (4.3.54)$$

$$\equiv A_{16}\mathcal{D}_n V^{\frac{1}{2}}.$$

From $(4.3.33) - (4.3.54)$ it follows $\mathcal{N}_n \leq \sum_{i=4}^{16} A_i \mathcal{D}_n V^{\frac{1}{2}}$, so $(4.3.32)$ reads

$$\tfrac{d}{dt} V(t) \leq - C\mathcal{D}_n \Big(1 - AV^{\frac{1}{2}} \Big), \text{ where } A = \Big(B + b \sum_{i=4}^{16} A_i \Big)/C.$$

All these allow us to prove, by recursive arguments similar to those in [MuloR89], the following theorem

Theorem 4.3.1. *If* $0 < \lambda < 1$, *i.e.* $R < R_H$, *and* $V(0) < A^{-2}$, *the conduction-diffusion solution* m_0 *is conditionally nonlinearly asymptotically stable with respect to the Lyapunov function* $(4.3.4)$.

Remark 4.3.1. From $(4.3.33) - (4.3.54)$ one easily sees the existence of a positive constant m such that $\inf_{R\in(0,R_H)} A^{-2} > m$.

Remark 4.3.2. The norm occurring in the Lyapunov function $V(t)$ is stronger than the L^2 norm and implies, as seen from $(4.3.19)$, a pointwise nonlinear asymptotic stability.

Remark 4.3.3. In the parameters space R_H is the critical hypersurface of the linear instability obtained by solving (by the classical normal mode technique) the two point problem derived from the eigenvalue problem governing the linear instability of the conduction-diffusion solution [MaiP84].

Chapter 5

Applications of direct methods based on Fourier series to linear Bénard convections

Throughout this chapter the principle of exchange of stabilities is assumed to hold and the perturbations are normal modes.

The flows studied in Sections 5.1 and 5.2 are rarely dealt with in literature, mainly due to their boundary conditions, high order of the governing equations and the presence of a large number of parameters. The linear stability of those complicated thermal vertical convections in unbounded horizontal layer, in ordinary or electromagnetic fluids, is treated by the B-D method (Appendix 5), directly applied to the linear equations governing the normal mode perturbations. These equations are deduced from the perturbation models (Section 1.4) for the approximate N-S-F models describing the perturbed flows (Sections 1.2, 1.4). The basic mechanical equilibrium is (1.4.2). The basic temperature gradient and magnetic field are constant and vertical. The most suitable series expansions are used and the symmetry properties are taken advantage of. The corresponding even and odd two-point eigenvalues problems have the same order as the given problem and half this order when in each equation and boundary condition either only odd or only even derivatives occur. The even and odd problems are solved separately, leading to two exact secular equations defined by a series and, thus, two neutral manifolds. It is found that, unlike the nonmagnetic case, in the electromagnetic fluids the odd case is the most relevant. In Section 5.1 we deal with the most complicated case when the temperature, magnetic field, the anisotropic currents of Hall and ion-slip are present. The fluid is situated in a layer bounded by two free surfaces. Section 5.2 deals with the same problem as Section 5.1 but here the ion-slip is absent. Most results are deduced from Section 5.1. Additional investigation of the neutral curves and of the singularities in the secular equations are performed. The fluid considered in Section 5.3 is micro-polar and the temperature and outer magnetic field are taken into account. The governing equations depend on four parameters and have the eighth order while for the even and odd problems this order is four. Numerical results are available. Particular classical cases are regained.

Section 5.4 is concerned with convections described by ode's with non-constant coefficients. Due to convergence reasons of the double series occurring as a result of the presence of variable coefficients, so far this case was, practically, not treated at all from the theoretical point of view. In all these convections the B-D method still applies, but the reasonings are complicated by the necessity of writing additional series expansions for the coefficients in the equations. Two main methods are applied. Comparison with the results obtained by means of some other methods can be easily done because our formulae are explicit. This is why we do not quote tables or graphs.

5.1 Magnetic Bénard convection in a partially ionized fluid for stress-free perfectly thermoelectrically conducting boundaries

By closely following [PalG04b], the governing two-point eigenvalue problem for a twelfth order system of ode's, containing derivatives of even as well as odd orders and four physical parameters, is solved by means of the B-D method. The given problem was split into an *even* and an *odd* problems, each of them having the same order as the given problem, namely 12, and containing even as well as odd parts of the unknown functions. In order to introduce a minimum number of constraints in the even problem, the Fourier expansion functions $\{E_{2n-1}\}$ and $\{F_{2n-1}\}$ are used, while in the odd case the functions $\{E_{2n}\}$ and $\{F_{2n}\}$ prove to be the most convenient involving two constraints, and not four corresponding to the first expansion functions. For these choices, in both even and odd cases, only two constraints occurred. The singularities in the secular equations are treated separately. By truncating to one and two terms the series defining the secular equation, the coupled action of Hall and ion-slip currents is found to be destabilizing, while the action of the ion-slip effect on the Bénard convection subject to Hall current depends on the Hartmann number and wave number. The closed form of the Rayleigh number as a function of the other parameters, defining the neutral curves up the two terms, is obtained. For two terms this neutral curve is situated under the neutral curve corresponding to one term. This was to be expected because the involved series were convergent at least as n^{-2} as $n \to \infty$. In the first *even case* it was found that if the sums were truncated to a single term, the stability bounds correspond to the case $\beta_H = 0$. In the case of two terms the neutral curves for $\beta_H \neq 0$ were situated below the neutral curves from the case $\beta_H = 0$. When trying to compare our results with the case when the Hall effect is absent, we found that our calculations implied a singularity and so we treated this case separately. The comparison between the case of absence and presence of the Hall current showed a destabilizing effect of this one.

A suitably chosen expression containing the eigenvalues make it possible to carry out the cumbersome computations especially in proving that the secular equation has real eigenvalues R (Rayleigh number). It is shown that the *odd* problem corresponding to odd velocity and temperature perturbations with respect to the vertical coordinate is more appropriate than the even one: for one term in the series, in the even case the neutral curve contains no influence of β_I and β_H, while in the odd case these influences do occur. In addition, better neutral curves correspond to the case of two terms. For the case of the absence of the Hall and ion-slip effects the existing results are regained [Chan].

In Section 5.1.5 the supplementary influence of porosity, concentration and compressibility is considered.

5.1.1 *Mathematical results for the magnetic Bénard problem*

By the magnetic Bénard problem we understand the problem of the stability of the conduction state of a horizontal fluid layer subject to a temperature gradient and in the presence of a magnetic field. Since this chapter is mostly concerned with this problem, we first present a survey of the existing results. Due to its relevance in many astrophysical and geophysical applications, the magnetic Bénard problem received considerable attention in the literature (for a list of references we recommend [PalG04b], [GeoPalR06]. We treated this problem in several physical situations: in the linear and nonlinear cases; without Hall and ion-slip effects; with Hall effect; with Hall and ion-slip effects; for rigid walls; for free surfaces; for one free and one rigid surface; with additional effects (*e.g.* porosity, fluid mixture). Our methods were analytical or numerical. The analytical methods were direct, variational or based on series. In this last case the B-D method was proved to be the most appropriate. However, we still have the open problem of the eigenvalues in the case when the Cramer determinant of the system in the Fourier coefficients vanishes, and, thus, the secular equation contains singular terms. This question is treated in Sections 5.2.1.2, 5.2.1.3 for the even problem and in Sections 5.2.2.2, 5.2.2.3 for the odd problem.

The Bénard problem of the stability of the thermodiffusive equilibrium of a viscous fluid layer heated from below has been extensively investigated. The problem of existence, continuous dependence, uniqueness, linear and nonlinear stability of this equilibrium was studied in the hydrodynamic case [Chan], [J76], [MuloR89], [MuloR94], in the magnetohydrodynamic case [Chan], [RioM88c], for a fluid characterized by a scalar electrical conductivity, *i.e.* in the case when the standard Ohm's law is supposed to hold and in the magnetohydrodynamic anisotropic case [AbS], [EbS1], [EbS2], [GeoPalPasB], [GeoPal96c], [MaiP84], [MaiPL], [MuloS88], [Pal97], [PalG03], [PalGPashB], [SharS], [SharR], [SharT], [SolM]. In the case when the generalized Ohm's law holds, the linear instability of the thermal equilibrium for the magnetic Bénard problem has been studied [MaiP84], [MaiPL] by supposing the layer delimited by two thermally but not electrically conducting free boundaries.

Due to the (stress-free) boundary conditions, in [MaiP84], [MaiPL], the eigenvalue problem governing the linear stability was solved by the Chandrasekhar-Galerkin method, yielding the smallest eigenvalue, defining the neutral curve, and showing the destabilizing effect either of the Hall current only, or of both Hall and ion-slip currents. The stabilizing-destabilizing effect of the Hall and ion-slip currents, compared with the case where only the Hall current is present, is investigated in [GeoPalPasB]; in [GeoPalPasB] there is a first numerical investigation of the electroanisotropic effects on the stability domain, showing the dependence of the stability results on the given physical parameters of the problem. In [GeoPalR06] we study another linear magnetic anisotropic Bénard problem in the presence of the Hall current, in a layer with rigid boundaries. The governing equations are suitably reformulated, then the energy method is applied and an associate variational problem is found. With it we associate the Euler-Lagrange equations and to it we apply the B-D method to obtain the neutral curve as a determinant the entries of which are some series involving the Fourier coefficients of the unknown perturbations. Retaining only one or two terms in the series in the secular equation we determine the approximate neutral curves, showing the destabilizing effect of the Hall current. In [GeoPalR06] we reconsider a linear magnetic anisotropic Bénard problem with other (namely rigid) boundary conditions, regaining the destabilizing effect of the Hall current and, in the case when the Hall effect is absent, we prove that the singularities of the secular equation are eigenvalues of the given boundary-value problem. In other words, the smallest singular value of the secular equation defines the neutral curve of the linear instability.

In Section 5.1 we consider the same problem as in [GeoPalR06] for a partially ionized fluid. Like in [GeoPalR06], [PalG03], we recover the result that the presence of both Hall and ion-slip effects is destabilizing.

5.1.2 *Splitting of the two-point eigenvalue problem*

Let us consider a homogeneous electrically conducting fluid in a horizontal layer bounded by two planes both stress-free, perfectly thermally and electrically conductors. Assume the layer subject to a vertical constant temperature gradient $\beta > 0$ and immersed in an external magnetic field $\mathbf{H_0} = H_o\mathbf{k}$, in an orthonormal frame reference $\{O, \mathbf{i}, \mathbf{j}, \mathbf{k}\}$. The electrical conductivity is supposed a tensorial one, *i.e.* the generalized Ohm's law holds, in the presence of both Hall and ion-slip currents. In the framework of continua and if the O-B approximation holds, the dimensionless equations governing the perturbation $\mathbf{v}(u, v, w)$, $\mathbf{h}$, θ, p of the thermodiffusive equilibrium (4.1.2) are (1.4.17$'$), where $\mathbf{v}$, $\mathbf{h}$, θ, p are the velocity, magnetic, temperature and pressure field, respectively. The positive coefficients P_r, P_m, M^2, R, β_H and β_I are the Prandtl, magnetic Prandtl, Hartmann and Rayleigh numbers and, Hall and ion-slip coefficients, respectively. The boundary conditions appropriate to stress-free, perfectly thermoelectrically conductors, are (1.4.4), (1.4.7).

Assume that the perturbations are doubly periodic of period $2\pi/\alpha$ and $2\pi/\beta$ in the x and y direction respectively and use the variables $w = \mathbf{k} \cdot \mathbf{v}$, $\zeta = \mathbf{k} \cdot \nabla \times \mathbf{v}$, $h_3 = \mathbf{k} \cdot \mathbf{h}$, $j = \mathbf{k} \cdot \nabla \times \mathbf{h}$ and the notation $\Delta_1 = \frac{\partial}{\partial x^2} + \frac{\partial}{\partial y^2}$ [J76]. From equations $(1.4.4')$, linearized about the equilibrium m_0, we have

$$\frac{\partial}{\partial t} P_m \Delta \Delta w + P_m M^2 \frac{\partial}{\partial z} \Delta h_3 + \mathcal{R} \frac{P_m^2}{P_r} \Delta_1 \theta,$$

$$\frac{\partial}{\partial t} \zeta = P_m \Delta \zeta + P_m M^2 \frac{\partial}{\partial z} j,$$

$$\frac{\partial}{\partial t} h_3 = \frac{\partial}{\partial z} w + (1 + \beta_I) \Delta h_3 - \beta_H \frac{\partial}{\partial z} j,$$

$$\frac{\partial}{\partial t} j = \frac{\partial}{\partial z} \zeta + \Delta j + \beta_H \frac{\partial}{\partial z} \Delta h_3 + \beta_I \frac{\partial^2}{\partial z^2} j,$$

$$\frac{\partial}{\partial t} \theta = w + \frac{P_m}{P_r} \Delta \theta. \qquad (5.1.1)$$

Equations $(5.1.1)_{1,2,4}$ are obtained by applying the operators $\mathbf{k} \cdot \nabla \times \nabla\times$, $\mathbf{k} \cdot \nabla\times$ to equation $(1.4.4')_1$ and $\mathbf{k} \cdot \nabla\times$ to equation $(1.4.4')_2$ respectively. For vanishing ion-slip coefficient this problem becomes that considered in [GeoPalR06]. The boundary conditions obtained by linearizing $(1.4.7')$ are

$$w = \frac{\partial^2}{\partial z^2} w = h_3 = \frac{\partial}{\partial z} j = \frac{\partial}{\partial z} \zeta = \theta = 0 \qquad z = \pm 0.5. \qquad (5.1.2)$$

Assume that the perturbations are normal modes, *i.e.*

$$(w, h_3, j, \zeta, \theta) = \{W(z), K(z), X(z), Z(z), \Theta(z)\} exp[i(a_x x + a_y y) + ct], \qquad (5.1.3)$$

and introduce $(5.1.3)$ into $(5.1.1)$ and $(5.1.2)$, to obtain

$$(D^2 - a^2)[P_m(D^2 - a^2) - c]W + P_m M^2 D(D^2 - a^2)K - \mathcal{R}\frac{P_m^2}{P_r} a^2 \Theta = 0,$$

$$[P_m(D^2 - a^2) - c]Z + P_m M^2 DX = 0,$$
$$[(1 + \beta_I)(D^2 - a^2) - c]K + DW - \beta_H DX = 0, \qquad (5.1.4)$$
$$[D^2(1 + \beta_I) - a^2 - c]X + DZ + \beta_H D(D^2 - a^2)K = 0,$$

$$[\frac{P_m}{P_r}(D^2 - a^2) - c]\Theta + W = 0,$$

where $a^2 = a_x^2 + a_y^2$,

$$W = D^2 W = K = DX = DZ = \Theta = 0 \qquad z = \pm 0.5. \qquad (5.1.5)$$

Since the principle of exchange of stabilities is supposed to hold, that is instability occurs as a stationary convection, we must assume $c = 0$. In order to solve the eigenvalue problem $(5.1.4)$, $(5.1.5)$ we first split the unknown functions in their even and odd parts, [Geo85], *e.g.* $W = W^e + W^o$ and then separate the even and the odd parts of the equations (Lemma 4.3). Due to the occurrence of odd and even derivatives in equations $(5.1.4)_{1-4}$, we obtain two problems containing some even and some odd parts of the unknown functions. The first is the so-called even problem, in W^e, Θ^e, Z^o, X^e, K^o,

$$(D^2 - a^2)^2 W^e + M^2 D(D^2 - a^2)K^o - \mathcal{R}a^2 \frac{P_m}{P_r}\Theta^e = 0,$$

$$(D^2 - a^2)Z^o + M^2 D X^e = 0,$$

$$(1 + \beta_I)(D^2 - a^2)K^o + DW^e - \beta_H D X^e = 0, \qquad (5.1.4')$$

$$[(1 + \beta_I)D^2 - a^2]X^e + DZ^o + \beta_H D(D^2 - a^2)K^o = 0,$$

$$\frac{P_m}{P_r}(D^2 - a^2)\Theta^e + W^e = 0,$$

$$W^e = D^2 W^e = K^o = DX^e = DZ^o = \Theta^e = 0 \qquad z = \pm 0.5 \qquad (5.1.5')$$

and the second is the so-called odd problem in W^o, Θ^o, Z^e, X^o, K^e,

$$(D^2 - a^2)^2 W^o + M^2 D(D^2 - a^2)K^e - \mathcal{R}a^2 \frac{P_m}{P_r}\Theta^o = 0,$$

$$(D^2 - a^2)Z^e + M^2 D X^o = 0,$$

$$(1 + \beta_I)(D^2 - a^2)K^e + DW^o - \beta_H D X^o = 0, \qquad (5.1.4'')$$

$$[(1 + \beta_I)D^2 - a^2]X^o + DZ^e + \beta_H D(D^2 - a^2)K^e = 0,$$

$$\frac{P_m}{P_r}(D^2 - a^2)\Theta^o + W^o = 0,$$

$$W^o = D^2 W^0 = K^e = DX^o = DZ^e = \Theta^o = 0 \qquad z = \pm 0.5. \qquad (5.1.5'')$$

In the following we solve these problems by B-D method (Appendix 5).

5.1.3   *Neutral curves for the even case*

Here we expand the even unknown functions on the set $\{E_{2n-1}\}$ total in the subspace of $L^2(-0.5, 0.5)$ consisting of even functions, where $E_{2n-1} = \sqrt{2}\cos([(2n-1)\pi z])$. For the odd unknown functions we use the total set $\{F_{2n-1}\}$, where $F_{2n-1} = \sqrt{2}\sin([(2n-1)\pi z])$. For instance put $W^e(z) = \sum_{n=1}^{\infty} W^e_{2n-1} E_{2n-1}$, where W^e_{2n-1} are the Fourier coefficients. By the backwards integration technique the Fourier coefficients of the derivatives of the unknown functions are determined in terms of the Fourier coefficients of the functions themselves. Let

$$D^{2k+1}W^e(z) = \sum_{n=1}^{\infty} W^{e(2k+1)}_{2n-1} F_{2n-1}(z), \qquad D^{2k}W^e(z) = \sum_{n=1}^{\infty} W^{e(2k)}_{2n-1} E_{2n-1}(z).$$

Then, by $(5.7)_1$ and $(5.8)_1$ we have

$$W^{e(2k+1)}_{2n-1} = 2\sqrt{2}(-1)^{n+1} D^{2k}W^e(0.5) - (2n-1)\pi W^{e(2k)}_{2n-1},$$

$$W^{e(2k)}_{2n-1} = (2n-1)\pi W^{e(2k-1)}_{2n-1}.$$

Similarly, $W^o(z) = \sum_{n=1}^{\infty} W^o_{2n-1} F_{2n-1}$, $D^{2k+1}W^o(z) = \sum_{n=1}^{\infty} W^{o(2k+1)}_{2n-1} E_{2n-1}(z)$,

$D^{2k}W^o(z) = \sum_{n=1}^{\infty} W_{2n-1}^{o(2k)} F_{2n-1}(z)$, where, by $(5.7)_2$ and $(5.8)_2$,

$$W_{2n-1}^{o(2k+1)} = (2n-1)\pi W_{2n-1}^{o(2k)},$$

$$W_{2n-1}^{o(2k)} = 2\sqrt{2}(-1)^{n+1} D^{2k-1}W^o(0.5) - (2n-1)\pi W_{2n-1}^{o(2k-1)}.$$

If the velocity and temperature fields are even functions of z, *i.e.* in the *even case*, the eigenvalue problem $(5.1.4')$, $(5.1.5')$ becomes

$$A_n^2 W_{2n-1}^e - M^2 A_n(2n-1)\pi K_{2n-1}^o - \frac{P_m}{P_r}Ra^2\Theta_{2n-1}^e = 2\sqrt{2}(-1)^n(2n-1)\pi\alpha_6 M^2,$$

$$-A_n Z_{2n-1}^o - M^2(2n-1)\pi X_{2n-1}^e = 2\sqrt{2}(-1)^n\alpha_4 M^2,$$

$$-(1+\beta_I)A_n K_{2n-1}^o - (2n-1)\pi W_{2n-1}^e + \beta_H(2n-1)\pi X_{2n-1}^e$$
$$= 2\sqrt{2}(-1)^n[(1+\beta_I)\alpha_6 - \beta_H\alpha_4],$$

$$-[(1+\beta_I)(2n-1)^2\pi^2 + a^2]X_{2n-1}^e + (2n-1)\pi Z_{2n-1}^o - \beta_H(2n-1)\pi A_n K_{2n-1}^o$$
$$= 2\sqrt{2}(-1)^n(2n-1)\pi[\alpha_6\beta_H + (1+\beta_I)\alpha_4],$$

$$-A_n\frac{P_m}{P_r}\Theta_{2n-1}^e + W_{2n-1}^e = 0,$$

$$(5.1.6)$$

where $\alpha_6 = DK^o(0.5)$, $\alpha_4 = X^e(0.5)$ and $A_n = a^2 + (2n-1)^2\pi^2$. Taking into account that $X_{2n-1}^{e^{(1)}} = 2\sqrt{2}(-1)^{n+1}\alpha_4 - (2n-1)\pi X_{2n-1}^e$, the constraints $DX^e = 0$, $K^o = 0$ at $z = \pm 0.5$ read

$$\sum_{n=1}^{\infty}[2\sqrt{2}(-1)^{n+1}\alpha_4 - (2n-1)\pi X_{2n-1}^e](-1)^{n+1} = 0,$$

$$(5.1.5')^*$$

$$\sum_{n=1}^{\infty}(-1)^{n+1}K_{2n-1}^o = 0.$$

Remark 5.1.1. In deriving (5.1.6), we took into account that equalities $(5.1.4')$ are understood in $C(-0.5, 0.5)$. Therefore their right-hand side is the continuous null function, the Fourier coefficients of which are null. In deriving the constraints, we used formulae (5.5) and $(5.6')$ respectively.

Denote $X_{2n-1}^{e^{(1)}} = \Delta_{4n}/\Delta_n$ and $K_{2n-1}^o = \Delta_{2n}/\Delta_n$. By solving this system we have

$$\Delta_n = \frac{P_m}{P_r}A_n\Big\{(Ra^2 - A_n^3)\big[G_n + \beta_I(H_n + L_n) + \beta_I^2 L_n\big] - M^2 L_n(H_n + \beta_I L_n)\Big\},$$

$$\Delta_{4n} = 2\sqrt{2}(-1)^n\frac{P_m}{P_r}(2n-1)\pi\alpha_4 A_n\Big\{-(Ra^2 - A_n^3)[A_n + M^2 + A_n\beta_H^2$$
$$+ \beta_I(M^2 + 2A_n) + \beta_I^2 A_n] + M^2 A_n L_n + M^2 A_n L_n\beta_I + M^4 L_n\Big\},$$

$$\Delta_{2n} = 2\sqrt{2}(-1)^n\frac{P_m}{P_r}\alpha_4 a^2\beta_H A_n(Ra^2 - A_n^3) - 2\sqrt{2}(-1)^n\frac{\Delta_n}{A_n}\alpha_6,$$

where $L_n = A_n(A_n - a^2)$, $H_n = A_n^2 + (A_n - a^2)M^2$, $G_n = H_n + \beta_H^2 L_n$.

So far α_4 and α_6 are arbitrary. Assume now that $\Delta_n \neq 0$ and introduce the expressions of $X_{2n-1}^{e}{}^{(1)}$ and K_{2n-1}^{o} into $(5.1.5')^*$ to obtain a linear algebraic system in α_4 and α_6. If its roots $(\alpha_4, \alpha_6) = (0,0)$, then all roots $(W_{2n-1}^{e}, K_{2n-1}^{o}, \Theta_{2n-1}^{e}, X^{e}, Z^{e})$ of (5.1.6) vanish, that is no eigenvalue of the problem $(5.1.4')$, $(5.1.5')$ exists. Therefore we must require the existence of a nonzero root (α_4, α_6), which occurs if the determinant $\det \begin{vmatrix} \alpha_4 C_1 & 0 \\ \alpha_4 C_2 & \alpha_6 C_3 \end{vmatrix}$ vanishes. Here C_1, C_2, C_3 are the coefficients of α_4 and α_6 in $(5.1.5')^*$ and they are converging series the term of which depend on the five parameters a, R, M, β_H and β_I. Since $C_3 = 2\sqrt{2}\sum_{n=1}^{\infty} A_n^{-1} > 0$, we must have $C_1 = 0$, which defines the secular equation, namely

$$\sum_{n=1}^{\infty} \frac{1}{A_n} \cdot \sum_{n=1}^{\infty} \frac{A_n^2[(Ra^2 - A_n^3)(1 + \beta_I) - M^2 L_n]}{\Delta_n} = 0, \tag{5.1.7}$$

or, equivalently [GrR],

$$\sum_{n=1}^{\infty} \frac{A_n[V_n(1 + \beta_I) + \beta_I]}{V_n[H_n + (\beta_H^2 + \beta_I^2)L_n + \beta_I(H_n + L_n)] + (\beta_H^2 + \beta_I^2)L_n + \beta_I H_n} = 0,$$

where $V_n = (Ra^2 - A_n H_n)M^{-2}L_n^{-1}$ and, therefore, $Ra^2 - A_n^3 = (V_n + 1)M^2 L_n$. With the notation $N_n = H_n + \beta_I L_n + P_n$, $P_n = (\beta_H^2 + \beta_I^2)L_n + \beta_I H_n$ this secular equation becomes

$$\sum_{n=1}^{\infty} \frac{A_n[V_n(1 + \beta_I) + \beta_I]}{V_n N_n + P_n} = 0 \tag{5.1.8}$$

and its terms tend to zero like n^2 as $n \to \infty$, therefore the series in (5.1.8) is convergent.

The expressions of Δ_n, Δ_{2n} and Δ_{4n} show that $K_{2n-1}^{o} \sim n^{-2}$ and $X_{2n-1}^{e} \sim n^{-2}$ as $n \to \infty$. Then $(5.1.6)_1$, $(5.1.6)_6$, $(5.1.6)_2$ imply that $W_{2n-1}^{e} \sim n^{-3}$, $\Theta \sim n^{-5}$, $Z \sim n^{-3}$ as $n \to \infty$, ensuring the convergence of the implied series.

Among all the expressions occurring in (5.1.8), the only one containing R is V_0^n. In fact, a better notation for the value of R corresponding to V_0^n would be R_n but, for the sake of simplicity, we keep the notation R. As the secular equation provides the eigenvalues R, it follows that we must solve (5.1.8) with respect to V_0^n.

Retaining only one term in (5.1.8) we have the solution $\overline{V}_1 = 0$, *i.e.* $R = A_1 H_1/a^2$. This eigenvalue does not depend on β_I and β_H.

Further truncate equation (5.1.8) to two terms. Since $V_1 = \dfrac{V_2 M^2 L_2 + Q}{M^2 L_1}$, where $Q = H_2 A_2 - H_1 A_1$ (therefore it does not depend on R), it becomes

$$\begin{aligned}
V_2^2 M^2 L_2 (1 + \beta_I)(A_1 N_2 + A_2 N_1) &+ V_2\Big\{Q(A_1 N_2 + A_2 N_1) \\
&+ M^2(A_1 L_2 P_2 + A_2 L_1 P_1) + \beta_I\big[M^2(A_1 L_1 N_2 + A_2 L_2 N_1) \\
&+ Q(A_1 N_2 + A_2 N_1) + M^2(A_1 L_2 P_2 + A_2 L_1 P_1)\big]\Big\} \\
&+ A_1 Q P_2 + \beta_I\big[M^2 L_1(A_1 P_2 + A_2 P_1) + Q(A_1 P_2 + A_2 N_1)\big] = 0.
\end{aligned} \tag{5.1.9}$$

This equation has two negative solutions and so, $Ra^2 - A_n H_n < 0$.

Since $R = A_1 H_1 / a^2$ is the neutral curve for $\beta_H = \beta_I = 0$, it follows that the Hall and ion-slip effects are destabilizing. For $\beta_I = 0$ we regain the results from [GeoPalR06].

Summing-up, we have proved

Theorem 5.1.1 [PalG04b]. *For even velocity and temperature vector fields and odd magnetic field the mechanical equilibrium has the following neutral curve*

$$R = \min_{V_2^{(1),(2)}} \left\{ \frac{(4\pi^2 + a^2)^3 + 4M^2\pi^2(4\pi^2 + a^2)}{a^2} + V_2^{(1),(2)} \frac{4M^2\pi^2(4\pi^2 + a^2)}{a^2} \right\},$$

$$(5.1.10)$$

where $V_2^{(1),(2)}$ are the quoted solutions of the secular equation (5.1.9).

Remark 5.1.2. Since in $(5.1.4')$ the operators D and $D^2 - a^2$ correspond in $(5.1.6)$ to the multiplication by $(2n-1)\pi i$ and $-A_n$ respectively, the properties of equation $(5.1.4')$, their solutions and involved operators are easier to detect by the inspection of their correspondents in $(5.1.6)$ (see also Section 3.4.3.1). For instance $(5.1.6)_5$ shows the invertibility of the operator $D^2 - a^2 : C^\infty(-0.5, 0.5) \cap L^2(-0.5, 0.5) \to C^\infty(-0.5, 0.5) \cap L^2(-0.5, 0.5)$.

Similarly, equation $(5.1.6)_2$ shows that $(D^2 - a^2)Z = M^2 DX$, i.e. $(5.1.4')_2$, and the fact that the constraint $(5.1.5')_1^*$ is $(D^2 - a^2)Z(\pm 0.5) = 0$.

Each unknown in $(5.1.6)$, e.g. K_{2n-1}^o, is a quotient of two determinants: the denominator is the Cramer determinant Δ_n of the system $(5.1.6)$, while the numerator is a linear combination of α_4 and α_6. It follows that if $\Delta_n = 0$, then, in order for the unknowns in $(5.1.6)$ be bounded and not all equal to zero, it is necessary that all those five determinants be equal to zero. The case $\Delta_n = 0$ is separately treated in Sections 5.2.1.2, 5.2.1.3, 5.2.2.2, 5.2.2.3, for the case $\beta_I = 0$.

If in Δ_n we replace $(2n-1)\pi$ by $-\lambda i$, where $\lambda \in \mathbf{C} \setminus \mathbf{R}$, then $\Delta_n = 0$ becomes a polynomial equation in λ

$$P(\lambda) \equiv \frac{P_m}{P_r}(a^2 - \lambda^2)\left\{\left[Ra^2 - (a^2 - \lambda^2)^3\right]\left[(a^2 - \lambda^2)^2 - \lambda^2 M^2 - \beta_H^2\lambda^2(a^2 - \lambda^2)\right.\right.$$

$$+ \beta_I\left((a^2 - \lambda^2)^2 - \lambda^2 M^2 - \lambda^2(a^2 - \lambda^2)\right) - \beta_I^2\lambda^2(a^2 - \lambda^2)\Big]$$

$$+ M^2\lambda^2(a^2 - \lambda^2)\left[(a^2 - \lambda^2)^2 - \lambda^2 M^2 - \beta_I\lambda^2(a^2 - \lambda^2)\right]\right\} = 0.$$

For $\lambda = D \equiv \frac{d}{dz}$, its left-hand side becomes just the operator, say L,

$$\frac{P_M}{P_r}(a^2 - D^2)\left\{\left[Ra^2 - (a^2 - D^2)^3\right]\left[(a^2 - D^2)^2 - D^2 M^2 - \beta_H^2 D^2(a^2 - D^2)\right.\right.$$

$$+ \beta_I\left((a^2 - D^2)^2 - D^2 M^2 - D^2(a^2 - D^2)\right) - \beta_I^2 D^2(a^2 - D^2)\Big]$$

$$+ M^2 D^2(a^2 - D^2)\left[(a^2 - D^2)^2 - D^2 M^2 - \beta_I^2 D^2(a^2 - D^2)\right]\right\} = 0,$$

defining the equation $L^e W = 0$ in W^e, which can be obtained from $(5.1.4')$ by eliminating the other unknown functions. Therefore, $P(\lambda) = 0$ is the characteristic equation corresponding to $LW^e = 0$.

The ode $LW^e = 0$ reads in the form $AW^e + RBW^e = 0$, where the operators A and B are

$$
\begin{aligned}
A =&(D^2 - a^2)^6\big[\beta_H^2 + (1 + \beta_I^2)^2\big] + (D^2 - a^2)^5\big[\beta_H^2 + \beta_I^2 + \beta_I) - 2M^2(1 + \beta_I)\big] \\
&+ (D^2 - a^2)^4(M^4 - 3M^2\beta_I a^2 - 2a^2 M^2) + (D^2 - a^2)^3(2M^4 a^2 - M^2 a^4 \beta_I) \\
&+ (D^2 - a^2)^2 M^4 a^4,
\end{aligned}
$$

$$
\begin{aligned}
B =&a^2\Big\{(D^2 - a^2)^3\big[\beta_H^2 + (1 + \beta_I^2)^2\big] + (D^2 - a^2)^2\big\{a^2\big[\beta_H^2 + \beta_I(1 + \beta_I)\big] \\
&- M^2(1 + \beta_I)\big\} - a^2 M^2(1 + \beta_I)(D^2 - a^2)\Big\}.
\end{aligned}
$$

Recall that R is taken as an eigenvalue in the sense used in applied papers namely as that value for which the equation LW^e has at least a nontrivial solution (Section 3.2.6).

The parameter R occurs in the expression of the roots λ of the characteristic equation and, therefore, in the general solution of the equation $LW^e = 0$. Imposing to this solution to satisfy all boundary conditions we get the secular equation the solutions R of which are the eigenvalues. Take $\lambda = (2n - 1)\pi i$. Then the equation $\Delta_n = 0$ takes some values for R which are not eigenvalues of the problem (5.1.4$'$), (5.1.5$'$). Indeed, $\Delta_n = 0$ is derived from equations (5.1.6) which do not take into account conditions (5.1.5$'$)$_{3,4}$. Consequently, $\Delta_n = 0$ will be a secular equation and its solutions R will be eigenvalues if they satisfy the constraints (5.1.5$'$)*, expressed by some relationships between the physical parameters.

Since $\Delta_n = 0$ reads $A\big((2n - 1)\pi\big) + RB\big((2n - 1)\pi\big) = 0$, where $A\big((2n - 1)\pi\big)$ and $B\big((2n-1)\pi\big)$ are obtained from A and B by, formally, putting $D^2 - a^2 = -A_n$, we have

$$
\begin{aligned}
R_n =&\Big\{A_n^5\big[\beta_H^2 + (1 + \beta_I^2)^2\big] - A_n^4\big[a^2\beta_H^2 + \beta_I + \beta_I^2) - 2M^2(1 + \beta_I)\big] + A_n^3(M^4 \\
&- 3M^2\beta_I a^2 - 2a^2 M^2) - A_n^2(2M^4 a^2 - M^2 a^4 \beta_I) + A_n M^4 a^4\Big\}a^{-2}\Big\{A_n^2\big[\beta_H^2 \\
&+ (1 + \beta_I)^2\big] - A_n\big[a^2(\beta_H^2 + \beta_I + \beta_I^2) - M^2(1 + \beta_I)\big] - a^2 M^2(1 + \beta_I)\Big\}^{-1}.
\end{aligned}
$$

In the absence of the electromagnetic effects, *i.e.* for $\beta_H = \beta_I = M = 0$, we obtain the classical result $R_n = A_n^3 a^{-2}$, with a minimum for $n = 1$, therefore the neutral curve is $R_1 = (a^2 + \pi^2)^3 a^{-2}$.

For the non-hydrodynamic case we must introduce into equations (5.1.6) the above expression for R_n and discuss the decoupling of the equations. For $\beta_I = 0$, this question is studied in Sections 5.2.1.2 and 5.2.1.3. For the case of multiple solutions of the characteristic equation $\Delta_n = 0$, the study follows the lines in Chapter 7. So far these are open problems.

In the equation $\Delta_n = 0$ the presence of the factor A_n shows that equations (5.1.6) decouple. Since $A_n \neq 0$, the corresponding operator $D^2 - a^2$ is invertible, therefore the decoupled second order equation has no eigenvalue. In the case $\beta_I = 0$ one finds that this is indeed the case.

Since Δ_n is common to all unknowns W^e_{2n-1}, Θ^e_{2n-1} X^e_{2n-1} Z^o_{2n-1} and K^o_{2n-1}, from (5.1.6) it follows that the ode equation $LW^e = 0$ is satisfied by all other unknown functions from (5.1.4').

In addition, the simplest way to construct it by eliminating all other unknown functions in (5.1.4') is suggested by the corresponding operations in (5.1.6). The operations are multiplications and divisions, therefore their correspondents in (5.1.4') are differentiations and inversions of differential operators. However, we must take care of the fact that the application of some differential operators, *e.g.* D, $D^2 - a^2$, in (5.1.4') corresponds, in general, in (5.1.6) not only to multiplication by $(2n-1)\pi i$ and $-A_n$ respectively, but also to the appearance of new arbitrary boundary values, which cannot occur in (5.1.6). On the other hand, the division in (5.1.6) by some expressions containing $-A_n$ and $(2n-1)\pi i$ can correspond to non-invertible operators in (5.1.4').

5.1.4 *Neutral curves for the odd case*

Let us suppose the velocity and temperature fields being odd functions of z, *i.e.* consider the *odd case* of problem (5.1.4''), (5.1.5''), and apply to it the Budiansky-DiPrima method. If the unknown functions are expanded on the sets $\{E_{2n-1}\}$ and $\{F_{2n-1}\}$ then four constraints are introduced leading to cumbersome computations. This is why in the *odd case* we use the total sets $\{1, E_2, E_4, \ldots\}$, $\{F_2, F_4, \ldots\}$ implying only two constraints.

We first remind the relationships between the Fourier coefficients of the successive derivatives

$$D^{2k+1}Z^e(z) = \sum_{n=1}^{\infty} Z^{e(2k+1)}_{2n} F_{2n}(z), \qquad D^{2k}Z^e(z) = Z^{e(2k)}_0 + \sum_{n=1}^{\infty} Z^{e(2k)}_{2n} E_{2n}(z),$$

for an even function, *e.g.* $Z^e(z) = Z^e_o + \sum_{n=1}^{\infty} Z^e_{2n} E_{2n}(z)$, where, by (5.1.5),

$$Z^e_o = \int_{-0,5}^{0.5} Z^e(z)dz, \qquad Z^{e(2k)}_0 = 2D^{2k-1}Z^e(0.5),$$

$$Z^{e(2k+1)}_{2n} = -2n\pi Z^{e(2k)}_{2n}, \qquad Z^{e(2k)}_{2n} = 2\sqrt{2}(-1)^n D^{2k-1}Z^e(0.5) + 2n\pi Z^{e(2k-1)}_{2n}.$$

Similarly, for an odd function, *e.g.* $W^o(z) = \sum_{n=1}^{\infty} W^o_{2n} F_{2n}(z)$, its derivatives are

$$D^{2k+1}W^o(z) = W^{o(2k+1)}_0 + \sum_{n=1}^{\infty} W^{o(2k+1)}_{2n} E_{2n}(z), \quad D^{2k}W^o(z) = \sum_{n=1}^{\infty} W^{o(2k+1)}_{2n} F_{2n}(z),$$

where, by (5.17),

$$W^{o(2k+1)}_0 = 2D^{2k}W^o(0.5), \quad W^{o(2k+1)}_{2n} = 2\sqrt{2}(-1)^n D^{2k}W^o(0.5) + 2n\pi W^{o(2k)}_{2n},$$

$$W^{o(2k)}_{2n} = -2n\pi W^{o(2k-1)}_{2n}.$$

Then set of equations (5.1.4″) imply

$$B_n^2 W_{2n}^o + M^2 B_n 2n\pi K_{2n}^e - \frac{P_m}{P_r} Ra^2 \Theta_{2n}^o = -2\sqrt{2}(-1)^{n+1} 2n\pi\alpha_6 M^2,$$

$$-B_n Z_{2n}^e + M^2 2n\pi X_{2n}^o = 2\sqrt{2}(-1)^{n+1}\alpha_4 M^2,$$

$$-(1+\beta_I)B_n K_{2n}^e + 2n\pi W_{2n}^o - \beta_H 2n\pi X_{2n}^o = 2\sqrt{2}(-1)^{n+1}[(1+\beta_I)\alpha_6 - \beta_H\alpha_4],$$

$$-[(1+\beta_I)(B_n - a^2) + a^2]X_{2n}^o - 2n\pi Z_{2n}^e + \beta_H 2n\pi B_n K_{2n}^o$$
$$= -2\sqrt{2}(-1)^{n+1}2n\pi[\beta_H\alpha_6 + (1+\beta_I)\alpha_4],$$

$$-\frac{P_m}{P_r}B_n\Theta_{2n}^o + W_{2n}^o = 0, \quad (1+\beta_I)(2\alpha_6 - a^2\alpha_2) - 2\beta_H\alpha_4 = 0, \quad -a^2\alpha_1 + 2M^2\alpha_4 = 0,$$

while the boundary conditions (5.1.5″) imply the constraints $DX^o(\pm 0.5) = K^e(\pm 0.5) = 0$, which, taking into account that $X_{2n}^{o^{(1)}} = 2\sqrt{2}(-1)^n\alpha_4 + 2n\pi X_{2n}^o$, have the form

$$2\alpha_4 + \sum_{n=1}^{\infty}[2\sqrt{2}(-1)^n\alpha_4 + 2n\pi X_{2n}^o]\sqrt{2}(-1)^n = 0, \quad \alpha_2 + \sum_{n=1}^{\infty}K_{2n}^e\sqrt{2}(-1)^n = 0,$$

where $B_n = 4n^2\pi^2 + a^2$, $\alpha_2 = \int_{-0.5}^{0.5} K^e(z)dz = K_o^e$, $\alpha_1 = \int_{-0.5}^{0.5} Z^e(z)dz = Z_o^e$, $\alpha_4 = X^o(0.5) = X^{o^{(1)}}/2$, $\alpha_6 = DK^e(0.5) = K^{e^{(2)}}/2$. Putting, for $n \geq 1$, $X_{2n}^o = \Delta_{4n}^o/\Delta_n^o$ and $K_{2n}^e = \Delta_{2n}^o/\Delta_n^o$, by solving the above system we have

$$\Delta_n^o = \frac{P_m}{P_r}B_n\Big\{(Ra^2 - B_n^3)\big\{B_n^2 + (B_n - a^2)(M^2 + B_n\beta_H^2) + \beta_I[B_n^2$$
$$+ M^2(B_n - a^2) + B_n(B_n - a^2)] + \beta_I^2 B_n(B_n - a^2)\big\}$$
$$- M^2 B_n(B_n - a^2)[B_n^2 + (B_n - a^2)M^2] - \beta_I M^2 B_n^2(B_n - a^2)^2\Big\},$$

$$\Delta_{4n}^o = 2\sqrt{2}(-1)^{n+1}\frac{P_m}{P_r}2n\pi\alpha_4 B_n\Big\{(Ra^2 - B_n^3)[B_n + M^2 + B_n\beta_H^2 + \beta_I(M^2 + 2B_n)$$
$$+ \beta_I^2 B_n] - M^2 B_n^2(B_n - a^2)(M^2 + B_n) - \beta_I M^2 B_n^2(B_n - a^2)\Big\},$$

$$\Delta_{2n}^o = 2\sqrt{2}(-1)^{n+1}\frac{P_m}{P_r}\Big\{\alpha_4 a^2\beta_H B_n(Ra^2 - B_n^3) - \alpha_6\frac{P_r}{P_m}\frac{\Delta_n}{B_n}\Big\}.$$

Remark 5.1.3. Formally, Δ_n^o, Δ_{4n}^o, Δ_{2n}^o can be derived from Δ_n, Δ_{4n}, and Δ_{2n} respectively from the even case by simply replacing $(2n-1)\pi$ by $2n\pi$ and, therefore A_n by B_n, and $(-1)^{n+1}$ by $(-1)^n$. This is obvious by the system in the Fourier coefficients. Indeed, if in the odd case we perform the changes $W_{2n}^o = \overline{W}_{2n}^o$, $X_{2n}^o = -\overline{X}_{2n}^o$, $\Theta_{2n}^o = \overline{\Theta}_{2n}^o$, and replace $2n\pi$ by $(2n-1)\pi$ and $(-1)^{n+1}$ by $(-1)^n$, then we obtain just the system in Fourier coefficients from the even case. Consequently $\overline{X}_{2n}^o$ becomes $-X_{2n-1}^e$ and K_{2n}^e becomes $-K_{2n}^o$. Hence X_{2n}^o corresponds to X_{2n-1}^e.

With the above expressions, the constraints lead to the following secular equation

$$\left\{1 + 2a^2\sum_{n=1}^{\infty}\frac{[(Ra^2 - B_n^3)(1+\beta_I) - M^2 B_n(B_n - a^2)B_n^2]}{\Delta_n\frac{P_r}{P_m}}\right\}\left(\frac{2}{a^2} + 4\sum_{n=1}^{\infty}\frac{1}{B_n}\right) = 0.$$

Introduce the notation similar to the even case, namely $L_n^o = B_n(B_n - a^2)$, $H_n^o = B_n^2 + (B_n - a^2)M^2$, $G_n^o = H_n^o + \beta_H^2 L_n^o$, take into account that $H_n^o B_n = B_n^3 + M^2 L_n^o$ and use the expressions $V_n^o = (Ra^2 - B_n H_n^o)M^{-2}L_n^{o^{-1}}$, hence $(Ra^2 - B_n^3) = (V_n^o + 1)M^2 L_n^o$. Remark that among all these expressions only V_n^o contains the eigenvalue R (in fact, R_n but we keep the notation R). Then we have

$$\Delta_n^o = \frac{P_m}{P_r} B_n M^2 L_n^o \left[V_n^o N_n^o + P_n^o \right],$$

where $N_n^o = H_n^o + \beta_I L_n^o + P_n^o$, $P_n^o = (\beta_H^2 + \beta_I^2)L_n^o + \beta_I L_n^o$. Consequently, the secular equation reads

$$1 + 2a^2 \sum_{n=1}^{\infty} \frac{B_n \left[V_n^o(1 + \beta_I) + \beta_I \right]}{V_n^o N_n^o + P_n^o} = 0. \tag{5.1.11}$$

For $n = 1$ we obtain

$$V_1^o = \overline{V}_1^o = -\frac{P_1^o + 2a^2 B_1 \beta_I}{N_1^o + 2a^2 B_1(1 + \beta_I)} \tag{5.1.12}$$

or, equivalently,

$$\begin{aligned}
R^o = &\frac{(4\pi^2 + a^2)^3 + 4\pi^2 M^2(4\pi^2 + a^2)}{a^2} - 4\pi^2 M^2(4\pi^2 + a^2) \\
&\left\{ (\beta_H^2 + \beta_I^2)4\pi^2(4\pi^2 + a^2) + \beta_I \left[(4\pi^2 + a^2)^2 + 4\pi^2 M^2 \right.\right. \\
&\left.\left. + 2a^2(4\pi^2 + a^2) \right] \right\} a^{-2} \left\{ (4\pi^2 + a^2)^2 + 4\pi^2 M^2 + 2a^2(4\pi^2 + a^2) \right. \\
&+ \beta_I \left[(4\pi^2 + a^2)^2 + 4\pi^2 M^2 + (4\pi^2 + 2a^2)(4\pi^2 + a^2) \right] \\
&\left. + (\beta_H^2 + \beta_I^2)4\pi^2(4\pi^2 + a^2) \right\}^{-1}.
\end{aligned} \tag{5.1.13}$$

Since $\overline{V}_1^o < 0$, the neutral curve $R = R^o$ for β_H, $\beta_I \neq 0$ is situated under the neutral curve $R = R_0^o$ from the case when the Hall and ion-slip currents are neglected.

The case $\beta_I = 0$ has been considered in [GeoPalR06]. From (5.1.13) it follows that the ion-slip effect can be stabilizing or destabilizing (with respect to the case when we consider only the Hall current) in dependence on various values of a and M^2, in agreement with the results in [GeoPalPasB], [MaiPL].

If the ion-slip effect is absent, *i.e.* $\beta_I = 0$, (5.1.13) becomes

$$\begin{aligned}
R_0^o = &\frac{(4\pi^2 + a^2)^3 + 4\pi^2 M^2(4\pi^2 + a^2)}{a^2} \\
&- \frac{16M^2\pi^4\beta_H^2(4\pi^2 + a^2)^2}{a^2 \left[(4\pi^2 + a^2)^2 + 4\pi^2 M^2 + 2a^2(4\pi^2 + a^2) + 4\pi^2\beta_H^2(4\pi^2 + a^2) \right]}
\end{aligned} \tag{5.1.14}$$

and agrees with the computations from [GeoPalR06].

If the series in (5.1.11) is truncated to two terms then, taking into account that $V_2^o = \dfrac{V_1^o M^2 L_1^o - Q^o}{M^2 L_2^o}$, where $Q^o = B_2 H_2^o - B_1 H_1^o$, we obtain the following secular equation in $\mathcal{V}_1^o = V_1^o - \overline{V}_1^o$

$$\mathcal{V}_1^{o^2} M^2 L_1^o T_1^2 \big[N_2^o T_1 + 2a^2 B_2 N_1^o (1 + \beta_I) \big] + \mathcal{V}_1^o T_1 \Big\{ M^2 L_2^o P_2^o T_1^2$$

$$- (\mathcal{P}_1 M^2 L_1^o + Q^o T_1) \big[N_2^o T_1 + 2a^2 B_2 N_1^o (1 + \beta_I) \big] + (M^2 L_2^o T_1 N_1^o$$

$$- 4a^4 B_1 B_2 M^2 L_1^o N_1^o) \beta_I + 4a^4 B_1 B_2 M^2 L_1^o P_1^o (1 + \beta_I)^2 \Big\}$$

$$+ 4a^4 B_1 B_2 L_1^o \beta_H^2 \big[M^2 L_2^o T_1 \beta_I - (\mathcal{P}_1 M^2 L_1^o + Q^o T_1)(1 + \beta_I) \big] = 0, \tag{5.1.15}$$

where $\mathcal{P}_1 = P_1^o + 2a^2 B_1 \beta_I$ and $T_1 = N_1^o + 2a^2 B_1 (1 + \beta_I)$. For $\beta_I = 0$ (5.1.15) coincides with the corresponding equation from [GeoPalR06].

If in the sum from (5.1.11) only the first two terms are left, then the eigenvalue R occurs only in V_1^0 and V_2^0, where $V_1^0 \neq \overline{V}_2^0$. Expressing V_2^0 in terms of V_1^0 means to reduce the determination of R to that of V_1^0. Moreover, we wrote the truncated secular equation not in V_1^0 but in the difference $\mathcal{V}_1^0$ of V_1^0 and $\overline{V}_1^0$ (the solution from the case $n = 1$), because it was much more simple. In addition, $\overline{V}_1^0$, given by (5.1.12), does not depend on R, therefore the eigenvalues could be immediately deduced from V_1^0.

Since $+4a^4 B_1 B_2 L_1^o \beta_H^2 \big[M^2 L_2^o T_1 \beta_I - (\mathcal{P}_1 M^2 L_1^o + Q^o T_1)(1 + \beta_I) \big] = T_1 [-Q^o + \beta_I (B_1^3 - B_2^3)] - \beta_H^2 M^2 L_1^{o^2} < 0$ because $B_1^3 - B_2^3 < 0$, it follows that equation (5.1.15) has two real opposite solutions. For the negative solution $\mathcal{V}_1^o$ it follows that $V_1^o < \overline{V}_1^o$. Therefore, the neutral curve for two terms in the series in (5.1.11) is situated under the neutral curve corresponding to one term, whence, taking into account the expressions $(Ra^2 - B_n^3) = (V_n^o + 1) M^2 L_n^o$, the result

Theorem 5.1.2 [PalG04b]. *For odd velocity and temperature fields and even magnetic field the mechanical equilibrium has the following approximate neutral curve*

$$R = \frac{(4\pi^2 + a^2) \big[(4\pi^2 + a^2)^2 + 4\pi^2 M^2 \big]}{a^2} + \frac{V_1^o 4\pi^2 M^2 (4\pi^2 + a^2)}{a^2} \tag{5.1.16}$$

where $V_1^o = \overline{V}_1^o + \mathcal{V}_1^o$, $\overline{V}_1^o = -\mathcal{P}_1 T_1^{-1}$ *and* $\mathcal{V}_1^o$ *is the negative solution of* (5.1.15).

Remark 5.1.4. The secular equation truncated to n terms is an equation of n-th degree in R (or some expression of R). Therefore, there are n secular manifolds defined by the n eigenvalues R, roots of that equation, if they are real. With every new approximation a new manifold emerges and the others are modified. As $n \to \infty$ their number increases to infinity and the neutral manifold is the lowest among all other secular manifolds. In fact, in any numerical procedure to solve an eigenvalue problem, the number of eigenvalues increases with the order of approximation.

5.1.5 *Thermosolutal instability of a compressible Soret-Dufour mixture with Hall and ion-slip currents in a porous medium*

This is the most complex type of convective flow studied in this book. The fluid is a binary mixture which is compressible, thermally and electrically conducting and is subject to a vertical magnetic field. The effects of Hall, ion-slip, Soret and Dufour currents are also considered. Moreover, the fluid is situated in a horizontal layer of a porous medium. Correspondingly, a supplementary unknown field (the concentration) occurs and many other parameters are involved. By the C-G. method, in this section the secular equation for the neutral stationary convection is derived. From it, in [GeoPalPasB], [PalGPashB] the critical values were deduced numerically for the first time in the literature. The influence of the various effects on the stability is obtained. Necessary conditions for overstability are also found.

5.1.5.1 *Perturbation problem*

Consider the linear stability against normal modes of the thermodiffusive equilibrium, of a compressible viscous electrically conducting mixture acted upon by a constant magnetic field $\mathbf{H}(0, 0, H)$, situated in a horizontal layer S of a porous medium, in the presence of the electroanisotropic Hall and ion-slip currents, and the thermoanisotropic Soret and Dufour currents. The layer is heated and salted from above, its bounding surfaces are free, they are electrically conducting and the temperature and concentration are kept constant. Small change in density destabilize the equilibrium and give rise to a convective motion governed by an initial and boundary-value problem for the balance equations for the mass, momentum, internal energy, concentration, and electric charge. We use these equations in the unknown functions $\mathbf{v}(v_1, v_2, v_3), \mathbf{H}(H_1, H_2, H_3), T, C, p$ (the velocity, magnetic, temperature, concentration and pressure fields). They depend on $\mathbf{x}(x_1, x_2, x_3) \in S$ and $t \in (0, \infty)$). The equations depend on **26** parameters: the positive numbers $\Phi < 1, \alpha, \beta_c, \mu_e, \eta, \beta_1, \beta_2$ and they stand for coefficient of medium porosity, thermal expansion, solute expansion, magnetic permeability, electrical resistivity and anisotropic Hall and ion-slip currents, respectively; $\mathbf{g}$ is the gravitation acceleration, $s = \pm 1$ according to whether the solute density is larger or less than the solvent density, $N > 0$ is the thermodynamic coefficient describing, together the thermodiffusive coefficient λ, the effect of the diffusive thermal conductivity; ν is the coefficient of kinematic viscosity, $E = \Phi + (1 - \Phi)\dfrac{\rho_{0s} c_{0s}}{\rho_{0f} c_{0f}}, \ K_m = \left[\Phi K_f + (1 - \Phi) K_s\right] \rho_{0f} c_{0f}$ $K_{c_m} = \left[\Phi K_{c_f} + (1 - \Phi) K_{c_s}\right] \rho_{0f} c_{0f}$, where $\rho_{0f}, c_{0f}, K_f, K_{c_f}, \rho_{0s}, c_{0s}, K_s K_{c_s}$ stand for the density, specific heat, thermal diffusivity, solute diffusivity of the mixture and solid matrix, respectively. Finally, let d be the thickness of the layer. Other dimensional parameters are introduced later. The thermodiffusive equilibrium has the particular solution form

$$\overline{m} \equiv (\overline{\mathbf{v}}, \overline{p}, \overline{T}, \overline{C}, \overline{\mathbf{H}}) \equiv (0, \overline{p}(x_3), (T_d - T_0)d^{-1}x_3 + T_0, (C_d - C_0)d^{-1}x_3 + C_0, \mathbf{H}_0),$$
$$(5.1.17)$$

where the subscripts o, d stand for quantities taken at the lower and upper surface respectively. It is stable up to certain values of certain dimensionless parameters related to the above-mentioned physical constant characteristics. If these thresholds are crossed, then this solution, although still exists, loses its stability on the account of the emergence of a new solution, corresponding to the convective motion in the form of the following perturbed equilibrium $(\mathbf{v}, p, T, C, \mathbf{H}) = (\mathbf{u}, \overline{p} + p, \overline{T} + \theta, \overline{C} + \gamma, \overline{\mathbf{H}} + \mathbf{h})$. We consider only infinitesimal perturbations $(\mathbf{u}, p, \theta, \gamma, \mathbf{h})$, which satisfy the following equations, deduced from the governing equations linearized about $\overline{m}$:

$$\left(\frac{\partial}{\partial t} - \nu\Delta + \frac{\nu\Phi}{K_1}\right)\Delta u_3 = g\Phi\left(\frac{\partial^2}{\partial x_1^2} + \frac{\partial^2}{\partial x_2^2}\right)(\alpha\theta - s\beta_c\gamma)$$

$$+ \frac{\mu_e H_0}{4\pi\rho_{of}}\frac{\Phi}{\nu}\frac{\partial}{\partial x_3}\nabla h_3,$$

$$\left(E\frac{\partial}{\partial t} - (K_m + N\lambda^2 K_{c_f})\Delta\right)\theta = N\lambda K_{c_f}\Delta\gamma + \left(\beta - \frac{g}{c_p}\right)u_3,$$

$$\left(E\frac{\partial}{\partial t} - K_{c_m}\Delta\right)\gamma = \lambda K_{c_f}\Delta\theta + \beta' u_3, \qquad (5.1.18)$$

$$\left(\frac{\partial}{\partial t} - \nu\Delta + \frac{\nu\Phi}{K_1}\right)\zeta = \frac{\mu_e H_0}{4\pi\rho_{of}}\frac{\Phi}{\nu}\frac{\partial}{\partial x_3}\xi,$$

$$\left(\Phi\frac{\partial}{\partial t} - (\eta + \beta_I H^2)\Delta\right)h_3 = H\frac{\partial}{\partial x_3}u_3 - \Phi\beta_H H\frac{\partial}{\partial x_3}\xi,$$

$$\left(\Phi\frac{\partial}{\partial t} - \eta\Delta\right)\xi = H\frac{\partial}{\partial x_3}\zeta + \Phi\beta_H H\frac{\partial}{\partial x_3}\Delta h_3 + \Phi\beta_I H^2\frac{\partial^2}{\partial x_3^2}\xi.$$

In (5.1.18),

$$\zeta = \frac{\partial u_2}{\partial x_1} - \frac{\partial u_1}{\partial x_2}, \qquad \xi = \frac{\partial h_2}{\partial x_1} - \frac{\partial h_1}{\partial x_2}$$

denote the x_3-components of vorticity and current density perturbation respectively; $(\beta - g/c_p)$ is the adiabatic temperature gradient of the equilibrium, where c_p and β stand for the specific heat of mixture at constant pressure and the static temperature gradient respectively. β' is the (constant) concentration gradient of one component of the mixture.

If the perturbations are assumed in the form of normal modes

$$(u_3, \theta, \gamma, \zeta, \xi, h_3)$$
$$= \{W(x_3), \Theta(x_3), \Gamma(x_3), Z(x_3), X(x_3), K(x_3)\} \cdot exp\left[i(K_1 x_1 + K_2 x_2) + ct\right],$$

and introduce the nondimensional parameters

$$\frac{a}{d} = K, \quad \sigma = cd^2/\nu = \sigma_r + \sigma_i, \quad \sigma_r, \sigma_i \in \mathbf{R}, \quad i = \sqrt{-1}, \quad P_r = \nu/K_m,$$

$$S_c = \nu/K_{c_m}, \quad P_2 = \nu/\eta_e, \quad \beta_H = \beta_1 H_0/\eta_e, \quad \beta_I = \beta_2 H_0^2/\eta_e$$

in (5.1.18), where $K^2 = K_1^2 + K_2^2$, then we obtain the following two-point eigenvalue problem governing the linear stability of the equilibrium

$$
\begin{cases}
\left(D^2 - a^2 - \sigma - \dfrac{\Phi d^2}{K_1}\right)(D^2 - a^2)W = \dfrac{g\Phi a^2 d^2}{\nu}(\alpha\Theta - s\beta_c\Gamma) \\[2mm]
\qquad\qquad\qquad\qquad\qquad\qquad - \dfrac{\mu_e H d}{4\pi\rho_{of}}\dfrac{\Phi}{\nu}D(D^2 - a^2)K, \\[4mm]
\left(D^2 - a^2 - \sigma - \dfrac{\Phi d^2}{K_1}\right)Z = -\dfrac{\mu_e H d}{4\pi\rho_{of}}\dfrac{\Phi}{\nu}DX, \\[4mm]
\left[(1 + N\lambda^2 r_D)(D^2 - a^2) - EP_r\sigma\right]\Theta = -\left(\beta - \dfrac{g}{c_p}\right)\dfrac{d^2}{K_m}W - N\lambda r_D(D^2 - a^2)\Gamma, \\[4mm]
(D^2 - a^2 - ES_c\sigma)\Gamma = -\dfrac{\lambda K_{c_f}}{K_{c_m}}(D^2 - a^2)\Theta - \dfrac{\beta' d^2}{K_{c_m}}W, \qquad (5.1.19)\\[4mm]
\left[(1 + \beta_I)(D^2 - a^2) - P_2\sigma\right]K = -\dfrac{H d}{\Phi\eta_e}DW + \beta_H dDX, \\[4mm]
\left[(1 + \beta_I)(D^2 - a^2) - P_2\sigma\right]X = -\dfrac{H d}{\Phi\eta_e}DZ - \dfrac{\beta_H}{d}D(D^2 - a^2)K, \\[4mm]
W = D^2 W = \Theta = \Gamma = DZ = D^2 X = X = K = 0, \qquad z = 0, 1
\end{cases}
$$

where $z = x_3 d^{-1}$ and

$$
r_D = \frac{K_{c_f}}{K_m} = \frac{K_{c_f}}{\Phi K_f + (1-\Phi)K_s}, \frac{K_{c_f}}{K_{c_m}} = \frac{K_{c_f}}{\Phi K_{c_f} + (1-\Phi)K_{c_s}}
$$

$(W, \Theta, \Gamma, Z, X, K)$ represents the eigenvector while for $a, \sigma, P_r, \dots$ fixed, R (defined in the next section) is the eigenvalue.

5.1.5.2 *Critical curves*

Assume that the instability occurs as stationary convection *i.e.* $\sigma = 0$. Then eliminating Z, K, X, Θ, Γ between equations (5.1.19) we obtain the following two-point eigenvalue problem

$$
\Bigg((D^2 - a^2 - P)(D^2 - a^2)^2 - \frac{Ra^2\Phi}{1 + N\lambda^2 r_D(1-\omega)}\left[\frac{1-G}{G} + N\lambda r_D\frac{\beta'}{\beta}\frac{S_c}{P_r}\right]
$$

$$
+ R_c a\Phi a^2\left\{-\frac{\lambda\omega}{1 + N\lambda^2 r_D(1-\omega)}\left[\frac{1-G}{G}\frac{\beta}{\beta'}\frac{P_r}{S_c} + N\lambda r_D\right] - 1\right\}\Bigg) \cdot
$$

$$
\left\{\left[(1 + \beta_I)D^2 - a^2\right](D^2 - a^2 - P) - M^2 D^2 + \frac{\beta_H^2}{1 + \beta_I}D^2(D^2 - a^2 - P)\right\}
$$

$$
- \frac{M^2}{1 + \beta_I}D^2(D^2 - a^2)\left\{\left[(1 + \beta_I)D^2 - a^2\right](D^2 - a^2 - P) - M^2 D^2\right\}W = 0, \quad (5.1.20)
$$

$$
W = D^2 W = D^4 W = D^6 W = D^8 W = 0, \qquad z = 0, \qquad z = 1 \qquad (5.1.21)
$$

which depends on 17 nondimensional parameters: $R = \dfrac{g\alpha\beta d^4}{\nu K_m}$ and $R_c = \dfrac{g\beta'\beta_c d^4}{\nu K_{c_m}}$, are the thermal and solute Rayleigh numbers, respectively;

$$\frac{1}{\omega} = \frac{K_{c_m}}{K_{c_f}} = \frac{\Phi K_{c_f} + (1-\Phi)K_{c_s}}{K_{c_f}} = \Phi + (1-\Phi)\frac{K_{c_s}}{K_{c_f}}; \quad P = \frac{\Phi d^2}{K_1}, \quad M^2 = \frac{\mu_e H^2 d^2}{4\pi\rho\nu\eta_e}$$

(M is the Hartmann number), $G = \dfrac{c_p\beta}{g}$.

The other unknown functions can be expressed in terms of W, therefore the eigenvalue problems (5.1.19) and (5.1.20), (5.1.21) are equivalent. Here W is the eigensolution while R is the eigenvalue (for any other parameter fixed). Hence the eigenvalue R is a function of $a, \omega, \Phi, G, R_c, N, \lambda, P_r, S_c, s, M, P, r_D, \frac{\beta'}{\beta}, \beta_I, \beta_H$, in the parameter space it represents a hypersurface with an infinity of sheets. The sheet situated at the smallest distance to the hyperplanes of coordinates is the neutral hypersurface, separating the domains of stability and instability of equilibrium. The smallest value of R on the neutral hypersurface is the critical Rayleigh number. Our aim is to derive the numerical values of the critical Rayleigh numbers on the basis of certain exact formulae.

The C-G method based on the set $\{\sin(n\pi z)\}$ total in $L^2(0,1)$ and applied to problem (5.1.20), (5.1.21), yields, for fixed values of the parameters except for R and a, the following secular equation

$$\frac{R_1\Phi}{1 + N\lambda^2 r_D(1-\omega)}\left[\frac{G-1}{G} - N\lambda r_D\frac{\beta'}{\beta}\frac{S_c}{P_r}\right] = aR_{c_1}\Phi\left\{-\frac{\lambda\omega}{1 + N\lambda^2 r_D(1-\omega)}\right.$$

$$\left. \cdot\left[\frac{G-1}{G} - N\lambda r_D\frac{\beta'}{\beta}\frac{S_c}{P_r}\right]\frac{\beta}{\beta'}\frac{P_r}{S_c} + 1\right\} + \frac{(1+x)^2(1+x+P_1)}{x}$$

$$+ \frac{M_1^2(1+x)\{(1+x+P_1)(1+x+\beta_I) + M_1^2\}}{x(1+\beta_I)\left[\{(1+x+P_1)(1+x+\beta_I) + M_1^2\} + \frac{\beta_H^2}{1+\beta_I}(1+x+P_1)\right]}$$

$$\tag{5.1.22}$$

where $R_1 = R/\pi^4$, $R_{c_1} = R_c/\pi^4$, $P_1 = P/\pi^4$, $M_1^2 = M^2/\pi^2$, $x = a^2/\pi^2$. In the absence of Soret-Dufour thermoanisotropic effects ($N = \lambda = 0$), from (5.1.22) we obtain

$$R_1 = \frac{G}{G-1}\left\{sR_{c_1} + \frac{(1+x)^2(1+x+P_1)}{x}\right.$$

$$\left. + \frac{M_1^2(1+x)\{(1+x+P_1)(1+x+\beta_I) + M_1^2\}}{x(1+\beta_I)\left[\{(1+x+P_1)(1+x+\beta_I) + M_1^2\} + \frac{\beta_H^2(1+x+P_1)}{(1+\beta_I)}\right]}\right\} \tag{5.1.23}$$

which for $\beta_I = 0$ becomes (34) of [SharT] (with a small difference supposed to be a slip in [SharT]).

In the absence of the compressibility, similar reasonings lead, formally, to (5.1.23) where $\dfrac{G}{G-1}$ must be taken to be equal to 1. Therefore, denoting by R_{1i} the expression of R_1 for the incompressible case we have

$$R_1 = \frac{G}{G-1}R_{1i} \tag{5.1.24}$$

which means that, for $G > 1$, the compressibility has a stabilizing effect.

In the general compressible case, a more useful form of (5.1.22) reads

$$R_1 \frac{C}{A} = R_{c_1} \Phi \left[\frac{N\lambda r_D}{A} \frac{C}{\Phi} \frac{G}{G-1} \frac{\alpha}{\beta_c} + s \right] + \frac{(1+x)^2(1+x+P_1)}{x}$$
$$+ \frac{M_1^2(1+x)\{(1+x+P_1)(1+x+\beta_I)+M_1^2\}}{x(1+\beta_I)\left[\{(1+x+P_1)(1+x+\beta_I)+M_1^2\} + \frac{\beta_H^2(1+x+P_1)}{(1+\beta_I)}\right]} \tag{5.1.25}$$

where $A = 1 + N\lambda^2 r_D(1-\omega)$, $C = \dfrac{\Phi(G-1)}{G}(1+\lambda s\omega \dfrac{\beta_c}{\alpha})$, which, for $C \neq 0$, may also be written as

$$R_1 = R_{c_1}\left[N\lambda r_D \frac{G}{G-1}\frac{\alpha}{\beta_c} + s\frac{A}{C}\Phi \right] + \frac{A}{C}\left\{ \frac{y^2(y+P_1)}{y-1} \right.$$
$$\left. + \frac{M_1^2 y\{(y+P_1)(y+\beta_I)+M_1^2\}}{(y-1)(1+\beta_I)\left[\{(y+P_1)(y+\beta_I)+M_1^2\} + \frac{\beta_H^2(y+P_1)}{(1+\beta_I)}\right]} \right\} \tag{5.1.26}$$

where $y = 1 + x > 1$. In the sequel, taking into account experimental data, we assume $M_1^2 \in (10, 10^4)$. For fixed values of the parameters $\alpha, \beta_c, \omega, N, \lambda, \beta_I, \beta_H$ and M_1^2, (5.1.25) provides R_1 as a rational function of y, $\lim_{y\to\infty} R_1(y) = \infty$ and $\lim_{y\to 1} R_1(y) = \infty$. Let (y_{cr}, R_{cr}) be the critical point; y_{cr} is the solution of the equation $\dfrac{dR_1}{dy} = 0$, which, up to a positive factor, can be written in the following form

$$\sum_{i=0}^{7} A_i y^{7-i} = 0 \tag{5.1.27}$$

where

$$A_0 = 2\overline{\beta}^2, \qquad \overline{\beta} = 1 + \beta_I, \quad A_1 = 4\overline{\beta}^3 + \overline{\beta}^2(5P_1 - 7) + 4\overline{\beta}\beta_H^2,$$

$$A_2 = 2\overline{\beta}^4 + \overline{\beta}^3(8P_1 - 10) + \overline{\beta}^2(4P_1^2 - 8P_1 + 4M_1^2 + 8) + (10P_1\overline{\beta} - 10\overline{\beta} + 4\overline{\beta}^2)\beta_H^2 + 2\beta_H^4$$

$$\equiv A_{22} + A_{21}\beta_H^2 + A_{20}\beta_H^4,$$

$$A_3 = \overline{\beta}^4(5P_1 - 3) + \overline{\beta}^3(8P_1^2 + 26P_1 + 6 + 4M_1^2) + \overline{\beta}^2(P_1^3 - 15P_1^2 + 21P_1 + 6P_1 M_1^2 - 10M_1^2)$$

$$- M_1^2\overline{\beta} + \left(M_1^2 + (8P_1^2 - 26P_1 + 6 + 4M_1^2)\overline{\beta} + (10P_1 - 6)\overline{\beta}^2 \right)\beta_H^2 + (5P_1 - 3)\beta_H^4$$

$$\equiv A_{32} + A_{31}\beta_H^2 + A_{30}\beta_H^4,$$

$$A_4 = \overline{\beta}^4(4P_1^2 - 8P_1) + \overline{\beta}^3\left(2P_1^3 - 22P_1^2 + 16P_1 + 6M_1^2 P_1 - 6M_1^2 \right)$$

$$+ \overline{\beta}^2\left(-4P_1^3 - 4P_1^2 + 2P_1^2 M_1^2 - 8P_1^- 16P_1 M_1^2 + 4M_1^2 + 2M_1^4 \right)$$

$$+\overline{\beta}\Big(-P_1+1\Big)2M_1^2+\Big(-2M_1^2+2M_1^2P_1+(-6M_1^2+8P_1+6M_1^2P_1+2P_1^3)\overline{\beta}$$

$$+(-16P_1+8P_1^2)\overline{\beta}^2\Big)\beta_H^2+(-8P_1+4P_1^2)\beta_H^4$$

$$\equiv A_{42}+A_{41}\beta_H^2+A_{40}\beta_H^4,$$

$$A_5=(2P_1^3-7P_1^2)\overline{\beta}^4+(-6P_1^3+14P_1^2+2P_1^2M_1^2)\overline{\beta}^3+(5P_1^3-6P_1^2M_1^2-7P_1^2+3P_1M_1^2$$

$$+P_1M_1^4+2M_1^2-3M_1^4)\overline{\beta}^2+(-M_1^2P_1^2-6M_1^2P_1-M_1^2-6M_1^4)\overline{\beta}$$

$$+\Big(M_1^2-M_1^4-4M_1^2P_1+M_1^2P_1^2+(-M_1^2-7P_1M_1^2+14P_1^2+2M_1^2P_1^2-6P_1^3)\overline{\beta}$$

$$+(-14P_1^2+2P_1^3)\overline{\beta}^2\Big)\beta_H^2-7P_1^2\beta_H^4\equiv A_{52}+A_{51}\beta_H^2+A_{50}\beta_H^4,$$

$$A_6=P_1^3\overline{\beta}^4+(-2P_1^3+2P_1^2M_1^2+P_1M_1^2)\overline{\beta}^3+(P_1^3-P_1^2M_1^2-2P_1M_1^2$$

$$+M_1^4)\overline{\beta}^2+(-M_1^2P_1^2+M_1^2P_1+P_1M_1^4-M_1^4)\overline{\beta}$$

$$+\Big(+M_1^2P_1^2-M_1^2P_1+(+P_1M_1^2+2M_1^2P_1^2-2P_1^3)\overline{\beta}2P_1^3\overline{\beta}^2\Big)\beta_H^2+P_1^3\beta_H^4$$

$$\equiv A_{62}+A_{61}\beta_H^2+A_{60}\beta_H^4,$$

$$A_7=P_1^2M_1^2\overline{\beta}^3+2(-P_1^2M_1^2+2P_1M_1^4)\overline{\beta}^2+(-M_1^2P_1+M_1^4)\overline{\beta}$$

$$+\Big(-M_1^2P_1^2+M_1^4P_1+M_1^2P_1^2\overline{\beta}\Big)\beta_H^2$$

$$\equiv A_{72}+A_{71}\beta_H^2+A_{70}\beta_H^4. \tag{5.1.28}$$

Consider now that the porosity is absent (*i.e.* $P_1=0$). Then, for a homogeneous fluid, from (5.1.26) and (5.1.27) we regain (2.2) from [GeoPalPasB] while for a binary mixture, with Soret-Dufour thermoanisotropic currents, Hall and ion-slip electroanisotropic currents we regain the following formula from [MaiP89]

$$R_1=R_{c_1}\Big[N\lambda r_D\frac{G}{G-1}\frac{\alpha}{\beta_c}+s\frac{A}{C}\Big]$$

$$+\frac{A}{C}\Big\{\frac{y^3}{y-1}+\frac{M_1^2y\{y(y+\beta_I)+M_1^2\}}{(y-1)(1+\beta_I)\big[\{y(y+\beta_I)+M_1^2\}+\frac{\beta_H^2y}{(1+\beta_I)}\big]}\Big\}. \tag{5.1.29}$$

For this last case formula (5.1.24), relating the compressible and incompressible Rayleigh numbers, also holds. Hence, as in [SharT], the compressibility delays the onset of thermal instability.

Obviously, (5.1.26) implies

$$\frac{dR_1}{dR_{c_1}} = N\lambda r_D \frac{G}{G-1}\frac{\alpha}{\beta_c} + s\frac{A}{C}\Phi$$

so from (5.1.25) it is easy to see that if $G > 1$ and the Soret-Dufour effect is absent, in which case $\left(\dfrac{dR_1}{dR_{c_1}}\right)_{\lambda=N=0} = \dfrac{G}{G-1}s\Phi$, then

$$\left(\frac{dR_1}{dR_{c_1}}\right) - \left(\frac{dR_1}{dR_{c_1}}\right)_{\lambda=N=0} = \frac{G}{G-1}\left\{N\lambda r_D \frac{\alpha}{\beta_c} + \frac{s\Phi\left[N\lambda^2 r_D(1-\omega) - \lambda s\omega\beta_c\alpha^{-1}\right]}{\left[1 + \lambda s\omega\beta_c\alpha^{-1}\right]}\right\},$$

$$(5.1.30)$$

hence the stabilizing or destabilizing effect depends on the sign of the second factor in (5.1.30). Many other reasonings may be done but they do not show how vary R_{1c_r} with respect to y_{c_r}. This is why we apply to (5.1.27) the idea of [GeoPalPasB]: instead of solving the seventh-order algebraic equation (5.1.27) with respect to y and to obtain the critical points $y_{c_r} = y_{c_r}(M_1^2, P_1, \overline{\beta}, \beta_H^2)$, $R_{1c_r} = R_{1c_r}(y_{c_r}(M_1^2, P_1, \overline{\beta}, \beta_H^2), M_1^2, P_1, \overline{\beta}, \beta_H^2)$, we can think of (5.1.27) as a second order algebraic equation in β_H^2

$$B_0(\beta_H^2)^2 + B_1(\beta_H^2) + B_2 = 0, \qquad (5.1.31)$$

where $B_0 = A_{40}y^3 + A_{50}y^2 + A_{60}y + A_{70}$, $B_1 = A_{21}y^5 + A_{31}y^4 + A_{41}y^3 + A_{51}y^2 + A_{61}y + A_{71}$, $B_2 = \sum_{i=0}^{7} A_{i2}y^{7-i}$.

Equation (5.1.31) defines the same hypersurface (5.1.27), hence the solutions

$$\beta_H^2 = \beta_H^2(M_1^2, y, P_1, \overline{\beta})$$

correspond to y_{c_r}, showing that the critical points will consist of those y for which $M_1^2, P_1, \overline{\beta},\ \beta_H^2 > 0$ and of the corresponding $R_{1c_r} = R_{1c_r}(y, M_1^2, P_1, \overline{\beta}, \beta_H^2(M_1^2, y, P_1, \overline{\beta}))$, obtained by substituting in (5.1.22) the explicit solutions of (5.1.31).

The results of numerical calculations based on (5.1.26), (5.1.27) and (5.1.28) are plotted [PalGPashB] showing the influence of the following parameters: $\beta_H, \beta_I, P_1, M_1$, on the domain of stability.

In order to determine the critical point one considers in these figures the expression

$$R = \frac{y^2(y+P_1)}{y-1} + \frac{M_1^2 y\{(y+P_1)(y+\beta_I) + M_1^2\}}{(y-1)(1+\beta_I)\left[\{(y+P_1)(y+\beta_I) + M_1^2\} + \frac{\beta_H^2(y+P_1)}{(1+\beta_I)}\right]}$$

equivalent to (5.1.26). The graphical representation of the surface $R = R(y, \beta_H)$ for M_1, P_1, β_I fixed shows that the stability domain is considerably reduced as the parameter β_H increases. The projection of this surface on (y, R)-plane are plotted. The minimum points of curves $R = R(y), \beta_H = $ constant represent the critical values y_{c_r}, R_{c_r} which, for different parameters β_H determine a critical curve.

From the graphic of the surfaces $R = R(y, \beta_H)$ plotted for various values parameter M_1 it follows that the stability domain is considerably extended with the increasing of this parameter from 60 to 100.

From the graphic of the surfaces $R = R(y, \beta_H)$ plotted for different values of parameters β_I we deduce that with the increase of β_I the stability domain is considerably reduced.

The critical curves $R_{c_r} = R_{c_r}(y_{c_r})$ obtained by varying the parameter β_H in the range $[0, 50]$ for different values of parameters β_I, P_1, M_1, are represented too. With the increase in M_1 the (y_{c_r}, R_{c_r}) values representing the critical point increase, as is shown. For small values of parameters P_1 and β_I all critical curves have a maximum point for y_{c_r}. This means that, at values y_{c_r} greater than this point, the equilibrium is stable for any β_H.

When P_1 increases, the values and the range for y_{c_r} are considerably reduced, and R_{c_r} has a slight increase. At the same time these curves change their shape.

Due to the increase of the parameter β_I the (y_{c_r}, R_{c_r}) values decrease rapidly and the y_{c_r} range is reduced. At the same time the critical curves change their shape, this being more obvious for great values of the parameter P_1 when one observes the inversion of the curvature. In this case these curves have a minimum point for y_{c_r}. Finally, the influence of the parameters P_1 and β_I is stronger for small values of M_1 (curves (5) in these figures).

5.1.5.3   *The overstability problem*

In this section we investigate whether instability can arise as oscillatory motion of constant amplitude, that is as overstability. This means to deduce conditions such that $\sigma_r = 0$ implies $\sigma_i \neq 0$ (where $\sigma_i \in \mathbf{R}$). To this end, from (5.1.20) we deduce

$$\Phi R_1 \frac{G-1}{G}\left[\left(1 + x + \frac{\sigma S_c E}{\pi^2}\right) + \lambda s \omega \frac{\beta_c}{\alpha}(1+x)\right]x$$

$$-\Phi R_{c_1}\left\{N\lambda r_D(1+x)\frac{\alpha}{\beta_c} + s\left[(1+x)\left(1 + N\lambda^2 r_D\right) + \frac{\sigma P_r E}{\pi^2}\right]\right\}x$$

$$= \left\{\left[(1+x)\left(1 + N\lambda^2 r_D\right) + \frac{\sigma P_r E}{\pi^2}\right]\left(1 + x + \frac{\sigma S_c E}{\pi^2}\right) - N\lambda^2 \omega r_D(1+x)^2\right\}$$

$$(1+x)\left(1 + x + P + \frac{\sigma}{\pi^2}\right) + M_1^2\left\{\left[(1+x)\left(1 + N\lambda^2 r_D\right) + \frac{\sigma P_r E}{\pi^2}\right]\left(1 + x + \frac{\sigma S_c E}{\pi^2}\right)\right.$$

$$\left.- N\lambda^2 \omega r_D(1+x)^2\right\}(1+x)\left\{\left(1 + \beta_I + x + \frac{P\sigma}{\pi^2}\right)\left(1 + x + P + \frac{\sigma}{\pi^2}\right) + M_1^2\right\}$$

$$\left(\left\{\left(1 + \beta_I + x + \frac{P\sigma}{\pi^2}\right)\left(1 + x + P + \frac{\sigma}{\pi^2}\right) + M_1^2\right\}\left[(1+\beta_I)(1+x) + P\frac{\sigma}{\pi^2} + \beta_H^2(1+x)\right]\right.$$

$$\left.\left(1 + x + P + \frac{\sigma}{\pi^2}\right)\right)^{-1}. \tag{5.1.32}$$

Suppose that we are in the case of overstability, such that we denote $\sigma/\pi^2 = i\sigma_1$ where σ_1 must be real. If we take the real and imaginary parts of (5.1.32), in the case $s = 1$, we obtain

$$\mu^3 - v_1\mu^2 + v_2^2\mu - v_3$$

$$\equiv \mu^3 - \left(F_2\underline{A}_1' + F_3P_2 + F_4A_4 + F_4P_2M_1^2 - R_1B_3 + R_{c_1}C_3\right)(PF_4)^{-1}\cdot\mu^2$$

$$+\left(\underline{A}_1'F_1 + F_2\underline{A}_3 + F_3\underline{A}_4 + M_1^2Ay^3P + M_1^2\underline{A}_1F_4 + M_1^2By^2\underline{A}_2' - R_1B_2 - R_{c_1}C_2\right)(PF_4)^{-1}\mu$$

$$-\left(F_1\underline{A}_3 + M_1^2A\underline{A}_1y^3 - R_1B_1 - R_{c_1}C_1\right)(PF_4)^{-1} = 0,$$

$$\mu^2 - u_1\mu + u_2$$

$$\equiv \mu^2 - \left[P^2F_1 + F_2\underline{A}_4 + F_3\underline{A}_1' + \underline{A}_3F_4 + M_1^2(By^2P + \underline{A}_2'F_4) - R_1D_2 + R_{c_1}D_4\right]$$

$$(P^2F_2 + \underline{A}_1'F_4)^{-1}\mu + (F_1\underline{A}_4 + F_3\underline{A}_4 + M_1^2By^2\underline{A}_1 + M_1^2\underline{A}_2'Ay^3 - R_1D_1 - R_{c_1}\underline{D}_3)$$

$$(P^2F_2 + \underline{A}_1'F_4)^{-1} = 0, \tag{5.1.33}$$

where $\sigma_1^2 = \mu$,

$$\underline{A}_1 = (y + \beta_I)(y + P) + M_1^2, \underline{A}_3 = \underline{A}_1y(1 + \beta_I) + \beta_H^2y(y + P), \underline{A}_1' = P(y\underline{\beta} + \underline{A}_2'),$$

$$\underline{A}_2' = y + \beta_I + P(y + P), \underline{A}_4 = \underline{A}_2'y(1 + \beta_I) + P\underline{A}_1 + \beta_H^2y,$$

$$B_1 = y(y - 1)C\underline{A}_3, B_2 = (y - 1)[C\underline{A}_1'y + F\underline{A}_4], B_3 = P^2(y - 1)F,$$

$$C_1 = -Dy(y - 1)\underline{A}_3, C_2 = -Dy(y - 1)\underline{A}_1' - (y - 1)G\underline{A}_4, C_3 = (y - 1)GP^2,$$

$$D_1 = (y - 1)[C\underline{A}_4y + F\underline{A}_3], D_2 = (y - 1)[CP^2y + F\underline{A}_1'],$$

$$D_3 = -(y - 1)[D\underline{A}_4y + G\underline{A}_3], D_4 = (y - 1)[DP^2y + G_1\underline{A}_1'],$$

$$F_1 = y^3A(y + P), F_2 = E^2S_cP_ry(y + P) + By^2,$$

$$F_3 = (B + A)y^3 + By^2P, F_4 = E^2S_cP_ry,$$

$$B = (1 + N\lambda^2r_D)S_cE + P_rE, D = \Phi\left[N\lambda r_D\alpha\beta_c^{-1} + s(1 + N\lambda^2r_D)\right], F = \Phi\frac{G - 1}{G}S_cE.$$

As σ_1 is supposed to be real and different from zero, we must have $\mu > 0$. Hence the necessary and sufficient condition to have $\mu > 0$ reads

$$u_1 > 0, \quad u_2 > 0, \quad v_1 > 0, \quad v_2 > 0, \quad v_3 > 0, \quad u_1^2 - 4u_2 > 0,$$

$$\frac{v_1^4(v_2 + v_3)}{81} + \frac{v_1^2v_3^2}{27} + \frac{v_3^3}{27} + \frac{v_1v_2v_3}{6} - \frac{v_1^3v_3}{27} - \frac{v_3^2}{4} - \frac{v_1^2v_2^2}{36} > 0. \tag{5.1.34}$$

Let us remark that the necessary condition in order for the principle of exchange of stability to hold becomes $v_3 = 0$.

Sufficient conditions in order for $u_1 > 0$, $u_2 > 0$, $v_1 > 0$, $v_2 > 0$, and $v_3 > 0$ are

$$\frac{R_1}{R_{c_1}} < \frac{D_4}{D_2}, \quad \frac{R_1}{R_{c_1}} < \frac{-D_3}{D_1}, \quad \frac{R_1}{R_{c_1}} < \frac{C_3}{B_3},$$

$$\frac{R_1}{R_{c_1}} < -\frac{C_2}{B_2}, \quad \frac{R_1}{R_{c_1}} < -\frac{C_1}{B_1} \tag{5.1.35}$$

respectively. These conditions are expressed in terms of R_1.

Other sufficient conditions in order for some necessary conditions of overstability hold can be obtained by using Sharma's method to write equation (5.1.32) in the form $\sum_{i=0}^4 T_{4-i}\mu^{4-i} = 0$, then to separate its real and imaginary parts and eliminate R_1 between the obtained equations.

Necessary conditions of overstability $T_0/T_u > 0$ may be considered to get Sharma's type more general conditions. In this type of conditions R_1 will not occur.

5.2 Magnetic Bénard convection for a fully ionized fluid

The same problem as in Sections $5.1.1 - 5.1.4$ is considered, but here only the Hall effect is present, *i.e.* we study the stability of a mechanical equilibrium of a magnetic electroanisotropic fluid. Formally, all formulae from Section 5.1 remain valid and become simpler because in them $\beta_I = 0$. This enables us, by following [GeoPalR06], a more detailed analysis of the neutral curves and of the singularities of the secular equations. More exactly, in Section 5.2.2.1 we treat the case $\beta_H = 0$. In Section 5.2.2.2 we consider the case $\beta_H \neq 0$ and determine the approximate neutral curve obtained by retaining one or two terms in the Fourier series representing the secular equation. We find that the Hall current has a destabilizing effect in agreement with [GeoPalPasB] [MaiP84] [MaiPL]; in the absence of the Hall current we obtain the smallest eigenvalue leading to the neutral curve.

5.2.1 *Neutral curves for the even case*

5.2.1.1 *Regular case $\Delta_n \neq 0$*

Since, for $\beta_I = 0$, we have $N_n = H_n + P_n$, $P_n = \beta_H^2 L_n$, therefore introducing the notation $X_n = Ra^2 - H_n A_n = V_n^e M^2 L_n$, the secular equation (5.1.8) becomes

$$\sum_{n=1}^{\infty} \frac{A_n X_n}{X_n(H_n + \beta_H^2 L_n) + M^2 \beta_H^2 L_n^2} = 0, \tag{5.2.1}$$

yielding the eigenvalues R as functions of a^2 and physical parameters M and β_H.

In order to see if the Hall currents are stabilizing or destabilizing, we must compare the secular curve defined by (5.2.1) with the secular curve from the case when the Hall effect is absent, *i.e.* $\beta_H = 0$. Thus, taking $\beta_H = 0$ in (5.2.1) we obtain the secular equation

$$\sum_{n=1}^{\infty} \frac{A_n(Ra^2 - A_n H_n)}{(Ra^2 - A_n H_n)(H_n + \beta_H^2 L_n)} = 0, \tag{5.2.1$'$}$$

each term of which is singular at $X_n = 0$, $\quad n = 1, 2, \ldots$. Then, assuming that $X_n \neq 0$, (5.2.1$'$) implies the equality $\displaystyle\sum_{n=1}^{\infty} \frac{A_n}{H_n + \beta_H^2 L_n} = 0$, which cannot be fulfilled because all its terms are positive. It follows that (5.2.1$'$) fails to represent the secular equation for $\beta_H = 0$. This is why we consider this case separately in Section 5.2.1.2, where we show that, in fact, the solutions of the true secular equation are $X_n = 0$, $\quad n = 1, 2, \ldots$ and they correspond to the eigenvalues $Ra^2 = H_n A_n$, or, equivalently,

$$R = \frac{A_n^3 + (A_n - a^2)M^2 A_n}{a^2}. \tag{5.2.1$''$}$$

Consequently, the smallest eigenvalue $R = \dfrac{(\pi^2 + a^2)^3 + M^2\pi^2(\pi^2 + a^2)}{a^2}$ defining the neutral curve corresponds to $n = 1$ and it is equal to IV, (163) of [Chan].

If $\beta_H \neq 0$ and $n = 1$, the eigenvalue is still that from the case $\beta_H = 0$, *i.e.* $R = H_1 A_1 a^{-2}$, and it corresponds to $X_1 = 0$. If $\beta_H \neq 0$ and $n = 1$ and 2, from (5.2.1) we derive the equation in X_2 (but which contains the expression in $X_1 = Ra^2 - A_1 H_1$ too)

$$X_2^2 + X_2[Q + M^2\beta_H^2 B_+] + M^2\beta_H^2 QD = 0,$$

where $Q = X_1 - X_2 \equiv H_2 A_2 - H_1 A_1$, the solutions of which are

$$X_2^{(1),(2)} = \frac{-(Q + M^2\beta_H^2 B_+) \pm \sqrt{(Q + M^2\beta_H^2 B_-)^2 + 4M^4\beta_H^4 C}}{2}, \tag{5.2.2}$$

where $B_\pm = (A_2 L_1^2 \pm A_1 L_2^2)(A_1 G_2 + A_2 G_1)^{-1}$, $C = A_1 A_2 L_1^2 L_2^2 (A_1 G_2 + A_2 G_1)^{-1}$, $D = A_1 L_2^2 (A_1 G_2 + A_2 G_1)^{-1}$, $G_n = A_n^2 + (A_n - a^2)(M^2 + A_n \beta_H^2)$, whence the result

Theorem 5.2.1 [GeoPalR06]. *If the velocity and temperature fields are even functions while the magnetic field is an odd function of the vertical coordinate z and only two terms of the secular equation are retained, then the approximate neutral curve reads*

$$R_0^e = \frac{(\pi^2 + a^2)^3 + M^2\pi^2(\pi^2 + a^2)}{a^2},$$

if the Hall effect is absent and

$$R^e = \frac{A_2 H_2 + X_2^{(2)}}{a^2}, \tag{5.2.3}$$

or, equivalently,

$$R^e = \frac{A_1 H_1 + X_1^{(2)}}{a^2}, \tag{5.2.4}$$

if the Hall effect is present, where $X_2^{(2)}$ has the expression (5.2.2), $X_1^{(2)} = Q + X_2^{(2)}$. In the $(a^2; R)$-plane, the curves (5.2.4) are situated below the curve R_0^e, thus the Hall effect is destabilizing.

Proof. The equation in X_1 can be obtained from the equation in X_2 by simply replacing Q by $-Q$. Then it is immediate that $X_1^{(2)} < 0$, hence the values of R given by (5.2.4) are smaller than those given by R_0^e. Moreover, $X_1^{(1)} > 0$, therefore the corresponding values for $R = [H_1 A_1 + X_1^{(1)}]a^{-2} = [H_2 A_2 + X_2^{(1)}]a^{-2}$ are higher than those given by R_0^e. Hence, R_0^e is situated between the two curves from the case $\beta_H \neq 0$, but only the last one is of interest to us. Since $X_2^{(1),(2)} < 0$ it follows that both the corresponding curves $R = [H_2 A_2 + X_2^{(1),(2)}]a^{-2}$ are situated below the curve $R = A_2 H_2 a^{-2}$. Their expression computed by (5.2.2) show that as $\beta_H \to 0$ these curves tend to the curves $R = A_2 H_2 a^{-2}$ and $R = A_1 H_1 a^{-2}$ respectively from the case of the absence of the Hall effect. Therefore, for a sufficiently small β_H the lower curve is the closest to R_0^e. In addition, the neutral curve from the case $\beta_H = 0$

is situated between the two curves corresponding to the two solutions of the secular equation (10) with $\beta_H \neq 0$.

If more terms in (5.2.1) are retained, higher degree equations in X_i are obtained. Due to the decreasing order of magnitude [Geo85] of the additional terms, their contribution to the solution diminishes and, thus, we expect that a limit neutral curve exists under that for the case $\beta_H = 0$. In fact, all involved series converge at least like n^{-1} as $n \to \infty$. In (5.2.1) the terms in Ra^2 are of order n^{-8} while those which do not contain Ra^2 are of order n^{-2} as $n \to \infty$ and they are negative. Indeed, for M^2, β_H and Ra^2 not too large, $X_n G_n + M^2 \beta_H^2 L_n^2 \sim X_n G_n \sim -A_n H_n^2 \sim -A_n^5$, $A_n X_n = A_n[Ra^2 - A_n H_n] = A_n Ra^2 - A_n^2 H_n$, therefore the coefficient of Ra^2 is of order A_n^{-4} and that of $A_n H_n$ is of order A_n^{-1} as $n \to \infty$. We emphasize that we are interested in the smallest R, therefore in the solution $X_k^{(k)}$ which corresponds to the value Ra^2 smaller than R_0^e. This solution exists. Indeed, all the smallest negative solutions of the k-th degree equation in X_k, obtained by truncating (5.2.1) to k terms, has a continuous dependence on β_H^2.

Moreover for $\beta_H^2 \to 0$ the corresponding Ra^2 tend to $H_1 A_1, H_2 A_2, \ldots, H_k A_k$. In particular, the lowest corresponding curve $R = R(a^2)$ tends to the neutral curve defined by R_0^e.

5.2.1.2 *Neutral curve for the singular case $\beta_H = 0$, $\Delta_n = 0$*

Throughout this and the following section we assume that (5.1.6) is written for $\beta_I = 0$. If $\beta_H = 0$, the system (5.1.6) decouples in two systems [GeoPal03]

$$-A_n Z_{2n-1}^o - M^2(2n-1)\pi X_{2n-1}^e = 2\sqrt{2}(-1)^n \alpha_4 M^2,$$
$$-A_n X_{2n-1}^e + (2n-1)\pi Z_{2n-1}^o = 2\sqrt{2}(-1)^n(2n-1)\pi \alpha_4, \tag{5.1.6$'$}$$

and

$$-A_n K_{2n-1}^o - (2n-1)\pi W_{2n-1}^e = 2\sqrt{2}(-1)^n \alpha_6,$$
$$A_n^2 W_{2n-1}^e - M^2 A_n(2n-1)\pi K_{2n-1}^o - \frac{P_m}{P_r} Ra^2 \Theta_{2n-1}^e$$
$$= 2\sqrt{2}(-1)^n(2n-1)\pi \alpha_6 M^2, \tag{5.1.6$''$}$$
$$-A_n \frac{P_m}{P_r} \Theta_{2n-1}^e + W_{2n-1}^e = 0.$$

Correspondingly, $\Delta_n = H_n[\frac{P_m}{P_r} A_n(Ra^2 - A_n H_n)] = \Delta_n' \Delta_n''$, where Δ_n'' and $\Delta_n'(= H_n)$ are the Cramer determinants of these systems.

From the first system it follows $X_{2n-1}^e = -\frac{(M^2 + A_n)(2n-1)\pi}{H_n} 2\sqrt{2}(-1)^n \alpha_4$, so the first constraint (5.1.5$'$)* becomes $2\sqrt{2}\alpha_4 \sum_{n=1}^{\infty}\left[1 + \frac{(M^2 + A_n)(A_n - a^2)}{H_n}\right] = 0$ which cannot hold (because each term is positive) unless $\alpha_4 = 0$. But, if $\alpha_4 = 0$, it follows that X_{2n-1}^e and $Z_{2n-1}^o = 0$. Hence, in order to see if there is some eigenvalue we must study the system (5.1.6$''$).

Thus, assume $\Delta_n'' \neq 0$, i.e. $Ra^2 - A_n H_n \neq 0$, for every $n \in \mathbf{N}^*$. Then, for $\Delta_n'' = \frac{P_m}{P_r} A_n(R_{a^2} - A_n H_n) \neq 0$, it follows $W_{2n-1}^e = \Theta_{2n-1}^e = 0$, $K_{2n-1}^o = \frac{2\sqrt{2}(-1)^{n+1}\alpha_6}{A_n}$, so the second constraint $(5.1.5')^*$ becomes the relation $2\sqrt{2}\alpha_6 \sum \frac{1}{A_n} = 0$, implying $\alpha_6 = 0$, and, therefore, $W_{2n-1}^e = \Theta_{2n-1}^e = K_{2n-1}^e = 0$. Hence, any number R such that $R \neq A_n H_n a^{-2}$ for every $n \in \mathbf{N}^*$cannot be an eigenvalue. Therefore R can be an eigenvalue only if $R = A_m H_m a^{-2}$, for some $m \in \mathbf{N}^*$. In this case $\Delta_m'' = 0$, while $\Delta_n'' \neq 0$ if $n \neq m$. In addition, for $n \neq m$ we have $\Delta_n'' = \frac{P_m}{P_r} A_n(A_m H_m - A_n H_n)$, implying $W_{2n-1}^e = \Theta_{2n-1}^e = 0$, $K_{2n-1}^o = \frac{2\sqrt{2}(-1)^{n+1}\alpha_6}{A_n}$ and, again, $(5.1.5')^*$ implies $\alpha_6 = 0$.

In the case $n = m$ the equations in $(5.1.6'')$ are not linearly independent. Indeed, $M^2(2n - 1)\pi(5.1.6'')_1 + [A_n^2 + M^2(A_n - a^2)](5.1.6)_3 = (5.1.6)_2$. Therefore, for $n = m$ we consider the system $(5.1.6'')_{1,3}$ the solutions of which are $W_{2m-1}^e = \frac{P_m}{P_r} A_m \Theta_{2m-1}^e$, $K_{2m-1}^o = -\frac{P_m}{P_r}(2m - 1)\pi\Theta_{2m-1}^e - \frac{2\sqrt{2}(-1)^m \alpha_6}{A_m}$. In this way, the constraint $(5.1.5')_2^*$ becomes $\sum_{n=1}^{\infty}\left[\frac{2\sqrt{2}\alpha_6}{A_n}\right] + (-1)^m \frac{P_m}{P_r}(2m - 1)\pi\Theta_{2m-1}^e = 0$, implying $K_{2m-1}^o = 0$. Since for $n \neq m$ we already saw that $K_{2m-1}^o = 0$, it follows that $K^o(z) = 0$. Consequently, (unlike in [GeoPal03]) for any value $R = A_m H_m a^{-2}$ the system $(5.1.6'')$ has this nontrivial solution $W^e(z) = \sqrt{2}\frac{P_m}{P_r} A_m \Theta_{2m-1}^e \cos((2m - 1)\pi z)$, $\Theta_e(z) = \sqrt{2}\Theta_{2m-1}^e \cos((2m - 1)\pi z)$, $K^o(z) = 0$. Therefore, there exist an infinity of eigenvalues, namely $R = A_m H_m a^{-2}$, $m \in \mathbf{N}^*$. Of course, it would suffice the fact that $\Theta^e(z)$ is not vanishing; we gave the expressions of $W^e(z)$ and $K^o(z)$ since they are useful if we want to know the corresponding eigensolution.

Consequently, the neutral curve is $R_1 = A_1 H_1 a^{-2}$ i.e. $R_1 = (\pi^2 + a^2)[(\pi^2 + a^2)^2 + M^2\pi^2]a^{-2}$ and it corresponds to the perturbations $W^e(z) = \sqrt{2}\frac{P_m}{P_r}(\pi^2 + a^2)\Theta_1^e \cos(\pi z)$, $\Theta^e(z) = \sqrt{2}\Theta_1^e \cos(\pi z)$, $K^e(z) = 0$, where Θ_1^e is an arbitrary constant.

5.2.1.3 *Neutral curve for the singular case $\beta_H \neq 0$, $\Delta_n = 0$*

In this case $\Delta_n = 0$ reads

$$Ra^2 = A_m H_m - \beta_H^2 M^2 \frac{L_m^2}{H_m + \beta_H^2 L_m}. \qquad (5.2.1')$$

In addition, for $m = n$,

$$\frac{(5.1.6)_3 M^2(2m - 1)\pi H_m + (5.1.6)_2(2m - 1)\pi\beta_H M^2(A_m - a^2)}{H_m + \beta_H^2 L_m}$$

$$+\frac{\beta_H M^2 L_m}{H_m + \beta_H L_m}(5.1.6)_4 - (5.1.6)_1 \frac{\alpha_4 \beta_H a^2 A_m(2m - 1)\pi M^2}{H_m + \beta_H^2 L_m} = 0,$$

if in equations (5.1.6) the expressions in the right-hand side were passed in the left-hand side. It follows that equation $(5.1.6)_1$ is a linear combination of $(5.1.6)_{3,2,4,5}$ if $\alpha_4 = 0$. Therefore assume that $\alpha_4 = 0$. Then the system $(5.1.6)$ for $n = m$ is consistent and the system $(5.1.6)_{3,2,4,5}$ has the solutions $W^e_{2m-1} = \frac{P_m}{P_r} A_m \Theta^e_{2m-1}$, $Z^o_{2m-1} = \frac{\frac{P_m}{P_r}(2m-1)\pi\Theta^e_{2m-1}\beta_H M^2 L^2_m}{A_m(H_m+\beta^2_H L_m)}$, $X^e_{2m-1} = \frac{\frac{P_m}{P_r}\Theta^e_{2m-1}\beta_H L^2_m}{(H_m+\beta^2_H L_m)}$, $K^e_{2m-1} = -\frac{\frac{P_m}{P_r}(2m-1)\pi\Theta^e_{2m-1}H_m}{(H_m+\beta^2_H L_m)} - \frac{2\sqrt{2}(-1)^m\alpha_6}{A_m}$.

For $m \neq n$, formulae for Δ_n, Δ_{2n} and Δ_{4n} still hold, of course for Ra^2 given by $(5.2.1')$ and $\alpha_4 = 0$. In addition, $W^e_{2n-1} = \Theta^e_{2n-1} = X^e_{2n-1} = Z^o_{2n-1} = 0$, while $K^e_{2n-1} = -\frac{2\sqrt{2}(-1)^n\alpha_6}{A_m}$. The restriction $(5.1.5')^*_1$ implies $\Theta^e_{2m-1} = 0$ and $(5.1.5')^*_2$ implies $\alpha_6 = 0$, whence the solution of $(5.1.6)$ is trivial and, consequently, R given by $(5.2.1')$ is not an eigenvalue for the problem $(5.1.4')$ $(5.1.5')$.

The Fourier coefficients were determined separately for $n \neq m$ and for $n = m$ and then the results were introduced in the constraints. In this way we found that, in the absence of the Hall effect, $R = R_m$ was an eigenvalue, while in the other case, *i.e.* when the Hall effect was present, no such eigenvalue exist.

5.2.2 *Neutral curves for the odd case*

5.2.2.1 *Regular case*

In this case, formally, $(5.1.11)$ written for $\beta_I = 0$ is the secular equation. However, we recall that it was deduced by introducing $X^o_{2n} = \Delta_{4n}/\Delta_n$, $K^e_{2n} = \Delta_{2n}/\Delta_n$ into the constraints $DX^o(\pm 0.5) = K^e(\pm 0.5)$, where

$$\Delta_n = \frac{P_m}{P_r}B_n\Big\{(Ra^2 - B^3_n)\big[B^2_n + (B_n - a^2)(M^2 + B_n\beta^2_H)\big]$$
$$- M^2 B_n(B_n - a^2)\big[B^2_n + (B_n - a^2)M^2\big]\Big\},$$

$$\Delta_{4n} = 2\sqrt{2}(-1)^{n+1}\frac{P_m}{P_r}2n\pi\alpha_4\Big[(Ra^2 - B^3_n)(B^2_n + M^2 B_n + B^2_n\beta^2_H)$$
$$- M^2 B^2_n(B_n - a^2)(M^2 + B_n)\Big],$$

$$\Delta_{2n} = 2\sqrt{2}(-1)^{n+1}\frac{P_m}{P_r}\Big\{\alpha_4 a^2 \beta_H B_n(Ra^2 - B^3_n) - \alpha_6\frac{P_r}{P_m}\frac{\Delta_n}{B_n}\Big\}.$$

Thus, the constraints lead to the secular equation $(5.1.1)$ for $\beta_I = 0$. It can be written in the form of the following determinant containing infinite sums of series converging at least like n^{-1} as $n \to \infty$

$$\det\begin{vmatrix} a_{11} & a_{12} \\ a_{21} & a_{22} \end{vmatrix} = 0, \tag{5.2.5}$$

where

$$a_{11} = 1 + 2a^2 \sum_{n=1}^{\infty} B_n^2 \frac{[(Ra^2 - B_n^3) - M^2 B_n(B_n - a^2)]}{\Delta_n \frac{P_r}{P_m}}, \qquad a_{12} = 0,$$

$$a_{21} = \frac{-\beta_H}{a^2} - 2 \sum_{n=1}^{\infty} \frac{a^2 \beta_H B_n(Ra^2 - B_n^3)}{\Delta_n \frac{P_r}{P_m}}, \qquad a_{22} = \frac{1}{a^2} + 2 \sum_{n=1}^{\infty} \frac{1}{B_n}.$$

In the above we understood that $\Delta_n \neq 0$.

5.2.2.2 *Singular case* $\beta_H = 0$

In this case $\Delta_n = \frac{P_m}{P_r} B_n[(Ra^2 - B_n^3) - M^2 B_n(B_n - a^2)][B_n^2 + M^2(B_n - a^2)]$,

$\Delta_{4n} = 2\sqrt{2}(-1)^{n+1} \frac{P_m}{P_r} 2n\pi\alpha_4 B_n[(Ra^2 - B_n^3) - M^2 B_n(B_n - a^2)](B_n + M^2)$, $\Delta_{2n} = 2\sqrt{2}(-1)^n \alpha_6 \frac{\Delta_n}{B_n}$. Consequently, if $\Delta_n \neq 0$, the constraints imply the secular equation

$$\left(1 + 2a^2 \sum_{n=1}^{\infty} \frac{B_n}{B_n^2 + M^2(B_n - a^2)}\right)\left(\frac{1}{a^2} + 2 \sum_{n=1}^{\infty} \frac{1}{B_n}\right) = 0.$$

This relation cannot hold because all involved terms are positive. Hence, our problem has no eigenvalue.

Then assume that $\Delta_n = 0$. In this case $Ra^2 = B_n^3 + M^2 B_n(B_n - a^2)$, or, with the notation $L_n = B_n(B_n - a^2)$ and $H_n = B_n^2 + M^2(B_n - a^2)$, we have the eigenvalues $R_n = \frac{H_n B_n}{a^2}$. Let us prove

Theorem 5.2.2. *For* $R_n = \frac{H_n B_n}{a^2}$ *nontrivial solution of* (5.1.4''), (5.1.5'') *for* $\beta_I = 0$ *exist.*

Proof. First remark that for $\beta_H = 0$ system (5.1.4'') for $\beta_I = 0$ splits into two noncoupled systems

$$\begin{aligned}
(D^2 - a^2)Z^e + M^2 D X^o &= 0, \\
(D^2 - a^2)X^o + D Z^e &= 0,
\end{aligned} \tag{5.2.6}$$

$$(D^2 - a^2)K^e + D W^o = 0,$$

$$(D^2 - a^2)^2 W^o + M^2 D(D^2 - a^2)K^e - H_n B_n \frac{P_m}{P_r}\Theta^o = 0, \tag{5.2.7}$$

$$\frac{P_m}{P_r}(D^2 - a^2)\Theta^o + W^o = 0.$$

By eliminating Z^e between (5.2.6)$_1$ and (5.2.6)$_2$ it follows $(D^2 - a^2)^2 X^o - M^2 D X^o = 0$, while taking into account in (5.2.6)$_2$ the boundary conditions for Z^e it follows the supplementary boundary conditions $(D^2 - a^2)X^o = 0$ at $z = \pm 0.5$. If $\lambda_1 \neq \lambda_2$ are the roots of the corresponding characteristic equation $(\lambda^2 - a^2)^2 - M^2\lambda^2 = 0$, introducing the general odd solution $X^o = A \sinh \lambda_1 + B \sinh \lambda_2$ into these conditions, we obtain the secular equation

$$\frac{(\lambda_2^2 - a^2)\tanh\frac{\lambda_2}{2}}{\lambda_2} = \frac{(\lambda_1^2 - a^2)\tanh\frac{\lambda_1}{2}}{\lambda_1}.$$

The function $f(\lambda) = \lambda^{-1}(\lambda^2 - a^2)\tanh\frac{\lambda}{2}$ is monotone for $\lambda > 0$ and for $\lambda < 0$. Indeed, $\dfrac{df}{d\lambda} = 0$ reads $\sinh\lambda = -\lambda(\lambda^2 - a^2)(\lambda^2 + a^2)^{-1}$ and the graphs of the functions defined by the two sides of this equality are intersecting only for $\lambda = 0$. Therefore the secular equation has only the trivial solution. Consequently X^o and Z^e are trivial functions.

Consider now the system (5.2.7) with the corresponding boundary conditions from (5.1.5″) for β_I. Then, using the expansion of the unknown functions on the total sets $\{E_{2n}\}_{n\in\mathbf{N}}$ and $\{F_{2n}\}_{n\in\mathbf{N}^*}$ we have

$$-B_n K_{2n}^e + 2n\pi W_{2n}^o = 2\sqrt{2}(-1)^{n+1}\alpha_6,$$

$$B_n^2 W_{2n}^o + M^2 B_n 2n\pi K_{2n}^e - \frac{P_m}{P_r}B_n H_n \Theta_{2n}^o = -2\sqrt{2}(-1)^{n+1}2n\pi\alpha_6 M^2,$$

$$-\frac{P_m}{P_r}B_n\Theta_{2n}^o + W_{2n}^o = 0,$$

$$-2\alpha_6 + a^2\alpha_2 = 0.$$

(5.2.8)

Remark that $-2n\pi M^2(5.2.8)_1 + [B_n^2 + M^2(B_n - a^2)](5.2.8)_3 = (5.2.8)_2$, (we remind that $H_n = B_n^2 + M^2(B_n - a^2)$) therefore equations $(5.2.8)_{1,2,3}$ are not independent, in other words, equations $(5.1.4'')_{1,3,5}$ are not independent. This can be seen by performing $M^2 D(18)_1 + [(D^2 - a^2)^2 - M^2 D^2](5.2.7)_3$ and add to $(5.2.7)_2$ to obtain

$$(D^2 - a^2)^2 W^o + M^2 D(D^2 - a^2)K^e + \frac{P_m}{P_r}[(D^2 - a^2)^2 - M^2 D^2](D^2 - a^2)\Theta = 0. \quad (5.2.9)$$

On the other hand, the elimination of W^o and K^e between (5.2.7) leads to the following equation in Θ

$$[(D^2 - a^2)^2 - M^2 D^2](D^2 - a^2)\Theta + Ra^2\Theta = 0, \qquad (5.2.10)$$

therefore $\dfrac{P_m}{P_r}[(D^2 - a^2)^2 - M^2 D^2](D^2 - a^2)\Theta = -\dfrac{P_m}{P_r}Ra^2\Theta$ and so, for $Ra^2 = B_n H_n$ (5.2.9) becomes $(5.2.7)_2$.

Remark 5.2.1. The operators $M^2 D$ and $(D^2 - a^2)^2 - M^2 D^2$ were constructed by taking into account that a factor of $2n\pi i$ is generated by the application of the operator D and, so, B_n is generated by $-(D^2 - a^2)$. This type of reasoning is generally useful in order to express properties of the system in Fourier coefficients in terms of those in the corresponding system of differential equations.

The characteristic equation for (5.2.10) reads

$$[(\lambda^2 - a^2)^2 - M^2\lambda^2](\lambda^2 - a^2) + Ra^2 = 0. \qquad (5.2.11)$$

Therefore for every eigenvalue $R_n = B_n H_n/a^2 = a^{-2}[(2n\pi)^2 + a^2]\{[(2n\pi)^2 + a^2]^2 + M^2(2n\pi)^2\}$ we can find the six solutions of (5.2.11) such that the odd general solution Θ^o of (5.2.10) has the form $\Theta^o(z) = A_{1n}\sin(2n\pi z) + A_{3n}\sinh\lambda_3 z + A_{5n}\sinh\lambda_5 z$, where λ_3 and λ_5 are given in the following, Θ^o satisfies the boundary conditions given in (5.1.5'') or deduced from (5.1.4'') and (5.1.5'') and it is not identically equal to zero. The same can be said about W^o and K^e. Really, $R_n = H_n B_n/a^2$ reads, equivalently, as $R_n = B_n H_n/a^2 = a^{-2}[(2n\pi)^2 + a^2]\{[(2n\pi)^2 + a^2]^2 + M^2(2n\pi)^2\}$ and, thus, (5.2.11) can be written in the form

$$\lambda^6 - (3a^2 + M^2)\lambda^4 + a^2(3a^2 + M^2)\lambda^2 - a^6 \\ + [(2n\pi)^2 + a^2]\{[(2n\pi)^2 + a^2]^2 + M^2(2n\pi)^2\} = 0. \tag{5.2.11'}$$

Since this equation has two roots $\lambda_{1,2} = \pm 2n\pi i$ it follows that the other four roots, written as $\lambda_{3,4} = \pm(\alpha + i\beta)$ and $\lambda_{5,6} = \pm(\alpha - i\beta)$ satisfy the equation

$$\lambda^4 - (3a^2 + M^2 + 4n^2\pi^2)\lambda^2 + (3a^2 + M^2)(a^2 + 4n^2\pi^2) + 16n^4\pi^4 = 0, \tag{5.2.11''}$$

whence the above quoted form for Θ^o. By construction this function satisfies equation (5.2.10). It must also satisfy the following boundary conditions derived from (5.1.4'') and (5.1.5'')

$$\Theta^o = D^2\Theta^o = D^4\Theta^o = 0 \qquad z = \pm 0.5 \tag{5.2.12}$$

implying the secular equation in $\mathbf{C}$

$$\det \begin{vmatrix} \sin(n\pi) & \sinh(\frac{\lambda_3}{2}) & \sinh(\frac{\lambda_5}{2}) \\ -4n^2\pi^2\sin(n\pi) & \lambda_3^2\sinh(\frac{\lambda_3}{2}) & \lambda_5^2\sinh(\frac{\lambda_5}{2}) \\ 16n^4\pi^4\sin(n\pi) & \lambda_3^4\sinh(\frac{\lambda_3}{2}) & \lambda_5^4\sinh(\frac{\lambda_5}{2}) \end{vmatrix} = 0, \tag{5.2.13}$$

which is automatically satisfied for both real and complex values of λ_3 and λ_5. This determinant always vanishes. However, it is easy to check that there exists a 2×2 nonvanishing minor formed with the minor of the upper-right corner of (5.2.13). The corresponding equations in A_{1n}, A_{3n}, A_{5n} read

$$A_{1n}\sin(n\pi) + A_{3n}\sinh(\alpha/2)\cos(\beta/2) + A_{5n}\sin(\beta/2)\cosh(\alpha/2) = 0,$$

$$-4n^2\pi^2\sin(n\pi)A_{1n} + A_{3n}[(\alpha^2 - \beta^2)\sinh(\alpha/2)\cos(\beta/2) - 2\alpha\beta\sin(\beta/2)\cosh(\alpha/2)]$$

$$+A_{5n}[\sinh(\alpha/2)\cos(\beta/2) + (\alpha^2 - \beta^2)\sin(\beta/2)\cosh(\alpha/2)] = 0,$$

implying $A_{3n} = A_{5n} = 0$. Therefore $\Theta^o = A_{1n}\sin(2n\pi z)$, where A_{1n} are determined up to a constant factor: they are the Fourier coefficients corresponding to the expansion functions $\sin(2n\pi z)$. Consequently for every R_n we have one nonvanishing solution Θ^o of the above form, *i.e.* R_n is an eigenvalue, indeed.

The system (5.2.8) gives the same result: $R_m = H_m B_m/a^2$ represent the eigenvalues for the problem (5.2.8), (5.1.5''). Indeed, since the Cramer determinant for $(5.2.8)_{1,2,3}$ is null, we choose equations $(5.2.8)_{1,3}$ in W_{2n}^o and K_{2n}^e. The corresponding Cramer determinant, which is a 2×2 minor of that for $(5.2.8)_{1,2,3}$, is nonvanishing. Therefore W_{2n}^o and K_{2n}^e can be uniquely determined in terms of

Θ^o_{2n} and α_6, *i.e.* $W^o_{2m} = \dfrac{P_m}{P_r} B_m \Theta^o_{2m}$ and $K^e_{2m} = 2m\pi\Theta^o_{2m}\dfrac{P_m}{P_r} + \dfrac{2\sqrt{2}(-1)^m\alpha_6}{B_m}$,

while for $m \neq n$ we have $W^o_{2n} = \Theta^o_{2n} = 0$, $K^e_{2n} = \dfrac{2\sqrt{2}(-1)^n\alpha_6}{B_n}$. Since, by $(5.2.8)_4$,

$\alpha_2 = 2\alpha_6 a^{-2}$, the constraint $K^e(\pm 0.5) = 0$ becomes $\dfrac{2\alpha_6}{a^2} + \sqrt{2}(-1)^m 2m\pi\Theta^o_{2m}\dfrac{P_m}{P_r} +$

$\displaystyle\sum_{n=1}^{\infty} 4\frac{\alpha_6}{B_n} = 0$, or, because $\displaystyle\sum_{n=1}^{\infty}\frac{1}{4n^2\pi^2 + a^2} = \frac{1}{4a}\coth\frac{a}{2} - \frac{1}{2a^2}$, it follows that

$\alpha_6 = (-1)^{m+1}\sqrt{2}a\tanh\dfrac{a}{2} 2m\pi\dfrac{P_m}{P_r}\Theta^o_{2m}$.

Taking into account the formula [GrR]

$$\sum_{k=1}^{\infty}\frac{(-1)^k\cos kx}{k^2 + b^2} = \frac{\pi}{2b}\frac{\cosh bx}{\sinh b\pi} - \frac{1}{2b^2}, \quad -\pi\leq x\leq\pi,$$

we find $K^e(z) = 2\sqrt{2}m\pi\dfrac{P_m}{P_r}\Theta^o_{2m}\left[(-1)^{m+1}\dfrac{\cosh az}{\cosh a/2} + \cos(2m\pi z)\right]$. In addition,

$W^o(z) = \sqrt{2}\dfrac{P_m}{P_r}B_m\Theta^o_{2m}\sin(2m\pi z)$ and $\Theta^o(z) = \Theta^o_{2m}\sqrt{2}\sin(2m\pi z)$, hence $R_m = B_m H_m a^{-2}$ is an eigenvalue.

The neutral curve corresponds to $m = 1$, *i.e.* it has the equation

$$R = \frac{(4\pi^2 + a^2)^3 + 4M^2\pi^2(4\pi^2 + a^2)}{a^2}. \tag{5.2.14}$$

Let us prove that no $R_n \neq B_n H_n a^{-2}$, $n \in \mathbf{N}^*$, is an eigenvalue. Indeed, $(5.2.8)$ has only trivial solutions. In this case the Cramer determinant is

$\Delta''_n = \dfrac{P_m}{P_r}B_n(Ra^2 - B_n H_n)$ and $W^o_{2n} = \Theta^o_{2n} = 0$, $K^e_{2n} = \dfrac{2\sqrt{2}(-1)^n\alpha_6}{B_n}$, which introduced into the constraint implies $\alpha_6 = 0$, hence $K^e_{2n} = 0 = \alpha_2$, hence the desired result.

5.2.2.3   *Singular case $\Delta_n = 0$, $\beta_H \neq 0$*

In this case the secular equation $(5.1.11)$ for $\beta_I = 0$ becomes

$$1 + 2a^2\sum_{n=1}^{\infty}\frac{(Ra^2 - B_n H^o_n)B_n}{(Ra^2 - B^3_n)(H^o_n + \beta^2_H L^o_n) - M^2 L^o_n H^o_n} = 0 \tag{5.2.15}$$

or, equivalently

$$1 + 2a^2\sum_{n=1}^{\infty}\frac{(Ra^2 - B_n H^o_n)B_n}{(Ra^2 - B_n H^o_n)G^o_n + \beta^2_H M^2 L^{o^2}_n} = 0 \tag{5.2.15'}$$

where $L^o_n = B_n(B_n - a^2)$, $H^o_n = B^2_n + M^2(B_n - a^2)$ and $G^o_n = H^o_n + \beta^2_H L^o_n$.

If in the sums in $(5.2.15')$ a single term is retained we obtain

$$R = \frac{(4\pi^2 + a^2)^3 + 4M^2\pi^2(4\pi^2 + a^2)}{a^2}$$
$$- \beta^2_H\frac{16M^2\pi^4(4\pi^2 + a^2)^2}{a^2\{(4\pi^2 + a^2)^2 + 4M^2\pi^2 + 2a^2(4\pi^2 + a^2) + \beta^2_H 4\pi^2(4\pi^2 + a^2)\}} \tag{5.2.16}$$

showing the destabilizing effect of the Hall current if compared with (5.2.14).

Now let truncate (5.2.15′) up to terms corresponding to $n = 2$ and introduce the notation $V_n^o = (Ra^2 - B_n H_n^o)M^{-2}L_n^{o^{-1}}$. Then $V_2^o = \dfrac{V_1^o M^2 L_1 - Q^o}{M^2 L_2}$ and so, (5.2.15′) becomes

$$(V_1^o - \overline{V}_1^o)^2 M^2 L_1 T_1^2 (N_2^o T_1 + 2a^2 B_2 N_1^o) + (V_1^o - \overline{V}_1^o)T_1[M^2 L_2 P_2 T_1^2$$
$$- (P_1^o M^2 L_1^o + Q^o T_1)(N_2^o T_1 + 2a^2 B_2 N_1^o) + 4a^4 B_1 B_2 M^2 L_1^o P_1^o] \qquad (5.2.17)$$
$$- 4a^4 B_1 B_2 P_1^o (P_1 M^2 L_1^o + Q^o T_1) = 0,$$

where $N_n^o = H_n + \beta_H^2 L_n^o$, $P_n^o = \beta_H^2 L_n^o$, $T_1 = N_1^o + 2a^2 B_1$, $Q^o = H_2^o B_2 - H_1^o B_1$ and $\overline{V}_1^o$ is the solution of (5.2.15′) corresponding to $n = 1$, *i.e.* $\overline{V}_1^o = -P_1 T_1^{-1}$. Due to the fact that $Ra^2 - B_1 H_1^o = V_1^o M^2 L_1^o = \overline{V}_1^o M^2 L_1^o + (V_1^o - \overline{V}_1^o)M^2 L_1^2$ this means that the case with two terms shows destabilizing effect if $V_1^o - \overline{V}_1^o < 0$ and stabilizing one otherwise. Equation (5.2.17) shows that we have one positive solution $V_1^o - \overline{V}_1^o$ and other negative. It follows that the neutral curve corresponding to this negative solution is better than (5.2.16).

5.3 Convection in a micro-polar fluid bounded by rigid walls

Following [Drag05] the stability of the mechanical equilibrium of a micro-polar fluid is dealt with. The derivation of the secular surfaces is immediate. Classical particular cases are regained.

The eigenvalue problem governing the perturbation of conduction layer of the thermally conducting micro-polar fluid, situated between two horizontal rigid walls maintained at constant temperature and subject to an external magnetic field reads [Ram]

$$\begin{cases} (1 + R)[(D^2 - a^2)^2 - QD^2]W + R(D^2 - a^2)Z - \mathcal{R}a^2\theta = 0, \\ [A(D^2 - a^2) - 2R] - R(D^2 - a^2)W = 0, \\ (D^2 - a^2)\theta + W - \overline{\delta}Z = 0, \end{cases} \qquad (5.3.1)$$

$$W = DW = Z = \Theta = 0 \quad \text{at} \quad z = \pm 0.5, \qquad (5.3.2)$$

containing five real parameters, the Rayleigh number $\mathcal{R} > 0$ being taken for the eigenvalue, $a > 0$ is the wave number, A, R, $\overline{\delta}$ are micro-polar parameters, Q is the constant intensity of the magnetic field and the functions W, Θ, $Z : [-0.5, 0.5] \to \mathbf{R}$ characterize the amplitude of the of the vertical component of the velocity, temperature and the vertical component of the spin vorticity, respectively.

Due to the fact that in (5.3.1) only even derivatives occur, the even and odd problems have the same form as (5.3.1), (5.3.2). Further on, we present only the even case. Expanding the unknown functions in Fourier series on $\{E_{2n-1}\}$ and introducing them in (5.3.1), the following algebraic system in the Fourier coefficients is obtained

$$\begin{cases} (1+R)\{2\sqrt{2}(-1)^{n+1}\alpha(2n-1)\pi + [A_n^2 + Q(2n-1)^2\pi^2]W_{2n-1}\}, \\ \quad - RA_n Z_{2n-1} - R_a a^2 \Theta_{2n-1} = 0, \\ RA_n W_{2n-1} - (AA_n + 2R)Z_{2n-1} = 0, \\ W_{2n-1} - Z_{2n-1} - \bar{\delta}Z2n - 1 - A_n\Theta_{2n-1} = 0, \end{cases} \tag{5.3.3}$$

where $\alpha = D^2 W(0.5) \neq 0$. The single boundary condition which is not automatically satisfied by the Fourier series expansions is $DW(\pm 0.5) = 0$. They introduce the constraint

$$\sum_{n=1}^{\infty} (-1)^n \sqrt{2}(2n-1)\pi W_{2n-1} = 0. \tag{5.3.4}$$

For $Q = 0$ and $\bar{\delta} = 0$ (5.3.3), (5.3.4) leads to the secular equation

$$\sum_{n=1}^{\infty} \frac{(2n-1)^2 A_n(AA_n + 2R)}{(AA_n + 2R)[A_n^3 + RA_n^3 - R_a a^2] - R^2 A_n^3} = 0. \tag{5.3.5}$$

If the intensity of the magnetic field is absent ($Q = 0$), but the micro-polar parameter $\bar{\delta} \neq 0$, the secular equation has the form

$$\sum_{n=1}^{\infty} \frac{(2n-1)^2 A_n D_n}{A_n^3[D_n(1+R) - R^2] + R_a a^2(\bar{\delta}RA_n - D_n)} = 0, \tag{5.3.6}$$

while for $Q \neq 0$ and $\bar{\delta} \neq 0$ the secular equation reads

$$\sum_{n=1}^{\infty} \frac{(2n-1)^2 A_n D_n}{A_n D_n(1+R)[A_n^2 + Q(2n-1)^2\pi^2] - R^2 A_n^2 + R_a a^2(\bar{\delta}RA_n - D_n)} = 0, \tag{5.3.7}$$

where $D_n = (AA_n + 2R)$. For $\bar{\delta} = 0$ (5.3.6) reduces to (5.3.5), while for $Q = 0$, (5.3.7) reduces to (5.3.6). Numerical computations [Drag05] reveal the stabilizing effect of A and R for $\bar{\delta} = 0$, while $\bar{\delta} \neq 0$ has a destabilizing effect. They agree perfectly with those existing in the literature.

In the absence of the micro-polar structure, *i.e.* $A = R = \bar{\delta} = Z = 0$, problem (5.3.1) $-$ (5.3.2) becomes that from [Chan] and it is studied in Section 7.2.

If, in addition, $Q = 0$, (5.3.1) $-$ (5.3.2) becomes the classical Bénard problem. Its treatment in [Geo82a] and [GeoOP] shows the possible existence of neutral curves which do not depend on the boundary conditions and the presence of a false neutral curve. Since, usually, these properties are inherited by the more general models of which they are derived, the study of (5.3.5) and (5.3.6) must be continued with the investigation of the singularities of these secular equations, related to particular eigenvalues (Remark 5.1.2), and with the investigation of the associated characteristic equation by the direct method (Appendix 6, Chapter 7).

5.4   Convections governed by ode's with variable coefficients

In Chapter 4 and Sections 5.1, 5.2 and 5.3, we study convections emerging as a result of the loss of stability of certain equilibria which are quite simple leading to a constant coefficients in the perturbed equations. Moreover, no heat (or other kind of) source is supposed to act. In these conditions, in the Boussinesq approximation, the N-S-F model and its generalizations to the magnetic case (Section 1.4) are appropriate. In this section, physical conditions, of the deep convection Section 5.4.1, the variable gravity field (Section 5.4.2) and a heat source (Sections 5.4.3, 5.4.4) imposed some other modified models, belonging to the Müller general scheme but not written by us explicitly. This is why they are briefly described in each of these sections.

Two types of convections are treated: the variable coefficients arise in the internal energy conservation equation (Sections 5.4.1, 5.4.3, 5.4.4) or in the momentum conservation equation (Section 5.4.2). In Section 5.4.1 we treat the case of the free boundaries, in Sections 5.4.3, 5.4.4 the case of rigid boundaries, while in Section 5.4.2 both these cases. The methods used are based on expansions in Fourier series on sets total in $L^2[0,1]$. In addition, in Section 5.4.1 the symmetry with respect to the straight line $z = 0.5$ or to the point $z = -0.5$ is assumed.

In Section 5.4.1 and partially in Section 5.4.2 the problem was reduced to an eigenvalue two-point problem for a single equation. Further on, in Section 5.4.1 series expansions are introduced into that equation to produce an equation containing the variable coefficient term multiplied by the expansion functions. Next, each of these products is again expanded in Fourier series. Finally, imposing to the obtained equation to be orthogonal to the expansion functions an infinite-dimensional algebraic equation linear in the Fourier coefficients is obtained. The requirement that the Cramer determinant vanishes yields the secular equation.

In Section 5.4.2 other two approaches are used. The series expansion with variable coefficients of the unknown functions Θ is introduced and by the variation of coefficients method the other function is derived. The further requirement that the boundary conditions for W be fulfilled leads to a system of equations where all Fourier coefficients for W are expressed in terms of those for Θ. Then both expansions are introduced into the other equation to get the infinite algebraic linear system in the Fourier coefficients of Θ, the standard condition of existence of non-trivial such coefficients leading to the secular equation. In Section 5.4.2, a variant of this approach is checked: the governing problem is reduced to an equation in Θ, then a new unknown function is introduced as a convenient linear function of Θ, to obtain a new system to which the first approach is applied.

In the penetrative convection treated in Section 5.4.3, one bounding plane is thermally conducting and other thermally nonconducting, requiring a special choice of the total set of expansion functions.

So far, in Chapter 5 we treated convections in unbounded layers. In Section

5.4.4 we deal with convection in a rectangular box in the presence of a heat source. The method used is that from Section 5.4.2 based on the equation in Θ.

5.4.1 *Deep convection*

Following [GeoLP], a linear stability analysis of the Bénard problem for deep convection is performed. The convergence problem is minutely analyzed. An estimate of the critical Rayleigh number that reduces to the classical value for vanishing depth parameter is obtained.

Unlike the convections governed by ode's with constant coefficients, due to local physical influences, the convections governed by ode's with variable coefficients are no longer of global type. Consequently, they are more or less similar to local flows, *e.g.* in boundary layers or of the plane Couette-Poiseuille type, the linear stability of which is described by the Orr-Sommerfeld (O-S) equation. The classical mathematical treatment of this equation reflected the large differences in the local behavior of the solution. This is a reason why asymptotic series of powers of a small parameter were used. In spite of the fact that in a certain sense the Fourier series can be viewed as asymptotic series, only very rarely they were used to investigate this equation. In this respect we quote the pioneering paper [Goldste], where the Fourier series were used to study the stability of the plane Poiseuille flow. One can say that it opens a new trend in hydrodynamic stability theory, namely the investigation of convections with local features.

The method applied in [GeoLP] originates in this famous paper. It is the reason why we chose the set $\{F_n\}$ total in $L^2(0,1)$. In fact, its use goes back to Reynolds in his seminal book *Theory of sound*, with a big subsequent influence on stability studies, *e.g.* [Goldste], [Chan].

Consider a homogeneous viscous and thermo-conducting fluid confined in a horizontal layer, subject to an adverse temperature gradient β and with stress-free and perfectly thermo-conducting bounding planes. Then [Zey], [Er] the nondimensional N-S-F equations governing the deep convection are

$$\begin{cases} \dfrac{d}{dt}\mathbf{u} = -P_r\nabla p + P_r\Delta\mathbf{u} + RP_rT\mathbf{k}, \\[2mm] \dfrac{d}{dt}T = R u\mathbf{k} + \mu(z)\Delta T + 2\dfrac{\delta}{r}\mu(z)\mathbf{d}\cdot\mathbf{d}, \\[2mm] \nabla\cdot\mathbf{u} = 0, \end{cases} \tag{5.4.1}$$

where $\mu(z) = \frac{1}{1+\delta(1-z)}$ is the function of the deep convection and $\delta \in [0,1]$ is a parameter. In $(5.4.1)_2$ the last term stands for dissipation. As $\delta \to 0$ we regain the N-S-F equations in the Boussinesq approximation for the classical Bénard shallow convection.

Suppose that the mechanical equilibrium (1.4.11) is the basic state and the perturbation fields of the velocity, temperature, pressure $\mathbf{v}(u,v,w)$, θ, and p' respectively. Then by linearizing (5.4.1) about this equilibrium, we obtain the perturbation equations

$$\begin{cases} \dfrac{1}{P_r}\dfrac{\partial}{\partial t}\mathbf{v} = -\nabla p' + \Delta\mathbf{v} + R\theta\mathbf{k}, \\[2mm] [1+\delta(1-z)]\left(\dfrac{\partial}{\partial t}\theta - R\mathbf{v}\cdot\mathbf{k}\right) = \Delta\theta, \\[2mm] \nabla\cdot\mathbf{v} = 0, \end{cases} \tag{5.4.2}$$

while the boundary conditions at $z=0,1$ read (1.4.9) (Remark 1.4.1). Then the instability of (1.4.11) to normal mode perturbations (1.4.12) is governed by the two-point eigenvalue problem for the system, following from (5.4.2),

$$\begin{cases} (D^2-a^2)^3 W = -[1+\delta(1-z)]Ra^2 W, \\[2mm] W = D^2W = D^4W = 0, \quad \text{at} \quad z=0,1. \end{cases} \tag{5.4.3}$$

Let us write W as a Fourier series on the set $\{F_n\}_{n\in\mathbf{N}}$, $F_n(z)=\sqrt{2}\sin[(2n-1)\pi z]$ (total in the subspace of $L^2[0,1]$ of functions symmetric with respect to the $z=0.5$), i.e. $W=\sum_{n=1}^{\infty}W_{2n-1}F_{2n-1}(z)$, $z\in[0,1]$.

By using the backward integration technique [DiP61] we obtain

$$D^{2k+1}W(z) = \sum_{n=1}^{\infty}W_{2n-1}^{(2k+1)}E_{2n-1}(z), \quad D^{2k}W(z) = \sum_{n=1}^{\infty}W_{2n-1}^{(2k)}F_{2n-1}(z),$$

where $E_{2n-1}(z)=\sqrt{2}\cos[(2n-1)\pi z]$ and, for every $k\in\mathbf{N}$, by (5.7) and (5.8), we have $W_{2n-1}^{(2k+2)}=-(2n-1)n^2\pi^2 W_{2n-1}^{(2k)}$. The boundary conditions are automatically satisfied. Thus $(5.4.3)_1$ becomes

$$\sum_{n=1}^{\infty}A_n^3 W_{2n-1}^{(2k)}F_{2n-1}(z) = (1+\delta)Ra^2\sum_{n=1}^{\infty}W_{2n-1}^{(2k)}F_{2n-1}(z)-\delta Ra^2 z\sum_{n=1}^{\infty}W_{2n-1}^{(2k)}F_{2n-1}(z),$$

$$\tag{5.4.4}$$

where $A_n = a^2 + (2n-1)^2\pi^2$.

Consider the expansion $zF_{2n-1}(z)=\sum_{n=1}^{\infty}\alpha_{(2n-1)k}^{(2k)}F_{2n-1}(z)$, $z\in(0,1)$ where

$$\alpha_{(2n-1)k} = \begin{cases} \dfrac{1}{4} - \dfrac{1}{8(2n-1)^2\pi^2}, & k=2n-1 \\[4mm] \dfrac{2}{\pi^2}\dfrac{k(2n-1)}{\left((2n-1)^2-k^2\right)^2}, & k\neq 2n-1 \end{cases}$$

and impose the orthogonality of $F_{2n-1}(z)$ to (5.4.4), to obtain the system of linear algebraic equations in the Fourier coefficients W_{2n-1},

$$A_n^3 W_{2n-1} = (1+\delta)Ra^2 W_{2n-1} - \delta Ra^2\sum_{m=1}^{\infty}\alpha_{(2m-1)k}^{(2k-1)}W_{2m-1},$$

which, by denoting $b_{2n-1}=(2n-1)^3 W_{2n-1}$, becomes the relation

$$b_{2n-1} = \dfrac{(1+\delta)Ra^2}{A_n^3}b_{2n-1} - \dfrac{\delta Ra^2}{(2n-1)^3 A_n^3}\sum_{m=1}^{\infty}\dfrac{\alpha_{(2n-1)(2m-1)}}{(2m-1)^3}b_{2m-1},$$

that can be written as

$$(\delta_{(2n-1)(2m-1)} + a_{(2n-1)(2m-1)})b_{(2n-1)} = 0, \tag{5.4.5}$$

where

$$a_{(2n-1)(2m-1)} = \begin{cases} -\left(1 + \dfrac{3}{4}\delta + \dfrac{\delta}{8}\dfrac{1}{(2n-1)^2\pi^2}\right)\dfrac{Ra^2}{A_n^3}, & n = m \\[2ex] \dfrac{2\delta}{\pi^2}\dfrac{(2n-1)^4}{(2m-1)^2\big((2n-1)^2 - (2m-1)^2\big)^2}\dfrac{Ra^2}{A_n^3}, & n \neq m. \end{cases}$$

Introducing $\beta_n = \dfrac{2Ra^2}{n^2\pi^2}$ and $\gamma_m = \dfrac{1}{m^4\pi^4}$, we obviously have that $|a_{nm}| \leq \beta_n\gamma_m$ for all n, m.

Let

$$\Delta(N) = \det \begin{vmatrix} 1 + a_{11} & a_{12} & \cdots & a_{1N} \\ a_{21} & 1 + a_{22} & \cdots & a_{2N} \\ \cdots & \cdots & \cdots & \cdots \\ a_{N1} & a_{N2} & \cdots & 1 + a_{NN} \end{vmatrix} = 0$$

be the Cramer determinant associated with the system (5.4.5). Due to the fact that the series $\sum_{k=1}^{\infty} \beta_{2k-1}\gamma_{2k-1}$ is convergent, it follows [Ea] that $\lim_{N\to\infty} \Delta(N)$ is convergent too. Denote by $\sum_{(n,\infty)}$ the summation over n indexes $i_1, i_2, \ldots, i_n$ each one ranging from 1 to ∞. Then the previous limit is the sum of the absolutely convergent series (as $n \to \infty$)

$$\sum_{(n,\infty)} a_{i_1 i_1} + \frac{1}{2!} \sum_{(2,\infty)} \det \begin{vmatrix} a_{i_1 i_1} & a_{i_1 i_2} \\ a_{i_2 i_1} & a_{i_2 i_2} \end{vmatrix} + \cdots + \frac{1}{n!} \sum_{(n,\infty)} \det \begin{vmatrix} a_{i_1 i_1} & a_{i_1 i_2} & \cdots & a_{i_1 i_n} \\ a_{i_2 i_1} & a_{i_2 i_2} & \cdots & a_{i_2 i_n} \\ \cdots & \cdots & \cdots\cdots & \cdots \\ a_{i_n i_1} & a_{i_n i_2} & \cdots & a_{i_n i_n} \end{vmatrix} + \cdots$$

This series, truncated at the second term, gives an equation of the first order in the eigenvalue $\mathcal{R}$

$$1 - \left[\left(1 + \frac{3}{4}\delta\right) \sum_{n=1}^{\infty} \frac{a^2}{A_n^3} + \frac{\delta}{8} \sum_{n=1}^{\infty} \frac{a^2}{(A_n - a^2)A_n^3}\right]\mathcal{R} = 0. \tag{5.4.6}$$

For $\delta = 0$, (5.4.6) becomes

$$1 - \sum_{n=1}^{\infty} \frac{a^2}{A_n^3}\mathcal{R}_o = 0. \tag{5.4.7}$$

Taking into account the asymptotic expansions $\sum_{n=1}^{\infty} \frac{a^2}{A_n^3}$ and $\sum_{n=1}^{\infty} \frac{a^2}{(A_n - a^2)A_n^3}$ [GrR], from equations (5.4.6) and (5.4.7) we obtain the values of $\mathcal{R}_0$ and $\mathcal{R}(\delta)$, and, so, the secular curve $\mathcal{R}_0 = \mathcal{R}_0(a)$ and the secular surface $\mathcal{R} = \mathcal{R}(a, \delta)$ respectively.

From the first 2×2 minor of the infinite determinant, namely

$$\det \begin{vmatrix} 1 + a_{11} & a_{13} \\ a_{31} & 1 + a_{33} \end{vmatrix} = 1 + a_{11} + a_{33} + \det \begin{vmatrix} a_{11} & a_{13} \\ a_{31} & a_{33} \end{vmatrix},$$

we have the second order approximation of (5.4.7) in the form of a second degree equation in $\mathcal{R}$

$$(A_1 A_2 - B)\mathcal{R}^2 - (A_1 + A_2)\mathcal{R} + 1 = 0$$

where

$$A_1 = (1 + \frac{3\delta}{4} + \frac{\delta}{8\pi^2})\frac{a^2}{(\pi^2 + a^2)^3}, \quad A_2 = (1 + \frac{3\delta}{4} + \frac{\delta}{72\pi^2})\frac{a^2}{(\pi^2 + a^2)^3},$$

$$B = \frac{9}{2^{10}}\frac{\delta^2}{\pi^2}\frac{a^2}{(\pi^2 + a^2)^3}\frac{a^2}{(9\pi^2 + a^2)^3}$$

that satisfies the relation $A_1 A_2 > B$ and $A_1 > A_2$.

The minimum eigenvalue, *i.e.* the critical value, is then

$$\mathcal{R} = \frac{A_1 + A_2 - \sqrt{(A_1 - A_2)^2 + 4B}}{2(A_1 A_2 - B)},$$

that for $\delta = 0$ reduces to the classical value $\frac{(\pi^2 + a^2)^3}{a^2}$ [Chan], whence the values of $\mathcal{R}(\delta)$ and, so, two secular curves and surfaces respectively.

Numerical results [GeoLP] reveal a destabilizing effect of the bottom phenomena.

5.4.2 *Convection in a variable gravity field*

The presence of the variable gravity field $\mathbf{g}(z) = gr(z)\mathbf{k}$ produces differences in the buoyancy forces and, thus, part of the fluid layer tends to become unstable, the other stable. These differences in the local behavior of the mechanical equilibrium influence the emerging convection. As a result, the conducting-convection is governed by the N-S-F model for which we suppose that the O-B approximation does not hold. Consequently, in the case of the variable gravitation, the unique change in the equations of this model is that in (1.2.8) the body forces are now $\mathbf{g}(z)\alpha T$ instead of $\rho_0[1 - \alpha(T - T_0)\mathbf{k}]$. The nondimensional linearized equations in the normal mode perturbations read [Strau]

$$\begin{cases} (D^2 - a^2)^2 W = Rr(z)a^2\Theta, \\ (D^2 - a^2)\Theta = -RW. \end{cases} \tag{5.4.8}$$

By following [Drag06], have we take $r(z) = 1 - \epsilon kz$, $\epsilon \in [0, 1/k]$, $k \in \mathbf{N}^*$, hence for $z \in (0, 1)$, (5.4.8) becomes

$$\begin{cases} (D^2 - a^2)^2 W = R(1 - \epsilon kz)a^2\Theta, \\ (D^2 - a^2)\Theta = -RW, \end{cases} \tag{5.4.9}$$

while the boundary conditions at $z = 0, 1$ are

$$W = DW = \Theta = 0, \tag{5.4.10}$$

$$W = D^2 W = \Theta = 0, \tag{5.4.11}$$

for rigid and stress-free boundaries, respectively.

Remark 5.4.1. Problem (5.4.9), (5.4.11) is equivalent to (5.4.3), where $R(1+\delta)$ in (5.4.3) is equal to R in (5.4.9) and $\epsilon k = \delta(1+\delta)$. This enables us a quick comparison between the results in Sections 5.4.1 and 5.4.2.

Remark 5.4.2. An equivalent form for (5.4.9) reads

$$(D^2 - a^2)^3 \Theta = -Ra^2 r(z) N(z) \Theta, \tag{5.4.12}$$

showing that equation (5.4.12) is the same for the varying coefficient in the body forces, corresponding to $(5.4.9)_1$ or to the presence of heat sources, corresponding to (5.4.25) or to deep convection effects corresponding to $(5.4.2)_1$, these last two ones occurring in the internal energy equation.

First, following [Drag06], consider the problem (5.4.9), (5.4.10). Take for Θ the Fourier series $\sum_{m=1}^{\infty} C_m \sin(m\pi z)$, termwise satisfying the conditions $(5.4.10)_3$, and introduce it into $(5.4.9)_1$ to get

$$(D^2 - a^2)W = R(1 - \epsilon k z) \sum_{m=1}^{\infty} C_m \sqrt{2} \sin(m\pi z). \tag{5.4.13}$$

Remark 5.4.3. The expansion is performed for that unknown function Θ the coefficient of which is variable and this expansion is introduced just in that equation containing the variable coefficient.

The affine equation in W (5.4.13) has the solution, obtained by Lagrange method of variable coefficients,

$$W(z) = \sum_{m=1}^{\infty} \Big\{ A_1^m \sinh(az) + A_2^m \cosh(az) + A_3^m z \sinh(az) + A_4^m z \cosh(az)$$

$$+ Ra^2 C_m G_m^{-2} \big[(1 - \epsilon k z) \sin(m\pi z) + 4m\pi\epsilon k G_m^{-1} \cos(m\pi z) \big] \Big\}, \tag{5.4.14}$$

where $G_m = m^2 \pi^2 + a^2$. The sum of the four terms, the coefficients of which are $A_{1,2,3,4}^m$, form the general solution of the linear equation obtained from (5.4.13) by neglecting the right-hand side. The remaining part in W is a particular solution of (5.4.13), which can be alternatively obtained, in the simple case of (5.4.13), by assuming, for each m, that it is of the form $L_1(1 - \epsilon k z) \sin(m\pi z) + L_2 \cos(m\pi z)$ and then determining the constants L_1 and L_2 such that this expression satisfy (5.4.13).

Imposing to (5.4.14) to satisfy the boundary conditions $(5.4.10)_{1,2}$, four equations in $A_{1,2,3,4}^m$ are obtained. Solving them we find that $A_{1,2,3,4}^m$ are proportional to C_m, namely

$$A_1^m D_m^{-1} = [a + (-1)^m \sinh a] G_m + \epsilon k \Big\{ -4(a + \cosh a \sinh a) + (-1)^m \big[4(a \cosh a + \sinh a)$$

$$-G_m \sinh a]\Big\}, \quad A_2^m D_m^{-1} = 4\epsilon k(\sinh^2 a - a^2),$$

$$A_3^m D_m^{-1} = [-a + \cosh a \sinh a - (-1)^m(\sinh a - a\cosh a)]G_m + \epsilon k\Big\{-4a\sinh^2 a + (-1)^m$$

$$[4a^2 \sinh a + (\sinh a - a\cosh a)G_m]\Big\},$$

$$A_4^m D_m^{-1} = (-\sinh^2 a - (-1)^m a\sinh a)G_m + \epsilon k\Big\{4a(a + \cosh a \sinh a) + (-1)^m$$

$$[4a(\sinh a + a\cosh a) + a\sinh a]\Big\},$$

where $D_m = Ra^2 m\pi C_m G_m^{-3}(\sinh^2 a - a^2)$, a notation slightly different from that in [Drag06].

In this way, the Fourier coefficients of W are proportional to the Fourier coefficients of Θ. Substituting the Fourier series for W and Θ into $(5.4.9)_2$ and imposing to the obtained equation to be orthogonal to the expansion functions, we get the infinite system of linear algebraic equations in C_m

$$\sum_{m=1}^{\infty} 0.5 C_m G_m \delta_{nm} = Rn\pi G_n^{-1} \sum_{m=1}^{\infty} \Big\{ A_1^m(-1)^{n+1} \sinh a$$

$$+ A_2^m[1 + (-1)^{n+1}\cosh a] + A_3^m[(-1)^{n+1}\sinh a$$

$$- 2a[1 + (-1)^{n+1}\cosh a]G_n^{-1}] + A_4^m[(-1)^{n+1}\cosh a \qquad (5.4.15)$$

$$+ 2a(-1)^n \sinh a G_n^{-1}]\Big\}$$

$$+ Ra^2 \sum_{m=1}^{\infty} C_m G_m^{-2}[\delta_{nm}/2 + \epsilon k T_{mn} + 4\epsilon k m\pi G_m^{-1} U_{mn}],$$

where δ_{nm} is the Kronecker δ,

$$T_{nm} = \begin{cases} 0.25, & \text{if} \quad n = m \\ 2mn[(-1)^{m+1} - 1]\pi^{-2}(n^2 - m^2)^{-1}, & \text{if} \quad m \neq n \end{cases}$$

$$U_{nm} = \begin{cases} 0, & \text{if} \quad n = m \\ n[(-1)^{m+n} - 1]\pi^{-1}(n^2 - m^2)^{-1}, & \text{if} \quad m \neq n. \end{cases}$$

The condition that the Cramer determinant of (5.4.15) vanishes provides the secular equation.

For the stress-free boundaries the general form of W and of the secular equation are the same as in the rigid case but the expressions for $A_{1,2,3,4}^m$ are simpler, namely

$$A_1^m D_m^{-1} = -\epsilon k[(-1)^{m+1} + \cosh a][G_m(-1)^{m+1} + 4a\sinh a]a^{-1}\sinh^{-2} a,$$

$$A_2^m D_m^{-1} = 4\epsilon k,$$

$$A_3^m D_m^{-1} = -\epsilon k G_m/a, \quad A_4^m D_m^{-1} = \epsilon k G_m a^{-1}\sinh^{-1} a\big((-1)^{m+1} + \cosh a\big),$$

where $D_m = Ra^2 m\pi C_m G_m^{-3}$. Further the method proceeds along the above lines [Drag06].

The method applied in the above to solve this type of eigenvalue problems was applied in [DiP59] and then substantiated in [DiPS65].

Remark 5.4.4. In these two approaches the boundary conditions occur only in the expressions of $A^m_{1,2,3,4}$, all other expressions depend only on the equations.

In an alternative approach [Chan], the problem (5.4.9), (5.4.10) is written in Θ only in the form

$$(D^2 - a^2)^3\Theta = -R^2 a^2(1 - \epsilon kz)\Theta, \qquad (5.4.16)$$

$$\Theta = (D^2 - a^2)\Theta = D(D^2 - a^2)\Theta = 0, \quad \text{at } z = 0, 1. \qquad (5.4.17)$$

Then a new function $\Psi = -R^2 a^2\Theta$ is introduced to transform (5.4.16) into

$$(D^2 - a^2)^3\Theta = (1 - \epsilon kz)\Psi, \qquad (5.4.18)$$

$$\Psi = -R^2 a^2\Theta. \qquad (5.4.19)$$

From now on, the first approach applies: since $\Psi(0) = \Psi(1) = 0$, we choose the expansion $\Psi(z) = \sum_{n=1}^{\infty} \Psi_m \sin(m\pi z)$ and take $\Theta(z) = \sum_{n=1}^{\infty} \Psi_n \Theta_n(z)$, where, by (5.4.18),

$$(D^2 - a^2)^3\Theta_m = (1 - \epsilon kz)\sin(m\pi z), \qquad (5.4.20)$$

implying

$$\Theta_m(z) = -G_m^{-3}\Big[(A^m_0 + A^m_1 z + A^m_2 z^2)\cosh az + (B^m_0 + B^m_1 z + B^m_2 z^2)\sinh az$$

$$+ (1 - \epsilon kz)\sin(m\pi z) - 6\epsilon km\pi G_m^{-1}\cos(m\pi z)\Big].$$

$$(5.4.21)$$

Introducing (5.4.21) in (5.4.17) we get the system in the coefficients in (5.4.21), namely

$$\begin{cases} A^m_0 = 6m\pi\epsilon k G_m^{-1}, \\ (A^m_0 + A^m_1 + A^m_2)\cosh a + (B^m_0 + B^m_1 + B^m_2)\sinh a = (-1)^m 6\epsilon km\pi G_m^{-1} \\ A^m_2 + aB^m_1 = -2\epsilon km\pi, \\ A^m_1 a\sinh a + A^m_2(\cosh a + 2a\sinh a) + B^m_1 a\cosh a + B^m_2(\sinh a + 2a\cosh a) \\ \quad = (-1)^{m+1} 2\epsilon km\pi, \\ A^m_1 a^2 + B^m_2 3a^2 = 0.5m\pi G_m, \\ A^m_1 a^2 \cosh a + A^m_2(2a^2\cosh a + 3a\sinh a) + B^m_1 a^2\sinh a \\ \quad + B^m_2(2a^2\sinh a + 3a\cosh a) \\ \quad = 0.5(-1)^m(1 - \epsilon k)m\pi G_m. \end{cases}$$

Substituting these expressions in (5.4.21) it follows that $\Theta_m(z)$ are determined in terms of a and ϵk. As a result, the coefficients of $\Theta(z)$ are expressed in terms of those of Ψ. Introducing the series for Ψ and the obtained series for Θ in (5.4.19) and imposing to the resulted equation to be orthogonal to the expansion functions, an infinite system of linear algebraic equations in the Fourier coefficients Ψ_n is obtained. The requirement that its Cramer determinant vanish leads to the secular equation

$$\det \left\| -0.5\delta_{nm} + Ra^2 G_m^{-3} \left[A_0^m J_{01}^k + A_1^m J_{11}^k + A_2^m J_{21}^k + B_0^m J_{02}^k + B_1^m J_{12}^k + B_2^m J_{22}^k + 0.5\delta_{nm} \right. \right.$$

$$\left. \left. -\epsilon k T_{nm} + 6\epsilon k m\pi G_m^{-1} U_{mn} \right] \right\| = 0,$$

where $J_{ij}^k = \int_0^1 z^i \cosh(kaz)\sin(\pi z)dz$ if $j = 1$ and $J_{ij}^k = \int_0^1 z^i \sinh(az)\sin(k\pi z)dz$ if $j = 2$ and $i = 0, 1, 2$.

For $\epsilon k = 0$, numerical computations based on the secular equations presented in this section reveal a very good agreement of the critical Reynolds number to the classical value. For $\epsilon k \neq 0$, the neutral curves obtained by both approaches are, practically, the same and they coincide with the neutral curves given in Section 5.4.1 [Drag06].

5.4.3 *Penetrative convection*

This motion emerges due to the internal instability of part of the fluid and its penetration into an adjacent stable part. Penetrative convections occur largely in engineering and geophysics. They have important consequences on the control of biological activities in oceans and lakes, in the dispersion of pollutants in the environment in astrophysics, the prediction of smoke motion inside the buildings etc. In the following we consider the penetrative convection caused by a radiation heating at the lower boundary of a horizontal layer of thickness d, and by a density which is quadratic in the temperature field, *i.e.* $\rho = \rho_m[1 - \alpha(T - T_m)^2]$. The governing dimensional N-S-F equations are

$$\begin{cases} \dfrac{\partial}{\partial t}\mathbf{u} + \mathbf{u}\cdot\nabla\mathbf{u} = -\dfrac{1}{\rho_m}\nabla p + \nu\Delta\mathbf{u} - \alpha g\mathbf{k}[1 - \alpha(T - T_m)^2], \\[2mm] \nabla\cdot\mathbf{u} = 0, \\[2mm] \dfrac{\partial}{\partial t}T + \mathbf{u}\cdot\nabla T = k\Delta T, \end{cases} \qquad (5.4.22)$$

and the boundary conditions read

$$\mathbf{u} = \mathbf{0} \text{ at } z = 0 \text{ and } z = d, \qquad \frac{d}{dz}T = \gamma,\ T = T_0 \text{ at } z = 0, \qquad T = T_1 \text{ at } z = d, \qquad (5.4.23)$$

where ρ_m and T_m are the maximum value of the density and the corresponding temperature respectively [Strau] and α is a constant. The basic equilibrium is

$$m_0 = \left\{ \mathbf{u} = 0, T = T_1 - \gamma(d - z), p = -\rho_m g z \int_0^z \left[1 - \alpha\big(T(s) - T_m\big)^2 \right]ds, \right.$$

$$\left. \gamma = (T_1 - T_0)d^{-1} \right\}. \qquad (5.4.24)$$

The heat flux through the lower boundary is γk.

The nondimensional normal mode perturbations satisfy the two-point problem

$$\begin{cases} (D^2 - a^2)^2 W = -2Ra^2(\zeta - z)\Theta, \\ (D^2 - a^2)\Theta = RW, \end{cases} \tag{5.4.25}$$

$$W = DW = 0 \text{ at } z = 0,1; \qquad \Theta = 0 \text{ at } z = 1; \qquad D\Theta = 0 \text{ at } z = 0, \tag{5.4.26}$$

where $\zeta = 1 + (T_m - T_0)/(T_1 - T_0)$; W and Θ are the intensities of the vertical component of the perturbation $\mathbf{v}$ and perturbation temperature θ respectively and they depend on z only.

In [Drag07a], this problem is solved by using the first method from Section 5.4.2. and $\{\sqrt{2}\cos[\frac{(2m+1)\pi z}{2}]\}$ as the set total in the subspace of $L^2(0,1)$ of functions g satisfying the conditions $Dg(0) = 0$, $g(1) = 0$. $\big($Indeed, $\cos[\frac{(2m+1)\pi z}{2}]$ are the eigenfunctions of the positive symmetric operator $-D^2$ (Appendix 5).$\big)$ Therefore, the Fourier series expansion $\Theta = \sum_{n=1}^{\infty} C_m \sqrt{2}\cos[\frac{(2m+1)\pi z}{2}]$ is used, so the boundary conditions for Θ are automatically satisfied. Introducing it into $(5.4.25)_1$ it follows

$$W(z) = \sum_{m=0}^{\infty} \Big\{ A_1^m \cosh(az) + A_2^m \sinh(az) + A_3^m z \cosh(az) + A_4^m z \sinh(az)$$

$$+ D_m Ra^2 \Big[32\sqrt{2}(z - \zeta)\cos[\frac{(2m+1)\pi z}{2}] - 256\sqrt{2}(2m+1)\pi G_m^{-1}\sin[\frac{(2m+1)\pi z}{2}] \Big] \Big\}$$

and taking into account the boundary conditions we obtain

$$\begin{cases} A_1^m = 32\sqrt{2}Ra^2\zeta D_m, \\ A_1^m \cosh a + A_2^m \sinh a + A_3^m \cosh a + A_4^m \sinh a \\ \quad = (-1)^m 256\sqrt{2}Ra^2(2m+1)\pi D_m G_m^{-1}, \\ aA_2^m + A_3^m = 32\sqrt{2}Ra^2[3(2m+1)^2\pi^2 - 4a^2]D_m G_m^{-1}, \\ aA_1^m \sinh a + aA_2^m \cosh a + A_3^m(\cosh a + a\sinh a) + A_4^m(\sinh a + a\cosh a) \\ \quad = (-1)^m 16\sqrt{2}Ra^2(2m+1)\pi(1 - \zeta)D_m, \end{cases}$$

where $G_m = (2m+1)^2\pi^2 + 4a^2$ and $D_m = C_m G_m^{-2}$, whence the system in D_m

$$\sum_{m=0}^{\infty} RA_1^m\sqrt{2}\frac{4a(-1)^n\sinh a + \pi(2n+1)}{G_n} + RA_2^m\sqrt{2}\frac{4a(-1)^n\cosh a}{G_n}$$

$$+ RA_3^m\sqrt{2}\frac{4\pi^2(-1)^n(2n+1)^2(\cosh a + a\sinh a) + 16a^2(-1)^n(a\sinh a - \cosh a)}{G_n^2}$$

$$+ RA_4^m\sqrt{2}\Big\{ \frac{4\pi^2(-1)^n(2n+1)^2(\sinh a + a\cosh a) + 16a^2(-1)^n(a\cosh a - \sinh a)}{G_n^2}$$

$$- \frac{16a\pi(2n+1)}{G_n^2} \Big\} + \frac{32\sqrt{2}R^2a^2 C_m}{G_m^2}X_{mn} - \frac{32\sqrt{2}R^2a^2\zeta C_m}{G_m^2}T_{mn}$$

$$-\frac{256\sqrt{2}R^2 a^2 C_m(2m+1)\pi}{G_m^3}\delta_{mn} + \frac{C_m}{2}G_m T_{mn} = 0,$$

where

$$X_{mn} = \begin{cases} \dfrac{1}{(2m+1)\pi} & \text{if } m = n, \\[2ex] \dfrac{(-1)^{m+n}(2m+1)}{\pi(m+n+1)(m-n)} & \text{if } m \neq n, \end{cases}$$

$$T_{mn} = \begin{cases} \dfrac{1}{(2m+1)\pi} & \text{if } m = n \\[2ex] \dfrac{(-1)^{m+n}(2m+1) - 2n - 1}{\pi(m+n+1)(m-n)} & \text{if } m \neq n, \end{cases}$$

leading to the secular equation [Drag07a]. Numerical computations are found to agree with the existing ones.

5.4.4 Convection with a heat source in a rigid box

Due to their important applications, the thermal convections in rigid boxes were continuously theoretically and numerically starting with the years '60 of the past century. For the case of different boundary conditions imposed on the lower and upper surfaces, we quote [Robe1a], [Tri]. For the presence of a uniform heat source we mention the numerical study [Velt]. For the case of both upper and lower rigid heat conducting walls, following [Drag07] here we investigate the mathematical model from [Velt] by using the method presented in Section 5.4.2 based on the equation in the perturbation temperature.

The conduction state in a horizontal layer of a viscous incompressible fluid with constant viscosity and thermal conductivity coefficients ν and χ is governed by the heat and hydrostatic transfer equations

$$-\frac{d}{dz}q_3 = \chi\frac{\partial^2}{\partial z^2}T_o, \tag{5.4.27}$$

$$\frac{d}{dz}p_0 = -\rho_0 g, \tag{5.4.28}$$

where $\eta = const$ is the heating rate, θ, p_0 and ρ_0 are the potential temperature, pressure and density at the bottom respectively. Assume a constant potential temperature difference between the lower and the upper boundaries $\Delta T = T_0 - T_1$, whence the potential temperature distribution $T_1 - T_0 = -\frac{\Delta T}{H}\left(z + \frac{H}{2}\right) + \frac{\eta}{2\chi}\left[z^2 - \left(\frac{H}{2}\right)^2\right]$.

In nondimensional coordinates the corresponding linearized perturbation N-S-F model reads

$$\begin{cases} \dfrac{d}{dt}\mathbf{v} = \nabla p + \Delta\mathbf{v} + G_r\theta\mathbf{k}, \\[2ex] \nabla \cdot \mathbf{v} = 0, \\[2ex] \dfrac{d}{dt}\theta = (1 - Nz)\mathbf{v}\mathbf{k} + Pr^{-1}\Delta\theta, \end{cases} \tag{5.4.29}$$

$$\mathbf{v} = \theta = 0, \quad \text{at} \quad z = \pm 0.5, \tag{5.4.30}$$

where N is a nondimensional parameter characterizing the heating (cooling) rate of the layer. For normal modes $(1.4.12')$ we can write $(5.4.29)$ and $(5.4.30)$ in W and Θ

$$\begin{cases} (D^2 - a^2)^2 W - a^2 G_r \Theta = 0, \\ (D^2 - a^2)\Theta + Pr(1 - Nz)W = 0, \end{cases} \tag{5.4.31}$$

$$W = DW = \Theta = 0, \quad \text{at} \quad z = \pm 0.5, \tag{5.4.32}$$

where $a^2 = m^2 + n^2$ and W and Θ are the intensity of the perturbation vertical velocity and temperature.

By the translation $x = z + 0.5$, denoting $N_1 = 1 + 0.5N$, and eliminating W between $(5.4.31)_{1,2}$, we obtain the two-point eigenvalue in Θ

$$(D^2 - a^2)^3 \Theta + Ra(N_1 - Nx)\Theta = 0, \tag{5.4.33}$$

$$\Theta = (D^2 - a^2)\Theta = D(D^2 - a^2)\Theta = 0 \quad \text{at} \quad x = 0, 1. \tag{5.4.34}$$

Introduce a new function Ψ to obtain the system in the unknown functions Θ and Ψ

$$\begin{cases} (D^2 - a^2)\Theta = (N_1 - Nx)\Psi, \\ \Psi = -Ra^2 \Theta, \end{cases} \tag{5.4.35}$$

to which the second (Chandrasekhar) approach from Section 5.4.2 is applied. Thus, taking into account the boundary conditions $(5.4.34)$, we set $\Psi = \sum_{m=1}^{\infty} \Psi_k \sin k\pi x$, $\Theta = \sum_{m=1}^{\infty} \Psi_k \Theta_k$, where Θ_k satisfies the equation $(D^2 - a^2)^3 \Theta_k = (N_1 - Nx)\sin k\pi x$, and, so,

$$\Theta_k = -\frac{1}{(k^2\pi^2 + a^2)^3}\Big[(A_0^k + A_1^k x + A_2^k x^2)\cosh(ax) + (B_0^k + B_1^k x + B_2^k x^2)\sinh(ax)$$

$$+(N_1 - Nx)\sin(k\pi x) - \frac{6k\pi N}{(k^2\pi^2 + a^2)}\cos(k\pi x)\Big].$$

Imposing the boundary conditions $(5.4.34)$, we obtain the following algebraic system in $A_0^k, A_1^k, A_2^k, B_0^k, B_1^k, B_2^k$

$$A_0^k = \frac{6k\pi N}{(k^2\pi^2 + a^2)},$$

$$(A_0^k + A_1^k + A_2^k)\cosh a + (B_0^k + B_1^k + B_2^k)\sinh a = \frac{6k\pi N}{(k^2\pi^2 + a^2)}(-1)^k,$$

$$A_2^k + aB_1^k = -2k\pi N,$$

$$a\sinh a A_1^k + A_2^k(\cosh a + 2a\sinh a) + a\cosh a B_1^k + B_2^k(\sinh a + 2a\cosh a) = 2k\pi N(-1)^{k+1},$$

$$a^2 A_1^k + 3aB_2^k = \frac{N_1}{2}k\pi(k^2\pi^2 + a^2),$$

$$a^2 \cosh a A_1^k + A_2^k(2a^2\cosh a + 3a\sinh a) + a^2\sinh a B_1^k + B_2^k(3a\cosh a + 2a^2\sinh a)$$

$$= \frac{N_1 - N}{2}k\pi(-1)^k(k^2\pi^2 + a^2).$$

Replacing the obtained solution Θ in equation $(5.4.35)_2$ and imposing the condition that the obtained equation be orthogonal to the functions $\sin(l\pi z)$, $l \in \mathbf{N}$ in $L^2(0,1)$, we get an infinite system of algebraic linear equations in Ψ_k, whence,

$$\det\|\frac{\delta_{kl}}{2} - \frac{a^2 Ra}{(k^2\pi^2 + a^2)}\left[A_0^k I_{01}^l + A_1^k I_{11}^l + A_2^k I_{21}^l + B_0^k I_{02}^l + B_1^k I_{12}^l + B_2^k I_{22}^l\right.$$

$$\left.+N_1\frac{\delta_{kl}}{2} - NT_{kl} + \frac{6k\pi N}{(k^2\pi^2 + a^2)}U_{kl}\right]\| = 0,$$

where

$$T_{kl} = \begin{cases} 1/4, & \text{if } k = l, \\ \dfrac{2kl[(-1)^{k+l} - 1]}{\pi^2(l-k)^2(l+k)^2}, & \text{if } k \neq l, \end{cases} \qquad U_{kl} = \begin{cases} 0, & \text{if } k = l, \\ \dfrac{l[(-1)^{k+l} - 1]}{\pi(l-k)(l+k)}, & \text{if } k \neq l, \end{cases}$$

$$I_{ij}^l = \begin{cases} \displaystyle\int_0^1 x^i \cosh(ax)\sin(l\pi x)dx, & \text{if } j = 1 \\ \displaystyle\int_0^1 x^i \sinh(ax)\sin(l\pi x)dx, & \text{if } j = 2, i = 0,1,2. \end{cases}$$

Keeping some parameters fixed, the neutral manifolds follow.

For $k = l = 1$ the neutral curve

$$Ra = \frac{(\pi^2 + a^2)^5(a + \sinh a)}{a^2[(a + \sinh a)(\pi^2 + a^2)^2 - 8a\pi^2(1 + \cosh a)]}$$

corresponds and it does not contain the parameter N. In [Dra07] numerical computations are performed for the second and the third approximation, *i.e.* $k = l = 2$ and $k = l = 3$ respectively, revealing the destabilizing influence of the heat source. For $N = 0$ a very good value for the critical Rayleigh number and the wave number are regained.

Chapter 6

Variational methods applied to linear stability

The involvement of three types of variational methods is presented. In the first, (Sections 6.1, 6.2) a (non-stationary) energy method is used to obtain a criterion expressed in terms of a minimum of a functional. With this functional the Euler-Lagrange equations are associated and the two-point problem for them is solved by the B-D method. As the involved functional represents only part of the rate of change of energy, its minimum value provides only a stability bound and not a limit of linear stability. Apart from these, the matrix differential operator associated with the Euler-Lagrange equations may happen to be non-symmetrizable. As a consequence, they give only a necessary condition for the minimum.

Due to the form of the boundary conditions and the presence of the basic conduction state, no variational principle of the type contained in Theorem 3.2.2 is expected to hold for the general stationary magnetohydrodynamic model corresponding to (1.4.17) when the Hall and ion-slip effects are present, or to (1.4.16) when only the Hall current is present. For the finite-dimensional models, derived from these for normal mode perturbations, we do not expect either that a variational principle holds, except for specific situations (Example 3.3.8). However, in quite particular cases, the validity of this principle can be proved (Sections 3.4, 3.5 and 6.4).

Yet in the finite-dimensional stationary case, for most particular models derived from (1.4.17) the associated operators are not symmetrizable mainly due to the boundary conditions which are different in the corresponding direct and adjoint two-point problems. Nevertheless, in these cases, usually variational principles can be shown to hold (Sections 3.4, 3.5 and 6.4) but they are expressed in terms of the given problem and its adjoint.

In Section 6.1 we consider a boundary-value problem for the linearized magnetohydrodynamic equations with a Hall current with boundary conditions different from those in Section 5.2. More exactly, here the bounding horizontal surfaces are rigid. This case is rarely treated in the literature because the standard expansion functions $\sin(n\pi z)$ are no longer appropriate and those used by Chandrasekhar turned out to yield too complicated expressions. Therefore, this is one of the cases where the B-D method is required and,

247

indeed, is successfully applied. More precisely, a preliminary reformulation of the perturbation problem as a vector integro-differential equation enables us to apply the energy method, to associate this equation with a variational problem and this one with the Euler-Lagrange equations. To these last equations the B-D method is applied. The destabilizing effect of the Hall current is regained. In this section, by the Lyapunov direct method, we study the linear Lyapunov stability of the conduction-diffusion solution of the anisotropic magnetic Bénard problem, for a fully ionized fluid in the case of stress-free boundaries. We introduce a sesquilinear Hermitian functional from which we derive the appropriate Lyapunov function and we regain the critical curves of the linear instability by applying the classical normal modes technique to the eigenvalue perturbation problem derived from the two-point problem governing the linear instability.

Section 6.3 is inserted in view of its relevance in applying energy method. It provides stability criteria for quasi-geostrophic zonal flows in the presence of lateral diffusion and bottom dissipation of the vertical vorticity.

Three general ideas for applying the energy relation are pointed out and exemplified.

The problem of the stability of the thermodiffusive equilibrium for the magnetic and nonmagnetic Bénard problem was widely investigated in both linear and nonlinear cases [Chan], [J76], [MuloR89], [MuloR94]. Thus, in [EbS1], [EbS2], [Geo85], [GeoPalPasB], [GeoPal96c], [MaiP84], [MaiPL], [Pal97], [SharR], [SharS], [SharT] the case of a fluid in the presence of a Hall current is considered.

In order to apply correctly the direct method to solve the two-point problem in linear anisotropic M.H.D. in the presence of Hall and ion-slip currents, in [GeoPal96c] false secular manifolds among the bifurcation characteristic manifolds are detected. In this case the catastrophe theory could not be used due to the high number of parameters. In [SolM] theorems of existence and uniqueness for M.H.D. linearized equations with Hall and ion slip currents are established. In [EbS1] the linear stability of a toroidal plasma with Hall current is studied, and in [EbS2] a linearization principle is obtained. Continuous dependence theorems are derived in [MuloS88]. The linear instability of the thermodiffusive equilibrium for the magnetic Bénard problem in the case of compressible fully ionized fluids through a porous medium is studied in [SharR], [SharS], [SharT] and necessary conditions for the existence of overstability are deduced in [SharS].

The Rayleigh-Taylor instability problem of a viscous plasma in the presence of a Hall current is investigated in [AbS] while in [MaiP84], [MaiPL] the linear instability of the thermodiffusive equilibrium for a partially ionized fluid in a free horizontal layer heated from below is considered. In [MaiP84] the exact solution of the characteristic value problem could be found by means of the C-G method due to the convenient boundary conditions.

6.1 Magnetic Bénard problem with Hall effect

In Section 6.1.1 the initial boundary-value problem for the system of linear pde's governing the evolution of perturbations for the Bénard magnetic problem with Hall effect is rewritten as a system associated with an integro-differential linear matrix operator, in a class convenient to the application of the energy method, leading to a variational problem with which the Euler-Lagrange equations are associated (Section 6.1.2). Due to the form of the boundary conditions the two-point problem for these equations is solved by the B-D method and the secular equation is obtained.

6.1.1 *Reformulation of the evolution equations of perturbations as an integro-differential equation*

Consider, following [PalG03], the problem $(1.4.16')$, $(1.4.5)$, $(1.4.7')$ governing the evolution of the perturbations around the conduction state $(1.4.2)$ of a homogeneous thermoelectrically conducting fluid with tensorial electrical conductivity, situated in a horizontal layer in which a constant vertical temperature gradient is maintained, in the presence of a uniform vertical magnetic field and in the O-B approximation. The layer is bounded by the planes $\pi_0 : z = 0$ and $\pi_1 : z = 1$, both rigid, thermally and electrically conductors.

Assume that the perturbation fields have the form $(1.4.12)$, *i.e.* they are doubly periodic of period $2\pi/\alpha_x$ and $2\pi/\alpha_y$ in x and y direction respectively and let us use the variables $\mathbf{k} \cdot \mathbf{u}$, $\mathbf{k} \cdot \nabla \times \mathbf{u}$, $\mathbf{k} \cdot \mathbf{h}$, $\mathbf{k} \cdot \nabla \times \mathbf{h}$ suitable to the formulation of the representation theorem for solenoidal fields in a plane layer [J76]. Denote by $\Omega = [0, \frac{2\pi}{\alpha_x}] \times [0, \frac{2\pi}{\alpha_y}] \times [0, 1]$ the periodicity cell and let $\partial\Omega$ be its boundary. Thus, equations $(1.4.16')$, linearized around the equilibrium m_0, become

$$\frac{\partial}{\partial t}w = -\frac{\partial}{\partial z}p + P_m\Delta w + P_m M^2\frac{\partial}{\partial z}h_3 + \mathcal{R}\frac{P_m^2}{P_r}\theta,$$

$$\frac{\partial}{\partial t}h_3 = \frac{\partial}{\partial z}w + \Delta h_3 - \beta_H\frac{\partial}{\partial z}j,$$

$$\frac{\partial}{\partial t}\zeta = P_m\Delta\zeta + P_m M^2\frac{\partial}{\partial z}j, \tag{6.1.1}$$

$$\frac{\partial}{\partial t}j = \frac{\partial}{\partial z}\zeta + \Delta j + \beta_H\frac{\partial}{\partial z}\Delta h_3,$$

$$\frac{\partial}{\partial t}\theta = w + \frac{P_m}{P_r}\Delta\theta,$$

where $w = \mathbf{k} \cdot \mathbf{u}$, $h_3 = \mathbf{k} \cdot \mathbf{h}$, $\zeta = \mathbf{k} \cdot \nabla \times \mathbf{u}$, $j = \mathbf{k} \cdot \nabla \times \mathbf{h}$. Equations $(6.1.1)_3$ and $(6.1.1)_4$ are obtained by applying [Chan] the operator $\nabla\times$ to the equations $(1.4.16)_1$ and $(1.4.16)_2$ respectively. The boundary conditions obtained by linearizing $(1.4.7')$ are

$$\frac{\partial^2}{\partial z^2}\theta = \frac{\partial}{\partial z}w = h_3 = \Delta h_3 = \frac{\partial}{\partial z}j = \zeta = \theta = 0 \qquad \text{at} \qquad z = 0, 1. \tag{6.1.2}$$

We also use the boundary conditions [Chan]

$$\frac{\partial}{\partial x}w = \frac{\partial}{\partial y}w = \frac{\partial}{\partial x}\theta = \frac{\partial}{\partial y}\theta = 0 \qquad \text{at} \qquad z = 0, 1. \tag{6.1.3}$$

For an easier mathematical handling, instead of differential system (6.1.1) we use the following system in the vector unknown function $\mathbf{U} \equiv (\nabla_1 w, \frac{\partial}{\partial z} w, h'_3, \frac{\partial}{\partial z} j, \zeta, \nabla_1 \theta)$

$$\frac{\partial}{\partial t} \nabla_1 w = -\frac{\partial}{\partial z} \nabla_1 p + P_m \Delta \nabla_1 w + P_m M^2 \frac{\partial}{\partial z} \nabla_1 h_3 + \mathcal{R} \frac{P_m^2}{P_r} \nabla_1 \theta,$$

$$\frac{\partial}{\partial t} \frac{\partial}{\partial z} w = -\frac{\partial^2}{\partial z^2} p + P_m \Delta \frac{\partial}{\partial z} w + P_m M^2 \frac{\partial^2}{\partial z^2} h_3 + \mathcal{R} \frac{P_m^2}{P_r} \frac{\partial}{\partial z} \theta,$$

$$\frac{\partial}{\partial t} h'_3 = \Delta \frac{\partial}{\partial z} w + \Delta h'_3 - \beta_H \Delta \frac{\partial}{\partial z} j,$$

$$\frac{\partial}{\partial t} \zeta = P_m \Delta \zeta + P_m M^2 \frac{\partial}{\partial z} j, \tag{6.1.4}$$

$$\frac{\partial}{\partial t} \frac{\partial}{\partial z} j = \frac{\partial^2}{\partial z^2} \zeta + \Delta \frac{\partial}{\partial z} j + \beta_H \frac{\partial^2}{\partial z^2} h'_3,$$

$$\frac{\partial}{\partial t} \nabla_1 \theta = \nabla_1 w + \frac{P_m}{P_r} \Delta \nabla_1 \theta,$$

where $h'_3 = \Delta h_3$ and $\nabla_1 = \frac{\partial}{\partial x}\mathbf{i} + \frac{\partial}{\partial y}\mathbf{j}$.

The system (6.1.4) is an integro-differential one. Indeed, $h_3 = h_3^* + h_\perp$, where h_3, h_3^* and $h_\perp$ are the unique solutions of the following Dirichlet problems for Laplace and Poisson equations

$$\begin{cases} \Delta h_3 = h'_3, \\ h_{3\|\partial\Omega} = h_{3w}, \end{cases} \qquad \begin{cases} \Delta h_3^* = 0, \\ h_{3\|\partial\Omega}^* = h_{3w}, \end{cases} \qquad \begin{cases} \Delta h_\perp = h'_3, \\ h_{\perp\|\partial\Omega} = 0, \end{cases}$$

where h_{3w} represents the value of h_3 on $\partial\Omega$. In fact, the boundary conditions $(6.1.2)_4$ imply $h_{3w} = 0$ on the planes $z = 0, 1$ while for the lateral parts of $\partial\Omega$ in general h_3 is not vanishing. It is immediate that in the class of functions vanishing on the $\partial\Omega$ the operator Δ is invertible whence

$$h_\perp = \Delta^{-1} h'_3, \qquad \Delta^{-1} h'_{3\|\partial\Omega} = 0$$

and therefore $h_3 = h_3^* + \Delta^{-1} h'_3$. Since the integral representation of harmonic functions implies that h_3^* is a linear function of h_{3w}, it follows that h_3 is a linear function of h'_3 [Mikh5]. Expressing $\mathbf{k} \cdot \frac{\partial}{\partial t}(\nabla \times \nabla \times \mathbf{u})$ firstly as the third component of $\nabla \times \nabla \times (1.4.16')_1$ and secondly as $\nabla_1(6.1.4)_1 + \frac{\partial}{\partial z}(6.1.4)_2$ we have

$$\mathcal{R} \frac{P_m^2}{P_r} \frac{\partial^2}{\partial z^2} \theta - \Delta \frac{\partial}{\partial z} p = 0. \tag{6.1.5}$$

In order to eliminate the pressure from (6.1.4), by using (6.1.5) we express it in terms of θ and $\nabla_1 \theta$. Namely $\frac{\partial}{\partial z} p = \frac{\partial}{\partial z} p^* + \frac{\partial}{\partial z} p_\perp$, where

$$\begin{cases} \Delta \frac{\partial}{\partial z} p = \mathcal{R} \frac{P_m^2}{P_r} \frac{\partial^2}{\partial z^2} \theta, \\ \frac{\partial}{\partial z} p_{\|\partial\Omega} = p_{zw}, \end{cases} \qquad \begin{cases} \Delta \frac{\partial}{\partial z} p^* = 0, \\ \frac{\partial}{\partial z} p_{\|\partial\Omega}^* = p_{zw}, \end{cases} \qquad \begin{cases} \Delta \frac{\partial}{\partial z} p_\perp = \mathcal{R} \frac{P_m^2}{P_r} \frac{\partial^2}{\partial z^2} \theta, \\ \frac{\partial}{\partial z} p_{\perp\|\partial\Omega} = 0, \end{cases}$$

whence

$$\frac{\partial}{\partial z}p_\perp = \mathcal{R}\frac{P_m^2}{P_r}\Delta^{-1}\frac{\partial^2}{\partial z^2}\theta, \qquad \Delta^{-1}\frac{\partial^2}{\partial z^2}\theta_{\|\partial\Omega} = 0.$$

Thus, in the components of $\mathbf{U}$ as unknowns, $(6.1.4)_{1,2}$ become

$$\frac{\partial}{\partial t}\nabla_1 w = P_m\Delta\nabla_1 w + P_m M^2\frac{\partial}{\partial z}\nabla_1\Delta^{-1}h_3' + \mathcal{R}\frac{P_m^2}{P_r}\Delta^{-1}\Delta_1\nabla_1\theta$$

$$+ P_m M^2\frac{\partial}{\partial z}\nabla_1 h_3^* - \frac{\partial}{\partial z}\nabla_1 p^*, \tag{6.1.6}$$

$$\frac{\partial}{\partial t}\frac{\partial}{\partial z}w = P_m\Delta\frac{\partial}{\partial z}w + P_m M^2\frac{\partial^2}{\partial z^2}\Delta^{-1}h_3' + \mathcal{R}\frac{P_m^2}{P_r}\frac{\partial}{\partial z}\Delta^{-1}\nabla_1\cdot\nabla_1\theta$$

$$+ P_m M^2\frac{\partial^2}{\partial z^2}h_3^* - \frac{\partial^2}{\partial z^2}p^*, \tag{6.1.7}$$

so system (6.1.6), (6.1.7), $(6.1.4)_3$-$(6.1.4)_6$ reads

$$\frac{d}{dt}\mathbf{U} = \mathbf{AU} + \mathbf{B(U)}, \tag{6.1.8}$$

where $\mathbf{B(U)} = (P_m M^2\dfrac{\partial}{\partial z}\nabla_1 h_3^* - \dfrac{\partial}{\partial z}\nabla_1 p^*, P_m M^2\dfrac{\partial^2}{\partial z^2}h_3^* - \dfrac{\partial^2}{\partial z^2}p^*, 0,0,0,0)^T$ and $\mathbf{A}$ is a matrix operator the entries of which are integro-differential operators.

6.1.2 *The associated functional and Euler equations*

Performing the inner product in $L^2(\Omega)$ of (6.1.8) by $\mathbf{U}$ and taking into account that direct computations yield $(\mathbf{B(U)}, \mathbf{U}) = 0$ we obtain

$$\left(\frac{d}{dt}\mathbf{U}, \mathbf{U}\right) = \left(\mathbf{A_s U}, \mathbf{U}\right), \tag{6.1.9}$$

where $(\cdot\cdot)$ stands for the inner product in $L^2(\Omega)$ and $\mathbf{A_s}$ is the symmetric part of $\mathbf{A}$. We have $\left(\mathbf{A_s U}, \mathbf{U}\right) = \Phi\left(\mathbf{U}\right)$, where the quadratic functional Φ has the form

$$\Phi\left(\mathbf{U}\right) = -\Big\{P_m\big[|\Delta_1 w|^2 + |\nabla_1\frac{\partial}{\partial z}w|^2 + |\nabla\frac{\partial}{\partial z}w|^2\big] + |\nabla h_3'|^2 + |\nabla\frac{\partial}{\partial z}j|^2 + P_m|\nabla\zeta|^2$$

$$+ \frac{P_m}{P_R}|\nabla\nabla_1\theta|^2\Big\} - \beta_H(\frac{\partial}{\partial z}h_3', \frac{\partial}{\partial z}\frac{\partial}{\partial z}j) - (\frac{\partial}{\partial z}\zeta, \frac{\partial}{\partial z}\frac{\partial}{\partial z}j) - (\nabla h_3', \nabla\frac{\partial}{\partial z}w)$$

$$+ \beta_H(\nabla h_3', \nabla\frac{\partial}{\partial z}j) + (1 + \mathcal{R}\frac{P_m^2}{P_r})(\nabla_1\theta, \nabla_1 w) + P_m M^2(\frac{\partial}{\partial z}j, \zeta) + P_m M^2(h_3', \frac{\partial}{\partial z}w).$$

Then the energy relation (6.1.9) reads

$$\frac{d}{dt}E(t) = \mathcal{I} - \mathcal{D}, \tag{6.1.10}$$

where

$$E(t) = \frac{1}{2}\Big\{|\frac{\partial}{\partial z}w|^2 + |\nabla_1 w|^2 + |h_3'|^2 + |\frac{\partial}{\partial z}j|^2 + |\zeta|^2 + |\nabla_1\theta|^2\Big\}$$

$$\mathcal{I} = -\beta_H\Big(\frac{\partial}{\partial z}h'_3, \frac{\partial}{\partial z}\frac{\partial}{\partial z}j\Big) - \Big(\frac{\partial}{\partial z}\zeta, \frac{\partial}{\partial z}\frac{\partial}{\partial z}j\Big) - \Big(\nabla h'_3, \nabla\frac{\partial}{\partial z}w\Big) + \beta_H\Big(\nabla h'_3, \nabla\frac{\partial}{\partial z}j\Big)$$

$$+ (1 + \mathcal{R}\frac{P_m^2}{P_r})(\nabla_1\theta, \nabla_1 w) + P_m M^2\Big(\frac{\partial}{\partial z}j, \zeta\Big) + P_m M^2\Big(h'_3, \frac{\partial}{\partial z}w\Big),$$

$$\mathcal{D} = P_m\Big\{|\Delta_1 w|^2 + |\frac{\partial}{\partial z}\nabla_1 w|^2 + |\nabla\frac{\partial}{\partial z}w|^2\Big\} + |\nabla h'_3|^2 + |\nabla\frac{\partial}{\partial z}j|^2$$

$$+ P_m|\nabla\zeta|^2 + \frac{P_m}{P_r}|\nabla\nabla_1\theta|^2.$$

From (6.1.10) we have

$$\frac{d}{dt}E(t) \le -\mathcal{D}(1 - \xi),$$

where

$$\xi = \max_{\mathcal{M}}\frac{\mathcal{I}}{\mathcal{D}}. \tag{6.1.11}$$

According to the class $\mathcal{M}$ of admissible functions for the variational problem (6.1.11), various extremal values for ξ can be obtained. The simplest choice is $\mathcal{M}$ consisting of vector elements $(w, h'_3, \frac{\partial}{\partial z}j, \zeta, \theta)$ the components of which are smooth functions satisfying the following conditions on $\partial\Omega$

$$w = \frac{\partial}{\partial z}w = h'_3 = \frac{\partial}{\partial z}j = \zeta = \theta = \frac{\partial^2}{\partial z^2}\theta = 0. \tag{6.1.12}$$

The corresponding Euler equations read

$$-2\xi P_m\Delta\Delta w - \frac{\partial}{\partial z}(P_m M^2 + \Delta)h'_3 - (1 + \mathcal{R}\frac{P_m^2}{P_r})\Delta_1\theta = 0,$$

$$\frac{\partial}{\partial z}\Big(\frac{\partial^2}{\partial z^2} + P_m M^2\Big)j + 2\xi P_m\Delta\zeta = 0,$$

$$\frac{\partial}{\partial z}(\Delta + P_m M^2)w + 2\xi\Delta h'_3 - \beta_H\Delta_1\frac{\partial}{\partial z}j = 0, \tag{6.1.13}$$

$$-\beta_H\Delta_1 h'_3 + 2\xi\Delta\frac{\partial}{\partial z}j + \Big(\frac{\partial^2}{\partial z^2} + P_m M^2\Big)\zeta = 0,$$

$$-(1 + \mathcal{R}\frac{P_m^2}{P_r})\Delta_1 w - 2\xi\frac{P_m}{P_r}\Delta_1\Delta\theta = 0.$$

6.1.3 *Stability criteria*

Apply the normal mode technique, *i.e.* assume that

$$(w, h_3', \frac{\partial}{\partial z}j, \zeta, \theta) = \{W(z), K(z), X(z), Z(z), \Theta(z)\}exp[i(\alpha_x x + \alpha_y y)]$$

and introduce these into (6.1.12), (6.1.13) and perform the change of variables $z \to z - 0.5$ to produce

$$-2\xi P_m (D^2 - a^2)^2 W - D(P_m M^2 + D^2 - a^2)K - Sa^2\Theta = 0,$$

$$(P_m M^2 + D^2)X + 2\xi P_m (D^2 - a^2)^2 Z = 0,$$

$$D(P_m M^2 + D^2 - a^2)W + 2\xi(D^2 - a^2)K + \beta_H a^2 X = 0, \qquad (6.1.14)$$

$$+\beta_H a^2 K + 2\xi(D^2 - a^2)X + (D^2 + P_m M^2)Z = 0,$$

$$Sa^2 W + 2\xi\frac{P_m}{P_r}a^2(D^2 - a^2)\Theta, = 0,$$

where $S = (1 + \mathcal{R}\frac{P_m^2}{P_r})$ and

$$W = DW = K = X = Z = \Theta = D^2\Theta = 0 \qquad z = \pm 0.5. \qquad (6.1.15)$$

Assume first that W and Θ are even while K, X and Z are odd functions and expand them in Fourier series with respect to the sets $\{E_1, E_3, \ldots\}$ and $\{F_1, F_3, \ldots\}$ respectively, where $E_{2n-1}(z) = \sqrt{2}\{\cos[(2n - 1)\pi z]\}$, $F_{2n-1}(z) = \sqrt{2}\{\sin[(2n - 1)\pi z]\}$. These sets are total in the subspaces of $L^2(-0.5, 0.5)$ consisting of even and odd functions respectively. Denote by f_{2n-1} the Fourier coefficients of a function f and put

$$A_n = (2n - 1)^2\pi^2 + \gamma^2, \quad B_n = P_m M^2 + \gamma^2 - A_n,$$

$$\alpha_1 = D^2 W(0.5), \quad \alpha_2 = DK(0.5), \quad \alpha_3 = DX(0.5), \quad \alpha_4 = DZ(0.5).$$

Then (6.1.14) become

$$-2\xi P_m A_n^2 W_{2n-1} + (2n - 1)\pi(a^2 - B_n)K_{2n-1} + Sa^2\Theta_{2n-1}$$

$$= 2\sqrt{2}(-1)^{n+1}(2n - 1)\pi \cdot (2\xi P_m \alpha_1 + \alpha_2),$$

$$(2n - 1)\pi(a^2 - B_n)W_{2n-1} - 2\xi A_n K_{2n-1} + \beta_H a^2 X_{2n-1} = 2\sqrt{2}(-1)^n(2\xi\alpha_2 + \alpha_1),$$

$$\beta_H a^2 K_{2n-1} - 2\xi A_n X_{2n-1} + B_n Z_{2n-1} = 2\sqrt{2}(-1)^n(2\xi\alpha_3 + \alpha_4), \qquad (6.1.16)$$

$$B_n X_{2n-1} - 2\xi P_m A_n Z_{2n-1} = 2\sqrt{2}(-1)^n(2\xi P_m \alpha_4 + \alpha_3),$$

$$Sa^2 W_{2n-1} - 2\xi\frac{P_m}{P_r}a^2 A_n \Theta_{2n-1} = 0,$$

the boundary conditions (6.1.15) introduce the constraints

$$\sum_{n=1}^{\infty}(-1)^{n+1}(2n-1)\pi W_{2n-1}=0, \quad \sum_{n=1}^{\infty}(-1)^{n+1}K_{2n-1}=0,$$

$$\sum_{n=1}^{\infty}(-1)^{n+1}X_{2n-1}=0, \quad \sum_{n=1}^{\infty}(-1)^{n+1}Z_{2n-1}=0,$$

implying a linear system in $\alpha_1,\ldots,\alpha_4$, namely $\sum_{i=1}^{4}\alpha_i C_{ji}=0,\ j=1,\ldots,4$ leading to

$$det\mathbf{C}=0 \tag{6.1.17}$$

where the entries C_{ij} of the matrix $\mathbf{C}$ are given in Section 6.1.4. Each C_{ij} is an infinite sum for n coming from 1 to ∞ and they converge to 0 at least like n^{-2} as $n\to\infty$. For $n=1$ (6.1.17) reduces to equation

$$x^2 A_1^4 T_{11}^{''} - x A_1^2 F_+ + S^2 a^4(B_1^2+D) + B_1^2 T_{31} = 0, \tag{6.1.18}$$

which has two positive solutions

$$x_{1,2}=\frac{F_+ \pm \sqrt{F_-^2 + 4T_{11}^{''} D T_{31} A_1}}{2A_1^2 T_{11}^{''}}, \tag{6.1.19}$$

where

$$F_{\pm}=T_{11}^{''}(B_1^2+D)\pm(S^2 a^4 + T_{31})$$

and the other symbols are defined in Section 6.1.4. Thus, in physical parameters we have

$$F_{\pm}=\frac{P_m}{P_r}a^2(\pi^2+a^2)\big[(P_m M^2-\pi^2)^2 + P_m\beta_H^2 a^4\big]$$

$$\pm\Big[(1+\mathcal{R}\frac{P_m^2}{P_r})^2 a^4 + \frac{P_m}{P_r}a^2\pi^2(\pi^2+a^2-P_m M^2)^2\Big],$$

$$4T_{11}^{''}D T_{31}A_1 = 4\pi^2(\pi^2+a^2)^2\frac{P_m^3}{P_r^2}\beta_H^2 a^8(\pi^2+a^2-P_m M^2)^2, \tag{6.1.20}$$

$$2A_1^2 T_{11}^{''} = 2(\pi^2+a^2)^3\frac{P_m}{P_r}a^2.$$

In fact, for a faster computation, instead of x we use the unknown $X=xA_1^2-B_1^2$.

Since, $\xi^2=\dfrac{x}{4P_m}$ and a sufficient condition for stability reads $\xi^2<1$, we obtained the following criteria.

Theorem 6.1.1. *The linear global stability of the fluid flow governed by (6.1.2) – (6.1.4) in the class of normal mode perturbations with even velocity and temperature and odd magnetic field holds if, for $\beta_H\neq 0$*

$$x_i < 4P_m \qquad i=1,2 \tag{6.1.21}$$

where each x_i stands for its expression from (6.1.19) − (6.1.20) *and if, for $\beta_H = 0$*

$$(P_m M^2 - \pi^2)^2 < 4(\pi^2 + a^2)^2 P_m \tag{6.1.22}$$

or

$$(1 + \mathcal{R}\frac{P_m^2}{P_r})^2 a^2 + \frac{P_m}{P_r}\pi^2(\pi^2 + a^2 - P_m M^2)^2 < 4(\pi^2 + a^2)^3 \frac{P_m^2}{P_r}. \tag{6.1.23}$$

Indeed, for $\xi^2 < 1$ the classical arguments in the energy method involving embedding theorems and Gronwall inequality imply Theorem 6.1.1.

Theorem 6.1.1 provides approximate criteria of linear stability. Better criteria, which can be derived from (6.1.17) only numerically, correspond to an increasing number n of terms in C_{ij}. In this case the unknown X can still be used because $xA_n^2 - B_n^2$ are affine functions of $xA_1^2 - B_1^2$.

Notice that $\mathcal{R}$ does not occur in (6.1.22) if $n = 1$, but this might happen if $n > 1$.

Assume now that W and Θ are odd while K, X and Z are even functions of z; this case will be referred to as the odd case. Then, similarly, we obtain a system which can be formally obtained from (6.1.16) by taking $\overline{W}_{2n-1}$, K_{2n-1}, X_{2n-1}, Z_{2n-1}, $\overline{\Theta}_{2n-1}$ instead of W_{2n-1}, K_{2n-1}, X_{2n-1}, Z_{2n-1}, Θ_{2n-1} respectively, and $\alpha, 0, 0, 0, \beta$ instead of $2\sqrt{2}(-1)^{n+1}(2n - 1)\pi(2\xi P_m \alpha_1 + \alpha_2)$, $2\sqrt{2}(-1)^n(2\xi\alpha_2+\alpha_1)$, $2\sqrt{2}(-1)^n(2\xi\alpha_3+\alpha_4)$, $2\sqrt{2}(-1)^n(2\xi P_m\alpha_4+\alpha_3)$ and 0 respectively, where $\overline{W}_{2n-1} = -W_{2n-1}$, $\overline{\Theta}_{2n-1} = -\Theta_{2n-1}$, $\alpha = 2\sqrt{2}(-1)^n(2\xi P_m\alpha_5 + \alpha_7)$ and $\beta = 2\sqrt{2}(-1)^{n+1}(2\xi\frac{P_m}{P_r}\gamma^2\alpha_6)$, $\alpha_5 = D^3 W_{2n-1}(0.5)$, $\alpha_6 = D\Theta_{2n-1}(0.5)$ and $\alpha_7 = D^2 K_{2n-1}(0.5)$. In this way the Cramer determinant is still Δ_n while

$$\overline{W}_{2n-1} = \frac{\overline{\Delta}_{1n}}{\Delta_n}, \; K_{2n-1} = \frac{\overline{\Delta}_{2n}}{\Delta_n}, \; X_{2n-1} = \frac{\overline{\Delta}_{3n}}{\Delta_n}, \; Z_{2n-1} = \frac{\overline{\Delta}_{4n}}{\Delta_n}, \; \overline{\Theta}_{2n-1} = \frac{\overline{\Delta}_{5n}}{\Delta_n},$$

where $\overline{\Delta}_{1n...5n}$ depend linearly on α and β and differ from $\Delta_{1n...5n}$. Moreover, for this odd case, the constraints corresponding to the boundary conditions $\overline{W} = \overline{\Theta} = 0$ read

$$\sum_{n=1}^{\infty}(-1)^{n+1}\overline{W}_{2n-1} = 0, \; \sum_{n=1}^{\infty}(-1)^{n+1}\overline{\Theta}_{2n-1} = 0, \text{ or, equivalently, } \alpha\overline{C}_{j1} + \beta\overline{C}_{j2} = 0,$$

$j = 1, 5$ leading to a determinantal equation similar to (6.1.17) for a 2×2 matrix $\overline{\mathbf{C}}$ the entries $\overline{C}_{ij}$ of which are an infinite sum for n ranging from 1 to ∞ and converges to 0 at least like n^{-2}. For $n = 1$ the odd analogue of (6.1.17) splits into the equation $X - D = 0$, where $D = P_m\beta_H^2\gamma^4$ and equation (6.1.18). Therefore, in the odd case, we have the extra solution $x = (D + B_1^2)A_1^{-2}$ which leads to the following

Theorem 6.1.2. *The linear global stability of the fluid flow governed by* (6.1.2) − (6.1.4) *in the class of normal mode perturbations with odd velocity and temperature and even magnetic field holds if* (6.1.21) − (6.1.23) *are fulfilled and, in addition, if, for $\beta_H \neq 0$,*

$$(P_m M^2 - \pi^2)^2 + P_m\beta_H^2 a^4 < 4P_m(\pi^2 + a^2)^2 \tag{6.1.24}$$

and if, for $\beta_H = 0$,

$$(P_m M^2 - \pi^2)^2 < 4P_m(\pi^2 + a^2)^2. \tag{6.1.25}$$

An interesting conclusion of the criterion (6.1.24) is that it implies the criterion (6.1.22) for the even case where no Hall effect is present. In addition, in the case when no Hall effect is present, the sufficient condition (6.1.24) for the stability from the case of the presence of the Hall effect becomes (6.1.25), which is just (6.1.22).

For $n > 1$ only numerical results can be deduced. However, since equation (6.1.17) from the odd case is of an odd order while that for the even case is of an even order, it follows that in the odd case at least a real solution X always exists. For boundary conditions containing second order upward derivative of the velocity the odd case is investigated in [MaiP84], [MaiPL] and the same destabilizing effect of the Hall current is found.

6.1.4   *Concluding remarks and a list of formulae used in Section 6.1.3*

The magnetic Bénard stability problem in the presence of the Hall effect and of rigid bounding surfaces, dealt with in this section, was not solved hitherto, even in the linear case, due to two reasons. Firstly, in order to associate the governing problem with a variational problem involving a symmetric functional, this problem had to be reformulated as an integro-differential equation. In addition, apart from the usual unknowns, we chose the first order derivatives as new unknown functions. It is only in this way that we could use the boundary conditions and, so, apply the classical energy method leading to a variational problem. With this we associated the Euler-Lagrange equations and the boundary conditions. Unlike the case of the free boundaries (when the velocity and its second order upward derivative vanish on the horizontal boundaries), in the case of rigid boundaries the velocity and its first order upward derivative vanish. In this case the standard Galerkin type method is not appropriate due to the extremely complicated form of the involved Fourier series expansion functions. In exchange, by means of the B-D method we were able to solve the extremum problem in the class of normal modes for even as well as for odd velocity and temperature fields. From (6.1.24) we deduced that the sufficient condition for linear global asymptotic exponential stability in the anisotropic magnetohydrodynamic case ensures the linear stability in the case when no Hall current is present in both odd and even cases. A similar conclusion was reached for the odd situation, *e.g.* for free boundaries, too [MaiP84], [MaiPL].

The solution of (6.1.16) is $\overline{W}_{2n-1} = \dfrac{\Delta_{1n}}{\Delta_n}$, $K_{2n-1} = \dfrac{\Delta_{2n}}{\Delta_n}$, $X_{2n-1} = \dfrac{\Delta_{3n}}{\Delta_n}$,

$Z_{2n-1} = \dfrac{\Delta_{4n}}{\Delta_n}$, $\Theta_{2n-1} = \dfrac{\Delta_{5n}}{\Delta_n}$, where

$$\Delta_n = 2\xi A_n[T_{1n}T_{4n} + T_{3n}(xA_n^2 - B_n^2)],$$

$$\Delta_{1n} = (-1)^n (2n-1)\pi 4\sqrt{2}\xi A_n \frac{P_m}{P_r}\gamma^2 \Big\{ \alpha_1[-xA_nT_{4n} + (\gamma^2 - B_n)(xA_n^2 - B_n^2)]$$

$$+2\xi\alpha_2[-A_nT_{4n} + (\gamma^2 - B_n)(xA_n^2 - B_n^2)] + \alpha_3\beta_H\gamma^2(\gamma^2 - B_n)(xA_n + B_n)$$

$$+2\xi\alpha_4 P_m\beta_H\gamma^2(\gamma^2 - B_n)(A_n + B_n) \Big\},$$

$$\Delta_{2n} = (-1)^{n+1} 2\sqrt{2}\Big\{ (xA_n^2 - B_n^2)[\alpha_1(T_{1n} + xA_nT_{2n}) + 2\xi\alpha_2(T_{1n} + A_nT_{2n})]$$

$$+[\alpha_3(xA_n + B_n) + 2\xi\alpha_4 P_m(A_n + B_n)]T_{1n} \cdot \beta_H\gamma^2 \Big\},$$

$$\Delta_{3n} = (-1)^{n+1} 2\sqrt{2}\Big\{ 2\xi\alpha_1 P_m\beta_H\gamma^2 A_n(T_{1n} + xA_nT_{2n}) + \alpha_2 x\beta_H\gamma^2 A_n(T_{1n} + A_nT_{2n})$$

$$+[2\xi\alpha_3 A_n(xA_n + B_n) + \alpha_4 xA_n(A_n + B_n)](T_{1n} + T_{3n}) \Big\},$$

$$\Delta_{4n} = (-1)^{n+1} 2\sqrt{2}\Big\{ \alpha_1 B_n\beta_H\gamma^2(T_{1n} + xA_nT_{2n}) + 2\xi\alpha_2\beta_H\gamma^2 B_n(T_{1n} + A_nT_{2n})$$

$$+[\alpha_3 4\xi^2 A_n(A_n + B_n) + 2\xi\alpha_4 A_n(xA_n + B_n)](T_{1n} + T_{3n}) - (\alpha_3 + 2\xi P_m\alpha_4)T_{1n}\beta_H^2\gamma^4 \Big\},$$

$$\Delta_{5n} = (-1)^n (2n-1)\pi 2\sqrt{2}S\gamma^2 \Big\{ \alpha_1[-xA_nT_{4n} + (\gamma^2 - B_n)(xA_n^2 - B_n^2)] + 2\xi\alpha_2$$

$$\cdot[-A_nT_{4n} + (\gamma^2 - B_n)(xA_n^2 - B_n^2)] - \beta_H\gamma^2(\gamma^2 - B_n)[\alpha_3(xA_n + B_n) - \alpha_4 2\xi P_m(A_n + B_n)] \Big\},$$

$T_{1n} = S^2\gamma^4 - xA_n^3\frac{P_m}{P_r}\gamma^2$, $T_{2n} = \frac{P_m}{P_r}\gamma^2(\gamma^2 - B_n)(A_n - \gamma^2)$, $T_{3n} = \frac{P_m}{P_r}\gamma^2(\gamma^2 - B_n)^2(A_n - \gamma^2)$, $T_{4n} = xA_n^2 - B_n^2 - D$. Introduce the constants C_{ijn} as follows

$$\frac{\Delta_{1n}(-1)^n(2n-1)\pi}{\Delta_n} = 2\xi\frac{P_m}{P_r}\gamma^2\sum_{i=1}^4 C_{1in}\alpha_i, \qquad \frac{\Delta_{2n}(-1)^n}{\Delta_n} = \sum_{i=1}^4 C_{2in}\alpha_i$$

$$\frac{\Delta_{3n}(-1)^n}{\Delta_n} = \sum_{i=1}^4 C_{3in}\alpha_i, \qquad \frac{\Delta_{4n}(-1)^n}{\Delta_n} = \sum_{i=1}^4 C_{4in}\alpha_i.$$

For an easier algebra, it is convenient to express the symbols containing x in terms of $xA_n^2 - B_n^2 = X_n$. Thus we have $T_{1n} = T'_{1n} - T''_{1n}X_n$, $T_{4n} = X - D$, $T'_{1n}S^2\gamma^4 - A_nB_n^2\frac{P_m}{P_r}\gamma^2$, $T''_{1n} = A_n\frac{P_m}{P_r}\gamma^2$, $C = P_mM^2 + \gamma^2$. We also note that $T_{3n} = (\gamma^2 - B_n)T_{2n}$ and $T'_{1n} = S^2\gamma^4 - B_n^2 T''_{1n}$.

For the case of $\overline{W}$ and $\overline{\Theta}$ odd functions we have

$$\overline{\Delta}_{1n} = \alpha 4\xi^2 A_n T''_{1n}T_{4n} + 2\xi\beta S\gamma^2 A_nT_{4n},$$

$$\overline{\Delta}_{5n} = 2\xi\alpha S\gamma^2 A_nT_{4n} + \beta\Big\{ X^2 A_n - X[A_n(D - B_n^2) - (A_n - \gamma^2)(\gamma^2 - B_n)^2] - A_nDB_n^2 \Big\},$$

$$\frac{\overline{\Delta}_{1n}(-1)^{n+1}}{\Delta_n} = C_{11n}\alpha + C_{12n}\beta, \qquad \frac{\overline{\Delta}_{5n}(-1)^{n+1}}{\Delta_n} = C_{51n}\alpha + C_{52n}\beta.$$

6.2   Lyapunov method applied to the anisotropic Bénard problem

The Bénard linear problem dealt with here consists of equations (1.4.16) and the boundary conditions (1.4.4) and (1.4.6). Therefore, up to the characteristic quantities used in the non-dimensionalization, the equations are the same as in Section 6.1, but the boundary conditions are different. In this section, they correspond to stress-free boundaries and prevent us from deriving an energy relation for the energy defined as in Section 6.1. Thus, we are forced to define a new energy, and to reformulate the problem in terms of some functions representing solenoidal fields in the plane layer. The reformulated equations are still pde's. Under the influence of Joseph's method of the parameters differentiation, in order to obtain the best possible stability criteria, we define a sesquilinear Hermitian functional defined on a suitable linear space of admissible vector fields. From it we derive a positive definite quadratic functional, *i.e.* the Lyapunov energy functional.

The coefficients in this functional are functions of the physical parameters of the problem. Due to the negative sign of some of these coefficients, the most difficult part of the analysis is the proof of the positive definiteness of the energy functional in the class of the normal modes, necessary in the direct Lyapunov method (Section 2.2). Further on, the study proceeds like in Section 6.1: the energy method involves the energy functional which is associated with the two-point eigenvalue problem for the Euler-Lagrange equations. Next these equations are solved by a Fourier series method. Then the maximum follows from the secular equation using the idea of the parameter differentiation. The requirement that this maximum corresponds to the best stability sufficient condition of linear global asymptotic Lyapunov stability permits us to determine the energy coefficients. The critical curves of linear stability from the literature [Mai84], [MaiPL] are regained.

Our presentation follows [PalG04a]. More detailed computations can be found in the first version from 2003 of this paper. Following the approach of [PalG04a], in [PalG04b], in the linear case and for free boundaries which are perfectly thermoelectrically conductors, a stability criterion for the most general case of $\beta_I \neq 0$, $\beta_H \neq 0$ it is deduced. In [Pal06], in the linear case, the situation from [PalG04b] was treated, but for electrically nonconducting boundaries.

6.2.1   *Energy relation for the Lyapunov (energy) functional*

Consider the problem (4.3.1), (4.3.2) in the class (4.3.3). By the Lyapunov direct method, we study the linear stability of the conduction state solution (1.4.2). The evolution equations, obtained by linearizing (4.3.1) about the equilibrium solution corresponding to the thermodiffusive state m_0, are

$$\begin{cases} -\dfrac{\partial}{\partial t}\Delta w = -\Delta\Delta w - M^2\Delta h_3' - \dfrac{\mathcal{R}}{P_r}\Delta_1\theta, \\[2mm] \dfrac{\partial}{\partial t}\zeta' = +\Delta\zeta' + M^2\partial_{zz}j, \\[2mm] \dfrac{\partial}{\partial t}j = \zeta' + \dfrac{P_m}{P_r}\Delta j + \beta_H\dfrac{P_m}{P_r}\Delta h_3', \\[2mm] \dfrac{\partial}{\partial t}h_3' = \partial_{zz}w + \dfrac{P_m}{P_r}\Delta h_3' - \beta_H\dfrac{P_m}{P_r}\partial_{zz}j \\[2mm] \dfrac{\partial}{\partial t}\theta = w + \dfrac{1}{P_r}\Delta\theta, \end{cases} \qquad (6.2.1)$$

with the boundary conditions (4.3.2). We choose as a Lyapunov function the expression:

$$E_l(t) = (\nabla w, \nabla w) + d_1(\zeta',\zeta') + d_2(j,j) + d_3(h_3',h_3') + d_4(\nabla_1\theta,\nabla_1\theta), \qquad (6.2.2)$$

where $< f, f > = \int_\Omega f\overline{f}d\Omega$ and the parameters d_i are, hitherto, arbitrary. In order to evaluate the temporal derivative of (6.2.2), along the solution of (6.2.1), multiply equations (6.2.1) by w, $d_1\zeta'$, d_2j, d_3h_3', $-d_4\Delta_1\theta$ respectively and integrate the result over Ω, by using the boundary conditions (1.4.4), (1.4.6), (4.3.2), to obtain the energy relation

$$\frac{d}{dt}E_l = \mathcal{J}_l - \mathcal{D}_l, \qquad (6.2.3)$$

where

$$\mathcal{J}_l = -M^2(w,\Delta h_3') - \left(\frac{R}{P_r} + d_4\right)(w,\Delta_1\theta) + d_1 M^2(\zeta',\partial_{zz}j) + d_2(j,\zeta')$$
$$+ d_3(h_3',\partial_{zz}w) + \beta_H\frac{P_m}{P_r}d_2(j,\Delta h_3') - \beta_H\frac{P_m}{P_r}d_3(h_3',\partial_{zz}j),$$

$$\mathcal{D}_l = (\Delta w,\Delta w) + d_1(\nabla\zeta',\nabla\zeta') + d_2\frac{P_m}{P_r}(\nabla j,\nabla j) + d_3\frac{P_m}{P_r}(\nabla h_3',\nabla h_3')$$
$$+ d_4\frac{1}{P_r}(\nabla\nabla_1\theta,\nabla\nabla_1\theta). \qquad (6.2.3')$$

Let

$$\lambda = \max_{\mathcal{M}}\frac{\mathcal{J}_l}{\mathcal{D}_l}. \qquad (6.2.4)$$

Then the energy relation (6.2.3) implies the condition

$$\frac{d}{dt}E_l \leq \mathcal{D}_l(\lambda - 1)$$

or, by using standard embedding theorems of Poincaré type,

$$\frac{d}{dt}E_l \leq E_l(\lambda - 1),$$

whence the sufficient condition

$$\lambda < 1 \qquad (6.2.5)$$

ensuring the linear exponential stability of the diffusion solution.

6.2.2   *Associated Euler-Lagrange equations and the secular equation*

In order to determine the critical parameters of the linear stability we study the variational problem (6.2.4). The Euler-Lagrange equations associated with the maximum (6.2.4) are

$$-M^2 \Delta h_3' - (\frac{R}{P_r} + d_4)\Delta_1 \theta + d_3 \partial_{zz} h_3' - 2\lambda \Delta \Delta w = 0,$$

$$-M^2 \Delta w + \beta_H \frac{P_m}{P_r} d_2 \Delta j + d_3 \partial_{zz} w - \beta_H \frac{P_m}{P_r} d_3 \partial_{zz} j + 2\lambda \frac{P_m}{P_r} d_3 \Delta h_3' = 0,$$

$$M^2 d_1 \partial_{zz} j + d_2 j + 2\lambda d_1 \Delta \zeta' = 0, \qquad (6.2.6)$$

$$M^2 d_1 \partial_{zz} \zeta' + d_2 \zeta' + \beta_H \frac{P_m}{P_r} d_2 \Delta h_3' - \beta_H \frac{P_m}{P_r} d_3 \partial_{zz} h_3' + 2\lambda \frac{P_m}{P_r} d_2 \Delta j = 0,$$

$$-(\frac{R}{P_r} + d_4)\Delta_1 w - 2\lambda d_4 \frac{1}{P_r}\Delta \Delta_1 \theta = 0.$$

They are equivalent to the equation in w

$$\left\{ \left[(d_2 + M^2 d_1 \partial_{zz})^2 P_r - 4\lambda^2 d_1 d_2 P_m \Delta \Delta \right] \left[(d_3 \partial_{zz} - M^2 \Delta)^2 \right.\right.$$

$$\left. -2d_3 \frac{P_m}{P_r}\lambda\left(\frac{\Delta_1 P_r}{2\lambda d_4}(\frac{R}{P_r} + d_4)^2 - 2\lambda\Delta\Delta\Delta\right)\right] \qquad (6.2.7)$$

$$\left. -\beta_H^2 (d_2 \Delta - d_3 \partial_{zz})^2 2\lambda d_1 \frac{P_m^2}{P_r}\left(\frac{\Delta_1 P_r}{2\lambda d_4}(\frac{R}{P_r} + d_4)^2 - 2\lambda\Delta\Delta\Delta\right)\right\} w = 0,$$

where w satisfies the boundary conditions (some of them are obtained from equations which are supposed to be valid up to the boundary of Ω)

$$w = \partial_{zz} w = \Delta w = \Delta w_{zz} = 0 \qquad \text{at} \qquad z = 0, 1. \qquad (6.2.8)$$

Assume that the perturbations are normal modes $w(x, y, z) = W(z)exp[i(a_x x + b_y y)]$ and expand $W(z)$ in a Fourier series in $L^2(0,1)$ upon the (total in $L^2(0,1)$) set of functions $\{\sin(n\pi z)\}$ satisfying all boundary conditions deduced from (6.2.8), namely

$$W = D^{(2k)}W = 0 \qquad \text{at} \qquad z = 0, 1 \qquad k \in N. \qquad (6.2.8')$$

Then (6.2.7) implies the secular equation

$$\mu^2 + \mu\left[-(M^2 B_n - d_3 n^2 \pi^2)^2 - P_m d_3 \frac{a^2(\frac{R}{P_r} + d_4)^2}{d_4}\right.$$

$$+ d_3 B_n \frac{(d_2 - M^2 d_1 n^2 \pi^2)^2}{d_1 d_2} + \beta_H^2 (d_3 n^2 \pi^2 - d_2 B_n)^2 \frac{B_n}{d_2}\frac{P_m}{P_r}\right] \qquad (6.2.9)$$

$$- (M^2 B_n - d_3 n^2 \pi^2)^2 \beta_H^2 (d_3 n^2 \pi^2 - d_2 B_n)^2 \frac{B_n}{d_2}\frac{P_m}{P_r} = 0,$$

where $B_n = n^2 \pi^2 + a^2$, $a^2 = a_x^2 + a_y^2$, $D = \frac{d}{dz}$, and

$$\mu = -4\lambda^2 d_3 B_n^3 \frac{P_m}{P_r} + (M^2 B_n - d_3 n^2 \pi^2)^2 + d_3 P_m \frac{a^2(\frac{R}{P_r} + d_4)^2}{d_4}.$$

In deriving (6.2.9) we used the backwards integration technique.

6.2.3 *Positive definiteness of the Lyapunov functional E_l*

Let us now prove that, for every fixed t, the Lyapunov energy functional (6.2.2) is positive definite in the class of normal modes.

Hypothesis: $d_1, d_3 < 0$, $d_2, d_4 > 0$.

Assume that w, ζ, j, h_3 and θ are normal modes, where their factors functions of z are real. Then equations $(1.4.16)_{4,5}$ show that h_3' and ζ' are purely imaginary. Introducing the normal forms in (6.2.1), the resulting system consists in ode's containing the wave numbers a_x and a_y or, more exactly, $a^2 = a_x^2 + b_y^2$, recording the presence in (1.4.16) of the derivatives with respect to x and y, and derivatives with respect to z. In addition, all terms of this system of ode's have the factor $e^{i(a_x x + a_y y)}$. Multiply the ode's by the normal modes $w^*, d_1\zeta'^*, d_2 j^*, d_3 h_3'^*, -d_4\Delta_1\theta^*$ respectively and add the resulting equations to obtain a single ode each term of which has the factor $e^{2i(a_x x + a_y y)}$. Then simplify by this factor, integrate this ode by parts with respect to z over $[0,1]$ and take into account the boundary conditions (4.3.2). In this way the left-hand side of the ode reads $\frac{d}{dt}E_{lnm}(\mathbf{V}, \mathbf{V}^*)$, where

$$E_{lnm}(\mathbf{V}, \mathbf{V}^*) = \; < DWDW^* + a^2 WW^* > + d_1 < Z'Z'^* > + d_2 < JJ^* >$$
$$+ \, d_3 < H_3 H_3'^* > + d_4 a^2 < \Theta\Theta^* >$$

and $< fg > = \int_0^1 f(z)g(z)dz$. The form $E_{lnm}(\mathbf{V}, \mathbf{V}^*)$ is a sesquilinear Hermitian functional on $\mathcal{V} \times \mathcal{V}$ [We2], [Mikh5]. Here $\mathcal{V}$ is a linear space of vector functions whose components belong to $C^4[0,1] \times C^2[0,1] \times C^2[0,1] \times C^2[0,1] \times C^2[0,1]$. In addition, the second and the fourth components are purely imaginary, while the others are real. Therefore, we have $E_{lnm}(\mathbf{V}, \mathbf{V}^*) = \overline{E_{lnm}(\mathbf{V}^*, \mathbf{V})}$, where the bar stands for the complex conjugation, because the terms WW^*, JJ^*, $\Theta\Theta^*$ are real while

$$< Z'Z'^* > = < Z'^* Z' > = < -Z'^* \cdot (-Z') > = \overline{< Z'^* Z' >},$$

and, similarly,

$$< H_3' \cdot H_3'^* > = \overline{< H_3'^* H_3' >}.$$

Moreover the quadratic functional $E_{lnm}(\mathbf{V}) = E_{lnm}(\mathbf{V}, \mathbf{V})$ is real and positive definite if and only if $d_1, d_3 < 0$, $d_2, d_4 > 0$.

6.2.4 *Stability criteria*

By hypothesis, in terms of physical parameters, expressing λ in terms of μ, where μ is the explicit solution of (6.2.9), the condition (6.2.5) reads

$$\frac{a^2 (\frac{R}{P_r} + d_4)^2}{4\frac{d_4}{P_r}} < B_n^3 + B_n^2 [M^2 B_n - d_3 n^2 \pi^2]^2 d_2.$$

$$\frac{\left[1 - \frac{(d_2 - M^2 d_1 n^2 \pi^2)^2 P_r}{4 d_1 d_2 P_m B_n^2}\right]}{-4 d_2 d_3 B_n^2 \frac{P_m}{P_r} + \beta_H^2 (d_3 n^2 \pi^2 - d_2 B_n)^2 \frac{P_m}{P_r} + \frac{d_3}{d_1}(d_2 - M^2 d_1 n^2 \pi^2)^2}, \tag{6.2.10}$$

where the denominator consists of three positive terms and the square brackets is positive.

So far, d_4 was arbitrary. For hydrodynamic stability reasons, let us determine it such that the expression bounding R derived from (6.2.10) be maximal, *i.e.* it corresponds to the largest stability domain.

Hence, denote by F the right-hand side of (6.2.10).

Then (6.2.10) implies $R < \left[\frac{2}{a}\sqrt{\frac{F}{P_r}}\sqrt{d_4} - d_4\right] P_r$. The right-hand side of this last inequality is maximal for $d_4 = \frac{F}{a^2 P_r}$. Thus, the best inequality (6.2.10) reads $R < \frac{F}{a^2}$, *i.e.*

$$Ra^2 < B_n^3 + \frac{B_n^2 [M^2 B_n - d_3 n^2 \pi^2]^2 d_2 \left[1 - \frac{(d_2 - M^2 d_1 n^2 \pi^2)^2 P_r}{4 d_1 d_2 P_m B_n^2}\right]}{-4 d_2 d_3 B_n^2 \frac{P_m}{P_r}\left[1 - \frac{(d_2 - M^2 d_1 n^2 \pi^2)^2 P_r}{4 d_1 d_2 P_m B_n^2}\right] + \beta_H^2 (d_3 n^2 \pi^2 - d_2 B_n)^2 \frac{P_m}{P_r}}. \tag{6.2.11}$$

Maximizing now the right-hand side of (6.2.11) with respect to $d_1 (< 0)$ and $d_2 (> 0)$ we find

$$Ra^2 < B_n^3 + \frac{(B_n^2 + M^2 n^2 \pi^2 \frac{P_r}{P_m})[M^2 B_n - d_3 n^2 \pi^2]^2 d_2}{-4 d_2 d_3 (B_n^2 + M^2 n^2 \pi^2 \frac{P_r}{P_m}) \frac{P_m}{P_r} + \beta_H^2 (d_3 n^2 \pi^2 - d_2 B_n)^2 \frac{P_m}{P_r}} \tag{6.2.12}$$

and

$$Ra^2 < B_n^3 - \frac{(B_n^2 + M^2 n^2 \pi^2 \frac{P_r}{P_m})[M^2 B_n - d_3 n^2 \pi^2]^2}{4 d_3 (B_n^2 + M^2 n^2 \pi^2 \frac{P_r}{P_m} + \beta_H^2 n^2 \pi^2 B_n) \frac{P_m}{P_r}}, \tag{6.2.13}$$

respectively.

The corresponding values for d_1 and d_2 are $d_1 = -\frac{d_2}{M^2 n^2 \pi^2}$ and $d_2 = -\frac{n^2 \pi^2 d_3}{B_n}$.

Maximizing the right-hand side of (6.2.13) with respect to d_3 we obtain

$$Ra^2 < B_n^3 + \frac{(B_n^2 + M^2 n^2 \pi^2 \frac{P_r}{P_m}) M^2 n^2 \pi^2 B_n \frac{P_r}{P_m}}{B_n^2 + M^2 n^2 \pi^2 \frac{P_r}{P_m} + \beta_H^2 n^2 \pi^2 B_n} \tag{6.2.14}$$

and this maximum is attained for $d_3 = -\frac{M^2 B_n}{n^2 \pi^2}$, whence

$$d_1 = -\frac{1}{n^2 \pi^2} \quad d_2 = M^2, \quad d_3 = -\frac{M^2 B_n}{n^2 \pi^2}. \tag{6.2.15}$$

The right-hand side of the inequality (6.2.14) is just (20) of [MaiP84] for a totally ionized fluid.

Like in [MaiP84], [MaiPL], the minimum value of (6.2.14) is obtained for $n = 1$. More exactly we have

Theorem 6.2.1. *If*

$$R < \frac{\pi^4}{x}\left\{(1+x)^3 + \frac{\frac{M^2}{\pi^2}\frac{P_r}{P_m}(1+x)[(1+x)^2 + \frac{M^2}{\pi^2}\frac{P_r}{P_m}]}{[(1+x)^2 + \beta_H^2(1+x) + \frac{M^2}{\pi^2}\frac{P_r}{P_m}]}\right\} \equiv R_H, \qquad (6.2.16)$$

where $x = \frac{a^2}{\pi^2}$, then, in the class of normal modes perturbations, the thermodiffusive equilibrium of a homogeneous thermoelectrically conducting fluid, characterized by a tensorial electrical conductivity, situated in a horizontal layer and subject to a constant vertical adverse temperature gradient and to a constant uniform vertical magnetic field, where the planes bounding S are both stress-free and thermally conducting but electrically non-conducting, is globally linearly exponentially asymptotically stable.

We remark that, for a non-conducting or isotropic electrically conducting fluid, from (6.2.16) we recover the conditions

$$R < \frac{\pi^4(1+x)^3}{x}, \qquad (6.2.17)$$

$$R < \frac{\pi^4}{x}\left\{(1+x)^3 + \frac{M^2}{\pi^2}\frac{P_r}{P_m}(1+x)\right\}, \qquad (6.2.18)$$

which ensure the linear stability [Chan] of the thermodiffusive equilibrium for the hydrodynamic and isotropic magnetohydrodynamic Bénard problem, respectively.

6.3 Stability criteria for a quasi-geostrophic forced zonal flow

For a problem in atmosphere dynamics, namely the stability of wind-driven flows, by following [GeoPal96c], linear stability criteria, implying the asymptotic vanishing of the perturbation kinetic energy, are deduced in terms of the maximum shear of the basic flow and/or its meridional derivative. The basic flow is assumed to be stationary and independent of the longitude. Some of these criteria extend those from [CrM], some others use results from this paper. The method is based on appropriate handling with integral and algebraic inequalities.

6.3.1 *Perturbation model*

Frequently, the quasi-geostrophic approximation is used to describe the dynamics of the planetary fluids (ocean and atmosphere) at synoptical scale. In the inviscid case, it is proved by C. Vamos and A. Georgescu that it represents the model of the fifth order asymptotic approximation of the primitive equations as Rossby number tends to zero. Thus, the main parameters remaining to govern the flow are the Reynolds number, related to the lateral dissipation, proportional to the parameter

A, and the parameter r, related to the bottom dissipation. The wind stress *curl* enters through the forcing term F. It is assumed to be longitude independent.

In spite of asymptotic and other simplifications involved, many results on the stability of flows of zonal interest were based on the initial and boundary-value problem for the balance equation for the local vertical vorticity $\nabla^2 \Psi$

$$\frac{\partial \nabla^2 \Psi}{\partial t} + \mathcal{J}(\Psi, \nabla^2 \Psi + \beta y) = F(y, t) - r\nabla^2 \Psi + A\nabla^4 \Psi, \qquad (6.3.1)$$

where β is the planetary vorticity gradient.

Non-stationary forcing, implying non-steady basic flows $\Psi_0(y, t)$, were used to study the case of barotropic flows. The corresponding type of stability seems to represent an appropriate explanation for the Antarctic Circumpolar Current. A more complicated forcing, and, correspondingly a two-dimensional basic flow $\Psi_0(x, y, t)$ proved to suit to baroclinic instability studies in channel geometry (*e.g.* for Antarctic flows) or in rectangular closed basins (*e.g.* for Northern Hemisphere flows). The linear as well as the nonlinear cases were considered.

With equation (6.3.1) we associate the condition of zero mass flux across the wall latitudes y_1 and y_2 ($\Psi_x = 0$) and the zero lateral vorticity diffusion ($\nabla^2 \Psi = 0$). The subscript indicates the differentiation. Thus, the basic flow, characterized by the local vorticity ($\nabla^2 \Psi_0 = q_0$), is the solution of the following two-point problem $q_0(y_1) = q_0(y_2) = 0$ for the ode $Aq_{0yy} - rq_0 + F(y) = 0$, where A and r are constants and $F(y)$ is a given continuous function. The explicit form of the solution q_0 can be immediately determined by means of the variation of coefficients method.

Let Ω be the closed basin $\Omega = \{(x, y, z) \in \mathbf{R}^3 \mid 0 \le x \le L, y_1 \le y \le y_2, 0 \le z \le D\}$, and $\mathbf{R}_+ = \{t \in \mathbf{R} \mid t \ge 0\}$. Then the linear perturbation $\phi(x, y, t) = \Psi - \Psi_0$, induced by the perturbation of the initial condition, satisfies the following equation

$$(\nabla^2 \phi)_t - \Psi_{0y}(\nabla^2 \phi)_x + (\Psi_{0yyy} + \beta)\phi_x + r\nabla^2 \phi - A\nabla^4 \phi = 0, \qquad (6.3.2)$$

for all $(x, y, z, t) \in \Omega \times R_+$, with boundary conditions

$$\phi_x = 0, \quad \nabla^2 \phi = 0 \qquad \text{at } y = y_1 \text{ and } y = y_2 \qquad (6.3.3)$$

and some initial condition $\phi = \phi_0$ at $t = 0$.

6.3.2 *Energy inequality*

Let $K(t) = \dfrac{1}{2} \displaystyle\int_\Omega (\phi_x^2 + \phi_y^2)dxdydz$ represent the corresponding perturbation kinetic energy. The basic flow Ψ_0 is asymptotically stable (in the mean) if $lim_{t \to \infty} K(t) = 0$. It is stable in the mean if $\dfrac{dK}{dt} < 0$. In order to deduce criteria for asymptotic stability we need inequalities of the form

$$\frac{dK}{dt} + aK \le b|g(t)|, \qquad (6.3.4)$$

where $a > 0$ and $b < 0$ are constants and g is a bounded function. For stability criteria we have $a = 0$ *i.e.*

$$\frac{dK}{dt} \leq b|g(t)|. \tag{6.3.5}$$

These inequalities are obtained by multiplying (6.3.2) by ϕ, integrating the resulted equation over Ω and taking into account (6.3.3). In this way the evolution equation for K follows. Then, appropriate Young, Schwarz and Poincaré type (or more general embedding) inequalities are used. The key point is to handle these inequalities in such a manner to get negative b. As the evolution equation for K contains positive as well as negative terms, it is easier to obtain (6.3.5) and, correspondingly, to deduce stability criteria, rather than asymptotic stability criteria. In exchange, for asymptotic stability sharper inequalities are necessary at every stage of the derivation of (6.3.4).

In the next sections we shall obtain criteria for asymptotic stability exploiting three ideas.

The first is to retain in the left-hand side of (6.3.4) as few as possible terms. This will imply larger negative terms in b.

The second concerns better correlation between the Schwarz and embedding inequalities and the following form of the Young inequality (1.12)

$$\alpha\beta < \frac{\alpha^2}{2\epsilon} + \frac{\epsilon}{2}\beta^2, \qquad \alpha, \beta, \epsilon > 0. \tag{6.3.6}$$

It proved to lead to the best criteria permitted by the existing embedding inequalities. In this respect we quote: [Geo76], where universal criteria better than Ladyzhenskaya's criteria were obtained; the paper by [GeoPal95] improving the famous Prodi's parabola, which bounds the spectrum of the linearized N-S equations; and [GeoPal96a] extending Joseph's criteria to the stability of some mixtures in the presence of temperature, concentration and thermodiffusive effects.

The third idea is to use, instead of (6.3.4) and (6.3.5), inequalities of the form

$$\frac{dK}{dt} + aK \leq b_1\big(g(t)\big)^2 + 2b_2|g(t)h(t)| + b_3\big(h(t)\big)^2, \tag{6.3.7}$$

where $a > 0, b_1 < 0, b_2$ and b_3 are real constants. In this case the asymptotic stability criterion follows from the condition of negatively definiteness of the quadratic form from the right-hand side of (6.3.7).

Thus, we start with the most general inequality for $\dfrac{dK}{dt}$. Assuming that $\phi(x, y, z, t) = \mathcal{Re}\{B(y, t)e^{ikx}\}$, where $\mathcal{Re}$ stands for the real part and $k \in \mathbf{R}$ is the x- wave number, we have

$$K(t) = \frac{1}{2}\int_\Omega (\phi_x\phi_x^* + \phi_y\phi_y^*)dxdydz = \frac{1}{2}DL\Big(\|B_y\|^2 + \|B\|^2\Big), \tag{6.3.8}$$

where $*$ stands for the complex conjugacy and $\|\cdot\|$ is the $L^2(y_1, y_2)$ norm. Therefore, the evolution equation for K reads

$$\frac{1}{DL}\frac{dK}{dt} + r(\|B_y\|^2 + k^2\|B\|^2) + A(\|B_{yy}\|^2 + 2k^2\|B_y\|^2 + k^4\|B\|^2)$$
$$= -k\int_{y_1}^{y_2} q_o\mathcal{Im}(B^*B_y)dy \tag{6.3.9}$$

and, by Schwarz inequality, the most general inequality for $\frac{dK}{dt}$ follows

$$\frac{1}{DL}\frac{dK}{dt} + r(\|B_y\|^2 + k^2\|B\|^2) + A(\|B_{yy}\|^2 + 2k^2\|B_y\|^2 + k^4\|B\|^2) \tag{6.3.10}$$
$$\leq |k|\mu_2\|B\|\|B_y\|,$$

where $\mu_2 = max_{y\in[y_1,y_2]}|q_0(y)|$ and $\mathcal{I}m$ indicates the imaginary part. In Section 3.6.3 from (6.3.10) we deduce the inequality of the form

$$\frac{1}{DL}(\frac{dK}{dt} + aK)\leq - b_1'\|B_y\|^2 + |k|\mu_2\|B\|\|B_y\| - b_2'\|B\|^2, \tag{6.3.11}$$

where b_1', b_2' are constants. Hence, in passing from (6.3.10) to (6.3.11) the only neglected term is $A\|B_{yy}\|^2$. In Sections 6.3.4 and 6.3.5, from (6.3.10) we derive several inequalities of the form

$$\frac{1}{DL}(\frac{dK}{dt} + a'K)\leq b'\|B_y\|^2, \tag{6.3.12}$$

where $a' > 0$, $b' > 0$, are constants. Consequently, some terms in $\|B\|$ will be expressed in terms of K and $\|B_y\|^2$ but other terms in $\|B\|$ and the term in $\|B_{yy}\|$ will be disregarded. This is imposed by the fact that the only integral functional inequality which can be taken into account in (6.3.12) is the Poincaré inequality $\|B\|\leq\alpha^{-1}\|B_y\|$, where $\alpha = \pi(y_2 - y_1)^{-1}$. The balance between the neglected and the maintained terms in $\|B\|$ was established by the first idea quoted in the above.

Our criteria will be expressed in terms of r, A, k, μ_2 and $\mu_3 = max_{y\in[y_1,y_2]}|q_{0y}|$.

6.3.3 *Criteria expressed in terms of the maximum basic vorticity μ_2 only. Use of (6.3.11)*

Inequality (6.3.10) implies $\dfrac{1}{DL}\left[\dfrac{dK}{dt} + 2(r + k^2 A)K\right]\leq|k|\mu_2\|B\|\|B_y\| - k^2 A\|B_y\|^2$, or, equivalently,

$$\frac{1}{DL}\left[\frac{dK}{dt} + 2(r + k^2 A)\epsilon K\right] + (r + k^2 A)(1 - \epsilon)(\|B_y\|^2 + k^2\|B\|^2) \tag{6.3.13}$$
$$\leq|k|\mu_2\|B\|\|B_y\| - k^2 A\|B_y\|^2,$$

where $0 < \epsilon < 1$ is an arbitrary number. Therefore (6.3.13) has the form (6.3.11) where

$$a = (r + k^2 A)\epsilon, b_1 = -\left[Ak^2(2 - \epsilon) + r(1 - \epsilon)\right], b_2 = \frac{|k|\mu_2}{2}, b_3 = -k^2(r + k^2 A)(1 - \epsilon).$$

Then, (6.3.13) is negatively defined if $-4k^4 A^2(1 - \epsilon)(2 - \epsilon) - 4k^2 Ar(3 - \epsilon) + \mu_2^2 - 4r^2(1 - \epsilon)^2 < 0$, which occurs if

$$\mu_2 < 2r(1 - \epsilon) \tag{6.3.14}$$

or if

$$\mu_2\geq 2r(1 - \epsilon) \quad\text{and}\quad k^2\geq\frac{-(3 - 2\epsilon)r + \sqrt{r^2 + \mu_2^2(2 - \epsilon)(1 - \epsilon)^{-1}}}{2A(2 - \epsilon)}. \tag{6.3.15}$$

Inequalities (6.3.14) represent criteria for asymptotic stability. They do not depend on k and the Reynolds number $\mathcal{R}$ (which is proportional to A). As it will be shown in Sections 4 and 5, they are the best for large $\mathcal{R}$ and μ_3. As $\epsilon \to 0$ the asymptotic stability criteria (6.3.14) become better and better and tend to the limit criterion

$$\mu_2 < 2r. \tag{6.3.16}$$

For $\epsilon = 0$ the above reasonings show that

$$\mu_2 \leq 2r \tag{6.3.17}$$

is, among the criteria (6.3.14), the best criterion for stability only.

Criteria (6.3.15) depend on k. In order to obtain k-independent criteria opposite inequalities for k are needed. Such sufficient conditions will be obtained in Section 6.3.6 by using the opposite inequalities deduced in Section 6.3.4.

6.3.4 *Criteria in terms of μ_2 only. Use of (6.3.12)*

Write (6.3.13) in the equivalent form

$$\frac{1}{DL}\left(\frac{dK}{dt} + 2k^2 A\epsilon K\right) \leq |k|\mu_2\|B\|\|B_y\| - \|B_y\|^2\left[r + k^2 A(2-\epsilon)\right]$$
$$- k^2\left[r + k^2 A(1-\epsilon)\right]\|B\|^2,$$

where $0 < \epsilon \leq 1$. Then in this inequality use the Poincaré inequality and neglect terms in $\|B\|^2$ to get (6.3.12), where $a' = 2k^2 A\epsilon$, $b' = |k|\mu_2\alpha^{-1} - r - k^2 A(2-\epsilon)$. Therefore b' is negative for all k^2 if

$$\alpha^{-1}\mu_2 < 2\sqrt{(2-\epsilon)rA} \tag{6.3.18}$$

and

$$\alpha^{-1}\mu_2 \geq 2\sqrt{(2-\epsilon)rA}$$
$$|k| \leq \frac{1}{2A(2-\epsilon)}\left[\alpha^{-1}\mu_2 - \sqrt{\alpha^{-2}\mu_2^2 - 4(2-\epsilon)rA}\right]. \tag{6.3.19}$$

The criteria expressed by inequalities (6.3.18) are independent of k; the best ones correspond to $\epsilon \to 0$. Among the criteria (6.3.18) the limit inequality

$$\alpha^{-1}\mu_2 < 2\sqrt{2rA} \tag{6.3.20}$$

represents the limit of criteria of asymptotic stability and is the best stability criterion.

For $\epsilon = 0$ the inequality

$$\alpha^{-1}\mu_2 \leq 2\sqrt{2rA} \tag{6.3.21}$$

is, among criteria (6.3.18), the best criterion for stability.

For $\epsilon = 1$ we obtain the criteria from [CrM] corresponding to the curve $\mu_2 = 2\sqrt{rA}$. Criterion (6.3.17) is better than (6.3.21) for $\frac{r}{A\alpha^2} > 0.5$ and worse for $\frac{r}{A\alpha^2} < 0.5$. They coincide for $\frac{r}{A\alpha^2} = 0.5$. Criteria (6.3.19) are treated in Section 6.3.5.

6.3.5   *Criteria in terms of μ_2 and μ_3*

Multiplying (6.3.2) by $B_{yy}^* - k^2 B^*$ and integrating over Ω, in [CrM] the following criteria

$$|k| \geq \frac{\mu_3}{r}, \tag{6.3.22}$$

$$|k| \geq \sqrt[3]{\frac{\mu_3}{A}} \tag{6.3.23}$$

were obtained. From (6.3.19) and (6.3.22) it follows

$$\frac{\mu_3}{r} \leq \frac{1}{2A(2-\epsilon)}\left[\alpha^{-1}\mu_2 - \sqrt{\alpha^{-2}\mu_2^2 - 4(2-\epsilon)rA}\right], \tag{6.3.24}$$

which, in the plane $(\mu_3, \alpha^{-1}\mu_2)$, represents a region of asymptotic stability. Define the curve $Q_\epsilon P_\epsilon$ by the equation $\alpha^{-1}\mu_2 = \frac{(2-\epsilon)A}{r}\mu_3 + \frac{r^2}{\mu_3}$, and let OP_ϵ be the straight line $\alpha^{-1}\mu_2 = \frac{2A}{r}\mu_3$, so that P_ϵ is located at the intersection of $Q_\epsilon P_\epsilon$, OP_ϵ and $S_\epsilon P_\epsilon$, where $S_\epsilon P_\epsilon$ is the straight line $\alpha^{-1}\mu_2 = 2\sqrt{(2-\epsilon)rA}$ defining criteria (6.3.18). At the same time, P_ϵ is the point of minimum for $Q_\epsilon P_\epsilon$ and its abscissa is $\frac{1}{\sqrt{(2-\epsilon)}}r\sqrt{\frac{r}{A}}$. Thus, the domain of asymptotic stability corresponding to (6.3.24) (*i.e.* (6.3.18), (6.3.19) and (6.3.22)) can be represented by (6.3.24) and these curves. The largest such region is obtained for $\epsilon = 0$ and it is bounded by the curves $Q_0 P_0$ and $S_0 P_0$

$$\alpha^{-1}\mu_2 = \frac{2A}{r}\mu_3 + \frac{r^2}{\mu_3}, \qquad 0 < \mu_3 \leq \frac{1}{\sqrt{2}}r\sqrt{\frac{r}{A}} \tag{6.3.25}$$

$$\alpha^{-1}\mu_2 = 2\sqrt{2rA}, \qquad \mu_3 \geq \frac{1}{\sqrt{2}}r\sqrt{\frac{r}{A}}. \tag{6.3.26}$$

The criterion defined by (6.3.25), (6.3.26) is independent of k^2, but k^2 must satisfy (6.3.19) and (6.3.22) for all points (μ_3, μ_2) from the region corresponding to (6.3.25), (6.3.26).

From (6.3.26) and (6.3.23) it follows

$$\sqrt[3]{\frac{\mu_3}{A}} \leq \frac{1}{2A(2-\epsilon)}\left[\alpha^{-1}\mu_2 - \sqrt{\alpha^{-2}\mu_2^2 - 4(2-\epsilon)rA}\right], \tag{6.3.27}$$

which, in the plane $(\mu_3, \alpha^{-1}\mu_2)$ represents a region of asymptotic stability. Define the curve $Q_\epsilon' P_\epsilon'$ by the equation $\alpha^{-1}\mu_2 = (2-\epsilon)\sqrt[3]{\mu_3 A^2} + r\sqrt[3]{\frac{\mu_3}{A}}$, and let OP_ϵ' be the curve $\alpha^{-1}\mu_2 = 2(2-\epsilon)\sqrt[3]{\mu_3 A^2}$ and denote by P_ϵ' the intersection of $Q_\epsilon' P_\epsilon'$, OP_ϵ', and $S_\epsilon P_\epsilon'$. In addition, P_ϵ' represents the point of minimum for $Q_\epsilon' P_\epsilon'$ and its abscissa is $\frac{1}{(2-\epsilon)\sqrt{(2-\epsilon)}}r\sqrt{\frac{r}{A}}$. Note that the straight line $S_\epsilon P_\epsilon$ is the same for both cases (6.3.24) and (6.3.27). Therefore, the asymptotic stability region corresponding to (6.3.27) and that one determined by these curves correspond to (6.3.18), (6.3.19) and (6.3.23). The largest such region is obtained for $\epsilon = 0$ and it is bounded by the curves $Q'_0 P'_0$ and $S_0 P'_0$. It reads

$$\alpha^{-1}\mu_2 = 2\sqrt[3]{\mu_3 A^2} + r\sqrt[3]{\frac{\mu_3}{A}}, \qquad \mu_3 \leq \frac{1}{2\sqrt{2}}r\sqrt{\frac{r}{A}}, \tag{6.3.28}$$

$$\alpha^{-1}\mu_2 = 2\sqrt{2rA}, \qquad \mu_3 \geq \frac{1}{2\sqrt{2}}r\sqrt{\frac{r}{A}}. \tag{6.3.29}$$

In [CrM] it was treated only the case $\epsilon = 1$, obtaining the curves $Q_1 P_1$ and $Q'_1 P'_1$.

The curve $Q_{\epsilon_1} P_{\epsilon_1}$ is below the curve $Q_{\epsilon_2} P_{\epsilon_2}$ if $\epsilon_1 > \epsilon_2$. So, for $0 \leq \epsilon \leq 1$, the family $Q_\epsilon P_\epsilon$ is situated between the disjoint curves $Q_0 P_0$ and $Q_1 P_1$. Similarly, the curve $Q'_{\epsilon_1} P'_{\epsilon_1}$ is below the curve $Q'_{\epsilon_2} P'_{\epsilon_2}$ if $\epsilon_1 > \epsilon_2$ and, for $0 \leq \epsilon \leq 1$, the family $Q'_\epsilon P'_\epsilon$ is situated between $Q'_0 P'_0$ and $Q'_1 P'_1$. In addition, the curve $Q_1 P_1$ is above the curve $Q'_0 P'_0$ and they do not intersect each other. Thus the best asymptotic stability region is bounded by the curves $Q_0 P_0$ and $S_0 P_0$ i.e. it is given by (6.3.25), (6.3.26).

Criterion given by (6.3.28), (6.3.29), is independent of k^2 but for the corresponding points (μ_2, μ_3) k^2 must satisfy (6.3.19) and (6.3.23).

6.3.6 *Criteria in terms of μ_2 only. Use of (6.3.11) and (6.3.12)*

In Section 6.3.5 we associated (6.3.19) with the opposite inequalities (6.3.22) and (6.3.23). Here we associate (6.3.19) for $\epsilon = 0$ with (6.3.15), *i.e.*

$$\frac{1}{16A^2}\left[\alpha^{-1}\mu_2 - \sqrt{\alpha^{-2}\mu_2^2 - 8rA}\right]^2 \geq \frac{-3r + \sqrt{r^2 + 2\mu_2^2}}{4A}.$$

This inequality is valid for

$$\mu_2 \geq 2r, \qquad \alpha^{-1}\mu_2 \geq 2\sqrt{2rA} \tag{6.3.30}$$

and leads to the following asymptotic stability criterion

$$\alpha^{-1}\mu_2 < \frac{1}{\alpha}\sqrt{2r^2 + rA\alpha^2 + \alpha^4 A^2 + \sqrt{(2r^2 + rA\alpha^2 + \alpha^4 A^2)^2 + 4Ar^3\alpha^2}}$$
$$= \alpha^{-1}\mu_2^* \tag{6.3.31}$$

which must be considered together with (6.3.30) and the restrictions $\frac{2}{3} < \frac{r}{\alpha^2 A} \leq \frac{7 + \sqrt{41}}{4}$. It follows that we are in the conditions where (6.3.17) is better than (6.3.21).

6.4 Variational principle for problem (5.3.1), (5.3.2)

By means of the approach presented in Section 3.3.3.4, it is found that for $Q = \bar{\delta} = 0$, the problem (5.3.1), (5.3.2) can be transformed into a selfadjoint problem for which a variational principle is shown to hold. The subsequent application of the B-D method reduces the stationarity of the associated functional to the stationarity of a function of the Fourier coefficients. This leads to the same secular equation (5.3.5) as in the case when the B-D method is applied directly to the two-point problem (5.3.1), (5.3.2).

In the particular case $R = A = 0$, when problem (5.3.1), (5.3.2) becomes the classical Bénard convection, the associate functional is positive. The comparison with the corresponding problem written in the temperature only is performed [GeoOP].

6.4.1   *Variational principle for the case $Q = \bar{\delta} = 0$, $A, R \neq 0$*

The problem (5.3.1), (5.3.2) can be written in the form $Lf = 0$, where $L : A \rightarrow [C^0(-0.5, 0.5)]^3$ is the matricial linear differential operator

$$L = \begin{pmatrix} (1+R)(D^2 - a^2 I)^2 & R(D^2 - a^2 I) & -R_a a^2 I \\ -R(D^2 - a^2 I) & A(D^2 - a^2 I) - 2RI & O \\ I & O & D^2 - a^2 I \end{pmatrix},$$

I is the identity operator on A, O is the null operator on A, the set A consists of all vector functions $f = (W, Z, \Theta)$ satisfying (5.3.2) and

$$f \in [C^\infty(-0.5, 0.5)]^4 \cap C^0(-0.5, 0.5)] \times C^0(-0.5, 0.5) \times C^0(-0.5, 0.5).$$

The domain of definition A is embedded in $[L^2(-0.5, 0.5)]^3$. Direct (and quite easy) computations show that L is not selfadjoint, however, taking into account the approach in Section 3.3.3.4, if we multiply equation (5.3.1) by (-1) and equation $(5.3.1)_3$ by $(R_a a^2)$, equation (5.3.1) becomes $L_1 f = 0$, where L_1 is the selfadjoint operator

$$L = \begin{pmatrix} -(1+R)(D^2 - a^2 I)^2 & -R(D^2 - a^2 I) & R_a a^2 I \\ -R(D^2 - a^2 I) & A(D^2 - a^2 I) - 2RI & O \\ R_a a^2 I & O & R_a a^2 (D^2 - a^2 I) \end{pmatrix}.$$

The eigenvalue problem corresponding to the selfadjoint operator L_1 reads

$$\begin{cases} (1+R)(D^2 - a^2)^2 W - R(D^2 - a^2)Z + R_a a^2 \Theta = 0, \\ -R(D^2 - a^2)W + [A(D^2 - a^2) - 2R]Z = 0, \\ R_a a^2 W + R_a a^2 (D^2 - a^2)\Theta = 0, \end{cases} \tag{6.4.1}$$

$$W = DW = Z = \Theta = 0 \quad \text{at} \quad z = \pm 0.5. \tag{6.4.2}$$

Define the functional $J : A_1 \rightarrow \mathbf{R}$ by $J(f) = -(L_1 f, f)$, where

$$A_1 = \{f = (W, Z, \Theta) \in [C^\infty(-0.5, 0.5)]^3 \mid W, Z, \Theta \text{ satisfy } (6.4.2)\}.$$

Then, we have

$$J(f) = -\int_{-0.5}^{0.5} \{-(1+R)[(D^2 W)^2 + 2a^2 (DW)^2 + a^4 W^2] + 2R[DZDW + a^2 WZ]$$

$$+ 2Ra^2 W\Theta - [A(DZ)^2 + Aa^2 Z^2 + 2RZ^2] - R_a a^2 (D\Theta)^2 - R_a a^4 \Theta^2\} dz. \tag{6.4.3}$$

Theorem 6.4.1 [Drag07a]. *$L_1 f = 0$ if and only if $\delta J f = 0$.*

Proof. By parts integrations imply

$$\delta J(f) = -2 \int_{-0.5}^{0.5} \left\{ -(1+R)(D^4 W - 2a^2 D^2 W + a^4 W] + R(D^2 Z - a^2 Z - R_a a^2)\Theta \right\} \delta W$$

$$+ \left\{ -R(D^2 W - a^2 W) + [A(D^2 Z - a^2 Z) - 2RZ] \right\} \delta Z + \left\{ W + (D^2\Theta - a^2\Theta) \right\} \delta\Theta = 0.$$

Since δW, δZ and $\delta\Theta$ are arbitrary, $\delta J(f) = 0$ implies (6.4.1). Conversely, if (6.4.1) holds, it follows that $\delta J(f) = 0$, whence Theorem 6.4.1.

In order to determine the stationarity conditions for the functional J we use the B-D method. More exactly, replacing in (6.4.3) the series expansions for W, DW, $D^2 W$, Z, DZ, Θ and $D\Theta$, upon the total sets F_{2n-1}, E_{2n-1}, J becomes a function of the coefficients W_{2n-1}, Z_{2n-1} and Θ_{2n-1}. Imposing to this function to be stationary, we obtain an infinite linear system in these coefficients. Eliminating Z_{2n-1} and Θ_{2n-1} between the equations of this algebraic system we get W_{2n-1} as a function of $\alpha = D^2 W(0.5)$. Substituting the obtained expression for W_{2n-1} in the constraint (5.3.4) we are led to (5.3.5). Therefore, as it was expected, we obtained the same secular equation like in the direct application of the B-D method (Section 5.3).

The use of the functional J diminishes the computations to the half because J contains at most second order derivatives of W and first order derivative for Z and Θ, *i.e.* half of the order of the derivatives occurring in the system (5.3.1).

Remark 6.4.1. In the case $R = A = Q = \bar{\delta} = 0$, (5.3.1), (5.3.2) becomes the classical Bénard problem which in Θ reads

$$(D^2 - a^2)^3\Theta - R_a a^2\Theta = 0, \quad z \in [-0.5, 0.5] \tag{6.4.4}$$

$$\Theta = D^2\Theta = D(D^2 - a^2)\Theta = 0 \quad \text{at} \quad z = \pm 0.5. \tag{6.4.5}$$

Its adjoint problem in Θ^* consists of the same equation (6.4.4) and the boundary conditions

$$\Theta^* = D^2\Theta^* = D(D^2 - 2a^2)\Theta^* = 0 \quad \text{at} \quad z = \pm 0.5, \tag{6.4.6}$$

therefore the operator corresponding to (6.4.4) is not selfadjoint (in fact, it is neither symmetrizable), even if the problems (5.3.1), (5.3.2) for $A = R = Q = \bar{\delta} = 0$ and (6.4.4), (6.4.5) are equivalent.

6.5 Taylor-Dean problem

6.5.1 *Eigenvalue problem*

By Taylor vortices [Tay36a], [Tay36b] we understand the secondary flow bifurcating from the basic Couette flow between two rotating cylinders at the threshold where the Couette flow loses its stability. The Dean [Dea] secondary flow is the secondary flow bifurcating from the stationary Poiseuille flow in a curved channel driven by an azimuthal pressure gradient. In engineering applications, sometimes, both fluid rotation and pumping round an annulus between cylinders are simultaneously acting, whence the interest in stability of the corresponding stationary flow. This is known as the Taylor-Dean problem [DiPS65], [Kac] and it is considered in the class of stationary perturbations.

We included the Taylor-Dean problem in a chapter devoted to convections due to the already well-known equivalence of the governing models. For the Taylor-Dean problem we put into evidence two types of variational principles quoting only a few papers strictly related to our topics of interest. We treat only the small-gap case.

Let us present the Taylor-Dean problem following [Kac], where the relationship between many others of its variants can be found. Denote by a star the dimensional quantities. Let R_1^*, $R_2^* > R_1^*$ be the radii of the two cylinders such that the gap $d^* = R_2^* - R_1^*$ is small, *i.e.* $d^*/R_1^* << 1$. Assume that the outer cylinder is at rest, the inner is rotating with the angular velocity Ω_1, and the flow is axisymmetric, *i.e.* the velocity components in cylindrical coordinates (U^*, V^*, W^*) do not depend on the azimuthal coordinate θ^*. The cylindrical coordinates are (r^*, θ^*, z^*). Let $V_T = \Omega_1 R_1^*/2$ be the (linear) velocity of inner cylinder (the index T indicates the Taylor problem contribution), and denote by $V_D = kd^{*2}/(12\nu R_1^*)$ velocity due to the circumferential drop pressure (the index D indicates the Dean problem). Introduce the parameter $\alpha = V_T/(V_T + V_D)$; $\alpha = 1$ if only the rotation is acting and $\alpha = 0$ if only the pressure drop is present. Then the basic state is

$$U = W = 0, \quad V = V_0(x) = 2\alpha(1 - x) + 6(1 - \alpha)(x - x^2),$$
$$P^* = P_0^* = \rho k\theta + f(x), \tag{6.5.1}$$

($k = \frac{1}{\rho}\frac{\partial P_0^*}{\partial \theta}$ is a measure of the azimuthal pressure gradient), where the non-dimensionalization was done by means of the characteristic quantities d^*, $V_T + V_D$, $\nu/(2d^*)$ and $t^*\nu d^{*-2}$ for the length, azimuthal velocity, radial and axial velocity, and time respectively. Denote by x and z the nondimensional radial and axial coordinates respectively. Let (U, V, W) be the perturbed velocity about the steady state (6.5.1), therefore $U = u(x, y, t)$, $V = V_0(x) + v(x, y, t)$, $W = w(x, y, z)$, where (u, v, w) is the perturbation velocity. Assume that the perturbations are stationary normal forms, *i.e.* $f(x, z) = f(x)e^{ikz} + \overline{f(x)}e^{-ikz}$, where the bar over quantities indicates the complex conjugation. Then the N-S model in the cylindrical coordinates in the small-gap approximation implies the Taylor-Dean eigenvalue problem

$$\begin{cases} (D^2 - k^2)^2 u - k^2 T V_0 v = 0, \\ (D^2 - k^2)v - \dfrac{1}{2}V_0'u = 0, \qquad \text{for} \qquad x \in [0, 1] \end{cases} \tag{6.5.2}$$

$$u = Du = v = 0, \quad \text{at} \quad x = 0, 1, \tag{6.5.3}$$

where $D = \frac{d}{dx}$ and $T = [(V_T + V_D)2d^*/\nu]^2 d^*/R_1^*$ is the Taylor nondimensional number depending on the characteristic length and azimuthal velocity. This problem has variable coefficients and is similar to problems of thermal convection in horizontal layers if the radial and axial velocities u and v in the Taylor-Dean problem are associated with the vertical component of the velocity and temperature respectively in a thermal convection problem (Section 5.4).

6.5.2 Reformulation of (6.5.2), (6.5.3) as a selfadjoint problem and the associated variational principle

In general, having variable coefficients, the matricial differential operator associated with (6.5.2), (6.5.3) cannot be expected to be selfadjoint. However, due to the invertibility property of part of this operator, a problem of the type more general than (6.5.2), (6.5.3), namely the one consisting of the ode's

$$\begin{cases} (D^2 - k^2)^2 u = f(x)v, \\ (D^2 - k^2)v = -a^2 T g(x)u, \qquad \text{for} \qquad x \in [0,1] \end{cases} \tag{6.5.4}$$

and the boundary conditions (6.5.3), is shown by DiPrima and Sani [DiPS65] to be reformulated as a selfadjoint problem and fulfilling Mikhlin's Theorem 3.2.2.

Before sketching the proof, we recall Remark 5.1.2, where the positivity of coefficients like A_n, A_n^2 of the Fourier coefficients suggested the invertibility of the corresponding operators $(D^2 - a^2)$ and $(D^2 - a^2)^2$ (easily to be proved directly too).

Remark 6.5.1. The possibility to reformulate a problem as a selfadjoint one is higher in the case of systems of ode's rather than in the case of an equivalent (higher order) ode (see also Remark 6.4.1 and Section 3.3.3).

DiPrima and Sani wrote (6.5.4) in the form

$$A\mathbf{w} = \lambda B\mathbf{w}, \tag{6.5.5}$$

where $\lambda = aT^{1/2}$, $\mathbf{w} = (u,v)^T$ and A and B are the following differential matricial operators

$$A = \begin{pmatrix} (D^2 - a^2)^2 & 0 \\ 0 & -(D^2 - a^2) \end{pmatrix}, \quad B(x) = \begin{pmatrix} 0 & f(x) \\ g(x) & 0 \end{pmatrix}, \tag{6.5.6}$$

$A : \mathcal{D}_A \subset L^2(0,1) \to L^2(0,1)$, $\mathcal{D}_A = \{\mathbf{w} \in L^2(0,1) \mid \mathbf{w} = (u,v)^T, u \in C^4(0,1), v \in C^2(0,1), u, v,$ satisfy (6.5.3)$\}$, $\mathcal{D}_B = \{\mathbf{w} \in L^2(0,1) \mid \mathbf{w} = (u,v)^T; u, v,$ satisfy (6.5.3)$\}$. Obviously, A is a matricial differential operator with constant coefficients, while $B(x)$ is an algebraic matrix operator with variable coefficients. In addition, $\mathbf{w}$ is a complex-valued vector, and, correspondingly, the inner product is $(\mathbf{w}_1, \mathbf{w}_2) = \int_0^1 (u_1 \bar{v}_1 + u_2 \bar{v}_2)dx$. By parts integration shows that $(A\mathbf{w}_1, \mathbf{w}_2) = (\mathbf{w}_1, A\mathbf{w}_2)$ and $(A\mathbf{w}, \mathbf{w}) = \int_0^1 [\mid (D^2 - a^2)u \mid^2 + \mid Dv \mid^2 + a^2 \mid v \mid^2]dx > 0$ for every $(\mathbf{w}_1, \mathbf{w}_2) \in \mathcal{D}_A$. Therefore, A is a selfadjoint positive definite operator. Here, $\mathcal{D}_A$ is a linear space. Moreover it is a pre-Hilbert space with respect to the inner product $(\mathbf{w}_1, \mathbf{w}_2)_A = (A\mathbf{w}_1, \mathbf{w}_2)$. Let $\|\mathbf{w}\|_A^2 = (A\mathbf{w}, \mathbf{w})$ be the corresponding norm. Then, the energy space corresponding to A is the energy space $H_A = \overline{\mathcal{D}_A}^{\|\cdot\|_A}$. By the general theory [Mikh5], $H_A \subset L^2(0,1)$ and it is a Hilbert separate space with respect to $(\cdot, \cdot)_A$. Since $\mathcal{D}_A \subset \mathcal{D}_B$, the space of the problem (6.5.4), (6.5.3) and of the corresponding generalized problem is H_A.

If defined in $L^2(0,1)$, B is bounded. Indeed,

$$\|B\mathbf{w}\|^2 = \int_0^1 \big(f(x)v(x) \cdot f(x)\overline{v}(x) + g(x)u(x) \cdot g(x)\overline{u}(x)\big)dx$$

$$= M \int_0^1 \{|\,v(x)\,|^2 + |\,u(x)\,|^2\}dx = M\|\mathbf{w}\|^2,$$

where $M = \max_{x \in [0,1]}\{f^2(x), g^2(x)\}$. Moreover, it is uniformly bounded because f and g are fixed.

Define on $L^2(0,1)$ a new operator $L = A^{-1}B$ by the relation

$$L\mathbf{w} = \int_0^1 G(x, x')B(x')\mathbf{w}(x')dx', \tag{6.5.7}$$

where G is the matrix Green function corresponding to A and boundary conditions (6.5.3) and it is uniformly bounded, and let us prove that it takes its values in $L^2(0,1)$, *i.e.* $\|L\mathbf{w}\| < \infty$. Since $B(x)$ is uniformly bounded in $L^2(0,1)$, it follows that $\|L\mathbf{w}\| \leq$ const. $\|\mathbf{w}\|$, *i.e.* $L : L^2(0,1) \to L^2(0,1)$ and it is completely continuous on $L^2(0,1)$.

Define L on H_A and let us prove that its values belong to H_A too. Indeed, using (3.2.31), the Hölder inequality (1.11) for $p = 2$, and the boundedness of $\|B\|$ and of $\|L\|(= \|A^{-1}B\|)$, we have successively

$$\|L\|_A^2 = (AL\mathbf{w}, L\mathbf{w}) = (AA^{-1}B\mathbf{w}, A^{-1}B\mathbf{w}) \leq (B\mathbf{w}, A^{-1}B\mathbf{w}) \leq \|B\mathbf{w}\|\|A^{-1}B\mathbf{w}\|$$

$$\leq \|B\|\|\mathbf{w}\|\|A^{-1}B\|\|\mathbf{w}\| \leq \frac{1}{\gamma^2}\|B\|\|A^{-1}B\|\|\mathbf{w}\|_A^2,$$

i.e. $L\mathbf{w} \in L^2(0,1)$. In older terms, L is defined on H_A (Section 3.2.5).

Let us write (6.5.5) in the standard form for an eigenvalue problem

$$\lambda^{-1}\mathbf{w} = L\mathbf{w}. \tag{6.5.8}$$

Then it is associated with the Rayleigh quotient functional $F(\mathbf{w}) = (L\mathbf{w}, \mathbf{w})/(\mathbf{w}, \mathbf{w})$. Indeed, let us write the variational problem $\min_{\mathbf{w} \in D_A} F(\mathbf{w})$ as an isoperimetric problem $\min_{\mathbf{w} \in D_A, \|\mathbf{w}\|=1} F(\mathbf{w})$, where $G(\mathbf{w}) = \int_0^1 \mathbf{w}\overline{\mathbf{w}}dx = (\mathbf{Iw}, \mathbf{w})$, or equivalently, as the variational problem without constraint $\min_{\mathbf{w} \in D_A}((L - \mu I)\mathbf{w}, \mathbf{w})$, where μ is the Lagrange multiplier. Then, by the Lagrange method, we have $\delta(F - \mu G) = \int_0^1 \delta\big[(L - \mu I)\mathbf{w}\overline{\mathbf{w}}\big]dx = \int_0^1 \{\delta\big[(L - \mu I)\mathbf{w}\big]\overline{\mathbf{w}} + (L - \mu I)\mathbf{w}\delta\overline{\mathbf{w}}\}dx = \int_0^1 (L - \mu I)\delta\mathbf{w}\overline{\mathbf{w}} + (L - \mu I)\mathbf{w}\delta\overline{\mathbf{w}} = \int_0^1 \delta\mathbf{w}(L^* - \mu I^*)\delta\overline{\mathbf{w}} + (L - \mu I)\mathbf{w}\delta\overline{\mathbf{w}}\}dx$, implying (6.5.8) because L and I are selfadjoint.

By Remark 3.2.13 and Appendix 2, λ_1^{-1}, the least eigenvalue, can be computed by means of minimizing sequences.

6.5.3 *Variational principle for the non-selfadjoint problem (6.5.2), (6.5.3)*

Let us prove that the convergence results of DiPrima and Sani substantiate the variational methods applied to problems of the form (6.5.5) where A and B have the main properties as for the Taylor-Dean problem, namely A is self-adjoint (or at least can be extended up to a selfadjoint operator) and positive definite.

Denote $\mu = \lambda^{-1}$. Then in (6.5.8) the first eigenvalue μ_1 of L is the limit of the values of the Rayleigh quotient, corresponding to a minimizing sequence $\{\mathbf{w}_n\}$, $\mathbf{w}_n \in D(A)$, tending weakly (in H_A) towards μ_1 when the sequence $\mathbf{w}_n$ converges towards the corresponding eigenvalue. By (3.2.3.4′) we have $\mu_1^{(n)} = (L\mathbf{w}_n, \mathbf{w}_n)/(\mathbf{w}_n, \mathbf{w}_n)$, or, equivalently,

$$\mu_1^{(n)}(\mathbf{w}_n, \mathbf{w}_n) = (L\mathbf{w}_n, \mathbf{w}_n). \tag{6.5.9}$$

As L is compact, the weak convergence implies the strong convergence in $L^2(0,1)$ corresponding to

$$\mu_1^{(n)}\mathbf{w}_n = L\mathbf{w}_n \tag{6.5.10}$$

and then projecting (6.5.10) on $\mathbf{w}_n$. But (6.5.10) reads equivalently as

$$\mu_1^{(n)}A\mathbf{w}_n = B\mathbf{w}_n, \tag{6.5.5′}$$

i.e. the approximate form for (6.5.5). This is the reason why, starting with (6.5.5) and not with (6.5.8), and applying Fourier series methods, the same results are obtained. The expansion functions $\mathbf{h}_n \in \mathcal{D}(A)$ must form a total set in H_A, so $\mathbf{w}_n(x) = \mathbf{W}_n\mathbf{h}_n(x)$ and, taking into account the linear independence of $\{\mathbf{h}_i\}$, (6.5.5′) implies

$$\mu_1^{(n)}\mathbf{W}_n A\mathbf{h}_n(x) = \mathbf{W}_n B\mathbf{h}_n(x), \tag{6.5.11}$$

where $\mathbf{W}_n$ are matrices. In order to solve this problem, we must have on both sides the same vector function $\mathbf{h}_k$, therefore we must develop $A\mathbf{h}_n$ and $B\mathbf{h}_n$ upon the total set $\{\mathbf{h}_i\}$, then match the obtained series, before imposing the Cramer determinant of the system in the Fourier coefficients (the entries of $\mathbf{W}_n$) to vanish.

Since $A\mathbf{h}_n \in \mathcal{D}(A^*)$ and $B\mathbf{h}_n \in \mathcal{D}(B^*)$, it is natural to use an alternative way to solve the eigenvalue problem (6.5.5) and, so, (6.5.11) in H_A taking into account these facts. In this alternative method (6.5.5), (6.5.3) is associated with a variational principle for the case of non-selfadjoint operators, frequently used in applications [Chan], [DiP59]. It involves the problem (6.5.5) and its adjoint [DidG]

$$A^*\mathbf{w}^* = \overline{\lambda}B^*\mathbf{w}^*, \tag{6.5.12}$$

and states: *the stationary value of the generalized Rayleigh quotient functional* $H = \dfrac{(\mathbf{w}^*, L\mathbf{w})}{(\mathbf{w}^*, B\mathbf{w})}$ *in* H_A *is an eigenvalue of* (6.5.5) (in general we do not know if it is the least). Indeed, like for the selfadjoint operator L in the above, the Lagrange method shows that $\delta H = 0$ if and only if (6.5.5) and (6.5.12) hold. Now, let $\mathbf{w} = \sum_{n=1}^{\infty} a_n \mathbf{s}_n$, $\mathbf{w}^* = \sum_{n=1}^{\infty} a_n^* \mathbf{r}_n$, where $\{\mathbf{s}_n\}$ and $\{\mathbf{r}_n\}$ are sets total in H_A, substitute these expansions into H to obtain the function $H(a_n, a_n^*)$, the stationarity points of which satisfy the system $\frac{\partial H}{\partial a_n} = 0$, $\frac{\partial H}{\partial a_n^*} = 0$, hence the vanishing of the associated Cramer determinant reads $\det(A_{ij} - \lambda B_{ij}) = 0$, where $A_{ij} = (\mathbf{r}_i, A\mathbf{s}_j)$, $B_{ij} = (\mathbf{r}_i, B\mathbf{s}_j)$.

The Taylor-Dean problem used in [DidG] consists in the ode's

$$\begin{cases} [(D^2 - q^2)^2 - ipTaD^2V]u - 2q^2TaVv = \sigma_r(D^2 - q^2)u, \\ (D^2 - q^2 - ipTaV)v + TaDVu = \sigma_r, \end{cases} \tag{6.5.13}$$

and the boundary conditions (6.5.3), where $\sigma_r = -c_i$ (in formula (1.4.12)) is the frequency of the perturbation mode, q and p are the axial and azimuthal wave numbers respectively, $V = 3(1 + \mu)x^2 - 2(2 + \mu)x + 1$, and T_a is the Taylor number defined as $T_a = \Omega R_1^{1/2}(R_2 - d_1)^{3/2}/\nu$. In this case

$$A = \begin{pmatrix} A_{11} & A_{12} \\ A_{21} & A_{22} \end{pmatrix} = \begin{pmatrix} (D^2 - q^2)^2 - ipTaD^2V & -2q^2TaV \\ TaDV & D^2 - q^2 - ipTaV \end{pmatrix},$$

$$B = \begin{pmatrix} B_{11} & B_{12} \\ B_{21} & B_{22} \end{pmatrix} = \begin{pmatrix} D^2 - q^2 & 0 \\ 0 & 1 \end{pmatrix}$$

$\mathbf{w}^T = (u, v)^T$, $\mathbf{w}^{*^T} = (u^*, v^*)^T$, and the following expressions

$$\mathbf{w} = \begin{pmatrix} \sum a_n f_n(x) \\ \sum b_n g_n(x) \end{pmatrix} \qquad \mathbf{w}^* = \begin{pmatrix} \sum a_n^* f_n(x) \\ \sum b_n^* g_n(x) \end{pmatrix}$$

are used, where

$$f_n = x^{2+n}(1 - x)^2, \qquad g_n = x^{1+n}(1 - x) \tag{6.5.14}$$

are expansion functions such that $\{f_n\}$, $\{g_n\}$ are total in the subspace of $L^2(0, 1)$ of functions satisfying (6.5.3). The reported computations agree very well with experimental [MutWHA] values even for one and two terms in these sets.

In [Drag07a] one direct and one variational methods based on Fourier series are applied to the two-point problem for (6.5.13) in the case $p = \sigma_r = 0$. The set of expansion functions was $\{\sin(n\pi x)\}$.

Other variational principles for non-selfadjoint operators related to our topics are to be found in Section 3.4 and [Chan], [DidG].

Chapter 7

Applications of the direct method to linear stability

Each section of this chapter deals with a single physical situation and is based on the already published papers of our research group. However, some results are improved and some new results are added. The presentation follows the lines of Appendix 6. The system of governing equations is written as a single equation for a single unknown function. All other unknown functions are expressed in terms of the coefficients of that function. As a consequence, the boundary conditions are expressed in terms of these coefficients too. The multiplicity of the roots of the characteristic equation is attentively investigated. Bifurcation sets of the involved characteristic manifolds are analyzed. Various forms of the corresponding transcendental secular equations are provided. Secular equations independent of the boundary conditions are studied. False secular points are deduced. Difficult points of the proofs and open problems are pointed out. First they are related to the determination, in the parameter space, of bifurcation manifolds corresponding to multiple characteristic roots. Second, the open problems concern the type of the involved bifurcation points and numerical procedures to construct the secular surface around them. Apart from the Couette flow (Section 7.1), all other flows considered in this chapter are convections. The direct method was also applied to two particular thermal convections in Section 3.4.3.1. In Section 7.5 we complete their study.

7.1 Couette flow between two cylinders subject to a magnetic field

The characteristic and secular equations for a special Couette flow are derived and discussed using results in [GeoPalR05]. Some new results on the secular manifolds which do not depend on boundary conditions are reported too. The secular equation for simple and some multiple eigenvalues were dealt with first in [Geo82a], [Geo 82b]. An older exhaustive study of the eigenvalue multiplicity and of geometric properties of the bifurcation surfaces and curves can be found in [GeoOO] together with the graphical representation of some neutral curves.

277

7.1.1 *Characteristic equation and its bifurcation sets*

Consider the Couette flow of a fluid between two rotating coaxial cylinders situated at a very small distance. The fluid is electrically conducting and subject to an axial magnetic field. The eigenvalue problem governing the linear stability of this flow against normal mode perturbation reads [Chan]

$$\left\{(D^2 - a^2)^2 + Qa^2\right\}^2 v = -Ta^2(D^2 - a^2)v, \quad -0.5 < z < 0.5 \qquad (7.1.1)$$

$$Dv = (D^2 - a^2)v = \left\{(D^2 - a^2)^2 + Qa^2\right\}v = D\left\{(D^2 - a^2)^2 + Qa^2\right\}v = 0, \quad (7.1.2)$$

where T, $Q > 0$ are the dimensionless Taylor and Chandrasekhar numbers respectively, a is the positive wave number, z is the vertical coordinate, $D = \frac{d}{dz}$ and v is the unknown stream function. The associated characteristic equation is

$$(\lambda^2 - a^2)^4 + 2Qa^2(\lambda^2 - a^2)^2 + Ta^2(\lambda^2 - a^2) + Q^2a^4 = 0. \qquad (7.1.3)$$

The form of the secular equation depends on the form of the eigensolutions of (7.1.1), therefore on the multiplicity of the roots of (7.1.3). Thus, first we must study this multiplicity. With the notation $\mu = \lambda^2 - a^2$ (7.1.3) becomes

$$\mu^4 + 2Qa^2\mu^2 + Ta^2\mu + Q^2a^4 = 0. \qquad (7.1.4)$$

Here we consider only the case $a, T, \Theta > 0$, *i.e.* the first octant without coordinate planes. For points (a, T, Θ) belonging to these planes we quote the study performed in [GeoOO].

Equation (7.1.3) has multiple roots only if either equation (7.1.4) has multiple roots or $\mu = -a^2$. Since (7.1.4) can have only double roots, let us introduce the surfaces

$$\mathcal{C} : T = T^* \equiv 16aQ\sqrt{Q}/(3\sqrt{3}), \qquad (7.1.5)$$

$$\mathcal{C}_1 : T = T^{**} \equiv (Q + a^2)^2, \qquad (7.1.6)$$

corresponding to a double root of (7.1.4) and to the root $\mu = -a^2$ respectively. Denote by $\mathcal{C}^*$ their intersection, *i.e.*

$$\mathcal{C}^* : Q = 3a^2, \quad T = 16a^4. \qquad (7.1.7)$$

The projection of $\mathcal{C}^*$ on the plane (a, Q) is

$$\mathcal{C}_*^* : Q = 3a^2. \qquad (7.1.8)$$

Consider $a, Q, T > 0$. Then, for $(a, Q, T) \in \mathbf{R}_+^{*\,3} \setminus (C \cup C_1)$, (7.1.3) has eight mutually disjoint roots $\lambda_{1,\dots,8}$ and (7.1.4) has four mutually distinct roots $\mu_{1,\dots,4}$, related to $\lambda_{1,\dots,8}$ by the relations: $\lambda_{1,5} = \pm\sqrt{\mu_1 + a^2}$, $\lambda_{2,6} = \pm\sqrt{\mu_2 + a^2}$, $\lambda_{3,7} = \pm\sqrt{\mu_3 + a^2}$, $\lambda_{4,8} = \pm\sqrt{\mu_4 + a^2}$, for $\mu_i > -a^2$; $\lambda_{1,5} = \pm\sqrt{\mu_1 + a^2}$, $\lambda_{2,6} = \overline{\lambda_{1,5}}$, $\lambda_{3,7} = \pm\sqrt{\mu_1 + a^2}$, $\lambda_{4,8} = \overline{\lambda_{3,7}}$ for $\mu_i \in \mathbf{C}$ or $\mu_i < -a^2$.

Since (7.1.4) is much easier to study than (7.1.3), let us relate the multiplicities of the roots of these two characteristic equations. Thus, $\mu_1 = \mu_2$ implies either $\lambda_1 = \lambda_2 > 0$, $\lambda_5 = \lambda_6$ if $\mu_1 = \mu_2 \in \mathbf{R}$, or $\lambda_1 = \lambda_2 = \lambda_5 = \lambda_6 = 0$ if $\mu_1 = \mu_2 = -a^2$, or λ_1 is purely imaginary and $\lambda_1 = -\lambda_2$, $\lambda_5 = -\lambda_6$, if $\mu_{1,2} < -a^2$.

Therefore, almost everywhere in $\mathbf{R}_+^{*\,3}$, more exactly for $(a, Q, T) \in \mathbf{R}_+^{*\,3} \setminus (C \cup C_1)$, (7.1.4) and (7.1.3) have mutually distinct roots; for $(a, Q, T) \in C \setminus C^*$, (7.1.4) has two equal real roots $\mu_1 = \mu_2 \neq -a^2$ and (7.1.3) has two pairs of equal and nontrivial roots $\lambda_1 = \lambda_2$, $\lambda_5 = \lambda_6$; for $(a, Q, T) \in C^*$, (7.1.4) has two roots equal to $-a^2$, i.e. $\mu_1 = \mu_2 = -a^2$ and (7.1.3) has four trivial roots $\lambda_1 = \lambda_2 = \lambda_5 = \lambda_6 = 0$; for $(a, Q, T) \in C_1 \setminus C^*$, (7.1.4) has one root equal to $-a^2$ and the others different, i.e. $\mu_1 = -a^2$ μ_2, μ_3, $\mu_4 \neq -a^2$. In this case $\lambda_1 = \lambda_5 = 0$, all other roots of (7.1.3) are non-vanishing and mutually disjoint [Geo82a], [Geo00].

In the space (μ, a, Q, T) the characteristic manifold defined by (7.1.4) has four sheets which coalesce along the surface $\mu_1 = \mu_2 = -a\sqrt{Q/3}$, $T = T^*$. In the space (λ, a, Q, T) the characteristic manifold defined by (7.1.3) has eight sheets, two of them coalescing along the surface $\lambda_1 = \lambda_5 = 0$, $T = T^{**}$, $Q \neq 3a^2$; two pairs of them coalescing along the surfaces $\lambda_1 = \lambda_2 = \sqrt{a^2 - a\sqrt{Q/3}}$, $T = T^*$, $Q \neq 3a^2$ and $\lambda_5 = \lambda_6 = -\sqrt{a^2 - a\sqrt{Q/3}}$, $T = T^*$, $Q \neq 3a^2$; four of them coalesce along the curve $\lambda_1 = \lambda_2 = \lambda_5 = \lambda_6 = 0$, $T = 16a^4$, $Q = 3a^2$. These results suggested to us the following

Theorem 7.1.1 [GeoPalR05]. *The surface C is a bifurcation set for the characteristic manifolds (7.1.3) and (7.1.4), the surface C_1 for (7.1.3), but not for (7.1.4), C^* for (7.1.3) C_*^* for (7.1.3) taken on C. In addition, C_*^* is the bifurcation set for $C \cup C_1$.*

Proof. If the characteristic manifold defined by (7.1.3) has bifurcation points, then equation (7.1.3) has multiple roots. For the case of a double root, this root satisfies (7.1.3) and the equation obtained by differentiating (7.1.3) with respect to λ, *i.e.*

$$2\lambda[4(\lambda^2 - a^2)^3 + 4Qa^2(\lambda^2 - a^2) + Ta^2] = 0. \tag{7.1.9}$$

If $\lambda = 0$ is the double root, then (7.1.3) implies $T = T^{**}$, therefore C_1 is the bifurcation set for (7.1.3). If the double root is one of the roots of (7.1.3) and the equation

$$4(\lambda^2 - a^2)^3 + 4Qa^2(\lambda^2 - a^2) + Ta^2 = 0, \tag{7.1.10}$$

then, by the Euclid algorithm, it follows that the double root can be either $\lambda = \sqrt{a^2 - \frac{4aQ}{3\sqrt{T}}}$ or $\lambda = -\sqrt{a^2 - \frac{4aQ}{3\sqrt{T}}}$, both leading to $T = T^*$. Whence C is the bifurcation set for (7.1.3).

If the characteristic manifold defined by (7.1.4) has bifurcation points, then equation (7.1.4) has multiple roots. For the case of a double root, this root satisfies (7.1.4) and the equation obtained by differentiating (7.1.4) with respect to μ, *i.e.*

$$4\mu^3 + 4Qa^2\mu + Ta^2 = 0. \tag{7.1.11}$$

Then, again by the Euclid algorithm, it follows that the double root is $\mu = -\frac{16}{9}\frac{Q^2 a^2}{T}$, which introduced into (7.1.4) or (7.1.11) leads to $T = T^*$. Hence C is the single bifurcation set for (7.1.4).

As expected, C_1 is not a bifurcation set for (7.1.4) because for the points of C_1 equation (7.1.4) has mutually disjoint roots, one of which being equal to $-a^2$ and leading to two equal solutions of (7.1.3).

Assume that (7.1.3) has a multiple root of multiplicity equal to 4. Then it must satisfy (7.1.3), (7.1.9) and

$$4(\lambda^2 - a^2)^3 + 4Qa^2(\lambda^2 - a^2) + Ta^2 + 2\lambda^2[12(\lambda^2 - a^2)^2 + 4Qa^2] = 0, \quad (7.1.12)$$

$$\lambda[12(\lambda^2 - a^2)^2 + 4Qa^2] + 16\lambda^3(\lambda^2 - a^2) = 0. \quad (7.1.13)$$

Since the multiple root satisfies (7.1.10), from (7.1.12) it follows that it must satisfy the equation $12(\lambda^2 - a^2)^2 + 4Qa^2 = 0$ and from (7.1.13) it follows that it satisfies the equation $16\lambda^3(\lambda^2 - a^2) = 0$. Supposing that this root is nontrivial, it follows that it can be either $\lambda_1 = a$ or $\lambda_2 = -a$. In both these cases from $12(\lambda^2 - a^2)^2 + 4Qa^2 = 0$ we have $Qa^2 = 0$, which contradicts the assumption $a, Q, T > 0$. Therefore the single root of (7.1.3) which can have multiplicity equal to four is $\lambda_1 = 0$ for which from (7.1.12) it follows $-4a^6 - 4Qa^4 + Ta^2 = 0$, while from (7.1.3) we have $a^8 + 2Qa^6 - Ta^4 + Q^2a^4 = 0$. These two relations imply $Q = 3a^2$, $T = 16a^4$, hence C^* is a bifurcation set for (7.1.3) (of a type different from those of $C \setminus C^*$ and $C_1 \setminus C^*$).

For (7.1.4), C^* is just part of the bifurcation set.

The surface $C \cup C_1$ has two sheets but for C^*, where the two sheets coalesce. The projection of C^* on the (a, Q) plane is C_*^*, therefore C_*^* is the bifurcation set for $C \cup C_1$. This can also be seen considering the equation

$$[T - 16aQ\sqrt{Q}/(3\sqrt{3})][T - (Q + a^2)^2] = 0, \quad (7.1.14)$$

which defines $C \cup C_1$ and which possesses a double root for $Q = 3a^2$. C_*^* is a bifurcation set for (7.1.3) taken on C, *i.e.* for

$$(\lambda^2 - a^2)^4 + 2Qa^2(\lambda^2 - a^2)^2 + 16a^3Q\sqrt{Q}/(3\sqrt{3})(\lambda^2 - a^2) + Q^2a^4 = 0, \quad (7.1.15)$$

because differentiating this equation with respect to λ and imposing to the solution $\lambda = 0$ of the obtained equation to satisfy (7.1.15) we obtain $Q = 3a^2$ defining C_*^*.

The detailed geometrical structure of the characteristic manifolds and bifurcation manifolds is given in [Geo82a], [Geo82b].

In the parameter (a, Q, T)-space, the bifurcation sets determine some regions, called the strata (Appendix 6). The bifurcation sets or part of them are strata too.

7.1.2   *Secular equations dependent on boundary conditions*

The secular equations must be written separately for each stratum determined by the bifurcation sets, because for each of them the form of the general solution of (7.1.1), (7.1.2) is different. We write this solution only for an even function v [Geo82a] (physical reasons show that odd v is not realistic). In the parameter (a, Q, T)-space, the strata are: $\mathbf{R}_+^{*\,3} \setminus (C \cup C_1)$, (they are of topological dimension equal to three); $C_1 \setminus C^*$ and $C \setminus C^*$ (two-dimensional) and C^* (one-dimensional). The types of general even solutions for (7.1.1), (7.1.2) corresponding to the strata and, therefore, to various multiplicities of λ_i, are

$$v(z) = \sum_{i=1}^{4} A_i \cosh(\lambda_i z), \quad \text{for} \quad (a, Q, T) \in \mathbf{R}_+^{*\,3} \setminus (C \cup C_1) \tag{7.1.16}$$

$$v(z) = A_1 \cosh(\lambda_1 z) + B_2 z \sinh(\lambda_1 z) + \sum_{i=1}^{3} A_i \cosh(\lambda_i z), \quad \text{for} \quad (a, Q, T) \in C \setminus C^* \tag{7.1.17}$$

$$v(z) = A_1 + \sum_{i=2}^{4} A_i \cosh(\lambda_i z), \quad \text{for} \quad (a, Q, T) \in C_1 \setminus C^* \tag{7.1.18}$$

$$v(z) = A_1 + A_2 z^2 + \sum_{i=1}^{3} A_i \cosh(\lambda_i z), \quad \text{for} \quad (a, Q, T) \in C^*. \tag{7.1.19}$$

Since (7.1.18) can be, simply, obtained from (7.1.16) for $\lambda_1 = 0$, we no longer consider it. In this case no secular points exist [GeoOO] and, so, the entire surface $C_1 \setminus C^*$ is indeed a false secular manifold.

The secular equation corresponding to (7.1.16) reads

$$\begin{vmatrix} \lambda_1 \sinh \frac{\lambda_1}{2} & \cdots \\ \mu_1 \cosh \frac{\lambda_1}{2} & \cdots \\ (\mu_1^2 + Qa^2) \cosh \frac{\lambda_1}{2} & \cdots \\ \lambda_1 (\mu_1^2 + Qa^2) \sinh \frac{\lambda_1}{2} & \cdots \end{vmatrix} = 0, \quad \text{for} \quad (a, Q, T) \in \mathbf{R}_+^{*\,3} \setminus (C \cup C_1). \tag{7.1.20}$$

The i-th lacking column in (7.1.20) is identical to the first column but with λ_1 and μ_1 replaced by λ_i and μ_i.

Formally, in the secular equation for $(a, Q, T) \in C \setminus C^*$ the first, third and fourth columns are identical with those from (7.1.20), while the second column is obtained by differentiating the second column in (7.1.20) with respect to λ_2 and then replacing λ_2 by λ_1. Similarly, in the secular equation for $(a, Q, T) \in C^*$, the second column is obtained by differentiating twice the second column in (7.1.20) and then replacing λ_2 by λ_1 [Geo82a], [GeoOO], the remaining columns being identical with those from (7.1.20).

7.1.3 *Secular equations and points; independence of boundary conditions*

For an easier solution of the secular equations the notation $t_j = \lambda_j \tanh(\lambda_j/2)$ is introduced. As a consequence, the new form of the secular equations will contain the product of $\cosh(\lambda_j/2)$, $j = 1, \ldots, 4$. If some of λ_j are purely imaginary then the corresponding $\cosh(\lambda_j/2)$ vanish, the equality $\prod_{j=1}^{4} \cos(i\lambda_j/2) = 0$ representing additional secular equations which do not depend on the boundary conditions. Hence, (7.1.3) must have purely imaginary roots $\lambda_j = i(2n-1)\pi$, $n \in \mathbf{N}$. Correspondingly, (7.1.4) must have negative roots of the form

$$\mu_j = -A_n, \tag{7.1.21}$$

where $A_n = a^2 + (2n-1)^2\pi^2$. Equation (7.1.4) always has two negative and two complex conjugate solutions. Therefore we focus our attention only on the negative roots of (7.1.4). Let $(a, Q, T) \in \mathbf{R}_+^{*\,3} \setminus (C \cup C_1)$. Then substituting in (7.1.4) the expression $-A_n$ for μ we obtain the infinite set of the secular equations

$$T = (A_n^2 + Qa^2)/(a^2 A_n). \tag{7.1.22}$$

If $Qa^2 < 3A_1^2$, i.e. $Q{\leq}3a^2[a^2 + \pi^2/a^2]^2$, then T increases with n and, so, the neutral manifold corresponds to $n = 1$ and reads

$$T = [Qa^2 + (a^2 + \pi^2)^2]^2/[a^2(a^2 + \pi^2)], \quad \text{for} \quad Qa^2 < 3A_1^2. \tag{7.1.23}$$

If $Qa^2 > 3A_n^2$, then T decreases with n, therefore the corresponding neutral equation is

$$T = (A_n^2 + Qa^2)^2/(a^2 A_n), \quad \text{for} \quad Qa^2 > 3A_n^2. \tag{7.1.24}$$

Let $(a, Q, T) \in C \setminus C^*$. In this case the negative roots of (7.1.4) are double, namely $\mu_1 = \mu_2 = -a\sqrt{Q}/\sqrt{3}$. Substituting them in (7.1.21) we obtain an infinity of secular equations

$$Q = 3A_n^2/a^2, \tag{7.1.25}$$

whence the neutral equation

$$Q = 3(a^2 + \pi^2)^2/a^2. \tag{7.1.26}$$

If $(a, Q, T) \in C_1 \setminus C^*$, then $\mu = -a^2$ and $\mu_2 \in [-1, 0)$ for $Q{\leq}3a^2$ and $\mu_2{\leq} -1$ for $Q > 3a^2$. The equality (7.1.21) can hold only for $\mu_2 < -1$, when (7.1.4) implies $Q = -\mu + (a^2 - \mu)\sqrt{-\mu}/a$, whence the secular equations

$$Q = A_n + (a^2 + A_n)\sqrt{A_n}/a, \tag{7.1.27}$$

implying the neutral equation

$$Q = (a^2 + \pi^2) + (2a^2 + \pi^2)\sqrt{a^2 + \pi^2}/a. \tag{7.1.28}$$

For the points of C^*, we have $\mu_1 = \mu_2 = -a^2$, therefore in this case no additional secular points exist. Consequently, taking into account all these we have

Theorem 7.1.2 [GeoPal03]. *The additional secular points for (7.1.1), which are independent of the boundary conditions (7.1.2), are those defined by (7.1.23), (7.1.24), (7.1.26), (7.1.28). In the last two cases the secular points are situated on $C \setminus C^*$ and $C_1 \setminus C^*$ respectively.*

Remark 7.1.1. The computations simplify considerably if we notice that (7.1.4) reads $Ta^2\mu = -(\mu^2 + Qa^2)^2$ and if we use the notation $\mu = a^2\gamma$, $Q = a^2S$. For the points of $C_1 \setminus C^*$ we could use the closed form of μ_2 (given by Cardano formula).

7.1.4 *Open problems for (7.1.1) and (7.1.2)*

The first is the existence of solutions of (7.1.20) and of the secular equation for points of $C \setminus C^*$. However, computations show that there exist infinitely many secular surfaces (sheets) defined by (7.1.20) and an infinity of spatial curves situated on $C \setminus C^*$ which consist of secular points. Among them we quote the secular manifold which are independent of the boundary conditions, studied in the above, namely (7.1.22), (7.1.25), (7.1.27) for points of $\mathbf{R}_+^{*3} \setminus (C \cup C_1)$, $C \setminus C^*$ and $C_1 \setminus C^*$ respectively.

In [GeoOO] it was shown that no secular points exist on C^*, therefore C^* is a FNM.

Thus, even for the very simple case of (7.1.1), (7.1.2), the geometry of the set of the secular points is complicated, this set consisting of surfaces and curves separated by C_1 and by C^*. A heuristic reasoning [GeoOO] shows that these curves (all of them belonging to $C \setminus C^*$) are limits for the secular surfaces of (7.1.20). A few numerical results reported in [GeoOO] lead to the following

Conjecture. *Except for the secular points which are independent of the boundary conditions and some other secular curves situated on $C \setminus C^*$, the curve $T = T^*$, $Q = const$ and the surface $C \setminus C^*$, are false secular manifolds. When they exist, the secular curves situated on $C \setminus C^*$ are limit sets of the secular surfaces defined by (7.1.20) and have some extremality properties.*

7.2 Soret-Dufour driven convection

The system of ode's is reduced to a single higher-order ode for each of the unknown functions. It is shown that these higher-order equations are different for different unknown functions. Then, by following [PalGPas], three cases of multiple characteristic values are studied.

7.2.1 *Equations satisfied by the unknown functions*

Consider a horizontal layer of a two-component fluid subject to an external thermal (ΔT) and concentrational (ΔN_1) gradients. The layer is bounded by two rigid and heat-conducting plates, impervious to mass transport. The stability of the mechanical equilibrium of the layer depends on the intensity and direction of these gradients: if $\Delta T < 0$ (the fluid is heated from above) and $\Delta N_1 > 0$ (at the top the solute density is greater than the solvent density), the equilibrium is stable. Indeed, in this case in the upper regions of the layer the fluid density is less than that in the bottom regions. The unstable equilibrium arises when $\Delta T > 0$, $\Delta N_1 < 0$ (the layer is heated from below and salted from above).

The other two cases, *i.e.* $\Delta T \cdot \Delta N_1 < 0$, correspond to the competing temperature and concentration effects. Therefore no *a priori* stability estimation may be done; these cases are investigated in the following.

If the principle of exchange of stabilities holds, the two-point problem governing the neutral stability against normal mode perturbations reads [Ve]

$$(D^2 - a^2)^2 W - s_T R a^2 \Theta + R_s a^2 r_D (\Delta N_1)^{-1} \Gamma = 0, \tag{7.2.1}$$

$$s_T W + (D^2 - a^2)\Theta + D_F(D^2 - a^2)\Gamma = 0, \tag{7.2.2}$$

$$\Delta N_1 r_D^{-1} W - s_T \Delta N_1 (D^2 - a^2)\Theta + (D^2 - a^2)\Gamma = 0, \tag{7.2.3}$$

for $z \in (-0.5; 0.5)$, and

$$W = DW = \Theta = 0, \quad D\Gamma = s_T \Delta N_1 D\Theta \tag{7.2.4}$$

at $z = \pm 0.5$. Here $D \equiv d/dz$, $s_T = |\Delta T|/\Delta T$; $W, \Theta, \Gamma : (-0.5, 0.5) \to \mathbf{R}$ are the unknown eigenfunctions which depend on z. The seven parameters $a > 0$, $\Delta T(> 0)$, $\Delta N_1(< 0)$, $r_D(> 0)$, $R = g\alpha \Delta T d^3/(\nu K)$, $R_s = g\gamma \Delta N_1 d^3/(\nu D_F)$, $D_F(> 0)$, represent the wave number, the difference of bottom and top temperatures, the difference of bottom and top concentrations, the Lewis number, the thermal Rayleigh number, the concentrational Rayleigh number and the coefficient of the concentration diffusivity respectively. The coefficients α and γ are associated with the density change with respect to temperature and concentration, *i.e.* $\rho = \rho_0[1 - \alpha s_T(T - T_0) + \gamma s_N(C - C_0)]$, where $s_N = \Delta N_1/|\Delta N_1|$.

Equations $(7.2.1) - (7.2.3)$ are related to equations (9) from [MaiP89] in some particular situations, namely

$$R = R^*, \ R_c^* = R_s r_D s^* F, \ \lambda^* = (\gamma^*/\beta^*)(K_T^*/K_C^*),$$

$$N^* = -(\beta^{*2}/\gamma^{*2})(K_C^*/K_T^*)/(1 + F), \ r_D = -F/(1 + F).$$

Here $F = D_F \Delta N_1$, $*$ indicates quantities from [MaiP89]. In [MaiP89] the characteristic quantities for Θ_1 and Γ_1 were taken to be β^* and $\gamma^* D_F$ respectively.

If $a, R, R_s, \neq 0$, then $(7.2.1) - (7.2.3)$ imply

$$(D^2 - a^2)\Gamma = -\alpha_1 W, \tag{7.2.5}$$

$$(D^2 - a^2)\Theta = \beta_1 W, \tag{7.2.6}$$

$$(D^2 - a^2)^3 W = a^2 k W, \tag{7.2.7}$$

where $\alpha_1 = \Delta N_1(1 + r_D^{-1})/(1 + F s_T)$, $\beta_1 = (F r_D^{-1} - s_T)/(1 + F s_T)$, $k = \beta_1 R s_T + \alpha_1 r_D R_s (\Delta N_1)^{-1}$, whence

$$(D^2 - a^2)^4 \Theta = a^2 k (D^2 - a^2)\Theta, \tag{7.2.8}$$

$$(D^2 - a^2)^4 \Gamma = a^2 k (D^2 - a^2)\Gamma. \tag{7.2.9}$$

Put $[(D^2 - a^2)^3 - a^2 k]\Theta = V$. Then equation (7.2.8) reads $(D^2 - a^2)V = 0$, while at $z = \pm 0.5$ we have $V(\pm 0.5) = -\beta_1 R_s a^2 r_D (\Delta N_1)^{-1} \Gamma(\pm 0.5)$. Since $\Gamma(\pm 0.5)$ are arbitrary, it follows that $V(\pm 0.5) \neq 0$. Moreover $DV(\pm 0.5) = -R_s a^2 D\Theta(\pm 0.5)$, where $D\Theta(\pm 0.5)$ are arbitrary. These two results show that, due to the boundary conditions $(7.2.4)_4$, it is impossible to write $(7.2.1)-(7.2.4)$ as an equivalent problem with the equation in Θ and Γ being of the form (7.2.7).

However, since (7.2.1) implies

$$(D^2 - a^2)^2 W = R_s a^2 r_D (\Delta N_1)^{-1} U, \tag{7.2.10}$$

where $U = \left(\dfrac{R s_T}{R_s r_D (\Delta N_1)^{-1}} \Theta - \Gamma \right)$, we perform a linear combination of (7.2.5) and (7.2.6) such that in the left-hand side stand $(D^2 - a^2)V$. Thus, $(7.2.6) \cdot \dfrac{R s_T}{R_s r_D \Delta N_1^{-1}} -$ (7.2.5) reads

$$(D^2 - a^2)U = \left(\frac{\beta_1 R s_T}{R_s r_D (\Delta N_1)^{-1}} + \alpha_1 \right) W = \frac{k}{R_s r_D (\Delta N_1)^{-1}} W. \tag{7.2.11}$$

Then, applying $(D^2 - a^2)^2$ to (7.2.11) and taking into account (7.2.10) we obtain

$$(D^2 - a^2)^3 U = a^2 k U. \tag{7.2.12}$$

This could be seen if we took into account that the general solution of equation (7.2.7) is

$$W(z) = \sum_{i=1}^{3} A_i \cosh(\lambda_i z),$$

and, applying the method of variation of coefficients to (7.2.5) and (7.2.6), we have

$$\Theta(z) = \sum_{i=1}^{3} \beta_1 A_i (\lambda_i^2 - a^2)^{-1} \cosh(\lambda_i z) + B \cosh(a z),$$

$$\Gamma(z) = -\sum_{i=1}^{3} \alpha_1 A_i (\lambda_i^2 - a^2)^{-1} \cosh(\lambda_i z) + C \cosh(a z)$$

where λ_i, $i = 1, \ldots, 6$ are the roots of the characteristic equation of (7.2.7), *i.e.*

$$(\lambda^2 - a^2)^3 - a^2 k = 0, \tag{7.2.13}$$

namely

$$\lambda_i = \sqrt{a^2 + \sqrt[3]{a^2 k}\,\epsilon_i}, \quad \lambda_{i+3} = -\lambda_i, \quad i = 1, 2, 3, \quad \epsilon_1 = 1, \quad \epsilon_{2,3} = (-1 \pm i\sqrt{3})/2.$$

Here A_1, A_2, A_3, B and C are still arbitrary constants. The boundary condition $(7.2.4)_3$ implies $B = -\beta_1 \sum_{i=1}^{3} A_i (\lambda_i^2 - a^2)^{-1} \cosh(\lambda_i/2)[\cosh(a/2)]^{-1}$, while (7.2.1) provides $C = B R \Delta N_1 s_T / (R_s r_D)$. In this way, the secular equation has the form

$$t_2 t_3 + \epsilon_2 t_1 t_2 + \epsilon_3 t_1 t_3 - E t_0 (t_1 + \epsilon_3 t_2 + \epsilon_2 t_3) = 0, \tag{7.2.14}$$

where $t_0 = a \tanh(a/2)$, $t_i = \lambda_i \tanh(\lambda_i/2)$, $i = 1, 2, 3$ and λ_i are the mutually distinct solutions of (7.2.13) (which implies $\Delta N_1 < \infty$, $a, k \neq 0$, $k \neq -a^4$) and

$$E = (F s_T - r_D)(S^{-1} - 1)(1 + F s_T)^{-1}, \qquad S = r_D R_s R^{-1}.$$

7.2.2   *Case $a = 0$ and case $k = 0$*

Although the problem $(7.2.1) - (7.2.4)$ contains seven parameters, the general characteristic equation $(7.2.13)$ has only two parameters (a, k), therefore we discuss the multiplicity of its solutions in terms of a and k. Differentiating $(7.2.13)$ we obtain $\lambda(\lambda^2 - a^2)^2 = 0$, therefore the possible multiple solutions are $\lambda = 0$, $\lambda = a$, $\lambda = -a$. For $\lambda = 0$, $(7.2.13)$ implies $a^6 + a^2 k = 0$, *i.e.* $a = 0$, or $k = -a^4$. For $\lambda = \pm a$ $(7.2.13)$ implies $a^2 k = 0$, *i.e.* either $a = 0$, or $k = 0$.

Case $a = 0$. Equation $(7.2.13)$ becomes $\lambda^6 = 0$, therefore $\lambda_1 = \ldots = \lambda_6 = 0$. However, in order to obtain the equation in W it is no longer suitable to apply the operator $D^2 - a^2$, the equation in W follows directly from $(7.2.1)$ and reads $D^4 W = 0$. Thus, instead of $(7.2.13)$ we have the particular characteristic equation $\lambda^4 = 0$. So, taking into account $(7.2.4)_{1,2}$ we get $W = 0$. Then $(7.2.2)$ and $(7.2.4)_3$ imply $\theta \equiv 0$ and $(7.2.1)$ and $(7.2.4)_4$ yield $\Gamma = const$. This means that in the case $a = 0$ (physically unrealistic because it corresponds to constant perturbation, but mathematically interesting because it is a limit situation) in the $(a, \Delta T, \Delta N_1, r_D, R, R_s, D_F)$ parameter space the manifold $a = 0$ consists of neutral points.

Case $k = 0, a \neq 0$. The terms in Θ and Γ are present in $(7.2.1)$ but their influence on the equation in W is null. Indeed, if $D^2 - a^2$ is applied to $(7.2.1)$, these terms no longer appear. Let us recall that k represents the joint thermal and concentrational influences on the dynamic quantity W. In spite of the fact that $k = 0$, the secular equation will contain these influences via the boundary conditions.

There are several possible subcases corresponding to $k = 0$. Let us consider the following: $R = R_s = 0, \beta_1 \neq 0$; $R_s = \beta_1 = 0, R \neq 0$; $R = \beta_1 = R_s = 0$. All of them lead to $W = \Theta = 0$, $\Gamma =$const, therefore the case similar to $a = 0$ in which all these points $(a, k, E) = (a, 0, E)$ are neutral points. In these subcases, equation $(7.2.3)$ was not suited because instead we had $W^4 = 0$.

If we avoid the case of films of fluids $(d \to 0)$ and feable concentration compression $(\gamma \to 0)$ and feable temperature compression $(\alpha \to 0)$, then $R_s = 0$ if and only if $\Delta N_1 = 0$ and $R = 0$ if and only if $\Delta T = 0$.

Thus the first and the third subcases correspond to the absence of concentration and temperature gradients whereas the second, to the absence of concentration gradient only.

In the cases $\Theta = 0$ or/and $\Gamma = const$, equations $(7.2.1)$ and $(7.2.3)$ are not entitled to represent energy and concentration balance equations (in fact, they reduce to the identity 0=0). However, as mentioned, these physically degenerated cases are interesting mathematically limit cases.

The last subcase of case $k = 0$ is $R, R_s, \beta_1 \neq 0$, $S \neq 0$ and corresponds to the even general solutions

$$
\begin{cases}
W(z) = A_1 \cosh(az) + A_2 z \sinh(az) + a_3 z^2 \cosh(az), \\[2mm]
\Theta(z) = B \cosh(az) + \dfrac{\beta_1}{4a^3}(2A_1 a^2 - A_2 a + A_3) z \sinh(az) \\[2mm]
\qquad + \dfrac{\beta_1}{4a^2}(A_2 a - A_3) z^2 \cosh(az) + \dfrac{\beta_1}{6a} A_3 z^3 \sinh(az), \\[2mm]
\Gamma(z) = C \cosh(az) - \dfrac{\alpha_1}{4a^3}(2A_1 a^2 - A_2 a + A_3) z \sinh(az) \\[2mm]
\qquad - \dfrac{\alpha_1}{4a^2}(A_2 a - A_3) z^2 \cosh(az) - \dfrac{\alpha_1}{6a} A_3 z^3 \sinh(az),
\end{cases}
\tag{7.2.15}
$$

which introduced in (7.2.4) and (7.2.1), provide the secular equation

$$
t_0^4 \left(\frac{1}{16a^5} - \frac{1}{48a^3} \right) + t_0^3 \left(-\frac{1}{4a^5} - \frac{1}{12a^3} - \frac{4}{aR_s} \right) + t_0^2 \left(\frac{1}{4a^5} + \frac{1}{8a^3} + \frac{1}{24a} + \frac{8}{aR_s} \right)
$$

$$
+ t_0 \left(\frac{1}{4a^3} + \frac{1}{12a} + \frac{4a}{R_s} \right) - \frac{a}{48} - \frac{3}{16a} = 0,
$$

whose solutions correspond to secular points of the form $(a, k, E) = (a, 0, E)$, where $R, R_s, \beta_1, S \neq 0$.

7.2.3 *Case* $k = -a^4$

The characteristic equation reduces to $(\lambda^2 - a^2)^3 + a^6 = 0$ and has the solutions

$$
\lambda_{1,4} = 0, \quad \lambda_{3,6} = \pm \frac{a\sqrt[4]{3}}{\sqrt{2}} \left(\frac{\sqrt{3}+1}{2} + i\frac{\sqrt{3}-1}{2} \right), \quad \lambda_{2,5} = \pm \frac{a\sqrt[4]{3}}{\sqrt{2}} \left(\frac{\sqrt{3}+1}{2} - i\frac{\sqrt{3}-1}{2} \right).
$$

Correspondingly, the general even solution of $(7.2.5) - (7.2.7)$ has the form given at the beginning of Section 7.2.2, where λ_i are given by these expressions. So, the secular equation will be the limit of equation (7.2.14) as $\lambda_1 \to 0$, i.e. $t_2 t_3 - E t_0 (\epsilon_3 t_2 + \epsilon_2 t_3) = 0$. This equation yields, for $E \neq 0, \infty$ (this last case corresponding to $S = 0$), the expression of E as a function of a $E(a) = t_2 t_3 [t_0 (\epsilon_3 t_2 + \epsilon_2 t_3)]^{-1}$, where t_2 and t_3 are computed using the λ_i corresponding to case $k = -a^4$. Then we have the expression

$$
S = \frac{F s_T - r_D}{F s_T - r_D + (1 + F s_T) t_2 t_3 [t_0 (\epsilon_3 t_2 + \epsilon_2 t_3)]^{-1}},
\tag{7.2.16}
$$

which depends on r_D, F, s_T, and a. Since $S = R_s r_D R^{-1}$, (7.2.16) is a linear relationship between R and R_s. Another relationship between R and R_s is obtained from the expressions of α_1, β_1 and k, taking into account that $k = -a^4$. It follows

$$
a^4 = \frac{[F(r_D^{-1} s_T - 1)R + R_s(1 + r_D)]}{1 + F s_T}.
\tag{7.2.17}
$$

In this way, for fixed r_D, F and s_T, (7.2.16) and (7.2.17) define the neutral curve given parametrically by $R = R(a), R_s = R_s(a)$. It is a limit curve for the neutral equation corresponding to the general case of α_1, β_1 and k. This curve consists of those points (a, R, R_s) of the surface $k = -a^4$ for which the right-hand side of (7.2.17) is positive. Graphical representations of the neutral hypersurface in the space $a, \Delta N_1, R$, and some neutral curves $R = R(a)$ for particular values of r_D, R_s and DN_1 are given in [PalGPas].

7.3   Magnetic Soret-Dufour driven convection

In the case of the even eigensolutions, for double roots of the characteristic equation, by arguments simpler than in [PalGM], the existence of secular curves is proved and false secular points are found. The nonmagnetic case is also treated.

7.3.1   *Eigenvalue problem*

Consider an orthonormal frame of reference $\{O, \mathbf{i}, \mathbf{j}, \mathbf{k}\}$ with $\mathbf{k}$ upwards positive and a two-component fluid in a horizontal layer bounded by the rigid walls $z = \pm 0.5$ in a constant magnetic field $\mathbf{H_0} = H_o \mathbf{k}$, subject to the action of external thermal and concentrational gradients. Assume that the walls are heat and diffusion conductors (and not impervious to mass transport as in Section 7.2) but electrical isolators. Suppose that the principle of exchange of stabilities holds. Then the eigenvalue two-point problem governing the neutral stability of the conduction state (1.4.18) against normal mode perturbations (1.4.12) consists of the equations for $z \in [-0.5, 0.5]$

$$(D^2 - a^2)^2 W - \mathcal{R}^2 a^2 \Theta + s\mathcal{C}^2 a^2 r_D \Gamma - M^2 D^2 W = 0, \tag{7.3.1}$$

$$W + (1 + N\lambda^2 r_D)(D^2 - a^2)\Theta + N\lambda \frac{\alpha}{\beta_c} \mathcal{C}^2 \mathcal{R}^{-2} r_D^2 (D^2 - a^2)\Gamma = 0, \tag{7.3.2}$$

$$W + \lambda \frac{\beta_c}{\alpha} \mathcal{R}^2 \mathcal{C}^{-2}(D^2 - a^2)\Theta + r_D(D^2 - a^2)\Gamma = 0, \tag{7.3.3}$$

$$DZ + (D^2 - a^2)X = 0, \tag{7.3.4}$$

$$(D^2 - a^2)Z + M^2 DX = 0, \tag{7.3.5}$$

and the boundary conditions

$$W = DW = \Theta = \Gamma = X = DZ = 0, \qquad \text{at} \qquad z = \pm 0.5\,. \tag{7.3.6}$$

The unknown amplitudes of the vertical velocity, temperature, concentration, magnetic and electric field $W, \Theta, \Gamma, X, Z : [-0.5; 0.5] \to R$ depend only on z. The eigenvector $(W, \Theta, \Gamma, X, Z)$ and its corresponding eigenvalue, say $\mathcal{R}$, depend on the remained seven parameters $a, M, \mathcal{C}, \tau, \sigma, \lambda, N$ defined in Section 1.1.1.

Remark 7.3.1. From (7.3.4), (7.3.5) follows that $[(D^2 - a^2)^2 - M^2 D^2](X, Z) = 0$ for $z \in [-0.5, 0.5]$ and, taking into account $(7.3.6)_5$ and $(7.3.6)_6$ respectively we deduce that X and Z are the null functions on $[-0.5, 0.5]$. Elimination of Γ and Θ between $(7.3.1) - (7.3.3)$ yields

$$[(D^2 - a^2)^2 - M^2 D^2](D^2 - a^2)W - b_1 a^2 W = 0, \quad z \in [-0.5, 0.5] \tag{7.3.7}$$

$$W = DW = [(D^2 - a^2)^2 - M^2 D^2]W = 0, \quad \text{at} \quad z = \pm 0.5 \tag{7.3.8}$$

where $b_1 = -\mathcal{R}^2 a_1 + s\mathcal{C}^2 r_D a_2$, $a_1 = 1 - N\lambda^2 r_D \mathcal{C}^2 \mathcal{R}^{-2}$, $a_2 = (1 + N\lambda^2 r_D - N\lambda \frac{\alpha}{\beta_c} \mathcal{R}^2 \mathcal{C}^{-2}) r_D^{-1}$. In this way, the eigenvalue $(7.3.1) - (7.3.6)$ is equivalent to the eigenvalue problem (7.3.7), (7.3.8), where b_1 is the eigenvalue, W is the eigensolution and they depend on the wave number a and Hartmann number M.

Remark 7.3.2. To $a = 0$, no eigenvalue corresponds. Indeed, in this case (7.3.7) reads $D^2 V = 0$, where $V = (D^4 - M^2 D^2)W$, and $(7.3.8)_3$, can be written as $V(\pm 0.5) = 0$. It is immediate that $V = 0$. Therefore $(D - M^2 D^2)W = 0$ for $z \in [-0.5, 0.5]$. Taking into account $(7.3.8)_{1,2}$, it follows that, up to a constant, for $M \neq 0$, the secular equation reads $M \sinh(M/2)[M/2 - \tanh(M/2)] = 0$. But this last equation has the unique root $M = 0$, therefore a contradiction; hence, for $M \neq 0$ no secular root exists. For $M = 0$ the last equation in W reads $D^4 W = 0$, implying $W = A_1 + A_2 z + A_3 z^2 + A_4 z^3$. Imposing $(7.3.8)_{1,2}$ yields $W = 0$. Consequently, in the $(a, M, \mathcal{C}, \tau, \sigma, \lambda, N)$ space, the points situated on the manifold $a = 0$ are false points.

Since secular eigenvalues exist only for $a > 0$, this is the unique case considered in the following.

7.3.2 *Case $M > 0$. Double roots of the characteristic equation*

The case of mutually distinct roots was treated theoretically and numerically in [PalGM]. The characteristic equation for (7.3.7) is

$$[(\lambda^2 - a^2)^2 - M^2 \lambda^2](\lambda^2 - a^2) - b_1 a^2 = 0, \tag{7.3.9}$$

or, equivalently, in terms of $\mu \equiv \lambda^2 - a^2$,

$$\mu^3 - M^2 \mu^2 - a^2 M^2 \mu - a^2 b_1 = 0. \tag{7.3.10}$$

Denote by $\lambda_1, \ldots, \lambda_6$ the roots of (7.3.9) and by μ_1, μ_2, μ_3 the roots of (7.3.10). We have

$$\lambda_{1,4} = \pm\sqrt{\mu_1 + a^2}; \quad \lambda_{2,5} = \pm\sqrt{\mu_2 + a^2}; \quad \lambda_{3,6} = \pm\sqrt{\mu_3 + a^2}. \tag{7.3.11}$$

The case $\mu_1 = \mu_2 = \mu_3$ is impossible. Indeed, the Viète relation $3\mu_1^2 = -a^2 M^2$ contradicts the fact that $\mu_1 \in \mathbf{R}$.

This is why in the following we consider the case $\mu_1 = \mu_2 \neq \mu_3$, where $\mu_1, \mu_3 \in \mathbf{R}$.

Case $\mu_1 = \mu_2 \neq \mu_3$. From the two Viète relations (therefore independently of b_1) it follows either

$$\mu_1 = \left(M^2 + \sqrt{M^4 + 3M^2 a^2}\right)/3 > 0, \quad \mu_3 = \left(M^2 - 2\sqrt{M^4 + 3M^2 a^2}\right)/3 < 0,$$

$$b_1 = -\left[2M^6 + 9a^2 M^4 + (2M^4 + 6M^2 a^2)\sqrt{M^4 + 3M^2 a^2}\right]/27 < 0, \qquad (7.3.12)$$

or

$$\mu_1 = \left(M^2 - \sqrt{M^4 + 3M^2 a^2}\right)/3 < 0, \quad \mu_3 = \left(M^2 + 2\sqrt{M^4 + 3M^2 a^2}\right)/3 > 0,$$

$$b_1 = \left[-(2M^6 + 9a^2 M^4) + (2M^4 + 6M^2 a^2)\sqrt{M^4 + 3M^2 a^2}\right]/27 > 0. \qquad (7.3.13)$$

The general even eigensolution of (7.3.7) reads

$$W(z) = A_1 \cosh(\lambda_1 z) + A_2 z \sinh(\lambda_2 z) + A_3 \cosh(\lambda_3 z)$$

and leads to the secular equation

$$\det \begin{vmatrix} \cosh(\lambda_1/2) & (1/2)\sinh(\lambda_1/2) & \cosh(\lambda_3/2) \\ \lambda_1 \sinh(\lambda_1/2) & \sinh(\lambda_1/2) + (\lambda_1/2)\cosh(\lambda_1/2) & \lambda_3 \sinh(\lambda_3/2) \\ \gamma_1 \cosh(\lambda_1/2) & (\gamma_1/2)\sinh(\lambda_1/2) + \gamma_1'^2 \cosh(\lambda_1/2) & \gamma_3 \cosh(\lambda_3/2) \end{vmatrix} = 0, \quad (7.3.14)$$

where we used the notation $\gamma_i = \mu_i^2 - M^2 \lambda_i^2$ and $\gamma_1' = \dfrac{d\gamma_1}{d\lambda_1} = 2\lambda_1(2\mu_1 - M^2)$ (we remind that in the case of a double root $\lambda_1 = \lambda_2$, formally, in the secular determinant the column corresponding to λ_2 is differentiated with respect to λ_2 and, then, in the result, λ_2 is replaced by λ_1).

Since $\lambda_1 \neq 0$, multiplying in (7.3.14) the first column by $\frac{1}{2}\tanh(\lambda_1/2)$ and subtracting it from the second column, yields the equivalent secular equation

$$(\gamma_3 - \gamma_1 + \gamma_1'\lambda_1)\sinh(\lambda_1/2)\cosh(\lambda_1/2)\cosh(\lambda_3/2)$$

$$+\frac{\lambda_1}{2}(\gamma_3 - \gamma_1)\cosh(\lambda_3/2) - \gamma_1'\lambda_3 \cosh^2(\lambda_1/2)\sinh(\lambda_3/2) = 0, \qquad (7.3.15)$$

where $\gamma_3 - \gamma_1 = \mu_1(\mu_1 - \mu_3) > 0$ and $\gamma_1' = -2\lambda_1 \mu_3$.

Subcase (7.3.12). We have $\gamma_1' > 0$, $\lambda_i^2 > 0$, $\lambda_1 > 0$ and $\lambda_3^2 = \sqrt{M^2 + 3a^2}[\sqrt{M^2 + 3a^2} - 2M]$. Therefore, $\lambda_3^2 > 0$, $\lambda_3 > 0$ for $M \in (0, a)$, $\lambda_3 = 0$ for $M = a$, $\lambda_3^2 < 0$ (hence, denote $\lambda_3 = i\delta_3$) for $M > a$.

For $M < a$ (3.7.15) implies $\tanh \frac{\lambda_3}{2} = h(M, a)$, where

$$h(M, a) = \left[(\gamma_3 - \gamma_1 + \gamma_1'\lambda_1)\sinh(\lambda_1/2)\cosh(\lambda_1/2) + \frac{\lambda_1}{2}(\gamma_3 - \gamma_1)\right] / \left[\gamma_1'\lambda_3 \cosh^2(\lambda_1/2)\right].$$

Taking into account the positivity of $\gamma_3 - \gamma_1$, γ_1', λ_1, λ_3 and since $\lambda_1 > \lambda_3$, it follows successively

$$h(M, a) > \frac{\gamma_3 - \gamma_1 + \gamma_1'\lambda_1}{\gamma_1'\lambda_3}\tanh\frac{\lambda_1}{2} > \frac{\gamma_1'\lambda_1}{\gamma_1'\lambda_3}\tanh\frac{\lambda_1}{2} > \tanh\frac{\lambda_1}{2},$$

implying $\tanh\frac{\lambda_3}{2} > \tanh\frac{\lambda_1}{2}$, therefore $\lambda_3 > \lambda_1$, contradicting (7.3.12).

For $M > a$, we have $\mu_3^2 < -a^2$, $\lambda_3 \in \mathbf{C} \setminus \mathbf{R}$, $\lambda_3 = i\delta_3$ (notation), $\cosh(\lambda_3/2)$ $= \cos(\delta_3/2)$, $\sinh(\lambda_3/2) = i\sin(\delta_3/2)$, $\gamma_1'\lambda_3\sinh(\lambda_3/2) = -\gamma_1'\delta_3\sin(\delta_3/2)$, so the secular equation (7.3.15) reads

$$\tan\frac{\delta_3}{2} = h_1(M, a), \tag{7.3.16}$$

where

$$h_1(M, a) = \frac{(\gamma_3 - \gamma_1 + \gamma_1'\lambda_1)\sinh(\lambda_1/2)\cosh(\lambda_1/2) + \frac{\lambda_1}{2}(\gamma_3 - \gamma_1)}{-\gamma_1'\delta_3\cosh^2(\lambda_1/2)},$$

showing that there is an infinity of secular cylinders defined by (7.3.16).

For $M = a$, we have $\mu_1 = a^2$, $\mu_3 = -a^2$, $\lambda_1 > 0$, $\lambda_3 = 0$. Even if, in this case, there are three pairs of equal roots of (7.3.9), the general even solution preserves its above form and, so, (7.3.15) reduces to the equation $3\sinh\frac{a\sqrt{2}}{2} + a\sqrt{2} = 0$, which has no solution.

Let C_1 be the surface defined by $(7.3.12)_3$. It is the bifurcation set for the characteristic manifolds. Its intersection with the cylinders (7.3.10) represents the set of secular curves, obtained from (7.3.15). Consequently, all points of C_1 which are not situated on these secular curves are false secular points. Their relationships with $\mathcal{R}$, $\mathcal{R}_c$ and the secular curves for mutually distinct roots of (7.3.9) are theoretically and numerically studied in [PalGM].

Subcase (7.3.13). We have $\gamma_1' < 0$, $\gamma_3 - \gamma_1 + \gamma_1'\lambda_1 < 0$, $\lambda_1, \lambda_3 > 0$, $\lambda_1 < \lambda_3$. Let us write $h(M, a)$ in the equivalent form

$$h(M, a) = \frac{[-\gamma_1'\lambda_1 - (\gamma_3 - \gamma_1)]\sinh(\lambda_1/2)\cosh(\lambda_1/2) - \frac{\lambda_1}{2}(\gamma_3 - \gamma_1)}{-\gamma_1'\lambda_3\cosh^2(\lambda_1/2)}.$$

We have

$$h(M, a) < \frac{-\gamma_1'\lambda_1 - (\gamma_3 - \gamma_1)}{-\gamma_1'\lambda_3}\tanh(\lambda_1/2) < \frac{-\gamma_1'\lambda_1}{-\gamma_1'\lambda_3}\tanh(\lambda_1/2) < \tanh(\lambda_1/2).$$

So, the corresponding secular equation $\tanh(\lambda_3/2) = h(M, a)$, (deduced from (7.3.15)) implies $\lambda_3 < \lambda_1$, which contradicts (7.3.13).

Consequently in this case the points of the surface $(7.3.13)_3$ are false secular points.

7.3.3 Case $M = 0$

For $b_1 \neq 0$, $b_1 \neq a^4$, equation (7.3.9) has no double roots. In this case the study of the secular equation corresponding to even eigensolutions is carried out in [PalGM]. For $b_1 = 0$, we have $\lambda_1 = \lambda_2 = \lambda_3 = a$, $\lambda_4 = \lambda_5 = \lambda_6 = -a$ and the general eigensolution is $W(z) = A_1\cosh(az) + A_2 z\sinh(az) + A_3 z^2\cosh(az)$. The corresponding secular equation reduces to $\sinh a + a = 0$ which has no solution. For $b_1 = a^4$, the general even eigensolution reads $W(z) = A_1 + A_2\cosh(\lambda_2 z) + A_3\cosh(\lambda_3 z)$, where $\lambda_{2,5} = \pm a\sqrt{\frac{3+i\sqrt{3}}{2}}$, $\lambda_{3,6} = \pm a\sqrt{\frac{3-i\sqrt{3}}{2}}$, and leads to the secular equation independent of the boundary conditions $\cos\beta/2 = 0$, implying $\beta = \pi + 2k\pi$ where $\beta = Im\lambda_2 = a\sqrt[4]{3}\frac{1-\sqrt{3}}{2\sqrt{2}} = (2-\sqrt{3})Re\lambda_2$. Therefore we get the neutral points $a = \pi\sqrt{2}\frac{(2k-1)(1+\sqrt{3})}{\sqrt[4]{3}}$, $k \in \mathbf{N}$. All other points of the curves $b_1 = 0$ and $b_1 = a^4$ are false secular points.

7.4   Convection in a porous medium

The equations governing the Bénard convection are supplemented with a term the coefficient of which is the dimensionless permeability. Two situations are considered: the bounding surfaces have different natures (Section 7.4.1) and both bounding surfaces are free (Section 7.4.2). The presence of false secular manifolds and their position with respect to the secular manifold from the case of simple eigenvalues is discussed, by following [GeoPal00], in terms of the three parameters involved.

7.4.1   *Lower rigid isothermal surface, upper free isolated surface*

The two-point eigenvalue problem [LebC]

$$(D^2 - k^2)^2 W - \mathcal{K}^{-1}(D^2 - k^2)W = k^2 R_a \Theta, \qquad z \in (0,1) \tag{7.4.1}$$

$$(D^2 - k^2)\Theta = -W, \tag{7.4.2}$$

$$W = D^2 W = D\Theta = 0, \quad \text{at } z = 1, \tag{7.4.3}$$

$$W = DW = \Theta = 0, \quad \text{at } z = 0, \tag{7.4.4}$$

governs the neutral stability of the mechanical equilibrium of a horizontal fluid layer in a porous medium heated from below, where W is the vertical component of the velocity, (W, Θ) is the vector eigenfunction, the thermal Rayleigh number $R_a \geq 0$ is the eigenvalue and $k, \mathcal{K}^{-1} \geq 0$ are given parameters, namely the wave number and the dimensionless permeability.

Several cases occur. Let us start with the most general one.

a) **Case** $k, R_a, \mathcal{K}^{-1} \neq 0$, $R_a \neq k^4 + k^2\mathcal{K}^{-1}$, $R_a \neq 4(27k^2\mathcal{K}^3)^{-1}$. Equations (7.4.1), (7.4.2) imply

$$(D^2 - k^2)^3 W - \mathcal{K}^{-1}(D^2 - k^2)^2 W + k^2 R_a W = 0, \tag{7.4.5}$$

whence, the characteristic equation has one of the following forms

$$\lambda^6 - \lambda^4(3k^2 + \mathcal{K}^{-1}) + \lambda^2(3k^4 + 2k^2\mathcal{K}^{-1}) - (k^6 + k^4\mathcal{K}^{-1} - k^2 R_a) = 0, \tag{7.4.6}$$

$$\mu^3 - \mu^2\mathcal{K}^{-1} + k^2 R_a = 0, \tag{7.4.7}$$

where $\mu = \lambda^2 - k^2$. Let λ_1, λ_2, λ_3, $\lambda_4(= -\lambda_1)$, $\lambda_5(= -\lambda_2)$ $\lambda_6(= -\lambda_3)$ be the solutions of (7.4.6). Since $R_a \neq k^4 + k^2\mathcal{K}^{-2}$, $\lambda_i, i = 1,\dots,6$ are nonnull. As $R_a \neq 4(27k^2\mathcal{K}^3)^{-1}$, (4.7.7) does not admit double roots μ_i, $i = 1,2,3$; it follows that (7.4.6) has no double solution too. Therefore all λ_i are distinct and the characteristic surface has six sheets

$$\mathcal{R}e\lambda_i = \mathcal{R}e\lambda_i(k, R_a, \mathcal{K}^{-1}), \qquad \mathcal{I}m\lambda_i = \mathcal{I}m\lambda_i(k, R_a, \mathcal{K}^{-1}),$$

and the general solution of (7.4.5) is $W(z) = \sum_{i=1}^{3} \big(A_i \cosh(\lambda_i z) + B_i \sinh(\lambda_i z)\big)$, which introduced in (7.4.1) leads to $\Theta(z) = -\sum_{i=1}^{3} \big(\frac{A_i}{\mu_i}\cosh(\lambda_i z) + \frac{B_i}{\mu_i}\sinh(\lambda_i z)\big)$, so imposing to these general solutions to satisfy (7.4.3), (7.4.4) we obtain the following secular equation

$$2\lambda_2\lambda_3(\mu_1 - \mu_3)(\mu_2 - \mu_1)\sinh\lambda_1 + 2\lambda_3\lambda_1(\mu_2 - \mu_1)(\mu_3 - \mu_2)\sinh\lambda_2$$

$$+2\lambda_1\lambda_2(\mu_3 - \mu_2)(\mu_1 - \mu_3)\sinh\lambda_3$$

$$+\frac{\lambda_2\lambda_3}{\mu_2\mu_3}(\mu_3 - \mu_1)(\mu_2 - \mu_1)(\mu_3^2 + \mu_2^2)\sinh\lambda_1\cosh\lambda_2\cosh\lambda_3$$

$$+\frac{\lambda_1\lambda_3}{\mu_1\mu_3}(\mu_2 - \mu_3)(\mu_2 - \mu_1)(\mu_3^2 + \mu_1^2)\sinh\lambda_2\cosh\lambda_1\cosh\lambda_3 \tag{7.4.8}$$

$$+\frac{\lambda_1\lambda_2}{\mu_1\mu_2}(\mu_1 - \mu_3)(\mu_2 - \mu_3)(\mu_2^2 + \mu_1^2)\sinh\lambda_3\cosh\lambda_1\cosh\lambda_3$$

$$-\big(9k^2 R_a + 2k^2\mathcal{K}^{-2}\big)\sinh\lambda_1\sinh\lambda_2\sinh\lambda_3 = 0,$$

defining the secular surface $R_a = R_a(k, \mathcal{K}^{-1})$ in the $(k, R_a, \mathcal{K}^{-1})$ space.

b) **Case** $k = 0, \mathcal{K}^{-1} \neq 0$. The elimination of Θ between (7.4.1) and (7.4.2) is unnecessary since the equations are decoupled. However from (7.4.1) we have $D^4 W - \mathcal{K}^{-1}D^2 W = 0$ and the corresponding characteristic equation reads $\lambda^4 - \mathcal{K}^{-1}\lambda^2 = 0$. Then the general solution of W has the form: $W(z) = A_1 + A_2 z + A_3 \cosh(\sqrt{\mathcal{K}^{-1}}z) + A_3 \sinh(\sqrt{\mathcal{K}^{-1}}z)$ and leads to the secular equation $\tanh\sqrt{\mathcal{K}^{-1}} = \sqrt{\mathcal{K}^{-1}}$, which admits the unique solution $\mathcal{K}^{-1} = 0$. This implies $W = 0$ and, so, equation (7.4.2) becomes $D^2\Theta = 0$. Imposing $(7.4.3)_3$ and $(7.4.4)_3$, we get $\Theta = 0$. Consequently, in case $b)$ no secular points exist, in spite of the fact that equation (7.4.8) is identically satisfied because four columns in the equivalent determinantal equation become identical. Indeed, for $k = 0$ equation (7.4.8) cannot be assimilated with the secular equation because it was derived using a general solution valid only for mutually distinct λ_i. Consequently, the points $(k, R_a) = (0, R_a)$ are false secular points and, so, they are not neutral points either.

Moreover, the points $(k, R_a, \mathcal{K}^{-1}, \mathcal{R}e\lambda, \mathcal{I}m\lambda)$ with $k = 0$ represent bifurcation points for the characteristic hypersurface defined by (7.4.6) [Col3]. In case a) this surface has six sheets whereas in case b) four among them (corresponding to the four zero solutions of (7.4.6)) coalesce. The corresponding bifurcation set consists of the points of the plane $k = 0$.

c) **Case** k, R_a, $\mathcal{K}^{-1} \neq 0$, $R_a = k^4 + k^2\mathcal{K}^{-1}$, $R_a \neq 4(27k^2\mathcal{K}^3)^{-1}$.

In the space $(k, R_a, \mathcal{K}^{-1}, \mathcal{R}e\lambda, \mathcal{I}m\lambda)$ equation $R_a = k^4 + k^2\mathcal{K}^{-1}$ represents a bifurcation manifold for the characteristic hypersurface defined by (7.4.6). At the points of this manifold the roots of (7.4.6) read

$$\lambda_{1,4} = 0, \quad \lambda_{2,3}^2 = \frac{3k^2 + \mathcal{K}^{-1} \pm \sqrt{(3k^2 + \mathcal{K}^{-1})^2 - 4k^2(3k^2 + 2\mathcal{K}^{-1})}}{2}, \quad \lambda_{5,6} = -\lambda_{2,3}$$

and, so, two of the sheets of the characteristic hypersurface coalesce. Moreover, the general solution of (7.4.5) reads

$$W(z) = A_1 + A_2 \cosh(\lambda_2 z) + A_3 \cosh(\lambda_3 z) + B_1 z + B_2 \sinh(\lambda_2 z) + B_3 \sinh(\lambda_3 z)$$

and (7.4.2) implies

$$\Theta(z) = \frac{A_1}{k^2} - \frac{A_2}{\lambda_2^2 - k^2} \cosh(\lambda_2 z) - \frac{A_3}{\lambda_3^2 - k^2} \cosh(\lambda_3 z)$$
$$+ \frac{B_1 z}{k^2} - \frac{B_2}{\lambda_2^2 - k^2} \sinh(\lambda_2 z) - \frac{B_3}{\lambda_3^2 - k^2} \sinh(\lambda_3 z),$$

whence the following secular equation

$$2(\mu_3 - \mu_2)\sinh\lambda_2 - \lambda_3(\lambda_2^2 + \lambda_3^2)\sinh\lambda_2\sinh\lambda_3 - 2\lambda_2\lambda_3^2$$
$$+\frac{2\lambda_3(\mu_2 - \mu_3)\sinh\lambda_3}{\lambda_2} + \frac{\lambda_3(\mu_2 - \mu_3)(k^4 + \mu_2^2)\sinh\lambda_3\cosh\lambda_2}{\lambda_2 k^2 \mu_2} \tag{7.4.9}$$
$$+\frac{(k^4 + \mu_3^2)(\mu_3 - \mu_2)\sinh\lambda_2\cosh\lambda_3}{k^2\mu_3} + \frac{\lambda_2\lambda_3^2(\mu_2^2 + \mu_3^2)\cosh\lambda_2\cosh\lambda_3}{\mu_2\mu_3} = 0.$$

Denote by S_1 the surface defined in the space $(k, R_a, \mathcal{K}^{-1})$ by the equation $k^2 R_a = k^6 + k^4\mathcal{K}^{-1}$. Then equation (7.4.9) defines a curve, say C_1, situated on S_1.

Numerical computations will decide if (7.4.9) possesses solutions. If this is the case, then the points of C_1 are secular points for the case c).

Theorem 7.4.1. *If the points of C_1 are secular, then they are limit points of the secular points of (7.4.8) as $\epsilon \to 0$ where ϵ is a smaller parameter defined by $k^2 R_a = k^6 + k^4\mathcal{K}^{-1} - \epsilon$.*

(In other words, the neutral surface defined by (7.4.8) is limited by the neutral curve C_1.)

Proof. The solutions of equation (7.4.6), written in the form $\lambda^6 - a_1\lambda^4 + a_2\lambda^2 - \epsilon = 0$, where $a_1 = 3k^2 + \mathcal{K}^{-1}$ and $a_2 = 3k^4 + 2k^2\mathcal{K}^{-1}$, have the following asymptotic behavior as $\epsilon \to 0$

$$\lambda_1 = \sqrt{\epsilon}\frac{1}{\sqrt{a_2}} + \epsilon\sqrt{\epsilon}\frac{a_1}{2a_2^2\sqrt{a_2}} + \epsilon^2\sqrt{\epsilon}(\dots)$$

$$\lambda_{2,3} = \sqrt{\frac{a_1 \pm \sqrt{a_1^2 - 4a_2}}{2}} \pm \epsilon\frac{1}{2\sqrt{a_1^2 - 4a_2}}\left(\frac{2}{a_1 \pm \sqrt{a_1^2 - 4a_2}}\right)^3 + \epsilon^2(\dots),$$

$$\mu_1 = -k^2 + \epsilon\frac{1}{a_2} + \epsilon^2(\dots),$$

$$\mu_{2,3} = \frac{a_1 - 2k^2 \pm \sqrt{a_1^2 - 4a_2}}{2} \pm \epsilon \frac{2}{\sqrt{a_1^2 - 4a_2}\left[a_1 \pm \sqrt{a_1^2 - 4a_2}\right]} + \epsilon^2(\dots).$$

Let us note that the first terms in these expansions are just the expressions for the corresponding eigenvalues from case c). Taking into account these expansions, the secular equation (7.4.8) reads $\sqrt{\epsilon}\lambda_2 g(\lambda_2, \lambda_3, k, \mathcal{K}^{-1}) + \epsilon\sqrt{\epsilon}(\dots) = 0$, where g stands for the left-hand side of equation (7.4.9). This proves Theorem 7.4.1.

In the above asymptotic expansions secular terms can occur if $a_1^2 - 4a_2 \equiv (\mathcal{K}^{-1} - 3k^2)(\mathcal{K}^{-1} + k^2)$ and $a_1 - \sqrt{a_1^2 - 4a_2} \equiv (\mathcal{K}^{-1} + 3k^2) - \sqrt{(\mathcal{K}^{-1} - 3k^2)(\mathcal{K}^{-1} + k^2)}$ are small [Geo95]. Therefore the above expansions are valid for $k >> \sqrt{\mathcal{K}^{-1}/3}$.

Let C_1' be the orthogonal projection of C_1 on the (k, R_a)-plane. Therefore C_1' is defined by (7.4.9) and it intersects the projection on the (k, R_a)-plane of the curve $k^2 R_a = k^6 + k^4 \mathcal{K}^{-1}$, $\mathcal{K}^{-1}$ assigned, at some point P_1. This point represents a limit for the projection on that plane of the neutral curve (7.4.8) corresponding to the assigned $\mathcal{K}^{-1}$. In this way, the part of C_1' corresponding to $k >> \sqrt{\mathcal{K}^{-1}/3}$ is a limit (possible envelope) of the family of projections of the neutral curves from (7.4.8). The remaining part of C_1' represents a bound for this family.

Similarly, in the $(k, R_a, \mathcal{K}^{-1})$-space, the curve C_1 provides a limit for the neutral surface (7.4.8). This curve consists of neutral points of problem $(7.4.1) - (7.4.4)$ and it is the intersection of the closure of the neutral surface from (7.4.8) and the surface S_1. The curve C_1 exists if equation (7.4.9) possesses real solutions $\mathcal{K}^{-1} = \mathcal{K}^{-1}(k)$.

Remark that only part of S_1, namely C_1, belongs to the neutral surface of problem $(7.4.1) - (7.4.4)$. The remaining part $S_1 \setminus C_1$ yields a bound for the neutral surface (7.4.8) and it consists of false neutral points.

d) **Case** $k, R_a, \mathcal{K}^{-1} \neq 0$, $R_a \neq k^4 + k^2 \mathcal{K}^{-1}$, $R_a = 4(27k^2\mathcal{K}^3)^{-1}$. We have

$$\mu_1 = -\mathcal{K}^{-1}/3, \ \mu_2 = \mu_3 - 2\frac{\mathcal{K}^{-1}}{3}, \ \lambda_1 = -\lambda_4 = \sqrt{\frac{(-\mathcal{K}^{-1} + 3k^2)}{3}},$$

$$\lambda_2 = \lambda_3 = -\lambda_5 = -\lambda_6 = \sqrt{\frac{(2\mathcal{K}^{-1} + 3k^2)}{3}},$$

and

$$W(z) = A_1 \cosh(\lambda_1 z) + A_2 \cosh(\lambda_2 z) + A_3 z \cosh(\lambda_2 z)$$
$$+ B_1 \sinh(\lambda_1 z)z + B_2 \sinh(\lambda_2 z) + B_3 z \sinh(\lambda_2 z),$$

$$\Theta(z) = -\frac{A_1}{\mu_1}\cosh(\lambda_1 z) - \frac{A_2}{\mu_2}\cosh(\lambda_2 z) + A_3\left[\frac{2\lambda_2(2\mu_2 - \mathcal{K}^{-1})}{-\mu_1\mu_2^2}\sinh(\lambda_2 z)\right.$$
$$\left. - \frac{z}{\mu_2}\cosh(\lambda_2 z)\right] - \frac{B_1}{\mu_1}\sinh(\lambda_1 z) - \frac{B_2}{\mu_2}\sinh(\lambda_2 z)$$
$$+ B_3\left[\frac{2\lambda_2(2\mu_2 - \mathcal{K}^{-1})}{-\mu_1\mu_2^2}\cosh(\lambda_2 z) - \frac{z}{\mu_2}\sinh(\lambda_2 z)\right].$$

The corresponding secular equation reads

$$\frac{3\lambda_1\lambda_2}{\mu_2^2}\left[\frac{-9\lambda_2^2\mu_2^2 + 7\mu_2^2 - 16k^2\mu_2 - 32k^4}{6\lambda_1\lambda_2\mu_2}\sinh\lambda_1(\sinh\lambda_2)^2\right.$$

$$-4\lambda_2\cosh\lambda_2 - 5\lambda_2\cosh\lambda_1 + \left(5 - \frac{2\lambda_2^2}{3\mu_2}\right)\sinh\lambda_2\cosh\lambda_1\cosh\lambda_2 \qquad (7.4.10)$$

$$\left.-\frac{4(\lambda_2^2 + 3k^2)}{3\mu_2}\sinh\lambda_2 + \frac{\lambda_2}{\lambda_1}\left(\mathcal{K}^{-1} + 6\frac{\lambda_2^2}{\mu_2}\right)\sinh\lambda_1(\cosh\lambda_2)^2\right] = 0.$$

In the $(k, R_a, \mathcal{K}^{-1})$ space, (7.4.10) defines a curve C_2, situated on the surface S_2, defined by $k^2 R_a = 4(27\mathcal{K}^3)^{-1}$. The points of C_2 are limit points of the neutral surface of (7.4.8). Indeed,

Theorem 7.4.2. *Equation (7.4.10) is a first asymptotic approximation of (7.4.8) as $\epsilon \to 0$, where $k^2 R_a = 4(27\mathcal{K}^3)^{-1} + \epsilon$.*

Proof. Let us write (7.4.6) in the form

$$\lambda^6 - a_1\lambda^4 + a_2\lambda^2 - a_3 + \epsilon = 0, \qquad (7.4.6)''$$

where a_1 and a_2 were introduced at point c) and $a_3 = k^6 + k^4\mathcal{K}^{-1} - 4(27\mathcal{K}^3)^{-1}$. From (7.4.6)″ it follows that the leading terms of the expansion of $\lambda_i, i = 1, \ldots, 6$ are just the expressions given at the beginning of point d). Let us denote them by λ_{io}. Then we have $\lambda_1 = \lambda_{1o} + \epsilon\lambda_{11} + o(\epsilon)$, $\lambda_{2,3} = \lambda_{20} \pm \sqrt{\epsilon}\lambda_{21} + \epsilon\lambda_{22} + o(\epsilon)$ as $\epsilon \to 0$, where λ_{io} are the solutions of equation (7.4.6)″ for $\epsilon = 0$ and λ_{i1} have the expressions $\lambda_{11} = -\left(6\lambda_{1o}^5 - 4a_1\lambda_{1o}^3 + 2a_2\lambda_{1o}\right)^{-1}$, $\lambda_{21}^2 = \lambda_{31}^2 = -\left(15\lambda_{20}^4 - 6a_1\lambda_{20}^2 + a_2\right)^{-1}$. It follows that, up to terms of order ϵ as $\epsilon \to 0$ and nonzero factors, the secular equation (7.4.8) reads $\epsilon g(\lambda_1, \lambda_2, k, \mathcal{K}^{-1}) = 0$, where g is the left-hand side of (7.4.10). Therefore, similar to point c), (7.4.10) defines the curve C_2. It represents a limit curve for the neutral surface (7.4.8) for $k \ll 1$ and it is the intersection of the closure of the neutral surface (7.4.8) and the surface S_2. The remaining part of S_2 provides a bound for (7.4.8).

Let C_2' be the orthogonal projection of C_2 on the (k, R_a)-plane. The properties of C_2' are similar to those of C_1'.

Thus the neutral surface of problem (7.4.1) − (7.4.4) is limited by C_1 for $k \gg \sqrt{\mathcal{K}^{-1}/3}$, by C_2 for $k \ll 1$, and it is bounded by the remaining parts of S_1 and S_2. The curve C_1' intersects C_2' at the point $(\sqrt{\mathcal{K}^{-1}/3}, 4\mathcal{K}^{-2}/9)$. About this point the bound yielded by S_1 and S_2 is the weakest.

In the classical case of nonporous medium, *i.e.* $\mathcal{K}^{-1} = 0$, the surface S_1 reads $R_a = k^4$ and S_2 degenerates into the plane $\mathcal{K}^{-1} = 0$. The surface S_1 is a bound for the neutral surface of the problem corresponding to (7.4.1) − (7.4.4) for $\mathcal{K}^{-1} = 0$.

e) **Case** $k, R_a, \mathcal{K}^{-1} \neq 0$, $R_a = k^4 + k^2\mathcal{K}^{-1}$, $R_a = 4(27k^2\mathcal{K}^3)^{-1}$. This case corresponds to $\lambda_1 = \lambda_4 = 0$, $\lambda_2 = \lambda_3 = k\sqrt{3}$, $\lambda_5 = \lambda_6 = -k\sqrt{3}$, $\mu_1 = -k^2$, $\mu_2 = \mu_3 = 2k^2$, $R_a = 4k^4$. It is obtained for $3k^2 = \mathcal{K}^{-1}$ (the only positive solution of equation $k^4 + k^2\mathcal{K}^{-1} = 4(27k^2\mathcal{K}^3)^{-1}$). This equation possesses also the double root $k^2 = -2\mathcal{K}^{-1}/3$, which is unacceptable because $k, \mathcal{K}^{-1} > 0$. Then the general solution of (7.4.5) has the form

$$W(z) = A_1 + A_2 \cosh(\lambda_2 z) + A_3 z \cosh(\lambda_2 z) + B_1 z + B_2 \sinh(\lambda_2 z) + B_3 z \sinh(\lambda_2 z),$$

while from (7.4.1) we have

$$\Theta(z) = \frac{A_1}{k^2} - \frac{A_2}{2k^2}\cosh(\lambda_2 z) + A_3\left[\frac{\sqrt{3}}{2k^3}\sinh(\lambda_2 z) - \frac{z}{2k^2}\cosh(\lambda_2 z)\right]$$
$$+ \frac{B_1 z}{k^2} - \frac{B_2}{2k^2}\sinh(\lambda_2 z) + B_3\left[\frac{\sqrt{3}}{2k^3}\cosh(\lambda_2 z) - \frac{z}{2k^2}\sinh(\lambda_2 z)\right].$$

In this way, the two boundary conditions for Θ are expressed in terms of the coefficients of W. The corresponding secular equation reads

$$\lambda_2^3 + 4\sinh\lambda_2(\cosh\lambda_2 - 1) + 4\lambda_2(\sinh\lambda_2)^2 + 4\lambda_2(\cosh\lambda_2)(\cosh\lambda_2 - 1) = 0.$$

As all terms in this equation are positive it follows that no secular points exist in case e).

f) **Case** $\mathcal{K}^{-1} = 0, R_a \neq 0$ corresponds to the classical Bénard problem, when the porous medium is absent [GeoOP]. Like in the case $\mathcal{K}^{-1} \neq 0$, some subcases must be considered. However, due to the fact that $\mu_i = -\sqrt[3]{k^2 R_a}\epsilon_i$, $\epsilon_1 = 1$, $\epsilon_{2,3} = \frac{-1\pm i\sqrt{3}}{2}$, here we have fewer possibilities for multiple solutions of (7.4.6) or (7.4.7). Moreover, the multiplicities of μ_i and λ_i are not always the same in the cases $\mathcal{K}^{-1} = 0$ and $\mathcal{K}^{-1} \neq 0$.

f_1) Subcase $k, R_a \neq 0, R_a \neq k^4$. The secular equation (7.4.8) is still valid and it becomes

$$2\frac{\sinh\lambda_1}{\lambda_1} + 2\frac{\sinh\lambda_2}{\epsilon_2\lambda_2} + 2\frac{\sinh\lambda_3}{\epsilon_3\lambda_3} + \frac{\sinh\lambda_1}{\lambda_1}\cosh\lambda_2\cosh\lambda_3 + \frac{\sinh\lambda_2}{\epsilon_2\lambda_2}\cosh\lambda_1\cosh\lambda_3$$
$$+ \frac{\sinh\lambda_3}{\epsilon_3\lambda_3}\cosh\lambda_1\cosh\lambda_2 + +3k^2\sqrt[3]{k^2 R_a}\frac{\sinh\lambda_1}{\lambda_1}\frac{\sinh\lambda_2}{\epsilon_2\lambda_2}\frac{\sinh\lambda_3}{\epsilon_3\lambda_3} = 0.$$

f_2) Subcase $k = 0$. We have $\lambda_i = 0$, $i = 1\ldots6$ and (7.4.1), (7.4.3)$_{1,2}$, (7.4.4)$_{1,4}$ imply $W = 0$. Then (7.4.2), (7.4.3)$_3$ and (7.4.4)$_3$ yield $\Theta = 0$. So, no secular points correspond to $k = 0$.

f_3) Subcase $k \neq 0$, $R_a = k^4$ is similar to case c). It can be obtained simply by letting $\mathcal{K}^{-1} = 0$ in (7.4.9).

g) **Case** $k, \mathcal{K}^{-1} \neq 0, R_a = 0$ corresponds to zero temperature gradients. The characteristic equation associated with (7.4.1) becomes $(\lambda^2 - k^2)(\lambda^2 - k^2 - \mathcal{K}^{-1}) = 0$ and it has four distinct solutions $\lambda_{1,3} = \pm k$, $\lambda_{2,4} = \pm\sqrt{k^2 + \mathcal{K}^{-1}}$. The general solution of (7.4.1) reads $W(z) = \sum_{i=1}^2 A_i \cosh(\lambda_i z) + B_i \sinh(\lambda_i z)$ and taking into account the boundary conditions we get the secular equation $\lambda_1^{-1}\tanh\lambda_1 = \lambda_2^{-1}\tanh\lambda_2$, which has no solution. So, $W = 0$ and (7.4.2) and the boundary conditions imply $\Theta = 0$. Hence, all points with $R_a = 0$ and, $k, \mathcal{K}^{-1} \neq 0$ are not secular.

h) **Case** $k \neq 0$, $\mathcal{K}^{-1} \to \infty$, $0 < R_a \mathcal{K} < \infty$. Equation (7.4.1) becomes $(D^2 - k^2)^2 W - k^2 R_a \mathcal{K} W = 0$, and leads to characteristic equation $(\lambda^2 - k^2)^2 - k^2 R_a \mathcal{K} = 0$ whose solutions are $\lambda_{1,3} = \pm\sqrt{k^2 + k\sqrt{R_a \mathcal{K}}}$, $\lambda_{2,4} = \pm\sqrt{k^2 - k\sqrt{R_a \mathcal{K}}}$. Then the secular equation reads $(\lambda_2^2 - \lambda_1^2)(\lambda_2 \sinh \lambda_1 \cosh \lambda_2 - \lambda_1 \sinh \lambda_2 \cosh \lambda_1) = 0$, or equivalently $\lambda_1^{-1} \tanh \lambda_1 = \lambda_2^{-1} \tanh \lambda_2$, which has no solution and implies $W = \Theta = 0$. Whence no secular points exist.

7.4.2   *Two free isothermal surfaces*

The above characteristic equations still hold (because they do not depend on the boundary conditions), whereas the secular equations are changed since now they are deduced from the following boundary conditions

$$W = D^2 W = \Theta = 0, \qquad \text{at } z = 0, 1. \tag{7.4.11}$$

Further on we consider the same cases as in Section 7.4.1. The eigenvalues are the same as these from Section 7.4.1, therefore, the general solutions W and Θ remain unchanged and will not be written any longer.

a) **Case** k, R_a, $\mathcal{K}^{-1} \neq 0$, $R_a \neq k^4 + k^2 \mathcal{K}^{-1}$, $R_a \neq 4(27 k^2 \mathcal{K}^3)^{-1}$. In this situation μ_i are mutually distinct and nonzero. If two roots, say μ_3 and μ_2, are complex-conjugate, then μ_1 is real and negative. If all μ_i are real, then one of them, say μ_1, is negative and the other two, positive. Since $\lambda_1 \lambda_2 \lambda_3 \lambda_4 \lambda_5 \lambda_6 = -\lambda_1^2 \lambda_2^2 \lambda_3^2 = -k^2(k^4 + k^2 \mathcal{K}^{-1} - R_a)$, it follows that for $R_a > k^4 + k^2 \mathcal{K}^{-1}$, two roots λ_1 and λ_4 are purely imaginary. This happens if $\mu_1 < -k^2$.

Introducing the general solutions W and Θ in (7.4.11) we obtain the secular equation

$$\sinh \lambda_1 \sinh \lambda_2 \sinh \lambda_3 \frac{(\lambda_1^2 - \lambda_2^2)^2 (\lambda_2^2 - \lambda_3^2)^2 (\lambda_3^2 - \lambda_1^2)^2}{\mu_1^2 \mu_2^2 \mu_3^2} = 0. \tag{7.4.12}$$

Since λ_1, λ_2 and λ_3 are mutually distinct, λ_2 and λ_3 are either positive or complex conjugate and $\mu_1, \mu_2, \mu_3 \neq 0$, then (7.4.12) implies $\sinh \lambda_1 = 0$, or, equivalently, $\sin(\lambda_1 \sqrt{-1}) = 0$, where λ_1 is purely imaginary, therefore the solutions of this last equation are $\lambda_1 \sqrt{-1} = n\pi$, $n \in \mathbf{Z}$. As λ_1 is a solution of (7.4.6), substituting this expression into (7.4.6) we obtain the secular equation

$$A_n^3 + A_n^2 \mathcal{K}^{-1} - k^2 R_a = 0, \tag{7.4.13}$$

where $A_n = n^2 \pi^2 + k^2$. In the (k, R_a)-plane this equation defines an infinity of nested curves. The neutral curve corresponds to $n = 1$ and, thus has the explicit form

$$R_a = \frac{(\pi^2 + k^2)^2}{k^2} (\mathcal{K}^{-1} + \pi^2 + k^2). \tag{7.4.14}$$

It was deduced in [LebC] by assuming that the eigenfunctions $W(z)$ have the form $W(z) = \sum_{n=1}^{\infty} a_n \sin(n\pi z)$ $n = 1, 2, \ldots$. Indeed, $\{\sin n\pi z\}_{n \in \mathbf{N}}$ is a total set in

$L^2(0,1)$. However, it is suitable only if $D^{2k}W(0) = D^{2k}W(1) = 0$ for every integer $k \geq 1$. So, this particular choice leads to simple calculations only in the case of free surfaces. In some other cases the expansion functions $\sin n\pi z$ do not satisfy all boundary conditions. Therefore by imposing W to satisfy them we obtain additional restrictions. For instance, $DW(1) = 0$ reads $\sum_{n=1}^{\infty} a_n n\pi(-1)^n = 0$ and $DW(0) = 0$ implies $\sum_{n=1}^{\infty} a_n n\pi = 0$. A posteriori it is necessary to prove the convergence of the involved series for the unknown functions and for their derivatives [Geo85]. This is the reason why the case of the free surfaces is frequently treated. In this paper we show how to treat analytically all types of surfaces, without recourse to function series.

b) **Case** $k = 0, \mathcal{K}^{-1} \neq 0$. In this case the secular equation reads $\lambda_2^4 \sinh \lambda_2 = 0$, where $\lambda_2 = \sqrt{\mathcal{K}^{-1}}$, and it has only vanishing roots, hence $W = 0$ and $\Theta = 0$ follows too. Consequently no eigenvalues exist in case b).

c) **Case** $k, R_a, \mathcal{K}^{-1} \neq 0$, $R_a = k^4 + k^2\mathcal{K}^{-1}$, $R_a \neq 4(27k^2\mathcal{K}^3)^{-1}$. This situation leads to the secular equation

$$\sinh \lambda_2 \sinh \lambda_3 \frac{\left(3k^2 + 2\mathcal{K}^{-1}\right)^2}{k^2 + \mathcal{K}^{-1}} (\mathcal{K}^{-1} - 3k^2) = 0, \tag{7.4.15}$$

which has no solution. Hence the surface $R_a = k^4 + k^2\mathcal{K}^{-1}$ contains no secular points.

d) **Case** $k, R_a, \mathcal{K}^{-1} \neq 0$, $R_a \neq k^4 + k^2\mathcal{K}^{-1}$, $R_a = 4(27k^2\mathcal{K}^3)^{-1}$. The secular equation reads

$$\lambda_2^2 \sinh \lambda_1 (\sinh \lambda_2)^2 (\mu_2 - \mu_1)^2 (3\mu_2 - \mathcal{K}^{-1})^2 \mu_1^{-2} \mu_2^{-4} = 0. \tag{7.4.16}$$

Its solution is a limit as $R_a \to 4(27k^2\mathcal{K}^3)^{-1}$ of the solutions of (7.4.12). Hence the points $(k, \mathcal{K}^{-1}, R_a)$ situated on the surface S_2 and corresponding to the solutions of (7.4.16) are secular points. They are situated on the curve C_2. The remaining part of S_2 represents a bound for the neutral surface of problem (7.4.1), (7.4.2), (7.4.11).

e) **Case** $k, R_a, \mathcal{K}^{-1} \neq 0$, $R_a = k^4 + k^2\mathcal{K}^{-1}$, $R_a = 4(27k^2\mathcal{K}^3)^{-1}$. Up to a non-null factor, the secular equation has the form $\sinh \lambda_2 = 0$, which has no solution. Correspondingly, in this case no secular points exist.

f) **Case** $\mathcal{K}^{-1} = 0$, $R_a \neq 0$.
$f_1)$ Subcase $k, R_a \neq 0$, $R_a \neq k^4$. The relation (7.4.14) where $\mathcal{K}^{-1} = 0$ still holds, and, so, we regain the classical result.
$f_2)$ Subcase $k = 0$ yields $W = \Theta = 0$.
$f_3)$ Subcase $k \neq 0$, $R_a = k^4$. The secular equation, obtained from (7.4.15) for $\mathcal{K}^{-1} = 0$, has no solution.

g) **Case** $k, \mathcal{K}^{-1} \neq 0$, $R_a = 0$. The secular equation $(\lambda_2^2 - \lambda_1^2)^2 \sinh \lambda_1 \sinh \lambda_2 = 0$, has no solution.

h) **Case** $k \neq 0$, $\mathcal{K}^{-1} \to \infty$, $0 < R_a\mathcal{K} < \infty$. We have the secular equation from the case g) of Section 7.4.1, which has no solution.

7.5   Convection in the presence of a dielectrophoretic force

The cases not covered by Section 3.4.3.1 are studied.

The problem $(3.3.1) - (3.3.3)$, $(3.3.7)$ governing this convection was studied by the direct method in [Geo 77] and [GeoC] in the general cases $k, R, El > 0$, $k^4 \neq R(1 + El)$ and $El = 0$, $k^4 \neq R$, from theoretical and numerical point of view respectively. In particular, the neutral curves were drawn in [GeoC] asserting that the secular equations $(3.4.3'')$ and $(3.4.3''')$ do have solutions. The case $El = 0$, $R \neq k^4$ is the well-known case of Bénard convection and it is regained as a limit case in all more complicated thermal convections. These two cases were considered in Section 3.4.3.1.

The excepted cases: 1)$k = 0$; 2)$El = 0$, $k > 0$, $R = 0$; 3)$El > 0$, $k > 0$, $R = 0$; 4)$El = 0$, $k > 0$, $R = k^4$; 5)$El > 0$, $k > 0$, $R(1 + El) = k^4$ were analyzed by the first author and D. Nica in a 1993 communication. In case 1) for general eigenfunctions and in cases 2) and 3) for even eigenfunctions, no secular points were found. In case 4), $\mu_1 = -k^4$ and, therefore, $\lambda_1 = \lambda_4 = 0$, we have $\lambda_{2,5}^2 = k^2(1 - \epsilon) = (u + iv)^2$, $\lambda_{3,6}^2 = k^2(1 - \epsilon') = (u - iv)^2$, whence the general even solution of $(3.3.8')$ is $\Theta = A_1 + A_2 \cosh(u + iv)x + A_3 \cosh(u - iv)x$. Introducing it in $(3.3.7')$ we obtain the secular equation

$$(u + iv)[(u - iv)^2 - k^2]\cosh\frac{u + iv}{2}\sinh\frac{u - iv}{2}$$
$$-(u - iv)[(u + iv)^2 - k^2]\sinh\frac{u + iv}{2}\cosh\frac{u - iv}{2} = 0 \qquad (7.5.1)$$

or, equivalently, $u(u^2 + v^2 + k^2)\sinh(iv) - iv(u^2 + v^2 + k^2)\sinh u = 0$, implying the transcendental equation $u \sin v - v \sinh u = 0$, which has no nontrivial solution, hence no point of the curve $R = k^4$ is secular. This curve is the bifurcation set for the characteristic manifold because, corresponding to these points, two sheets of this manifold coalesce.

In case 5), Lemma 3.4.1 asserts that μ_i cannot be multiple. Therefore λ_i can be multiple only if one μ_i, say μ_1, is equal to $-k^2$. In this situation equation $(3.3.9')$ implies $k^4 = R(1 + El)$ and $\lambda_{1,5} = 0$, while the other roots λ_i, $i = 2, 3, 4, 6, 7, 8$ satisfy the equation

$$\lambda^6 - 4k^2\lambda^4 + 6k^4\lambda^2 + k^2(R - 4k^4) = 0, \qquad (7.5.2)$$

deduced from the general characteristic equation $(3.3.9)$ in the hypothesis $k^4 = R(1 + El)$. In μ, $(7.5.2)$ reads

$$\mu^3 - k^2\mu^2 + k^4\mu + k^2(R - k^4) = 0$$

and leads to the secular equation, for the case of even eigensolutions,

$$CD(c^2 - d^2)(a - b) + BC(b^2 - c^2)(d - a) + DB(d^2 - b^2)(c - a) = 0, \qquad (7.5.3)$$

where $a = k^{-2}$, $b = (k^2 - \lambda_2^2)^{-1}$, $c = (k^2 - \lambda_3^2)^{-1}$, $d = (k^2 - \lambda_4^2)^{-1}$, $B = \lambda_2 \tanh(\lambda_2/2)$, $C = \lambda_3 \tanh(\lambda_3/2)$, $D = \lambda_4 \tanh(\lambda_4/2)$. Formally, (7.5.3) is obtained from (3.4.3″) by taking $\lambda_1 = 0$. Correspondingly, the secular curves defined by (7.5.3) are limits as $k^4 \to R(1 + El)$ of the curves defined by (3.4.3″).

The existence of solutions of (7.5.3) is an open problem. This is why we do not know if the points of the surface $k^4 = R(1 + El)$ are or not secular.

7.6 Convection in an anisotropic M.H.D. thermodiffusive mixture

The eigenvalue problem treated in this section governs the linear stability of some complex fluid flows. The order of the involved system of ode's is equal to 14 and they contain 10 parameters. By following [GeoPalR96c] we perform a detailed investigation of the multiplicity of the roots of the characteristic equation and detect the false secular points. Physically unrealistic cases are shown. Open problems are revealed.

7.6.1 *Formulation of the eigenvalue problem*

In the case of mutually distinct eigenvalues, we derived the secular equation for the following eigenvalue problem in $(-0.5; 0.5)$

$$(D^2 - a^2)^2 W - Ra^2 \Theta + sR_c a^2 r_D \Gamma + M^2 D(D^2 - a^2)K = 0, \qquad (7.6.1)$$

$$W + (1 + N_1 D_1 r_D)(D^2 - a^2)\Theta + N_1 R_c R^{-1} r_D^2 (D^2 - a^2)\Gamma = 0, \qquad (7.6.2)$$

$$W + D_1 RR_c^{-1}(D^2 - a^2)\Theta + r_D(D^2 - a^2)\Gamma = 0, \qquad (7.6.3)$$

$$[(1 + \beta_I)D^2 - a^2]X + DZ + \beta_H D(D^2 - a^2)K = 0, \qquad (7.6.4)$$

$$(D^2 - a^2)Z + M^2 DX = 0, \qquad (7.6.5)$$

$$(1 + \beta_I)(D^2 - a^2)K + DW - \beta_H DX = 0, \qquad (7.6.6)$$

$$W = DW = \Theta = \Gamma = X = DZ = K = 0, \qquad (7.6.7)$$

where $(W, \Theta, \Gamma, X, Z, K) \in (C^\infty[-0.5; 0.5])^6$ is the eigensolution and $\mathbf{P} \equiv (R, a, R_c, r_D, N_1, D_1, \beta_I, \beta_H, M, s) \in (\mathbf{R} - \{0\}) \times \mathbf{R}_*^+ \times (\mathbf{R} - \{0\}) \times \mathbf{R}_*^+ \times \mathbf{R}_*^+ \times \mathbf{R}_*^+ \times \mathbf{R}_*^+ \times \mathbf{R}_*^+ \times \mathbf{R}_*^+ \times \{-1, 1\}$ represents the eigenvalue and $D \equiv d/dz$, $W, \Theta, \Gamma, X, Z, K$

are the amplitudes of the perturbation of vertical velocity, temperature, concentration, vorticity vector, density current vector, magnetic field, in the 10-dimensional parameter space. From $(7.6.1) - (7.6.6)$ we deduce

$$[(D^2-a^2)^3 - b_1 a^2](\{(D^2-a^2)[(1+\beta_I)D^2-a^2] - M^2 D^2\}(1+\beta_I) + \beta_H^2 D^2(D^2-a^2))W$$

$$-M^2 D^2(D^2-a^2)\{[(1+\beta_I)D^2 - a^2](D^2-a^2) - M^2 D^2\}W = 0, \qquad (7.6.1')$$

$$(D^2-a^2)\Theta = -a_1 W, \qquad (7.6.2')$$

$$(D^2-a^2)\Gamma = -a_2 W, \qquad (7.6.3')$$

$$D(D^2-a^2)K = -M^{-2}\{(D^2-a^2)^2 W - Ra^2\Theta + sR_c a^2 r_D \Gamma\}, \qquad (7.6.4')$$

$$DX = (1+\beta_I)\beta_H^{-1}(D^2-a^2)K + \beta_H^{-1}DW, \qquad (7.6.5')$$

$$DZ = -\beta_H D(D^2-a^2)K - [(1+\beta_I)D^2 - a^2]X \qquad (7.6.6')$$

where $a_1 = 1 - N_1 R_C r_D R^{-1}$, $a_2 = 1 + N_1 D_1 r_D - D_1 RR_C^{-1}$, $b_1 = -Ra_1 + sR_C r_D a_2$. Equations $(7.6.1') - (7.6.6')$ are equivalent to $(7.6.1) - (7.6.6)$ if with them we associate equation $(7.6.5)$, so far used only to write $(7.6.1')$. Indeed, we took into account only part of the information contained in $(7.6.5)$, we use only its derivative, substituted it into $(7.6.4)$ and then have eliminated X between $(7.6.6)$ and this equation obtained from $(7.6.4)$. Therefore $(7.6.1) - (7.6.6)$ is equivalent to $(7.6.1') - (7.6.6')$ and $(7.6.5)$.

The general solution of $(7.6.1') - (7.6.6')$ leads to the general characteristic equation

$$\begin{aligned}
f(\lambda) \equiv &A\lambda^{10} - [4a^2 A + (a^2 + 2M^2)(1+\beta_I)]\lambda^8 + [6a^4 A \\
&+ (4a^4 + 5a^2 M^2)(1+\beta_I) + M^4 + a^2 M^2]\lambda^6 - [4a^6 A + a^2 b_1 A \\
&+ (6a^6 + 4a^4 M^2)(1+\beta_I) + 2a^4 M^2 + a^2 M^4]\lambda^4 + [a^8 A + a^4 b_1 A \qquad (7.6.8)\\
&+ (4a^8 + a^6 M^2 + b_1 a^4 + b_1 a^2 M^2)(1+\beta_I) + a^6 M^2]\lambda^2 \\
&- a^6(a^4 + b_1)(1+\beta_I) = 0,
\end{aligned}$$

where $A = (1+\beta_I)^2 + \beta_H^2$ and $\mathbf{P}$ contain no limit (*i.e.* physically unacceptable) component (*e.g.* $\alpha = 0$, $M = 0$). This equation is to be used in the case of mutually disjoint roots λ_i.

Further on we treat several cases where multiple roots of $(7.6.8)$ exist. In some of these cases some parameter vanishes (*e.g.* $M = 0$), showing that the corresponding effect and the physical quantity K are not present. Moreover, the equation expressing the balance law and the boundary conditions for that quantity must no longer be considered. As a consequence, the system of governing equations and the boundary conditions as well as the characteristic equation are changed and they are to be studied separately. Consequently, we must study some cases where all physical quantities are present as well as the physically unacceptable values of the involved parameters. These cases can be significant for the bifurcation study.

In the following we suppose that W, Θ, Γ, X are even functions and Z and K are odd functions.

As an eigensolution is a 6-dimensional vector function, it follows that the eigenvalues exist if at least one component of this unknown vector is non-null. Thus, we must study the eigenproblem for W as well as for all other unknown functions.

7.6.2 *Case $a = 0$ of the perturbations depending only on the vertical coordinate z and time*

Subcase 1). $M > 0$, $\beta_I \geq 0$, $\beta_H > 0$, $b_1 \in \mathbf{R}$. Correspondingly (7.6.8) becomes

$$A\lambda^{10} - 2M^2(1 + \beta_I)\lambda^8 + M^4\lambda^6 = 0, \tag{7.6.9}$$

and hence $\lambda_1 = \lambda_4 = \lambda_3 = \lambda_6 = 0$. In order to obtain the equation in W we must no longer eliminate Θ and Γ, while the elimination of X and Z between $(7.6.4) - (7.6.6)$ is simpler. By direct calculation, from $(7.6.1)$, $(7.6.4) - (7.6.6)$ we obtain the equation in W

$$D^2[AD^4 - 2M^2(1 + \beta_I)D^2 + M^4]W = 0. \tag{7.6.10}$$

(Taking $a = 0$ in $(7.6.1')$ we would have been led to $D^6[AD^4 - 2M^2(1 + \beta_I)D^2 + M^4]W = 0$, the higher power of D showing that in the case $a \neq 0$, in order to obtain an equation in W only, additional differentiations were necessary. Similar situations hold for the equations formally derived from $(7.6.1') - (7.6.6')$; they will no longer be specified. Instead, either equations $(7.6.1) - (7.6.6)$ or the general balance equations will be used.) The characteristic equation reads

$$\lambda^2[A\lambda^4 - 2M^2(1 + \beta_I)\lambda^2 + M^4] = 0$$

and has the roots $\lambda_1 = 0$, $\lambda_2 = u + iv$, $\lambda_3 = u - iv$, $\lambda_4 = -\lambda_1$, $\lambda_5 = -\lambda_2$, $\lambda_6 = -\lambda_3$. Here $u = MA^{-1/2}\sqrt{(\sqrt{A} + 1 + \beta_I)/2}$, $v = MA^{-1/2}\sqrt{(\sqrt{A} - 1 - \beta_I)/2}$. The general even solution of (7.6.10) reads $W(z) = A_1 + A_2 \cosh \lambda_2 z + A_3 \cosh \lambda_3 z$. Equations for Θ, Γ, X, Z, K are obtained from $(7.6.2'), (7.6.3'), (7.6.5'), (7.6.6')$ and $(7.6.4')$ respectively by simply letting $a = 0$. They have the general solutions

$$\Theta(z) = B_1 - a_1\left[\frac{A_1}{2}z^2 + A_2\lambda_2^{-2}\cosh\lambda_2 z + A_3\lambda_3^{-2}\cosh\lambda_3 z\right],$$

$$\Gamma(z) = B_2 - a_2\left[\frac{A_1}{2}z^2 + A_2\lambda_2^{-2}\cosh\lambda_2 z + A_3\lambda_3^{-2}\cosh\lambda_3 z\right],$$

$$X(z) = B_4 + A_2[\beta_H^{-1} - \lambda_2^2\beta_H^{-1}M^{-2}(1 + \beta_I)]\cosh\lambda_2 z + A_3[\beta_H^{-1}$$
$$- \lambda_3^2\beta_H^{-1}M^{-2}(1 + \beta_I)]\cosh\lambda_3 z,$$

$$Z(z) = (B_5 - B_4M^2)z - M^2 A_2\lambda_2^{-1}[\beta_H^{-1} - \lambda_2^2\beta_H^{-1}M^{-2}(1 + \beta_I)]\sinh\lambda_2 z$$
$$- M^2 A_3\lambda_3^{-1}[\beta_H^{-1} - \lambda_3^2\beta_H^{-1}M^{-2}(1 + \beta_I)]\sinh\lambda_3 z,$$

$$K(z) = B_3 z - A_2\lambda_2 M^{-2}\sinh\lambda_2 z - A_3\lambda_3 M^{-2}\sinh\lambda_3 z.$$

However, equation (7.6.5) implies $B_5 = M^2 B_4$. Correspondingly, the secular equation is $\frac{\sin v}{v} = \frac{\sinh u}{u}$ and has only the trivial solution $u = v = 0$. This implies $\beta_I < 0$ which contradicts the hypothesis. Hence in this case there are no secular points.

Subcase 2). $M > 0$, $\beta_I \geq 0$, $\beta_H = 0$, $b_1 \in \mathbf{R}$, *i.e.* Hall effect is absent, implying $\beta_I = 0$ too. In this case, since $\beta_H = 0$, the elimination of K between (7.6.1) and (7.6.6) leads to an equation in W, $D^4 W - M^2 (1 + \beta_I)^{-1} D^2 W = 0$, and the corresponding characteristic equation reads $\lambda^4 - M^2 (1 + \beta_I)^{-1} \lambda^2 = 0$. It has the roots $\lambda_1 = 0$, $\lambda_2 = M/\sqrt{1 + \beta_I}$, $\lambda_3 = 0$, $\lambda_4 = -\lambda_2$, so the general even solution W reads $W(z) = A_1 + A_2 \cosh \lambda_2 z$. The equations for Θ, Γ, K, can be formally obtained from (7.6.2′), (7.6.3′), (7.6.4′), whereas X and Z satisfy equations (7.6.4) and (7.6.5) where $a = \beta_H = 0$ (easily using conditions $DZ = X = 0$ we find $X = Z = 0$). By reasoning as in case 1), we find that case 2) corresponds to false secular points.

Subcase 3). $M = 0$, $b_1 \in \mathbf{R}$. This case corresponds to the absence of the magnetic field and correspondingly to the absence of Hall and ion-slip effects, *i.e.* $\beta_H = \beta_I = 0$. In addition equations (7.6.5), (7.6.4) must be disregarded. From strictly mathematical viewpoint we can consider the case β_H, $\beta_I > 0$ and must keep these equations and see what happen if $M \to 0$. Direct very simple computations lead to the conclusion that neither in this case we have secular points.

7.6.3   *Case $a \neq 0$, $M = 0$, $b_1 = 0$*

This case corresponds to the absence of magnetic field and the physical comment done for subcase 3) holds too. Further on we deal only with the mathematically limit case introducing the notation $\lambda = a\lambda^*$, $-b_1 = a^4 R^*$, $M = aM^*$, $M^{*2}/A = M_i^2$, $(1 + \beta_I)/A = \beta^*$, $\delta = \lambda^{*2}$. Then, formally, (7.6.8) reads

$$\delta^5 + (-4 - \beta^* - 2\beta^* M^{*2})\delta^4 + (6 + 4\beta^* + 5\beta^* M^{*2} + M^{*2} M_i^2 + M_i^2)\delta^3$$
$$+ (-4 + R^* - 6\beta^* - 4M^{*2}\beta^* - 2M_i^2 - M^{*2} M_i^2)\delta^2$$
$$+ (1 - R^* + 4\beta^* + M^{*2}\beta^* - R^*\beta^* - R^*\beta^* M^{*2} + M_i^2)\delta + \beta^*(R^* - 1) = 0$$
$$\tag{7.6.11}$$

and it is valid also for the magnetic case ($M \neq 0$) and for $b_1 \neq 0$. The limit case $b_1 = 0$ is a physically limit case when $R = R_C = 0$ (*i.e.* no temperature and concentration gradients in the basic state exist and equation (7.6.3) for concentration loses its meaning, one of its coefficients being unbounded) and a purely mathematical case where R, $R_C \neq 0$. Anyhow, $b_1 = 0$ shows that if the operator $D^2 - a^2$ is applied to (7.6.1), in order to eliminate Θ and Γ, then no contribution of Θ and Γ remains.

Subcase 4). $\beta^* = 1$ which means $\beta_H = \beta_I = 0$. Direct calculations show that $\lambda_1 = \cdots = \lambda_5 = a$ and $\lambda_6 = \cdots = \lambda_{10} = -a$.

Subcase 5). $\beta^* \neq 1$. In this case (7.6.11) reads $(\delta - 1)^4 (\delta - \beta^*) = 0$. Elimination of Θ and Γ between (7.6.1) − (7.6.3) leads to the equation $(D^2 - a^2)^3 W = 0$. But (7.6.1) implies the boundary conditions $(D^2 - a^2)^2 W = 0$ at $z = \pm 0.5$ so denoting $(D^2 - a^2)^3 W = Y$ we have the problem $D^2 Y = 0$, $z \in (-0.5, 0.5)$ and $Y = 0$ at $z = \pm 0.5$, which immediately gives $Y = 0$, *i.e.* $(D^2 - a^2)^2 W = 0$ for $z \in (-0.5, 0.5)$. The corresponding characteristic equation is $(\lambda^2 - a^2)^2 = 0$ and has the roots $\lambda_1 = \lambda_2 = a$, $\lambda_3 = \lambda_4 = -a$, hence the general even solution reads $W(z) = A_1 \cosh az + A_2 z \sinh az$ and the secular equation becomes $\sinh a + a = 0$. Since this equation has only the solution $a = 0$, it follows that case 5) corresponds to false secular points.

Remark that in all these cases, either Θ, Γ, X, Y, Z are expressed in terms of W, so $W = 0$ implies $\Theta = \Gamma = X = K = Z = 0$, or there are some unknown functions which cannot be expressed in terms of W, but satisfy some eigenvalue problems which have no eigenvalue (as in case 2)).

In case 5) problem $(7.6.1) - (7.6.7)$ decouples into one problem for W, Θ, Γ, one for Z which leads to $Z \equiv 0$ and another for X and K, which implies also $X = K = 0$. The decoupling concerns first equation $(7.6.5)$ for Z and the remaining ones (connected with W) and after finding $W = 0$ we have a system of equations for Θ and Γ and another for X and K. In this subcase we have, in addition, a decoupling of the system for X and K.

7.6.4   *Case $a \neq 0$, $M^* = 0$, $R^* \neq 0$*

Equation $(7.6.11)$ becomes $(\delta - 1)(\delta - \beta^*)[(\delta - 1)^3 + R^*] = 0$

Subcase 6). $\beta^* \neq 1$, $b_1 \neq -a^4$. We have mutually distinct characteristic roots and direct computations give $[(D^2 - a^2)^3 - b_1 a^2]W = 0$ and, correspondingly, the characteristic equation reads $[(\lambda_i^2 - a^2)^3 - b_1 a^2] = 0$, or, equivalently, $(\delta - 1)^3 + R^* = 0$. Then $W(z) = \sum_{i=1}^{3} A_i \cosh \lambda_i z$, where $\lambda_i^2 = a^2 + r\epsilon_i$, $r = \sqrt[3]{b_1 a^2}$, ϵ_i are the cubic roots of unity. Writing $\lambda_{2,3} = u \pm iv$, where $u, v = \sqrt{[\sqrt{a^4 + r^2 - a^2 r} \pm (a^2 - r^2)]/2}$, the secular equation becomes $t_1 + \epsilon_3 t_2 + \epsilon_2 t_3 = 0$, where $t_i = \lambda_i \tanh(\lambda_i / 2)$, and it is just the secular equation for purely thermal case if we let $b_1 = -R$ (Rayleigh number). Therefore the points corresponding to subcase 6) are secular. Direct computations give $Z \equiv 0$, whereas X and Z are well-determined functions of W (linear combinations of the fundamental system of solutions which defines W, with coefficients proportional to the coefficients in the expression of W). If $\beta_H = 0$ then, in addition, $X \equiv 0$.

Subcase 7). $\beta^* \neq 1$, $b_1 = -a^4$. In this situation the characteristic equation, which is the same as in case 6), has two null roots $\lambda_1 = \lambda_4 = 0$. Reasonings similar to case 5) show that the points corresponding to case 7) are false secular points.

Cases 6) and 7) are physically unrealistic because they correspond to the presence of Hall and ion-slip effects in the absence of a magnetic field.

Subcase 8). $\beta^* = 1$, $b_1 \neq -a^4$ is similar to case 6), with the only difference that in this case $X = Z \equiv 0$ whereas $K \not\equiv 0$. The presence of K in the case of $M = \beta_I = \beta_H = 0$ is physically unrealistic.

Subcase 9). $\beta^* = 1$, $b_1 = -a^4$ corresponds to false secular points.

7.6.5   *Case $a \neq 0$, $M^* > 0$, $R^* = 0$*

This represents the magnetic case in the absence of thermal and concentration gradients or in their presence but such that $b_1 = 0$. It corresponds to the characteristic equation, formally deduced from (7.6.11)

$$(\delta - 1)[\delta^4 + (-3 - \beta^* - 2\beta^* M^{*2})\delta^3 + (3 + 3\beta^* + 3\beta^* M^{*2}$$
$$+ M^{*2} M_i^2 + M_i^2)\delta^2 + (-1 - 3\beta^* - \beta^* M^{*2} - M_i^2)\delta + \beta^*] = 0. \tag{7.6.12}$$

However, some factors, *e.g.* $\delta - 1$, showing unnecessary differentiation, will not occur in the true characteristic equation. Thus, if $R = R_c = 0$, then the true characteristic equation reads

$$\delta^4 + (-3 - \beta^* - 2\beta^* M^{*2})\delta^3 + (3 + 3\beta^* + 3\beta^* M^{*2}$$
$$+ M^{*2} M_i^2 + M_i^2)\delta^2 + (-1 - 3\beta^* - \beta^* M^{*2} - M_i^2)\delta + \beta^* = 0, \tag{7.6.13}$$

i.e. it is (7.6.12) without the factor $\delta - 1$.

In this very complicated case, the limiting situation $R^* = 0$ is itself a difficult case, which requires an extended study. Here the most important thing is not the fact that $R^* = 0$ is a limit value, but that the roots δ_i of the characteristic equation are multiple. We shall investigate the multiplicity of δ_i by using the relations

$$\delta_1 + \delta_2 + \delta_3 + \delta_4 = 3 + \beta^* + 2\beta^* M^{*2};$$
$$\delta_1\delta_2 + \delta_1\delta_3 + \delta_1\delta_4 + \delta_2\delta_3 + \delta_2\delta_4 + \delta_3\delta_4 = 3 + 3\beta^* + 3\beta^* M^{*2}$$
$$+ M^{*2} M_i^2 + M_i^2; \tag{7.6.14}$$
$$\delta_1\delta_2\delta_3 + \delta_1\delta_2\delta_4 + \delta_1\delta_3\delta_4 + \delta_2\delta_3\delta_4 = 1 + 3\beta^* + \beta^* M^{*2} + M_i^2;$$
$$\delta_1\delta_2\delta_3\delta_4 = \beta^*.$$

Subcase $\delta_1 = \delta_2 = \delta_3 = \delta_4 = \delta$. Relations (7.6.14) imply $4\delta = 3 + \beta^* + 2\beta^* M^{*2}$, $6\delta^2 = (3 + 3\beta^* + 3\beta^* M^{*2} + M^{*2} M_i^2 + M_i^2)$, $4\delta^3 = (1 + 3\beta^* + \beta^* M^{*2} + M_i^2)$, $\delta^4 = \beta^*$. Deriving M^{*2} from $(7.6.14)_1$ and M_i^2 from $(7.6.14)_3$, introducing the obtained expressions in $(7.6.14)_2$ and taking into account $(7.6.14)_4$ we get the consistency relation

$$(\delta - 1)^4(\delta + 1)^2(\delta^2 + 10\delta - 3) = 0. \tag{7.6.15}$$

By definition, $M_i \leq M$, $\beta^* \leq 1$, while from $(7.6.14)_{1,4}$ it follows that $3/4 < \delta \leq 1$. If $\delta = 1$, then $\beta^* = 1$ and $(7.6.14)_1$ implies $M^* = 0$, which contradicts the hypothesis $M > 0$. The other three solutions of (7.6.15) do not belong to the interval $(3/4, 1)$. In conclusion (7.6.13) cannot have a solution of multiplicity 4.

Subcase $\delta_1 = \delta_2 = \delta_3 = \delta \neq \delta_4$. Relations (7.6.14) imply

$$\delta + \delta_4 = 3 + \beta^* + 2\beta^* M^{*2}, \quad 3\delta^2 + 3\delta\delta_4 = 3 + 3\beta^* + 3\beta^* M^{*2} + M^{*2} M_i^2 + M_i^2,$$

$$\delta^3 + 3\delta^2\delta_4 = 1 + 3\beta^* + \beta^* M^{*2} + M_i^2, \delta^3\delta_4 = \beta^*. \qquad (7.6.15')$$

Relations $(7.6.15')_1$, $(7.6.15')_4$ and the fact that equation $(7.6.13)$ has real coefficients imply that $\delta_1\delta_4 > 0$. If $\delta = 1$, then $\delta_4 = \beta^*$ and $(7.6.15')_1$ gives $M^{*2} = 0$. Hence $\delta \neq 1$. From $(7.4.15')_1$ and $(7.4.15')_2$ we obtain

$$\begin{aligned}
M^{*2} &= (\delta - 1)[3\delta^3 - \beta^*(\delta^2 + \delta + 1)](2\beta^*\delta^3)^{-1}, \\
M_i^2 &= 3\beta^*(\delta - 1)[2\delta^4 - \delta^3 - \beta^*(\delta^2 + \delta - 1)][3\delta^4 + \delta^3(\beta^* - 3) + \beta^*]^{-1}.
\end{aligned} \qquad (7.6.16)$$

If $\beta^* = 1$ then, by definition, $M^{*2} = M_i^2$, and $(7.6.15')_{1,3}$ read $(\delta - 1)^3(\delta + 1)^3 = 0$, *i.e.* $\delta = 1$, $\delta = -1$. As these values are unacceptable, it follows that for $\beta^* = 1$ equation $(7.6.13)$ cannot have a triple characteristic root.

Therefore consider $\beta^* \neq 1$. Because $M^{*2} > 0$, it follows that $3\delta^3(\delta - 1) > \beta^*(\delta^3 - 1)$ and, hence, $3\delta^4 + \delta^3(\beta^* - 3) + \beta^* > 0$. Hence, in order for M^{*2} and M_i^2 to be positive, for $\delta < 1$, β^* must satisfy the inequalities

$$\frac{3\delta^3}{\delta^2 + \delta + 1} < \beta^*, 2\delta^4 - \delta^3 < \beta^*(\delta^2 + \delta - 1). \qquad (7.6.17)$$

On the other hand, introducing $(7.6.16)_{1,2}$ in $(7.6.15')_3$, we have the consistency relation

$$\begin{aligned}
\beta^{*2}(\delta^4 + 8\delta^3 + 3\delta^2 - 2\delta - 1) + \beta^*(2\delta^7 - 8\delta^6 - 18\delta^5 + 2\delta^4 + 4\delta^3) \\
+ 3\delta^6(2\delta^2 + 2\delta - 1) = 0.
\end{aligned} \qquad (7.6.18)$$

This equation in β^* has real solutions if the discriminant $\Delta(\delta) \equiv (\delta - 1)^2(\delta + 1)^3(\delta^3 - 9\delta^2 + 3\delta + 1)$ is positive. But $\Delta > 0$ if $0 < \delta \leq \delta_{01}$, $\delta \geq \delta_{02}$, where $\delta_{01} \in (0.5, 0.6)$ and $\delta_{02} \in (8, 9)$. But, for $\delta > 6$ the coefficients of equation $(7.6.18)$ are positive, hence it has no positive root β^* of $(7.6.18)$. Therefore we must find some contradiction for $0 < \delta \leq \delta_{01}$. To this purpose we come back to $(7.6.17)$. Finally, $\delta^2 + \delta - 1 < 0$ for $0 < \delta < (\sqrt{5} - 1)/2$. As $\Delta((\sqrt{5} - 1)/2) < 0$, it follows that $\delta_{01} < (\sqrt{5} - 1)/2$, whereas for $\delta < (\sqrt{5} - 1)/2$ we have $\delta^2 + \delta - 1 < 0$, so $(7.6.17)$ reads $\frac{3\delta^3}{\delta^2 + \delta + 1} < \beta^* < \frac{2\delta^4 - \delta^3}{\delta^2 + \delta - 1}$, and implies $3\delta^3(\delta^2 + \delta - 1) > (2\delta^4 - \delta^3)(\delta^2 + \delta + 1)$, that is $(\delta - 1)^2(\delta + 1) < 0$, which is false. Consequently, equation $(7.6.13)$ cannot have a triple root either.

Subcase $\delta_1 = \delta_2 \neq \delta_3 = \delta_4$**.** Letting $S = \delta_1 + \delta_3$, $P = \delta_1\delta_3$ relations $(7.6.14)$ read

$$\begin{aligned}
2S &= 3 + \beta^* + 2\beta^* M^{*2}; \\
S^2 + 2P &= (3 + 3\beta^* + 3\beta^* M^{*2} + M^{*2} M_i^2 + M_i^2); \\
2SP &= (1 + 3\beta^* + \beta^* M^{*2} + M_i^2); \\
P^2 &= \beta^*.
\end{aligned} \qquad (7.6.19)$$

Relation $(7.6.19)_1$ implies that $S > \frac{3 + \beta^*}{2}$, while $(7.6.19)_3$ shows that $P > 0$, hence from $(7.6.19)_4$ it follows that $P = \sqrt{\beta^*}$, thus $0 < P \leq 1$. If $\beta^* = 1$, then $P = 1$, $M^* = M_i$, while $(7.6.19)$ implies that $S = M^{*2} + 2$ and, therefore, equation $(7.6.12)$ reads

$$(\delta - 1)[\delta^2 - (2 + M^{*2})\delta + 1]^2 = 0. \qquad (7.6.13')$$

However, in this case (7.6.12) cannot stand for the characteristic equation because, due to the fact that $\beta_H = \beta_I = 0$, the equation for W reads $[(D^2-a^2)^3 - M^2 D^2(D^2 - a^2)]W = 0$. Consequently the true characteristic equation becomes

$$(\lambda^2 - a^2)[\lambda^4 - (2a^2 + M^2)\lambda^2 + a^4] = 0. \tag{7.6.20}$$

Comparing (7.6.20) with (7.6.13$'$), we notice the occurrence of a parasitic factor $\lambda^4 - (2a^2 + M^2)\lambda^2 + a^4$, as it was already remarked at the beginning of Section 7.6.5. This factor is due to the fact that certain differentiations done to deduce the equation for W are no longer necessary in the case $\beta_H = \beta_I = 0$. Hence, the limit points corresponding to this subcase are not bifurcation points for the particular characteristic equation (7.6.20). This subcase where $\beta^* = 1$ has been studied in [Geo82a] and it was shown that all points are secular. If $\beta^* < 1$, from $(7.6.19)_{1,2}$ we deduce the expressions

$$M^{*2} = \frac{2S - 3 - \beta^*}{2\beta^*}, \quad M_i^2 = \frac{\beta^*(2S^2 - 6S - 3\beta^* + 4\sqrt{\beta^*} + 3)}{(2S - 3 + \beta^*)},$$

which substituted in $(7.6.19)_3$ lead to the consistency condition

$$4S^2 - 4(2 + \sqrt{\beta^*})S + (-\beta^* + 6\sqrt{\beta^*} + 3) = 0.$$

As the solutions of this relation in S do not satisfy the condition $S > \frac{3+\beta^*}{2}$ (necessary for $M^{*2} > 0$), it follows that, for $\beta^* < 1$, (7.6.13) cannot have a pair of double roots.

Subcase $\delta_1 = \delta_2 = \delta$ and δ_1, δ_3, δ_4 are mutually distinct. Denote $\delta_3 + \delta_4 = S$, $\delta_3\delta_4 = P$. Equation (7.6.12) has real coefficients, hence $\delta \in \mathbf{R}$. Equations (7.6.14) become

$$\begin{aligned}
2\delta + S &= 3 + \beta^* + 2\beta^* M^{*2}, \\
\delta^2 + 2\delta S + P &= 3 + 3\beta^* + 3\beta^* M^{*2} + M^{*2}M_i^2 + M_i^2, \\
\delta^2 S + 2\delta P &= 1 + 3\beta^* + \beta^* M^{*2} + M_i^2, \\
\delta^2 P &= \beta^*.
\end{aligned} \tag{7.6.21}$$

Relation $(7.6.21)_1$ implies that δ, $S < 0$ cannot hold, whereas $(7.6.21)_4$ shows that $P > 0$. Adding $(7.6.21)_1$ to $(7.6.21)_3$, subtracting the obtained equation from $(7.6.21)_2$ and taking into account $(7.6.21)_4$ we get

$$M^{*2}M_i^2 = (\delta - 1)^2(P + 1 - S) \tag{7.6.22}$$

and, therefore, $S < P + 1$. From $(7.6.21)_1$ we have $2\delta + S > 3 + \beta^* = 3 + P\delta^2$ and, since $S < P + 1$, it follows $2\delta + P + 1 > 3 + P\delta^2$, therefore $2(\delta - 1) > P(\delta - 1)(\delta + 1)$, which cannot be satisfied if $\delta \leq -1$. From $(7.6.21)_3$ we have $\delta^2 S + 2\delta P > 1 + 3\beta^* = 1 + 3P\delta^2$ and, since $S < 1 + P$, it follows $(\delta - 1)(\delta + 1) > 2\delta P(\delta - 1)$, which cannot be satisfied if $-1 < \delta \leq 0$. From $(7.6.21)_3$ we have $\delta^2 S + 2\delta P > 1 + 3\beta^*$ and, taking into account $(7.6.21)_4$ and (7.6.22) we obtain $\delta^2 P + \delta^2 + 2\delta P > 1 + 3\beta^*$, therefore $\beta^* + \delta^2 + \frac{2\beta^*}{\delta} > 1 + 3\beta^*$, consequently, $\delta^2 - 1 > 2\beta^*(1 - \frac{1}{\delta})$ and, hence

$$(\delta - 1)(\delta + 1) > 2\beta^*\frac{\delta - 1}{\delta}. \tag{7.6.23}$$

If $0 < \delta < 1$, then (7.6.23) is not satisfied if $1 > \delta > \delta_2$, where $\delta_2 = \frac{4\beta^*}{(1+\sqrt{1+8\beta^*})}$ is the positive root of $\delta^2 + \delta - 2\beta^* = 0$. But, as β^* runs over $(0,1)$, the root δ_2 runs over $(0,1)$, hence (7.6.23) is not satisfied for $\delta \in (0,1)$. But for $\delta > 1$ we have $P \in (0,\frac{1}{2})$. Hence, except, possibly, for $P \in (0,\frac{1}{2})$, the subcase of a single double root of (7.6.13) cannot occur.

7.6.6   *General case $a \neq 0$, $M^* > 0$, $R^* \neq 0$*

This case takes into account all effects from problem $(7.6.1) - (7.6.7)$. We fix our attention on multiple roots of (7.6.11) by considering the relationships

$$\delta_1 + \delta_2 + \delta_3 + \delta_4 + \delta_5 = 4 + \beta^* + 2\beta^* M^{*2};$$

$$\delta_1\delta_2 + \delta_1\delta_3 + \delta_1\delta_4 + \delta_1\delta_5 + \delta_2\delta_3 + \delta_2\delta_4 + \delta_2\delta_5 + \delta_3\delta_4$$
$$+ \delta_3\delta_5 + \delta_4\delta_5 = 6 + 4\beta^* + 5\beta^* M^{*2} + M^{*2}M_i^2 + M_i^2;$$

$$\delta_1\delta_2\delta_3 + \delta_1\delta_2\delta_4 + \delta_1\delta_2\delta_5 + \delta_1\delta_3\delta_4 + \delta_1\delta_3\delta_5 + \delta_1\delta_4\delta_5 + \delta_2\delta_3\delta_4 + \delta_2\delta_3\delta_5$$
$$+ \delta_2\delta_4\delta_5 + \delta_3\delta_4\delta_5 = 4 - R^* + 6\beta^* + 4M^{*2}\beta^* + 2M_i^2 + M^{*2}M_i^2; \qquad (7.6.24)$$

$$\delta_1\delta_2\delta_3\delta_4 + \delta_1\delta_2\delta_3\delta_5 + \delta_1\delta_2\delta_4\delta_5 + \delta_1\delta_3\delta_4\delta_5 + \delta_2\delta_3\delta_4\delta_5$$
$$= 1 - R^* + 4\beta^* + M^{*2}\beta^* - R^*\beta^* - R^*\beta^* M^{*2} + M_i^2;$$

$$\delta_1\delta_2\delta_3\delta_4\delta_5 = \beta^*(1 - R^*).$$

Subcase $\delta_1 = \ldots = \delta_5 = \delta$. In this case (7.6.24) become, after eliminating R^* between $(7.6.24)_{3,4,5}$,

$$
\begin{aligned}
M^{*2} &= (5\delta - \beta^* - 4)(2\beta^*)^{-1}, \\
10\delta^2 &= 6 + 4\beta^* + 5\beta^* M^{*2} + M^{*2}M_i^2 + M_i^2, \\
10\delta^3 &= 3 + 6\beta^* + 4M^{*2}\beta^* + M_i^2(2 + M^{*2})\delta^5\beta^{*-1}, \\
5\delta^4 &= 3\beta^* + \delta^5 + \delta^5\beta^{*-1} + \delta^5 M^{*2} + M_i^2.
\end{aligned}
\qquad (7.6.25)
$$

Relation $(7.6.25)_1$ implies $\beta^* < 5\delta - 4$ and, since $\beta^* > 0$, it is necessary to have $\delta > 0.8$. Subtracting $(7.6.25)_{2,4}$ from $(7.6.25)_3$ and replacing in the obtained equation M^{*2} by its expression $(7.6.25)_1$ we obtain

$$f(\beta^*) \equiv \beta^{*2} + \beta^*(\delta^5 - 10\delta^4 + 20\delta^3 - 20\delta^2 + 5\delta + 2) + 5\delta^6 - 4\delta^5 = 0 \qquad (7.6.26)$$

Finally, replacing $(7.6.25)_1$ into $(7.6.25)_4$ a necessary condition for the positiveness of M_i^2 reads

$$6\beta^{*2} + \beta^*(\delta^5 - \delta^4) + 5\delta^6 - 2\delta^5 < 0. \qquad (7.6.27)$$

If $\delta \geq 10$ all coefficients in (7.6.27) are positive, hence (7.6.27) cannot hold. Since, for $10 > \delta > \frac{4}{5}$, we have $\delta^5 - 10\delta^4 < 0$, and using $\beta^* < 5\delta - 4$, we obtain

$$\beta^*(\delta^5 - \delta^4) + 5\delta^6 - 2\delta^5 > 2\delta^4(5\delta^2 - 28\delta + 20),$$

which shows that (7.6.27) cannot take place for $0.8 < \delta \leq \frac{14-4\sqrt{6}}{5} (\simeq 0.82)$ and $\delta \geq \frac{14+4\sqrt{6}}{5} (\simeq 4.759)$. In addition, the discriminant of the equation associated with (7.6.27) is negative for $0.8 < \delta \leq 1.1$ and $\delta \geq 3.5$, therefore it remains to investigate the interval $1.1 < \delta < 3.5$. Similarly, for $0.82 < \delta < 5$, we have $m(\delta) = \delta^5 - 10\delta^4 + 20\delta^3 - 20\delta^2 + 5\delta + 2 < 0$ and, therefore

$$\beta^*(\delta^5 - \delta^4 + 20\delta^3 - 20\delta^2 + 5\delta + 2) + 5\delta^6 - 2\delta^5 > (5\delta - 4)(2\delta^5 - 10\delta^4 + 20\delta^3 - 20\delta^2 + 5\delta + 2).$$

Simple reasonings show that the right-hand side of this inequality is strictly positive for $\delta \geq 2.2$. Moreover, (at least) for $1.1 < \delta < 2.2$ the roots of (7.6.26) are greater than 1, because $f(0) > 0$, $f(1) < f(0)$, (which reads $m < -1$) and $\frac{df}{d\beta^*} < 0$ at $\beta^* = 1$. Summing up all the above results it follows that the subcase we study cannot hold.

Remark that in the chosen subcase relation $(7.6.25)_5$ implies $R^* < 1$. Hence, for $R^* \geq 1$, this case cannot take place. The situation $\beta^* = 1$ also cannot occur because in this case $(7.6.25)_{1,2}$ imply $\delta = 1$ (and, hence $M^* = M_i = 0$) and $\delta = \frac{-1}{3}$ (which leads to negative M^*).

Subcase $\delta_1 = \cdots = \delta_4 = \delta \neq \delta_5$. In this case, (7.6.24) become

$$4\delta + \delta_5 = 4 + \beta^* + 2\beta^* M^{*2};$$
$$6\delta^2 + 4\delta\delta_5 = 6 + 4\beta^* + 5\beta^* M^{*2} + M^{*2} M_i^2 + M_i^2;$$
$$4\delta^3 + 6\delta^2\delta_5 = 4 - R^* + 6\beta^* + 4M^{*2}\beta^* + 2M_i^2 + M^{*2}M_i^2; \qquad (7.6.28)$$
$$\delta^4 + 4\delta^3\delta_5 = 1 - R^* + 4\beta^* + M^{*2}\beta^* - R^*\beta^* - R^*\beta^* M^{*2} + M_i^2;$$
$$\delta^4\delta_5 = -\beta^*(R^* - 1).$$

Relations $(7.6.28)_{1,5}$ show that $\delta, \delta_5 \in \mathbf{R}$. If $R^* = 1$ this subcase cannot occur. Indeed, $(7.6.28)_5$ implies either $\delta_5 = 0$ or $\delta = 0$. But $\delta = 0$ contradicts $(7.6.28)_2$. Hence $\delta_5 = 0$. Then (7.6.28) gives

$$M^{*2} = \frac{4\delta - 4 - \beta^*}{2\beta^*} \qquad (7.6.29)$$

implying $0 < \beta^* < 4(\delta - 1)$ whence $\delta > 1$. Then (7.6.28) gives

$$4\delta^3 - 3 - 6\beta^* - M_i^2(M^{*2} + 2) > 0$$

and, taking into account the expressions of M_i^2 and M^{*2} given by $(7.6.28)_4$ and (7.6.29), we obtain

$$3\beta^{*2} + \beta^*(3\delta^4 - 8\delta^3 - 12\delta + 18) + 4\delta^4(\delta - 1) < 0. \qquad (7.6.30)$$

Since $3\delta^4 - 8\delta^3 - 12\delta + 18 \geq 0$ for $1 \leq \delta \leq \delta_0 < 1.1$ and $\delta > \delta_1 \in (2,3)$ it follows that in these intervals (7.6.30) fails to hold. For $\delta_0 < \delta < \delta_1$ we have $3\delta^4 - 8\delta^3 - 12\delta + 18 < 0$ and, hence,

$$\beta^*(3\delta^4 - 8\delta^3 - 12\delta + 18) + 4\delta^4(\delta - 1) > 4(\delta - 1)(3\delta^4 - 8\delta^3 - 12\delta + 18) + 4(\delta - 1)$$

$$= 4(\delta - 1)(4\delta^4 - 8\delta^3 - 12\delta + 18).$$

But the function $4\delta^4 - 8\delta^3 - 12\delta + 18$ is positive for $\delta \in (\delta_0; \delta_3)$, where $\delta_3 \in (1.2; 1.3)$, and for $\delta \geq 3$, hence again (7.6.30) cannot take place for these ranges. Finally, adding $(7.6.28)_1$ and $(7.6.28)_4$ and then subtracting the sum of $(7.6.28)_3$ and the relation $(7.6.28)_1$ divided by 2, we find $\beta^* = 2\delta^4 - 8\delta^3 + 12\delta^2 - 4\delta - 2$. Substituting this expression into $(7.6.28)_4$ we find $M_i^2 = -5\delta^4 + 24\delta^3 - 36\delta^2 + 12\delta + 6$. It can be proved that $M_i^2 < 0$ for $\delta > 1.2$ which contradicts the fact that $M_i^2 > 0$. All these results shows that for $R^* = 1$ the corresponding subcase fails.

7.6.7 *Open problems*

In Sections $7.6.2 - 7.6.6$ we mainly dealt with the detection of cases of multiple roots of the characteristic equation. In simpler cases we use Viète relations, Young inequality, minima and maxima, positiveness and boundedness of some physical quantities, solutions of transcendental equations obtained by graphical representation, non-existence of solutions of transcendental equations written as $f(x_1) = f(x_2)$, where f is injective and $x_1 = x_2$ is physically unrealistic case etc. We could do a more systematic analysis by starting with the cases $a = 0$ and $b_1 = -a^4$, when there are the multiple roots $\lambda_{1,6} = 0$. There, for $a \neq 0$, taking into account the derivative of the characteristic equation, we must investigate the case of double roots of the equation in δ. Indeed, if there do not exist double roots, no root with higher multiplicity exists. We did not choose this alternative because we intended to show several tricks which could lead to the solution in a simpler way.

Several cases have not been treated yet. For instance, in the case $R^* \neq 1$ some theoretical results can be obtained but this involves cumbersome algebraic calculations which must be yielded by specialized computer codes. We must treat the cases $\delta_1 = \ldots = \delta_5$; $\delta_1 = \ldots = \delta_4 \neq \delta_5$; $\delta_1 = \delta_2 = \delta_3 = \delta$ where $\delta, \delta_4, \delta_5$ are mutually distinct or $\delta_4 = \delta_5 \neq \delta$. Although algebraic and containing only hyperbolic functions, the secular equation was not determined nor solved in several cases. This is due to the fact that it involves bifurcation questions still unsolved.

7.7 Inhibition of the thermal convection by a magnetic field

On the basis of a detailed analysis of the multiplicity of the characteristic roots [OpG], for each multiplicity the corresponding neutral stability surfaces are determined. False secular surfaces are determined. A construction of the secular curves is given by using a continuation algorithm from numerical bifurcation theory.

7.7.1   *Multiplicity of the characteristic roots*

The mathematical problem governing the neutral linear stability of the conduction state of a horizontal fluid layer, situated between two rigid walls $x = \pm 0.5$, heated from below in the presence of an external magnetic field parallel to the gravitational field, reads [DiP61], [Chan]

$$(D^2 - a^2)\big[(D^2 - a^2)^2 - QD^2\big]W = -Ra^2W, \quad x \in (-0.5, 0.5) \tag{7.7.1}$$

$$W = DW = \big[(D^2 - a^2)^2 - QD^2\big]W = 0, \quad x = \pm 0.5 \tag{7.7.2}$$

where $D = d/dx$, W is the amplitude of the vertical component of the perturbation velocity, $a(>0)$ is the dimensionless wave number, $Q(>0)$ is the dimensionless field strength (Chandrasekhar number) and $R(>0)$ represents the dimensionless Rayleigh number which depends on the temperature gradient.

To various extents this problem was dealt with: in [Geo82a], [OpG], [GeoO90], by using the direct method; in [DiP61] by a direct method based on Fourier series; in [GeoOP] by direct and variational methods based on Fourier series. The secular equation derived by the direct method was solved numerically in [GeoOO] and it was found that the bifurcation set for the characteristic manifold consists of false secular points and it is situated beneath the neutral curve, the first numerical (and graphical) confirmation of the existence of a false neutral curve $R = R(a)$ for some fixed values of Q. The numerical method used was a continuation method.

The characteristic equation for $(7.7.1) - (7.7.2)$ has the form

$$(\lambda^2 - a^2)^3 - Q\lambda^2(\lambda^2 - a^2) + Ra^2 = 0, \tag{7.7.3}$$

or, equivalently, putting $\mu = \lambda^2 - a^2$, the form

$$\mu^3 - Q\mu^2 - Qa^2\mu + a^2R = 0. \tag{7.7.4}$$

Let $\lambda_1, \ldots, \lambda_6$ be the roots of equation (7.7.3), which depend on the three parameters a, Q, R. The detailed discussion of the multiplicity of λ_i is carried out in [OpG].

The roots λ_i can be multiple in two cases: 1) if the roots of (7.7.4) are multiple, which happens if and only if the points (a, Q, R) of the parameter space are situated on the surface $C_1 = \{(a, Q, R) \in \mathbf{R}^3 \mid -27a^2R = (Q + \sqrt{Q^2 + 3a^2Q})^2 \times (Q - 2\sqrt{Q^2 + 3a^2Q})$ (the corresponding explicit function R of a and Q is denoted by $R = R^*(a, Q))$ for $a > 0$, $Q \geq 0$, and $a = 0$ for $Q = 0$, $R \geq 0\}$; 2) if one of the roots of (7.7.4) is $\mu_i = -a^2$, i.e. $\lambda_i = \lambda_{i+3} = 0$, which happens if and only if the points (a, Q, R) are situated on the cylinder $C_2 = \{(a, Q, R) \in \mathbf{R}^3 \mid R = a^4 \text{ for } a \geq 0, Q \geq 0\}$. The situation 1) was deduced by applying the Euclid algorithm to (7.7.4).

Let us now provide a geometrical and bifurcational interpretation of the sets of the points corresponding to multiple λ_i. In the $(Re\mu, Im\mu, a, Q, R)$ space, the set of solutions $\mu_i(a, Q, R)$ of equation (7.7.4) is a 3-dimensional (possibly singular) manifold M, and similarly, the set of solutions $\lambda_i(a, Q, R)$ of (7.7.3) is a 3-dimensional manifold Λ. The value and multiplicity of μ_i and λ_i and, correspondingly, the number of sheets of M and Λ, depend on the position of the points (a, Q, R) with respect to the surfaces C_1 and C_2 in the (a, Q, R) parameter space. The origin belongs to $C_1 \cap C_2$; C_1 contains the open half a-axis $a > 0$, $Q = 0$; $R = 0$ and the open half R-axis $R > 0$, $a = 0$; $Q = 0$.

Equation (7.7.4) can have: mutually distinct solutions for points (a, Q, R) not situated on C_1; $\mu_1 = \mu_2$ for (a, Q, R) belonging to C_1, but not to the R or a-halfaxis and the origin $(0, 0, 0)$; a triple solution $\mu_1 = \mu_2 = \mu_3$ when (a, Q, R) is situated at the origin or on these half-axes. This means that, in the (a, Q, R) parameter space, C_1 represents the bifurcation (catastrophe) surface for M; for $(a, Q, R) \notin C_1$ M has three distinct sheets, for (a, Q, R) on C_1 (for $a > 0$, $R > 0$) two sheets coalesce while for points on the half R- and a-axis the three sheets coalesce into a single one. Since, by definition, $\lambda_{1,4} = \pm\sqrt{\mu_1^2 + a^2}$, $\lambda_{2,5} = \pm\sqrt{\mu_2^2 + a^2}$, $\lambda_{3,6} = \pm\sqrt{\mu_3^2 + a^2}$, it follows that C_1 is a bifurcation surface for Λ too.

The manifold Λ may have any number of sheets from 1 to 6, namely, for $(a, Q, R) \notin C_1 \cup C_2$, Λ has six sheets not mutually intersecting; if $(a, Q, R) \in C_1 \setminus C_2$, $a > 0$, $R > 0$, four sheets of Λ coalesce (namely λ_1 with λ_2 and λ_4 with λ_5) so Λ has 4 distinct sheets; for $(a, Q, R) \notin C_1 \cap C_2$, $a > 0$, $R > 0$, (*i.e.* at points $(a, a^2, a^4) \equiv (\sqrt{Q}, Q, Q^2))$, Λ has only three distinct sheets, λ_1 coalesce with λ_2, λ_4 with λ_5 and λ_3 with λ_6. In this particular case $\lambda_3 = 0$; for points situated on the open half a-axis, Λ has two distinct sheets $\lambda_1 = \lambda_2 = \lambda_3 = a$, $\lambda_4 = \lambda_5 = \lambda_6 = -a$; at the points on the half R-axis, $R \geq 0$, $a = 0$, $Q = 0$, including the origin (which belongs to C_1) all the sheets of Λ coincide and are $\lambda_1 = \lambda_2 = \ldots = \lambda_6 = 0$; finally, for points of C_2 two sheets of Λ coalesce and they are $\lambda_3 = \lambda_6 = 0$. in this case, when $R \neq R^*$, Λ has five distinct sheets. The difference from the case of M arises from the fact that on C_2 we have $\mu_3 = -a^2$ implying $\lambda_3 = \lambda_4 = 0$.

Remark that although the physical problem is defined for $a, R > 0$, continuation reasons imposed the consideration of the cases $a = 0$, $R = 0$ too.

7.7.2 Secular equations

For $(a, Q, R) \notin C_1 \cup C_2$ $a, Q, R > 0$, Λ has six distinct sheets, so the general even solution (7.7.1) reads $W(x) = \sum_{i=1}^{3} A_i \cosh(\lambda_i x)$, leading to the secular equation

$$\det \begin{vmatrix} \cosh\frac{\lambda_1}{2} & \frac{1}{2}\cosh\frac{\lambda_2}{2} & \cosh\frac{\lambda_3}{2} \\ \lambda_1\sinh\frac{\lambda_1}{2} & \lambda_2\sinh\frac{\lambda_2}{2} & \lambda_3\sinh\frac{\lambda_3}{2} \\ (\mu_1^2 - Q\lambda_1^2)\cosh\frac{\lambda_1}{2} & (\mu_2^2 - Q\lambda_2^2)\cosh\frac{\lambda_2}{2} & (\mu_3^2 - Q\lambda_3^2)\cosh\frac{\lambda_3}{2} \end{vmatrix} = 0. \quad (7.7.5)$$

Denoting $t_i = \lambda_i \tanh(\lambda_i/2)$, it simply becomes

$$\mu_1\mu_2(t_1 - t_2) + \mu_2\mu_3(t_2 - t_3) + \mu_3\mu_1(t_3 - t_1) = 0 \quad (7.7.6)$$

and represent the implicit equation

$$f(a, Q, R) = 0 \quad (a, Q, R) \notin C_1 \cup C_2, \quad a, Q, R > 0 \quad (7.7.6')$$

of the secular surface S_1 corresponding to points of the first octant which do not belong to the bifurcation surfaces $R = R^*$ (referred to as C_1) and $R = a^4$ (denoted by C_2). Let us distinguish S_1 from the larger surface S defined by the equation $f(a, Q, R) = 0$ (a, Q, R), $a, Q, R \geq 0$.

For $(a, Q, R) \in C_1 \setminus C_2$, $a, R > 0$, we have $\mu_1 = \mu_2 = (Q + \sqrt{Q^2 + 3a^2 Q})/3$, $\mu_3 = (Q - 2\sqrt{Q^2 + 3a^2 Q})/3$, $\mu_1, \mu_3 \in \mathbf{R}$, $\lambda_1 = \lambda_2$, $\lambda_4 = \lambda_5$. Hence Λ has four distinct sheets and the general even solution of (7.7.1) reads $W_e(x) = A_1 \cosh \lambda_1 x + A_2 x \sinh \lambda_1 x + A_3 \cosh \lambda_3 x$, which lead to the secular equation

$$\det \begin{vmatrix} \cosh \frac{\lambda_1}{2} & \frac{1}{2} \sinh \frac{\lambda_1}{2} & \cosh \frac{\lambda_3}{2} \\ \lambda_1 \sinh \frac{\lambda_1}{2} & \sinh \frac{\lambda_1}{2} + \frac{\lambda_1}{2} \cosh \frac{\lambda_1}{2} & \lambda_3 \sinh \frac{\lambda_3}{2} \\ (\mu_1^2 - Q\lambda_1^2) \cosh \frac{\lambda_1}{2} & (4\lambda_1 \mu_1 - 2\lambda_1 Q) \cosh \frac{\lambda_1}{2} & (\mu_3^2 - Q\lambda_3^2) \cosh \frac{\lambda_3}{2} \\ & +[(\mu_1^2 - Q\lambda_1^2)]/2 \sinh \frac{\lambda_1}{2} & \end{vmatrix} = 0. \quad (7.7.7)$$

Let us remark that the general solution when λ_1 is a double root may be formally derived differentiating the general solution from the case when all roots λ_i are simple with respect to λ_2 and then letting $\lambda_2 \to \lambda_1$. Likewise (7.7.7) is obtained by differentiating (7.7.5) with respect to λ_2 and then letting $\lambda_2 \to \lambda_1$.

An equivalent alternative way of deriving (7.7.7) from (7.7.5) is the following: subtract the first column from the second, divide the so-obtained second column by $\lambda_2 - \lambda_1$ and let $\lambda_2 \to \lambda_1$.

A third approach to derive (7.7.7) is to write $\lambda_2 = \lambda_1 + \epsilon$, develop the determinant in (7.7.7) in power series of ϵ and then tend ϵ at zero.

All these procedures show that the secular curve S_2 consisting of points $(a, Q, R^*(a, Q))$ where (a, Q) satisfies (7.7.7) is a limit of the secular surface N_1 from the case of simple λ_i extended for $R \to R^*$, $R \neq a^4$. (Indeed, the first two operations implied in the second approach invariated the roots of equation (7.7.5).) This property is basic for pointing out those manifolds which do not belong to the secular surface N_1 of (7.7.1), (7.7.2). So, for all points of the bifurcation surface C_1 (except the points with $a = 0$ or $R = a^4$), we have $\lambda_1 = \lambda_2$, implying that (7.7.5) vanishes identically; hence all these points are formally secular, *i.e.* they belong to S. But among all these pints only the points of S_2 (satisfying (7.7.5)) of the neutral curve N_2 belong to the true secular surface S_t. Hence these points of C_1 are bifurcation points for the surface S to which S_t and C_2 belong. This enables us to construct the true secular surface S_t by starting with these points and applying a continuation algorithm.

Equation (7.7.7) can also be written as

$$\mu_1(\mu_3 - \mu_1)(\lambda_1^2 + 2t_1 - t_1^2) - 4\lambda_1^2 \mu_3(t_3 - t_1) = 0, \quad (7.7.8)$$

hence it is of the form

$$f_1\big(a, Q, R^*(a, Q)\big) = 0, \quad a > 0, \quad R^* > 0, \quad R^* \neq a^4. \quad (7.7.9)$$

It may be proved [OpG] that no point (a, Q, R), for which the number of sheets of Λ is 1, 2, 3 or 5 belongs to the true secular surface S_t. It follows that $S_t = S_1 \cup S_2$ or, equivalently, S_t consists of points of S and the limit points of S_1 when $R \to R^*$, $a > 0$, $R^* > 0$. To construct S_t we start from (7.7.8) which has $(0, 0)$ as bifurcation point and is satisfied for points $(a, 0)$ of the positive a-halfaxis. With the point $(0, 0)$ of the (a, Q)-plane as a departure point, we use a continuation algorithm to construct the curve S_2' which is the projection of S_2 on this plane. Then S_2 is immediately constructed as the curve of points $(a, Q, R^*(a, Q))$ with $(a, Q) \in S_2'$. Next, in planes $Q = $ constant, starting from the points of S_2 which belong to $S_1 \cap C_1$ (and are bifurcation points for S) the neutral curve in these planes is constructed also by a continuation procedure. In this way, various sections in the true secular surface are drawn.

This bifurcation analysis of (7.7.9) and (7.7.6)$'$ completes our method. The study herein belongs to those quoted by Collatz [Col3].

7.8 Microconvection in a binary layer subject to a strong Soret effect

By means of the direct method the bifurcation set of the characteristic manifolds are found and the false secular on them are determined following mainly [Drag07a].

7.8.1 *Eigenvalue problem*

The conduction-convection in a viscous binary fluid situated in an infinite horizontal layer, bounded by impermeable rigid walls, in the presence of a strong Soret effect and for low constant gravity g reads [GapZ]

$$\begin{cases} \nabla \mathbf{v} = S\Delta T + L_e \Delta C \\[2mm] P_r^{-1} \epsilon \mathbf{v} \cdot \nabla \mathbf{v} = -\nabla p'' + \Delta \mathbf{v} - \dfrac{G(T + C)}{1 + \epsilon(T + C)} \mathbf{k}, \\[2mm] \epsilon \mathbf{v} \nabla \cdot T = \Delta T, \\[2mm] \epsilon \mathbf{v} \nabla \cdot C = L_e \Delta(C - \sigma T) \end{cases} \qquad (7.8.1)$$

$$u = w = 0, \quad T_z = -1, \quad C_z = \sigma T_z, \quad S = 1 - L_e \sigma \quad \text{at} \quad z = 0, 1 \qquad (7.8.2)$$

where $p'' = p - \rho_0 g \mathbf{k} x - (\eta/3)\nabla \cdot \mathbf{v}$, $\mathbf{v} = (u, w)$ is the velocity field, T is the temperature, C is the concentration, $\epsilon > 0$ is the Boussinesq parameter, $L_e > 0$ is the Lewis number, $\sigma \in [0, 1]$ is the separation ratio, G stands for the Galileo number, $\mathbf{k}$ is the unit vector pointing vertically upwards and the index z indicates the differentiation.

Take the mechanical equilibrium $\overline{\mathbf{v}} = \mathbf{0}$, $\overline{T}_z = -1$, $\overline{C}_z = -\sigma$ as the basic state. Then the normal mode perturbations (denoted by primes) of the form

$f'(x, z) = f'(z) \, exp(-iax)$ satisfy the eigenvalue two-point problem

$$\begin{cases} (D^2 - a^2)^2 \Psi' + iaG'(T' + C') = 0, \\ -\epsilon(ia\Psi' + SDT' + L_e DC') = (D^2 - a^2)T', \\ \epsilon\sigma(ia\Psi' + SDT' + L_e DC') = L_e(D^2 - a^2)(C' - \sigma T'), \end{cases} \tag{7.8.3}$$

$$D\Psi' = ia(ST' + L_e C'), \quad \Psi' = DT' = DC' = 0 \quad \text{at} \quad z = 0, 1 \tag{7.8.4}$$

where $D = \frac{d}{dz}$, $G' = \frac{G}{1+\epsilon(\overline{T}+\overline{C})^2}$. We further assume that G' is a strictly positive constant. In this problem (Ψ', T', C') is the eigenvector corresponding to the eigenvalue L_e, which depends on the positive parameters $a, \epsilon, G', S, \sigma$. Using $(7.8.3)_{2,3}$ and the boundary conditions $(7.8.4)_{3,4}$, we have $(D^2 - a^2)U = 0$, $DU = 0$ at $z = 0, 1$, where $U = \sigma(1 - L_e)T' + L_e C'$. This problem has the trivial solution $U = 0$ for $z \in [0, 1]$, whence the relationship between the unknown functions T' and C', namely $C' = \sigma(L_e - 1)T'L_e^{-1}$. This reduces (7.8.3) to

$$L_e(D^2 - a^2)^3 T' + \epsilon L_e(1 - \sigma)D(D^2 - a^2)T' + a^2\epsilon G'[L_e - \sigma(L_e - 1)]T' = 0,$$

or, since, from physical reasons, $L_e \neq 0$, introducing the new parameters $a_1 = \epsilon(1 - \sigma)\geq 0$, $a_z = a^2\epsilon G'[1 + \sigma(1 - L_e^{-1})]$, to

$$(D^2 - a^2)^3 T' + a_1 D(D^2 - a^2)T' + a_2 T' = 0. \tag{7.8.6}$$

The boundary conditions (7.8.4) written in T' only read

$$DT' = (D^2 - a^2)T' = D^3 T' = 0 \quad \text{at} \quad z = 0, 1. \tag{7.8.7}$$

The eigenvalue two-point problem (7.8.6), (7.8.7) depends only on three parameters, namely a, a_1, a_2.

7.8.2   *Characteristic equation and its bifurcation set*

The characteristic equation associated with equation (7.8.6) reads

$$(\lambda^2 - a^2)^3 + a_1\lambda(\lambda^2 - a^2)^2 + a_2 = 0,$$

or, using the notation $\mu = \lambda/a$, $6b = a_1/a\geq 0$, $d = (a_2 - a^6)/a^6$,

$$P \equiv \mu^6 + 6b\mu^5 - 3\mu^4 - 12b\mu^3 + 3\mu^2 + 6b\mu + d = 0, \tag{7.8.8}$$

or, equivalently,

$$(\mu^2 - 1)[\mu^3 + 5b\mu^2 - \mu - b] = 0. \tag{7.8.9}$$

Case $b \neq 0$, $d \neq 0, -1$. The roots $\mu = \pm 1$ of (7.8.9) are not roots for (7.8.8) (because introduced in (7.8.8) imply $d = 0$). The other roots of (7.8.9) are all real and distinct because the discriminant of the equation

$$P_1 \equiv \mu^3 + 5b\mu^2 - \mu - b = 0 \tag{7.8.10}$$

is $\Delta = -(125b^4 + 22b^2 + 1)/27 < 0$. Moreover, equation (7.8.10) cannot have the roots $\mu = \pm 1$ because this would imply $b = 0$ and $\mu = 0$, in which case (7.8.8) would imply $d = -1; 0$. Therefore, all roots of (7.8.9) are simple, implying the fact that no root of (7.8.8) can have multiplicity greater than or equal to three.

The polynomials P and P_1 cannot have three common roots. Indeed, this would imply that $P = P_1^2$, which is impossible because the coefficients of μ^5 are $6b$ and $10b$ respectively. Therefore, assume that P and P_1 have at most two common roots. Then, applying Euclid's algorithm we have $P = P_1 Q_1 + P_2$, $P_1 = P_2 Q_2 + R_1$, where $P_2 = -A\mu^2 + (20b^3 + 4b)\mu + 25b^4 + d$, $Q_1 = \mu + b\mu^2 - (2 + 5b^2)\mu + 25b^3$, $A = 125b^4 + 4b^2 - 1$, $Q_2 = -\frac{1}{A}[\mu + (625b^4 + 40b^2 - 1)bA^{-1}]$, $R_1 = -A^{-2}\{\mu[2400b^6 + 349b^4 + 4b^2 - 1 + dA] + b[209b^4 + 8b^2 - 1 + d(625b^4 + 40b^2 - 1)]\}$. Therefore P and P_1 have in common two roots if both square brackets in the expression of R_1 vanish. Eliminating d between the two vanishing brackets it follows $b^2(375b^4 + 6b^2 - 1)\Delta = 0$, implying $b = b^* = (8\sqrt{6} - 3)/375$, the common roots being $\mu_{1,2} = \frac{-(2+3\sqrt{6})}{50b^*} \pm \frac{1}{\sqrt{2}}$. In the space (b, d, μ), equation (7.8.8) defines a surface, for (b^*, d^*) two pairs of its sheets coalescing. If $R_1 = 0$ but the square brackets are not vanishing, then (7.8.8) has a single double root, *i.e.* only two sheets coalesce. This happens if

$$\mu = -b[209b^4 + 8b^2 - 1 + d(625b^4 + 40b^2 - 1)]/[2400b^6 + 349b^4 + 4b^2 - 1 + dA].$$

Introducing this double root in $P_2 = 0$ it follows the restriction

$$d^3 + d^2(2 - 3b^2 - 375b^4 - 3125b^6)$$
$$+d(1 - 6b^2 - 366b^4 - 3050b^6) + (-3b^2 + 9b^4 + 1099b^6 + 9216b^8) = 0. \tag{7.8.11}$$

The discriminant of this equation is $D = -1024 \cdot 27^2 b^6 A^2 \Delta^3 < 0$, hence (7.8.11) has three roots $d = d_i(b)$, $i = 1, 2, 3$ defining in the (b, d) plane three curves $\gamma_1, \gamma_2, \gamma_3$, symmetric with respect to the straight line $b = 0$. Each of them is the bifurcation set of two sheets of the surface defined by (7.8.8). The bifurcation set for d_1, d_2, d_3 is (b^*, d^*). The points $(b, d) = (0, 0); (0, -1)$ are only apparent bifurcation points for γ_1, γ_2 and γ_3 respectively. In fact, at these points the two symmetric branches of d_1, d_2 and d_3 respectively, coalesce.

Case $b = 0$. Equation (7.8.8) becomes $(\mu^2 - 1)^3 + d + 1 = 0$. For $d = -1$ (7.8.8) has three pairs of double roots $\mu_{1,2,3} = 1$, $\mu_{4,5,6} = -1$, corresponding to $\lambda_{1,2,3} = a$, $\lambda_{4,5,6} = -a$, for $d = 0$, the characteristic equation in λ has the double root $\lambda_1 = \lambda_4 = 0$ while for $d \neq -1, 0$, (7.8.8) has no multiple root.

Case $d = -1$, $b \in \mathbf{R}$. The straight line $d = -1$ of the (b, d) plane is also a bifurcation set for (7.8.8). Except for $b = 0$, it consists of points corresponding to which (7.8.8) has two pairs of double roots $\mu_{1,2} = 1$, $\mu_{3,4} = -1$.

Theorem 7.8.1. *The bifurcation set for (7.8.8) consists of the curves d_1, d_2 and d_3, which are defined for $b \in \mathbf{R}$ by the roots of (7.8.11) and the straight line $d = -1$, $b \in \mathbf{R}$.*

In order to investigate the multiplicity of roots $\mu_1, \ldots, \mu_6$, of (7.8.8) we mainly use either the Viète relations: 1° $\sum_{i=1}^{6} \mu_i = -6b$, 2° $\sum_{i,j=1 \, i<j}^{6} \mu_i \mu_j = -3$, 3° $\sum_{i,j,k=1 \, i<j<k}^{6} \mu_i \mu_j \mu_k = 12b$, 4° $\sum_{i,j,k,r=1 \, i<j<k<r}^{6} \mu_i \mu_j \mu_k \mu_r = 3$, 5° $\sum_{i,j,k,r,s=1 \, i<j<k<r<s}^{6} \mu_i \mu_j \mu_k \mu_r \mu_s = -6b$, $\prod_{i=1}^{6} \mu_i = d$, or we do a static bifurcation analysis using the first two derivatives of (7.8.8), namely

$$\mu^5 + 5b\mu^4 - 2\mu^3 - 6b\mu^2 + \mu + b = 0, \tag{7.8.12}$$

$$5\mu^4 + 20b\mu^3 - 6\mu^2 - 12b\mu + 1 = 0. \tag{7.8.13}$$

Of course, in the last approach we use the Euclid algorithm.

In the following we show that the use of the Viète relations as they stand can lead to more complicated reasonings than those in the above. This shows the importance of combining the information given by these relations and some remarks deduced from the equations. In fact, although it looks algorithmic and easy to apply, the inspired remarks transparent from equations themselves are of much importance in the direct method.

Since the coefficients of (7.8.8) are real, it follows that if there exist 6, 5 or 4 multiple roots of this equation, then they are necessarily real.

The case when all μ_i are equal cannot occur since 2° implies the impossible relation $15\mu_1^2 = -3$.

Assume $\mu_1 = \mu_2 = \mu_3 = \mu_4 = \mu_5 \neq \mu_6$. Then 2° and 4° imply $10\mu_1^2 + 5\mu_1\mu_6 = -3$ and $5\mu_1^4 + 10\mu_1^3\mu_6 = 3$ respectively leading to the impossible relation $-6\mu_1^2 - 15\mu_1^4 = 3$, because $\mu_{1,\ldots,5}$ must be real.

Assume $\mu_1 = \mu_2 = \mu_3 = \mu_4 \neq \mu_{5,6}$ and denote $\mu_5 + \mu_6 = s$, $\mu_5\mu_6 = p$. Then 1°, 2°, 4°, 5° imply $4\mu_1 + s = -6b$, $6\mu_1^2 + 4\mu_1 s + p = -3$, $\mu_1^4 + 4\mu_1^3 s + 6\mu_1^2 p = 3$ and $\mu_1^4 s + 4\mu_1^3 p = -6b$ respectively, leading to $80\mu_1^8 - 35\mu_1^7 + 51\mu_1^4 - 21\mu_1^3 + 21 = 0$. This last equation has only positive solutions. Moreover, it can be written in the form $80\mu_1^8 + 16\mu_1^4 + (35\mu_1^4 + 21)(1 - \mu_1^3) = 0$, showing that it has no solution for $\mu_1 \leq 1$ and in the form $41\mu_1^8 + 35\mu_1^7(\mu_1 - 1) + 30\mu_1^4 + 21\mu_1^3(\mu_1 - 1) + 21 = 0$, showing that it has no solution $\mu_1 \geq 1$. Also it can be written as $[45\mu_1^8 + 30\mu_1^6 + 18\mu_1^2 + 3]\mu_1^{-3} = 0$, showing that it has no real solution μ_1, because, in view of (7.8.13), $\mu_1 \neq 0$. Whence the case of four equal characteristic roots cannot occur.

So far no restriction on b and d was imposed.

Assume $\mu_1 = \mu_2 = \mu_3$, therefore this multiple root is a common root of (7.8.8), (7.8.12) and (7.8.13). In addition, since (7.8.8) cannot have roots of multiplicity four, it follows that no root of (7.8.13) is multiple.

In order to prove that an equation and its derivatives have a common root, we apply the Euclid algorithm. Sometimes, using some additional remarks deduced by the inspection of the equations itself, this algorithm can lead immediately to the desired result. Thus, in particular cases, *e.g.* for (7.8.12), the elimination of the parameter can lead faster to the result. Indeed, (7.8.12) reads in the form (7.8.9), which shows that $\mu = \pm 1$ cannot be a solution for (7.8.13) except for $b = 0$. Therefore, (7.8.12) reduces to (7.8.10). Elimination of b between (7.8.10) and (7.8.13) leads to the equation $5\mu^4 + 2\mu^2 + 1 = 0$ which has no real solution (we remark that (7.8.10) has only real solutions). Consequently, for $b \neq 0$, (7.8.8) has no solutions of multiplicity equal to three.

7.8.3 *False secular points*

Case $b = 0$, $d = -1$. The general solution of (7.8.6) reads

$$T'(z) = (A_0 + A_1 z + A_2 z^2)\cosh(az) + (B_0 + B_1 z + B_2 z^2)\sinh(az),$$

where the A'_s and B'_s are arbitrary constants. Imposing to it to satisfy the boundary conditions it follows the secular equation $a^6 \sinh a(\sinh^2 a - a^2) = 0$, which has no solution $a \neq 0$. Hence the points $(Le, a, \sigma) = (0.5, a, 1)$ corresponding to $(b, d) \neq (0, 0)$ are false secular points.

Case $b = 0$, $d = 0$. The general solution of (7.8.6) reads

$$T'(z) = A + Bz + \sum_{i=2}^{3} A_i \cosh(\lambda_i z) + B_i \sinh(\lambda_i z),$$

where $\lambda_{2,3} = a\sqrt{(3 \pm i\sqrt{3})/2}$, and the A'_s and B'_s are arbitrary constants, leading to the secular equation

$$\sinh(\lambda_2/2)\sinh(\lambda_3/2)[\lambda_3^2\mu_2 \cosh(\lambda_3/2)\sinh(\lambda_2/2) - \lambda_2^3\mu_2 \cosh(\lambda_2/2)\sinh(\lambda_3/2)] = 0,$$

the root of which is $a = 0$, implying $T' = C' = \Psi' = 0$. Hence the points $(b, d) = (0, 0)$ and, therefore, $(Le, \sigma, \epsilon, G') = (Le, 1, \epsilon, a^4 Le[\epsilon(2Le - 1)]^{-1})$ are false secular points.

Cases $b \in \mathbf{R}$, $d = -1$; $b = b^*$, $d = d^*$; $b \in \mathbf{R}$, $d = d_1$, d_2 or d_3 are treated analytically in a similar way in [Drag07a]; numerical calculations show that all corresponding secular points are false secular points. No secular points independent of the boundary conditions were found.

7.9 Convection in the layer between the sea bed and the permafrost

The direct method is applied to this convection using the results in [Drag07a], reframed along the lines of Appendix 6. A new study involving multiplicity of the characteristic roots is included too. A false neutral curve is found while additional secular points were proved to be absent.

A permafrost means a soil or rock that remains below $0°C$ for at least two consecutive years. Temperatures increase in the soil causes permafrost to thaw, creating a layer of salting sediments beneath the sea bed. Permafrost melting has severe consequences on climate change. In the horizontal layer between the sea bed and the permafrost the convection takes place. The convection motion is determined by the salty layer melting the fresh ice, which being less dense, rises through the porous thawing layer. Such convection flow is observed on the coast of Alaska.

It is governed by the equations [Strau]

$$\begin{cases} Rs\delta \cdot \mathbf{k} = \nabla p + \mathbf{v}, \\ \nabla \cdot \mathbf{v} = 0, \\ R\mathbf{v} \cdot \mathbf{k} + \Delta s = 0 \end{cases} \tag{7.9.1}$$

for $z \in (0,1)$ and the boundary conditions

$$\begin{cases} \mathbf{v} \cdot \mathbf{n} = 0 \quad \text{at} \quad z = 0,1, \\ s = 0 \quad \text{at} \quad z = 0, \\ \dfrac{\partial}{\partial z} s + \alpha s = 0 \quad \text{at} \quad z = 1, \end{cases} \tag{7.9.2}$$

where $\mathbf{k}$ is the unit vector pointing vertically upwards, $R_a = R^2$ is the Rayleigh number, s, $\mathbf{v}$, p are the perturbations of the salt, velocity and pressure field respectively, and $\alpha \geq 0$ is a parameter related to the melting of the permafrost interface.

For normal mode perturbations the eigenvalue problem (7.9.1), (7.9.2) becomes

$$\begin{cases} (D^2 - a^2)W + Ra^2 S = 0, \\ (D^2 - a^2)S + RW = 0, \end{cases} \tag{7.9.3}$$

$$\begin{cases} W = 0 \quad \text{at} \quad z = 0,1, \\ S = 0 \quad \text{at} \quad z = 0, \\ DS + \alpha S = 0 \quad \text{at} \quad z = 1, \end{cases} \tag{7.9.4}$$

where $z \in (0,1)$, $D = \frac{d}{dz}$, a is the wave number, W, S are the intensity of $\mathbf{v} \cdot \mathbf{k}$ and s in the normal modes. Consider (W, S) as the eigenfunction and R_a as the eigenvalue. Eliminating W between $(7.9.3)_{1,2}$ we obtain

$$(D^2 - a^2)^2 S - R^2 a^2 S = 0, \tag{7.9.5}$$

$$\begin{aligned} D^2 S = 0 \,\text{at}\, z = 0; &\qquad D^2 S - a^2 S = 0 \,\text{at}\, z = 1; \\ S = 0 \,\text{at}\, z = 0; &\qquad DS + \alpha S = 0 \,\text{at}\, z = 1, \end{aligned} \tag{7.9.6}$$

so the associated characteristic equation is $(\lambda^2 - a^2)^2 - R^2 a^2 = 0$ and has the distinct roots $\lambda_1 = \sqrt{a^2 + Ra}\ (= -\lambda_3)$, $\lambda_2 = \sqrt{a^2 - Ra}\ (= -\lambda_4)$, for $R \neq a$, and $\lambda_{1,3} = a\sqrt{2}$, $\lambda_{2,4} = 0$ for $R = a$.

For $R \neq 0$, the general solution of (7.9.5) reads $\displaystyle\sum_{i=1}^{3}[A_i \cosh(\lambda_i z) + B_i \sinh(\lambda_i z)]$, leading to the secular equation

$$\lambda_2 \sinh \lambda_1 \cosh \lambda_2 + \lambda_1 \sinh \lambda_2 \cosh \lambda_1 + 2\alpha \sinh \lambda_1 \sinh \lambda_2 = 0. \tag{7.9.7}$$

For $R < a$ we have $\lambda_1, \lambda_2 > 0$, therefore all terms in (7.9.7) are positive, hence (7.9.7) has no solution. For $R > a$ the neutral curves were determined numerically in [Drag07a].

For $R = a$ we have $\lambda_1 = -\lambda_3 = a\sqrt{2}$, $\lambda_2 = -\lambda_4 = 0$, therefore the general solution of (7.9.5), (7.9.6) reads $S(z) = A_1 \cosh(a\sqrt{2}z) + B_1 \sinh(a\sqrt{2}z) + A_2 + B_2 z$, leading to the secular equation $\tanh(a\sqrt{2}) = -a\sqrt{2}/(1+2\alpha)$, which has no solution, the left-hand side being positive and the other negative. (The case $a = 0$ is not physically realistic.)

Consequently, the problem (7.9.1), (7.9.2) has a single secular curve for $R < a$, a false neutral curve $R = a$ and no extra secular equation independent of the boundary conditions (7.9.4).

Appendix 1

Sets endowed with an algebraic, topological or algebraic-topological structure. Inequalities related to them

The core of this book is the energy method (in both stationary and non-stationary case), involving total sets and Fourier spaces, orthogonality and splittings of Banach and Hilbert spaces and variational settings of differential equations. Here we present those basic facts which are related to the above topics, mainly using the terminology in [Kre]. In older papers, some constructions, definitions and properties we use today appeared in the process of solving related concrete problems.

A *vector (linear) space* over the field $\mathbf{K}$ is usually denoted by $V/\mathbf{K}$, where V is a (unstructured) set endowed with two operations $+ : V \times V \to V$ and $\cdot : \mathbf{K} \times V \to V$, such that if $u, v, w \in V$ and $a, b \in \mathbf{K}$, then, $u + v = v + u$, $(u+v) + w = u + (v+w)$, there exists an element $0 \in V$ such that $0 + v = v + 0 = v$ for every $v \in V$, $0 \cdot v = 0$, for every $v \in V$, $(a + b)u = au + bu$, $a \cdot (u + v) = a \cdot u + a \cdot v$, $(ab)u = a \cdot (b \cdot u)$, $1 \cdot v = v$, for every $v \in V$. If $\mathbf{K}$ is understood, instead of $V/\mathbf{K}$ we write V.

A *vector subspace* W of a vector space V is a nonempty subset of V which is closed to the two operations on V. If M be a nonempty set of a vector space V, the set of all linear combinations of elements of M is called the *span* of M. It is a vector subspace and is denoted by *span M*. An arbitrary (possibly infinite) subset B of V is a *linearly independent set* of V if every nonempty *finite* subset of B is linearly independent. In the sequel, $\mathbf{K} = \mathbf{C}$ or $\mathbf{K} = \mathbf{R}$. On a given set, several vector space structures can be defined.

Assume that V is a finite- or infinite-dimensional vector space and let B be a linearly independent set of V which spans V. Then B is called a *Hamel basis* of V and every $v \in V \setminus \{0\}$ has a unique representation as a linear combination of *finitely* many elements of the Hamel basis with nonzero scalars as coefficients.

Let $W \subset V$ be a subspace of the linear space V and define the *coset* of $u \in V$ with respect to W (denoted $u + W$) by $u + W = \{v \in V \mid v = u + w, w \in W\}$. In [Kat] a subspace is referred to as a *linear manifold* and a coset as a linear variety of u parallel to V. The set $V/W \equiv \{u + W \mid u \in V\}$ consisting of all cosets with respect to W of all elements of V is a linear space and it is called the *quotient* or *factor space* of V by W or *modulo* W. The dimension of V/W is called the *codimension* of W and is denoted by *codim W*, *i.e.* $dim V/W = codim\,W$. The space V/W realizes

323

a special partition of V such that every element of V belongs to its coset, therefore to an element of V/W.

A *metric space* $\mathcal{X} = (X, d)$ is a set X endowed with a geometric structure defined by a function $d : \mathcal{X} \times \mathcal{X} \to \mathbf{R}^+$ called a *metric* or *distance* such that: $d(u, u) = 0$; $d(u, v) = d(v, u)$ (symmetry); $d(u, v) + d(v, w) \geq d(u, w)$ (triangle inequality) for every $u, v, w \in X$. For $r > 0$ and $u \in X$, the *open ball* centered at u and of radius r is the set $B(u, r) = \{v \in X \mid d(u, v) < r\}$. Sometimes, the space $\mathcal{X}$ is simply denoted by X. Every metric space has a topology defined by d, in which a subset $U \subset X$ is a neighborhood of $u \in X$ if and only if it contains an open ball centered at u.

Let $M \subset X$ be an arbitrary subset. We say that u_0 is an *accumulation point* of M in $\mathcal{X}$ if every neighborhood of u_0 contains at least one element of M different from u_0. The set consisting of the points of M and of all accumulation points of M is referred to as the *closure* or *adherence* of M in $\mathcal{X}$ and it is denoted by $\overline{M}$.

We say that a metric space $\mathcal{X}$ is *complete* if for every Cauchy sequence $\{u_n\}$ of its elements, there exists an element u_0 of this space such that the sequence converges to u_0. Every part M of a metric space is a metric space with respect to the metric induced by d on M. Every non-complete metric space $\mathcal{X}$ can be completed up to a complete space $\overline{\mathcal{X}}$. In general, the nature of the elements of $\mathcal{X}$ and $\overline{\mathcal{X}}$ is different; $\mathcal{X} \subset \overline{\mathcal{X}}$ means the embedding of $\mathcal{X}$ in $\overline{\mathcal{X}}$. We say that a metric space X_1 is embedded in X_2 if there exists a subset Y of X_2 and a continuous bijection $X_1 \to Y$.

A subset $M \subset \mathcal{X}$ is said to be dense in $\mathcal{X}$ if $\overline{M} = \mathcal{X}$. In other words, every element $u_0 \in \mathcal{X}$ (not necessarily from M) is as close to an element of M as we wish. Therefore, for every u_0 in $\mathcal{X}$, we can find a sequence of elements of M converging to u_0. The metric space $\mathcal{X}$ is said to be *separable* if there exists a *countable* subset $M \subset \mathcal{X}$ dense in X.

If A is a continuous operator on $\mathcal{X}$, and M is a dense subset of $\mathcal{X}$, then all properties of A in M are transferred to $\mathcal{X}$ through A. As a consequence, we work in the simpler space M as if we were in $\mathcal{X}$.

Sometimes, the same set can be endowed with both algebraic and geometric structures. In applications, a relationship between these structures is useful. This is realized by means of the functions called the norms.

A *normed space* [Day] is a vector space X with a real-valued function defined on it, satisfying certain axioms, called the norm, and denoted by $\| \cdot \|_X$ or, simply, by $\| \cdot \|$. Then the distance induced by the norm is $d(u, v) = \| u - v \|$, for every $u, v \in X$. A normed space is usually denoted by $(X, \| \cdot \|)$, or, simply, by X. On a given vector space X several norms can be defined, corresponding to different resulted normed spaces. (Each norm has its geometrical and physical meaning.) In order to distinguish them we put indices, *e.g.* $\| \cdot \|_X$, $\| \cdot \|_{l,p}$ or denote them by special symbols, *e.g.* $[\cdot]$, $\|\| \cdot \|\|$, $| \cdot |$, $| \cdot |_A$. In this case the relationship between them is expressed by embeddings (see above). Let X be a vector space and define on it the norms $\| \cdot \|_1$ and $\| \cdot \|_2$. Denote by X_1 and X_2 the corresponding normed spaces. If there exists a positive number k such that $\| u \|_1 \leq k \| u \|_2$, for every $u \in X$, then X_2 is embedded in X_1: $X_2 \subset X_1$. If there exist two positive real numbers k_1, k_2, such that $k_1 \| u \|_2 \leq \| u \|_1 \leq k_2 \| u \|_2$ for every $u \in X$, then $X_1 = X_2$ as normed spaces. The norms are then said to be *equivalent*.

Let $l \in \mathbf{N}$ and $p \geq 1$ and denote by $C^l(a,b)$ the space of real functions u continuous on (a,b) together with their l first derivatives. Since for some u the expressions $\|u\|$, $\|u\|_p$ and $\|u\|_{l,p}$ defined by

$$\|u\| = \max_{x \in (a,b)} |u(x)| + \max_{x \in (a,b)} |Du(x)| + \ldots + \max_{x \in (a,b)} |D^l u(x)|, \tag{1.1}$$

$$\|u\|_p = \left(\int_a^b |u(x)|^p dx \right)^{\frac{1}{p}}, \tag{1.2}$$

$$\|u\|_{l,p} = \left(\int_a^b \sum_{0 \leq k \leq l} |D^k u(x)|^p dx \right)^{\frac{1}{p}}, \tag{1.3}$$

can be infinite, these expressions cannot define norms on $C^l(a,b)$. In order for (1.3) to make sense, we have to diminish the space $C^l(a,b)$.

For instance, let $C^l[a,b]$ the subspace of $C^l(a,b)$ such that $D^k u$, $k = 0,1,\ldots,l$ can be continued up to $[a,b]$. Let $C_0^l(a,b)$ be the subspace of $C^l(a,b)$ such that $\overline{\operatorname{supp} u} \subset (a,b)$, *i.e.* the functions u (and, consequently, their derivatives) vanish on some segments near $x = a$, and $x = b$. The functions possessing this property are called the *functions with a compact support*. Similarly, define $C_0^\infty(a,b)$. In Soviet literature [Smir], [Lad69], [Mikh3], instead of the symbol $C_0^\infty(a,b)$ it is used J_0 (and it is called the space of finite functions). All $C^l[a,b]$, $C_0^l(a,b)$, $C_0^\infty(a,b)$ are normed spaces and the following embeddings hold

$$C_0^\infty(a,b) \subset C_0^l(a,b) \subset C^l[a,b] \subset C[a,b]. \tag{1.4}$$

The space of functions defined on a domain $\Omega \subset \mathbf{R}^n$ and of class C^{α_i} with respect to the variables x_i, $i = 1,\ldots,n$, for every $\mathbf{x} \equiv (x_1,\ldots,x_n) \in \Omega$ is denoted by $C^{\alpha_1,\ldots,\alpha_n}$. It is not a normed space.

A *subspace* Y of a normed space X is a subspace of X as a linear space, with the norm obtained by restricting the norm of X to Y. Let $u \in X$. Then [Mikh5], the coset $u + Y$ is referred to as a *linear manifold*. We adopt this terminology. Let $\{u_n\}$ be a sequence of elements of the normed space X and denote $s_k = \sum_{i=1}^k u_i$. We say that the (infinite) series $\sum_{i=1}^\infty u_i$ converges if the sequence $\{s_k\}$ of partial sums converges, *i.e.* there exists some $s \in X$ such that $\|s_k - s\| \to 0$ as $k \to \infty$. Assume that X contains a sequence $\{e_n\}$ such that for every $u \in X$ there exists a unique sequence of scalars $\{\alpha_n\}$ such that $\|u - \sum_{i=1}^n \alpha_i e_i\| \to 0$ as $n \to \infty$. Then $\{e_n\}$ is called a *(Schauder) basis* for X and the series $\sum_{i=1}^\infty \alpha_i e_i$ which has the sum u is called the *expansion* of u with respect to $\{e_n\}$. If a normed space has a Schauder basis then it is separable; the converse assertion is not true. Every finite dimensional subspace Y of a normed space X is closed in X.

A *Banach space* is a normed space $(B, \|\cdot\|)$, shortly denoted by B, which is complete with respect to the metric induced by the norm. A *subspace* of B is a vector subspace of B considered as a normed space, hence it is not assumed to be complete. A subspace Y of a Banach space is complete if and only if it is closed. Hence, every finite dimensional subspace of B is a Banach space.

Let $(M, \|\cdot\|)$ be a normed space. Its metric completion is a Banach space $B = \overline{M}^{\|\cdot\|}$. Then M is embedded in B. Useful examples of completions are: $L^p(a,b) = \overline{C[a,b]}^{\|\cdot\|_p}$, $W^{\circ l,p}(a,b) = \overline{C_0^\infty(a,b)}^{\|\cdot\|_{l,p}}$. For $\Omega \subset \mathbf{R}^n$, $L^p(\Omega)$, $W^{\circ l,p}(\Omega)$, and the norms $(1.1) - (1.3)$ can be defined similarly. For bounded Ω, the Sobolev space is defined by $W^{l,p}(\Omega) = \{u \in L^p(\Omega) \mid D^\alpha u \in L^p(\Omega), |\alpha| = 1, \ldots, l, \ \alpha \equiv (\alpha_1, \ldots, \alpha_n), \ \alpha_i \geq 0, \ i = 1, \ldots, n\}$ and we have $W^{\circ l,p}(\Omega) = W^{l,p}(\Omega)$. Of course, $W^{0,p}(\Omega) = L^p(\Omega)$. The Sobolev spaces are separable Banach spaces. A lot of other spaces obtained by completion are now called Sobolev spaces [Ad]. The embeddings of various Sobolev spaces are possible only if certain relationship between l, p and n hold [Sob2], [Smir]. These sufficient conditions for embeddings are known as *Sobolev-Kondrashev embedding theorems*. Here are a few of them with the corresponding relationships between their norms

$$lp > n, \quad p > 1, \quad W^{l,p}(\Omega) \subset C(\overline{\Omega}), \quad \sup_{x \in \overline{\Omega}} |u(x)| \leq c\|u\|_{l,p,}; \tag{1.5}$$

$$l = [n/2] + 1 + r > n/2, \quad p = 2, \quad W^{[n/2]+1+r,2}(\Omega) \subset C(\overline{\Omega}),$$
$$\sup_{x \in \overline{\Omega}} |u(x)| \leq c\|u\|_{[n/2]+1+r,2}; \tag{1.5'}$$

$$lp \leq n, \quad p > 1, \quad m > n - lp, \quad q^* < mp/(n-lp);$$
$$W^{l,p}(\Omega) \subset L^{q^*}(S), \quad \|u\|_{q^*,S} < c\|u\|_{l,p,\Omega}, \tag{1.6}$$

where the square brackets stand for the integer part, S is the intersection of Ω with an m-dimensional linear manifold of $\mathbf{R}^n$ and the index S shows that the norm is defined for functions defined on S;

$$l = 0, \quad p > 1, \quad m = n, \quad q^* < p, \quad L^p(\Omega) \subset L^{q^*}(\Omega); \tag{1.7}$$

$$lp > n, \quad 0 \leq |\alpha| \leq j < l - n/p, \quad u \in W^{l,p}(\Omega)$$
$$\implies \quad D^\alpha u \in C(\overline{\Omega}), \quad \sup_{x \in \overline{\Omega}} |D^\alpha u| < c\|u\|_{l,p}; \tag{1.8}$$

$$|\alpha| = j \geq 0, \quad j \geq l - n/p, \quad u \in W^{l,p}(\Omega) \quad \implies \quad D^\alpha u_{|S} \in L^{q_j}(S),$$
$$q_j < \frac{mp}{n - (l-j)p}, \quad m > n - (l-j)p, \quad \|D^\alpha u\|_{q_j,S} \leq c\|u\|_{l,p,\Omega}; \tag{1.9}$$

$$0 < t = l - \frac{n-m}{p} - \left(\frac{1}{p} - \frac{1}{q}\right)m, \quad 1 < p < q < \infty, \quad u \in W^{l,p}(\Omega)$$
$$\implies \quad u_{|S} \in W^{[t],q}(S), \quad \|u\|_{[t],q,S} \leq c\|u\|_{l,p,\Omega}. \tag{1.10}$$

These inequalities hold only on star-like domains Ω; therefore they can be applied to standard domains of motion but not to fractal ones, which appeared recently in applications. The constants c depend on l, p, m, n, r, q, q^*, α, j, Ω, $\partial\Omega$ and S but not on u. The best such constants correspond to the minimum of the associated functionals, *e.g.* in (1.6) $c = min\dfrac{\|u\|_{q^*,S}}{\|u\|_{l,p,\Omega}}$. For details concerning the properties of Ω and $\partial\Omega$ we send to [Ad], [Mau1]. The properties of embeddings are different, *e.g.* in the case of (1.5), (1.6) they are bounded; for $q^* = mp/(n-lp)$ (1.6) still holds, but the embedding is no longer completely continuous.

The importance of inequalities for our topics is crucial. This explains why a lot of papers have been devoted especially to integral inequalities [Ne], [Sob1], [Sob2], [Wey], [Mau1], [Mikh3], [Cat], [Pro], [FoiP], [Aub2], [Wal], [Os], [PayW], [PayW2], [Smir], [Wel], [J76], [Lad69], [Strau], [Mikh5], [MuloR98], [MuloR97], [KaX], [MuloR03]. Other references can be found in Section 3.2.6.

In the following we present, sometimes not in the most general form, but in the form more suitable to our proposes, some other inequalities useful in energy stability theory.

Hölder inequality. Let Ω be a domain (connected open set) of $\mathbf{R^n}$, $\mathbf{u} \in [L^p(\Omega)]^n$, $1 \leq p \leq \infty$, $\mathbf{v} \in [L^q(\Omega)]^n$, with $1/p+1/q = 1$. Then $\mathbf{u} \cdot \mathbf{v} \in L^1(\Omega)$ and the following *Hölder inequality* holds [Geo85], [Smir]

$$\int_\Omega |\mathbf{uv}|dx \leq \left(\int_\Omega |\mathbf{u}|^p dx\right)^{1/p}\left(\int_\Omega |\mathbf{v}|^q dx\right)^{1/q}. \tag{1.11}$$

Schwartz inequality. When $p = q = 2$ $\mathbf{u}, \mathbf{v} \in [L^2(\Omega)]^n$, the Hölder inequality in the Hilbert space $[L^2(\Omega)]^n$ becomes the *Schwartz inequality*

$$\int_\Omega |\mathbf{uv}|dx \leq \left(\int_\Omega |\mathbf{u}|^2 dx\right)^{1/2}\left(\int_\Omega |\mathbf{v}|^2 dx\right)^{1/2}.$$

Minkowski inequality. Let Ω be a domain of $\mathbf{R^n}$, $\mathbf{u}, \mathbf{v} \in [L^p(\Omega)]^n$, $1 \leq p < \infty$, the following *Minkowski inequality* holds [Smir]

$$\left(\int_\Omega |\mathbf{u} + \mathbf{v}|^p dx\right)^{1/p} \leq \left(\int_\Omega |\mathbf{u}|^p dx\right)^{1/p} + \left(\int_\Omega |\mathbf{v}|^p dx\right)^{1/p}.$$

Poincaré inequality. Let $\Omega \subset \mathbf{R^3}$ be a star-shaped domain with respect to every point of a sphere $S \subset \Omega$. Any function $\mathbf{u}$ which belongs to the space $L^2(\Omega)$, and has the first derivatives $\frac{\partial \mathbf{u}}{\partial x_k} \in L^2(\Omega)$, satisfies the following *Poincaré inequality* [Smir].

$$\int_\Omega |\mathbf{u}|^2 dx \leq B\left[\int_\Omega |\nabla\mathbf{u}|^2 dx + \left(\int_\Omega \mathbf{u}dx\right)^2\right].$$

If the boundary $\partial\Omega$ of Ω is piecewise smooth, the following [Smir]

Friedrichs inequality.

$$\int_\Omega |\mathbf{u}|^2 dx \leq C\left[\int_\Omega |\nabla\mathbf{u}|^2 dx + \int_{\partial\Omega} |\mathbf{u}|^2 dx\right],$$

holds.

In the previous inequalities B and C are constants depending on various conditions, such as the geometry and size of the domain, the boundary conditions. This is the reason why, in hydrodynamic stability theory, different inequalities of Poincaré type occur. In the following we listed some of them.

For a *function* defined in a cell $V = \{(x, y, z) \in \mathbf{R}^3 | 0 \le x \le 2a_1, 0 \le y \le 2a_2, 0 < z < 1\}$ and *periodic* in x and y direction, of period $2a_1$ and $2a_2$ respectively, satisfying the boundary conditions $\mathbf{u} = \mathbf{0}$ on $z = 0, 1$, we have the following inequalities [Strau]

$$\int_V |\mathbf{u}|^2 dx \le \frac{1}{\pi^2} \int_V |\nabla \mathbf{u}|^2 dx,$$

and [HarLP]

$$\int_V |\nabla \mathbf{u}|^2 dx \le \frac{1}{\pi^2} \int_V |\Delta \mathbf{u}|^2 dx.$$

For a function defined in a cell V, periodic in x and y direction, of period $2a_1$ and $2a_2$ respectively, satisfying the boundary conditions $\mathbf{u} = \mathbf{0}$ on $z = 0$, and $\frac{\partial \mathbf{u}}{\partial z} = \mathbf{0}$ on $z = 1$, we have [RioM88c], [MuloR97], [MuloR98]

$$\int_V |\mathbf{u}|^2 dx \le \frac{4}{\pi^2} \int_V |\nabla \mathbf{u}|^2 dx.$$

In particular, the Poincaré inequality, for a function $\mathbf{u}$ as above, vanishing on the planes $z = 0, 1$ implies [MuloR97]

$$\int_V |\nabla_1 \mathbf{u}|^2 dx \le k_1^2 \int_V |\Delta_1 \mathbf{u}|^2 dx,$$

where $\nabla_1 = \frac{\partial}{\partial x}\mathbf{i} + \frac{\partial}{\partial y}\mathbf{j}$, $\Delta_1 = \frac{\partial^2}{\partial x^2} + \frac{\partial^2}{\partial y^2}$, $k_1^2 = \max\{a_1^{-2}, a_2^{-2}\}$.

For a function $u(x, y, z)$ defined in a cell V, periodic in x and y direction, of period $2a_1$ and $2a_2$ respectively, $z \in (0, 1)$, satisfying the boundary condition $\mathbf{u} = \mathbf{0}$ on $z = 0$, we have the following *boundary estimate* [Strau]

$$\int_\Gamma |\mathbf{u}|^2 dx \le 2\left(\int_V |\mathbf{u}|^2 dx\right)^{1/2}\left(\int_V |\nabla \mathbf{u}|^2 dx\right)^{1/2},$$

where Γ is the intersection of V with the plane $z = 1$.

Wirtinger inequality. For a *function* defined in a cell $V = \{(x, y, z) \in \mathbf{R}^3 | 0 \le x \le 2a_1, 0 \le y \le 2a_2, 0 < z < 1\}$ and *periodic* in x and y directions, of period $2a_1$ and $2a_2$ respectively, satisfying the condition $\int_V \mathbf{u} dx = \mathbf{0}$, and the boundary conditions $\frac{\partial \mathbf{u}}{\partial z} = \mathbf{0}$ on $z = 0, 1$, we have [HarLP], [Strau]

$$\int_V |\mathbf{u}|^2 dx \le c_1 \int_V |\nabla \mathbf{u}|^2 dx,$$

where the constant c_1 is given by [KaX] $c_1 = \max\{\pi^{-2}, a_1^{-2}, a_2^{-2}\}$.

Some Sobolev inequalities for u^4. Let u be a smooth scalar function defined on $\mathbf{R}^2$, and let Ω be its compact support. In [Lad69] the following inequality is derived

$$\int_{\mathbf{R}^2} u^4 dx \le 2 \int_{\mathbf{R}^2} u^2 dx \int_{\mathbf{R}^2} |\nabla u|^2 dx.$$

For a periodic function $u(x,y)$, of period k in x direction and m in y direction, if $\Gamma = \{(x,y)|\overline{x} \le x \le \overline{x}+k, \overline{y} \le y \le \overline{y}+m\}$ with $\overline{x}, \overline{y}$ fixed, it follows that [Strau]

$$\int_{\Gamma} u^4 dx \le \left(1 + \frac{4}{\pi^2} + \frac{2(k+m)}{\pi\sqrt{km}}\right) \int_{\Gamma} u^2 dx \int_{\Gamma} |\nabla u|^2 dx,$$

provided that $\int_{\Gamma} u dx = 0$.

Let u be a smooth scalar function defined on $\mathbf{R}^3$, and let Ω be its compact support. In [Lad69] the following inequality is derived

$$\int_{\mathbf{R}^3} u^4 dx \le 4\left(\int_{\mathbf{R}^3} u^2 dx\right)^{1/2}\left(\int_{\mathbf{R}^3} |\nabla u|^2 dx\right)^{3/2}.$$

From the previous inequality, by using the Young inequality, in [Lad69] for a function u of compact support Ω, we get

$$\int_{\mathbf{R}^3} u^4 dx \le \frac{1}{\epsilon^3}\left(\int_{\mathbf{R}^3} u^2 dx\right)^2 + 3\epsilon\left(\int_{\mathbf{R}^3} |\nabla u|^2 dx\right)^2 \quad \forall \epsilon > 0.$$

Let u be a *function* defined in a cell $V = \{(x,y,z) \in \mathbf{R}^3 | 0 \le x \le 2a_1, 0 \le y \le 2a_2, 0 < z < 1\}$ and *periodic in x and y direction, of period a_1 and a_2 respectively, without boundary conditions to be specified on $z = 0$, $z = 1$* [Strau]. Then

$$\int_V u^4 dx \le (a_2 a_2)^{1/3}\left(\int_V u^6 dx\right)^{2/3}.$$

From this inequality, taking into account some Sobolev inequalities [Ad], [Strau] the following inequality

$$\left(\int_V u^4 dx\right)^{1/2} \le 32(a_2 a_2)^{1/6}\left(\int_V u^2 dx + \int_V |\nabla u|^2 dx\right)$$

holds for arbitrary boundary conditions.

For a more general cell V, even if no periodicity conditions on $z = 0, 1$ are assumed, the following inequality

$$\int_V u^4 dx \le \gamma^4 \int_V |\nabla u|^2 dx$$

holds [Strau], the constant γ is determined for arbitrary boundary conditions, either assuming that u vanishes at $z = 0$ and $z = 1$, or taking into account some periodicity conditions.

A Sobolev inequality for u^6. Let Ω be a bounded domain in $\mathbf{R}^3$, and let u be a function satisfying the boundary conditions $u = 0$ on $z = 0, 1$. Then [Strau]

$$\left(\int_{\Omega} u^6 dx\right)^{1/3} \le c \int_{\Omega} |\nabla u|^2 dx,$$

where $c = 2^{2/3}/(3^{1/2}\pi^{2/3})$, irrespective of the boundary conditions.

An inequality for the supremum of a function **u**. From (1.5), if Ω is a bounded domain and $\partial\Omega$ has bounded first and second derivatives, then, for $\mathbf{u} \in [W^{2,2}]^n(\Omega) \cap [W^{\circ 1,2}(\Omega)]^n$, the following inequality [Lad69]

$$\int_\Omega |\nabla\mathbf{u}|^2 dx + \int_\Omega \mathbf{u}^2 dx \leq \int_\Omega |\Delta\mathbf{u}|^2 dx$$

holds. Therefore,

$$\sup_{x\in\Omega} |\mathbf{u}(x)| \leq C(\Omega) \int_\Omega |\Delta\mathbf{u}|^2 dx \quad \forall \mathbf{u} \in [W^{2,2}(\Omega)]^n \cap [W^{\circ 1,2}(\Omega)]^n,$$

where the constant $C(\Omega)$ depends on various conditions.

For instance, let V be a periodicity cell, and let $\mathbf{u} \in [W^{2,2}(V)]^n \cap [W^{\circ 1,2}(V)]^n$ be vanishing on the planes $z = 0$, $z = 1$, and periodic in x and y directions. In this case the constant $C(\Omega)$ is [GaldS85], [MuloR03], [Strau]

$$C = \frac{\sqrt{3}}{[\pi^5 h^3 \sqrt{2}(\sqrt{2}-1)]^{1/2}} + \frac{2^{5/2}5(1+\pi^2)^{1/2}h^{3/5}}{3\pi},$$

where $h = \min\{1, a_1, a_2\}$. If V is the periodicity cell, $\mathbf{u} \in [W^{2,2}(V)]^n \cap [W^{\circ 1,2}(V)]^n$ satisfies the condition $\int_V \mathbf{u}dx = \mathbf{0}$, the boundary conditions $\frac{\partial\mathbf{u}}{\partial z} = \mathbf{0}$ on $z = 0, 1$, and it is periodic in x and y direction, [GaldS85] [Strau]

$$C = c_3 h^{-3/2} k_0^{-2} + c_4 h^{3/5}\sqrt{1 + k_0^{-2}},$$

where $k_0^2 = \min\{\pi^2, a_1^2, a_2^2\}$, and c_3 and c_4 are real numbers [GaldS85].

In dealing with the previous inequalities an important algebraic tool is the Young inequality [Kre]

$$ab < \frac{a^p}{p\epsilon^p} + \frac{b^q\epsilon^q}{q}, \tag{1.12}$$

where $a, b, \epsilon > 0$ are arbitrary numbers and p and q are conjugate.

Some integral identities of Green type are also frequently used to derive stability criteria. Here are some of them.

Let $u \in C^{(2)}(\overline{\Omega})$, where Ω is a bounded domain with a piecewise smooth boundary $\partial\Omega$, and denote $Lu = \left(A_{jk}\frac{\partial^2}{\partial x_j \partial x_k} + A_k\frac{\partial}{\partial x_k} + A_0\right)u$, then the first *Green identity* and its four particular cases read [Mikh5], [Mikh2]

$$\int_\Omega vLudx = -\int_\Omega A_{jk}\frac{\partial v}{\partial x_j}\frac{\partial u}{\partial x_k}dx + \int_\Omega v\left(A_k\frac{\partial u}{\partial x_k} + A_0 u\right)dx + \int_{\partial\Omega} vA_{jk}\frac{\partial u}{\partial x_k}\cos(\nu x_j)d\sigma,$$

$$\int_\Omega v\Delta u dx = -\int_\Omega \frac{\partial v}{\partial x_k}\frac{\partial u}{\partial x_k}dx + \int_{\partial\Omega} v\frac{\partial u}{\partial\nu}d\sigma,$$

$$\int_\Omega u\Delta u dx = -\int_\Omega |\frac{\partial u}{\partial x_k}|^2 dx + \int_{\partial\Omega} u\frac{\partial u}{\partial\nu}d\sigma,$$

$$\int_\Omega \Delta u dx = \int_{\partial\Omega} \frac{\partial u}{\partial \nu} d\sigma,$$

where ν is the outer normal to the Ω.

The Gauss formula of by parts integration for a function $\mathbf{u} \equiv (u_1, u_2, u_3)$ defined on a three-dimensional domain Ω of points $\mathbf{x} \equiv (x, y, z)$ is [Mikh3]

$$\int_\Omega \left(\frac{\partial u_1}{\partial x} + \frac{\partial u_2}{\partial y} + \frac{\partial u_3}{\partial z} \right) dx d\Omega = \int_{\partial\Omega} \Big[u_1 \cos(\nu, x) + u_2 \cos(\nu, y) + u_3 \cos(\nu, z) \Big] d\sigma.$$

For two-dimensional domains it becomes the Green formula, while for $u_1 = u_2 = 0$, $u_3 = uv$, it reads

$$\int_\Omega u \frac{\partial v}{\partial x} = - \int_\Omega v \frac{\partial u}{\partial x} + \int_{\partial\Omega} uv \cos(\nu, x) d\sigma.$$

Let $M^{(k)}(\overline{\Omega})$ consists of functions $\zeta \in C^{(k)}(\overline{\Omega})$ such that $\zeta_{|\partial\Omega} = \ldots = D^{(k-1)} \zeta_{|\partial\Omega} = 0$. Then [Mikh5] for $\mathbf{g} \in C^1(\overline{\Omega})$ and $\zeta \in M^{(1)}(\overline{\Omega})$ we have a formula relating *grad* and *div*

$$\int_\Omega \mathbf{g}(\mathbf{x}) grad \zeta(\mathbf{x}) dx = - \int_\Omega \zeta(\mathbf{x}) div \mathbf{g}(\mathbf{x}) dx.$$

Starting with this formula, we can define the generalized divergence of $\mathbf{g}$: let $\mathbf{g} \in [L^2(\Omega)]^n$, then we say that $f \in L^2(\Omega)$ represents the generalized divergence of $\mathbf{g}$ if [Mikh5]

$$\int_\Omega \mathbf{g}(\mathbf{x}) grad \zeta(\mathbf{x}) dx = - \int_\Omega \zeta(\mathbf{x}) f(\mathbf{x}) dx. \quad \forall \zeta \in M^{(1)}(\overline{\Omega}).$$

Theorem 1.1 [Mikh5]. *Let B a Banach space and let M_1 be a subset of B dense in B. Then every subset M_2 such that $M_1 \subset M_2 \subset B$ is dense in B too.*

This theorem and the inclusion (1.4) show that the normed spaces $C_0^l(a, b)$ and $C^l[a, b]$ are dense in $W^{l,p}(a, b)$ if completed in the norm $\| \cdot \|_{l,p}$. In the classical sense the homogeneous boundary value problem for ode's are defined on spaces of the form $C^l(a, b) \cup C^{l-1}[a, b]$. Since they included $C_0^\infty(a, b)$, by Theorem 1.1, they are dense in $L^2(a, b)$. With the notation $C^{l_1, l_2, \ldots, l_n}(a, b)$ for the Cartesian product $C^{l_1}(a, b) \times \ldots \times C^{l_n}(a, b)$, the solutions $(u_1, \ldots, u_n)$ of a homogeneous boundary value problem for a system of ode's belongs to $C^{l_1, l_2, \ldots, l_n}(a, b) \cup C^{l_1-1, l_2-1, \ldots, l_n-1}[a, b]$. These spaces contain the space $\left(C_0^\infty(a, b) \right)^n$ which is dense in $\left(L^2(a, b) \right)^n$. By Theorem 1.1, it follows that these spaces are dense in $\left(L^2(a, b) \right)^n$ too. Usually, $\left(L^2(a, b) \right)^n$ is denoted by $\mathbf{L}^2(a, b)$, or, simply, by $L^2(a, b)$, when the context does not allow any confusion.

In some concrete problems, *e.g.* the incompressible Navier-Stokes model [Te] the solutions $\mathbf{u}$ must satisfy some constraints, *e.g.* $div\mathbf{u} = 0$, $\mathbf{u}_{|\partial\Omega} = 0$. In this case the role of $C_0^\infty(\Omega)$ is played by $\mathcal{N}(\Omega) = \{ \mathbf{u} \in \left(C_0^\infty(\Omega) \right)^n \mid div\mathbf{u} = 0 \}$ and

$C_*^n(\Omega) = \{\mathbf{u} \in \left(C^n(\overline{\Omega})\right)^n \mid div\mathbf{u} = 0, \mathbf{u}_{|\partial\Omega} = 0\}$. Then, instead of $\mathbf{L}^2(\Omega)$, we use $\overline{\mathcal{N}(\Omega)}^{\|\cdot\|_2} = N(\Omega)$. We quote also the notation $N^l = \overline{\mathcal{N}(\Omega)}^{\|\cdot\|_{l,2}}$, $l \in \mathbf{N}^*$. These notations differ from those in the Russian or French literature, where many other types of metric completions are considered. This leads to completely different properties of the elements of the resulting Banach spaces.

In calculus of variations the set of admissible functions is not a linear space because they must satisfy nonhomogeneous boundary conditions. In order to use Banach spaces theory, the space M is defined as consisting of the functions with the same regularity, but satisfying homogeneous boundary conditions.

The *dual* or *adjoint* X^* (sometimes denoted by X', *e.g.* in [Kre]), of a normed space X consists of all linear bounded functionals $\mathcal{F}$ on X and it is a Banach space, (even when X is not). The use of dual spaces is easier theoretically and numerically due to the passage from functions to functionals, from differential to integral operators and so, the coming back to the more natural original form of equations governing phenomena. If $u \in X$ and u^* is a linear bounded functional, then we write $u^*(u)$ or, more usual, $< u, u^* >$, where $< \cdot, \cdot >$ stands for the duality pairing relation.

A Banach space is reflexive if $B^{**} = B$, more exactly if these spaces are isomorphic.

Let $M \subset X$ be a nonempty subset of the normed subspace X. The *annihilator* M° is the set of all bounded linear functionals on M, *i.e.*

$$M^\circ = \{\mathcal{F} \in X^* \mid < u, \mathcal{F} >= 0, \forall u \in M\}$$

(cf. [EdE]). The set M° is a closed vector subspace of X^*. If $< u, \mathcal{F} >= 0$ we say that the functional $\mathcal{F}$ and the vector u are orthogonal. Hence M° consists of all functionals orthogonal to u. Let B_1 be a finite-dimensional Banach subspace of the Banach space B. Then B splits into an orthogonal direct sum $B = B_1 \oplus B_2$ where B/B_1 is isomorphic to B_2, $dim\, B_2 = dim\, B_1^\circ$, $dim\, B = dim\, B_1 + dim\, B_1^\circ$ [Ede], [Kre]. The Banach space splitting is possible only if $dim\, B_1 < \infty$ or $codim\, B_1 < \infty$, whence the importance in spectral theory of Fredholm operators whose kernel and co-kernel have finite dimension.

The Banach spaces splitting is basic in solving equations by suitably projecting them on finite-dimensional subspaces generated by the eigenvectors of the linearized equations (defined by Fredholm operators), the complementary other projection leading to the bifurcation equation [Geo85].

An *inner product* (or *pre-Hilbert*) space X is a vector space over $\mathbf{K}$ with a Hermitian (if $\mathbf{K}$ is complex) or bilinear (if $\mathbf{K}$ is real) form $(\cdot, \cdot) : X \times X \to \mathbf{K}$, called the *inner* or *scalar product*. Two elements u, v of X are called *orthogonal* if $(u, v) = 0$. In this case we write $u \perp v$. The inner product generates a norm $\|\cdot\| = \sqrt{(\cdot, \cdot)}$ and X is a normed space. Let X be an inner product space. A vector subspace Y of X endowed with the inner product of X induced on $Y \times Y$ is said to be a *subspace* of X. The annihilator $S^\perp$ of a nonempty subset S of an inner product space X is a closed subspace of X. In a pre-Hilbert space the Schwarz inequality $|(u, v)| \leq \|u\|\|v\|$ holds.

A *Hilbert space H* is a complete pre-Hilbert space. Every pre-Hilbert space has a completion, unique up to an isomorphism. In older Soviet literature, *e.g.* in [Mikh2], a pre-Hilbert space is referred to as a Hilbert space.

A Hilbert space is in particular a Banach space with respect to the norm generated by the inner product. A *subspace* of a Hilbert space H is a vector subspace Y endowed with the natural pre-Hilbert space structure obtained by restriction of the scalar product of H to Y. A subspace of H is complete (therefore Hilbert) if and only if it is closed in H. A Hilbert space is reflexive.

Theorem 1.2. *Let $\mathcal{H}$ be a Hilbert space and let M be a total subset of $\mathcal{H}$. Then $(u, v) = 0$, $\forall v \in M$, implies $u = 0$.*

Let Y be a closed subspace of H. Then $H = Y \oplus Y^{\perp}$, where $Y^{\perp} = \{u \in H \mid u \perp Y\}$ and it is called the *orthogonal complement* of Y. Therefore in Hilbert spaces the splitting is not conditioned by the finite dimension of Y (as in Banach spaces).

One of the most important results in functional analysis is the following *representation theorem.*

Riesz-Fréchet theorem [Kre]. *Every bounded linear functional $\mathcal{F}$ on a Hilbert space H can be represented by an inner product, i.e. there exists a unique element $v_{\mathcal{F}} \in H$ such that $\mathcal{F}(u) \equiv < u, \mathcal{F} >= (u, v_{\mathcal{F}})$. If $\mathcal{F} = 0$, then $v_{\mathcal{F}} = 0$. If $\mathcal{F} \neq 0$, then $v_{\mathcal{F}} = \frac{\mathcal{F}(u_0)}{<u_0, u_0>} u_0$, where u_0 is any non-vanishing vector in $\mathcal{N}^{\perp}(\mathcal{F})$. Here $\mathcal{N}(\mathcal{F})$ stands for the null space of $\mathcal{F}$.*

The following theorem is fundamental for energy method in evolution equations.

Weyl Theorem [Wey]. *$L^2(\Omega) = G(\Omega) \oplus N(\Omega)$, where $G(\Omega)$ is the Hilbert space of generalized gradients and $N(\Omega)$ is the Hilbert space of solenoidal vectors $\mathbf{u}$, i.e. $div\,\mathbf{u} = 0$.*

The Hilbert space $L^2(\Omega)$ is the most important in hydrodynamics and hydromagnetics. It consists in $\mathbf{K}$-valued square Lebesgue integrable functions on Ω. If $\mathbf{K} = \mathbf{C}$, the inner product in $L^2(a, b)$ is $(u, v) = \int_a^b u(x)\overline{v(x)}dx$, and in $\mathbf{L}^2(\Omega)$ it is $(\mathbf{u}, \mathbf{v}) = \int_\Omega \sum_{i=1}^n u_i(x)\overline{v_i(x)}dx$, where the bar stands for the complex conjugation. If $\mathbf{K} = \mathbf{R}$, then the bar no longer occurs.

Let H be a separable Hilbert space. Since H is a particular complete metric space, it follows that all the above mentioned concepts can be defined in H too. Moreover, in this case one can introduce an important class of subsets of H which allow to approximate (represent) every element of H by linear combinations of their elements. More precisely, a subset $M \subset H$ is *total* [Kre] or *fundamental, complete* ([Mikh2], [Mikh3]) in H if its span (the set of all linear combinations of elements in M) is dense in H, *i.e.* $\overline{\mathrm{span}M} = H$.

The most useful total sets of Hilbert spaces consist of sets of orthonormal elements. Such sets are referred to as *total orthonormal sets* or *orthonormal bases* of H. This label is related to the fact that all total orthonormal sets in a nontrivial Hilbert space have the same cardinality called the *Hilbert dimension* or *orthogonal dimension* of H.

Let M be an orthonormal set dense in H. Every linear combination of elements of M is linearly independent, hence an orthonormal set is a linearly independent set. Therefore, M is a Hamel basis for $span\,M$. But, $M \subset span\,M$ and $\overline{M} = H$. Then, by Theorem 1.1, $\overline{span\,M} = H$, *i.e.* M is total in H. This is why total sets in Hilbert spaces are also called complete because in this case completeness implies totality. In every Hilbert space $H \neq \{0\}$ there exists a total orthonormal set. The role of M in Theorem 1.2 is taken by this orthonormal set, therefore

Theorem 1.3. *If $u \in H$ is orthogonal to every element of a total orthonormal set M of X, then u is the null element of H.*

In addition, every element $u \in H$ is approximated by a sequence of elements u_n of $span\,M$, where u_n has a representation as a linear combination of finitely many elements of M. In particular, this property is important in numerical methods of calculus of variations, when even if the solution is generalized, in computations only elements of C^n occur.

Let H be a separable Hilbert space and let the sequence $\{\phi_k\}$ be a total orthonormal set in H. A *Fourier series* with respect to ϕ_k is an infinite sum $\sum a_k \phi_k$. Every element $u \in H$ can be formally represented as a series $u = \sum_{k=1}^{\infty} a_k \phi_k$, called the *Fourier expansion of u* upon the total set $\{\phi_k\}$. The coefficients $a_k = (u, \phi_k)$ are called the *Fourier coefficients* of u.

The property in Theorem 1.3 is basic for Fourier series theory.

If H is a separable Hilbert space then every total orthonormal set in H is countable. If the Hilbert space H contains a countable total orthonormal set, then H is separable.

In particular, the Hilbert space $L^2(a, b)$ is separable because it contains (among many others) the total orthonormal set of Legendre polynomials; $L^2(-\infty, +\infty)$ is separable because it contains the total orthonormal set of Hermite polynomials; $L^2(-\infty, b)$, $L^2(a, +\infty)$ are separable because they contain total sets of Laguerre polynomials etc. [Kre].

In this book we are mainly interested in special total (and, so, countable) orthonormal sets $\{\phi_n\}$ of functions $\phi_n \in L^2(-0.5, 0.5)$ which satisfy certain boundary conditions at $x = \pm 0.5$. By means of them we expand in Fourier series the unknown functions u of the boundary-value problem or the admissible functions of the associated variational problem.

Thus, the use of Fourier series reduces the determination of a function to the determination of its Fourier coefficients. Therefore the solution of a differential equation is reduced to the solution of a system of algebraic equations in these coefficients.

The Fourier coefficients satisfy the Bessel inequality $\sum_{n=1}^{\infty} \mid u_n \mid^2 \leq \parallel u \parallel^2$, where $\parallel \cdot \parallel^2 = (\cdot, \cdot)$. Let H be a Hilbert space. Then for every $u \in H$ we have $\sum_{n=1}^{\infty} \mid u_n \mid^2 = \parallel u \parallel^2$ (Parseval relation), *i.e.* the orthonormal set ϕ_n is total in H. This equality is very important for the validation of the variational methods [Sob2]. The Fourier series converges (in the norm $\parallel \cdot \parallel$) if and only if $\sum_{n=1}^{\infty} \mid u_n \mid^2$ converges. These two properties imply the fact that the Fourier expansion of every $u \in H$ converges in the norm of H.

In a general (pre)Hilbert space H and for a more general orthonormal set (not necessary total), for every $u \in H$, the set of non-vanishing Fourier coefficients is at most countable. This powerful result together with the property of convergence of the Fourier series expansion of arbitrary elements of H justify the term of *Fourier series representation* of elements of Hilbert spaces. This is why the Fourier series for the unknown functions are called *the exact solutions* [Kre], [Mikh3].

The Fourier series methods apply either directly to the boundary value problems or to their associated variational formulation leading to the same results if the variational principle holds.

Appendix 2

Operators and functionals

Due to the non-uniformity in terminology in functional analysis, especially the differences between Soviet and Western concepts, we have decided to present both and to specify those adopted by us.

An *operator* is a function (transformation) $T : V_1 \to V_2$, where V_1 and V_2 are two abstract linear, in particular normed, spaces [Kre]. We are mainly concerned with function spaces; in this context an operator is a function of functions [Mikh3]. In spectral theory, concerning linear operators, by an operator it is meant a linear function A transforming V_1 into V_1 [Fri2], [Wel]. For the theory of adjoint operators, it is understood that the operator A is linear, densely defined, V_1 and V_2 are Hilbert spaces, denoted by H_1 and H_2 respectively, the domain of definition $\mathcal{D}(A)$ of A, and the range $\mathcal{R}(A)$ of A, are contained in H_1 and H_2 respectively [Bal]. In our paper we use the word of *mapping* for the nonlinear case and operator for the linear case as in [Mikh2]. Sometimes, in order to emphasize the linearity, we write (linear) operator. Usually we denote by A the operator and by T the mapping.

A *linear operator* over the field $\mathbf{K}$ is characterized by: $1°$ linear domain of definition $\mathcal{D}(A)$, $2°$ additivity, $A(u+v) = Au + Av$, $\forall u, v \in \mathcal{D}(A)$, $3°$ multiplicativity $A(\alpha u) = \alpha Au$, $\forall \alpha \in \mathbf{K}$, $\forall u \in \mathcal{D}(A)$.

The sets $\mathcal{D}(A)$ and $\mathcal{R}(A)$ are linear spaces over $\mathbf{K}$ and they belong to linear spaces over the same field.

Bounded operators. Let H_1, H_2 be two Hilbert spaces over the same field. $A : \mathcal{D}(A) \subset H_1 \to H_2$ is *bounded* if $\mathcal{D}(A) = H_1$ and there exists $M > 0$ such that $sup \parallel Au \parallel_{H_2} / \parallel u \parallel_{H_1} = M < \infty$ [Bal]. Let X, Y be two normed spaces. The operator $A : \mathcal{D}(A) \subset X \to Y$ is bounded if there exists a real number M such that $\parallel Au \parallel_Y \leq M \parallel u \parallel_X$, $\forall u \in \mathcal{D}(A)$ [Kre].

Bounded mappings. Let X_1, X_2 be two Banach spaces. A mapping $T : X_1 \to X_2$ is bounded if it transforms every bounded set of X_1 in a bounded set of X_2. This concept differs from the boundedness of a real-valued function defined on $\mathbf{R}$, which is bounded if its range is bounded. A linear operator defined on a finite-dimensional Banach space is bounded.

An *unbounded operator* is an operator which is not bounded, *e.g.* the operators

337

of differentiation.

Continuous operators and mappings. Let X and Y be two normed spaces. A (linear or nonlinear) mapping $T : \mathcal{D}(T) \subset X \to Y$ is *continuous* at some u_0 in $\mathcal{D}(T)$ if for every $\epsilon > 0$ there exists $\delta > 0$ such that $\| T(u) - T(u_0) \|_Y < \epsilon$ for all $u \in \mathcal{D}(T)$ satisfying $\| u - u_0 \|_X < \delta$. T is continuous if it is continuous at every $u_0 \in \mathcal{D}(T)$. Let A be a (linear) operator. Then A defined on the entire Banach space X is continuous if and only if A is bounded [Kre].

Some authors (especially in the older Soviet literature) call a continuous and linear operator a linear operator. We do not adopt this definition.

Completely continuous operators and mappings. Let X, Y be two reflexive Banach spaces and let $T : X \to Y$ be a mapping or an operator. We say that T is *completely continuous* if it transforms weakly (*i.e.* in the distributions sense) converging sequences into strongly (*i.e.* in the norm) converging sequences [Mau1].

Compact operators and mappings. Let X and Y be two reflexive Banach spaces and let $T : X \to Y$ be a (linear or nonlinear) mapping. If T is 1° continuous and 2° it transforms every bounded subset M of X into a relatively compact subset $T(M)$ of Y, *i.e.* $\overline{T(M)}$ is compact, then T is called a *compact mapping*. Every completely continuous mapping is compact, while the converse is not always true. If X is not reflexive, then the complete continuity does not necessary implies the compactness. The complete continuity can also be defined for X and Y normed spaces. Compact mappings are bounded but the inverse is not true. A compact linear operator is defined only by 2° because 2° implies the boundedness of A, then being defined on an entire Banach space, A is continuous, whence 1°. Every compact linear operator is completely continuous; it follows that in the case of linear operators the two concepts coincide. A compact operator, therefore, is continuous but a continuous operator is not always compact. For instance, if $\dim X_1 = \infty$, the identity operator $I : X_1 \to X_1$ is continuous but not compact because it transforms the closed unit ball (which is bounded) in itself (but the unit ball is not compact and so nor relatively compact).

Densely defined operators. Let H be a Hilbert space and let $A : \mathcal{D}(A) \subset H \to H$ be an operator. We say that A is densely defined if $\overline{\mathcal{D}}(A) = H$, *i.e.* the closure of the domain of A with respect to the norm of H is equal to H.

Adjoint, selfadjoint, symmetric, positive and positive definite operator.

Several definitions for such operators A exist. They depend on the fact whether A is bounded or unbounded, its domain and range belong to the same or different spaces, its domain is a Banach or a Hilbert space, its domain is an entire Hilbert space or only part of it, the Hilbert space is real or complex, the operator A is densely defined or not.

Banach space framework [Kre]. Let B_1, B_2 be two Banach spaces and let B_1^* and B_2^* their dual or adjoint spaces. Let $A : B_1 \to B_2$ be a bounded linear operator. The *adjoint* of A is an operator $A^* : B_2^* \to B_1^*$, satisfying $< Au, v^* > = < u, A^* v^* >$, for every $u \in B_1$ and for every $v^* \in B_2^*$.

Hilbert space framework and bounded operators [Kre]. Let H_1, H_2 be two real or complex Hilbert spaces and let $A : H_1 \to H_2$ be a bounded operator. The *Hilbert-adjoint* of A is the operator $A^* : H_2 \to H_1$ such that $(Au, v) = (u, A^*v)$, $\forall u \in H_1$, $\forall v \in H_2$. Let $H_1 = H_2 = H$. The operator $A : H \to H$ is *selfadjoint* if $A = A^*$, *i.e.* $(Au, v) = (u, Av)$ for every $u, v \in H$. For unbounded operators these definitions cannot be formulated, while the concepts exist but they have other definitions. Indeed, by Hellinger-Toeplitz theorem, an operator A defined on an entire complex Hilbert space H and satisfying the relation $(Au, v) = (u, Av)$ for every $u, v \in H$, is bounded. On the other hand, in order for the Hilbert-adjoint operator A^* of an operator A to exist, A must be densely defined on H. It is also interesting to remark that if $A : H \to H$ is only linear, its Hilbert adjoint is bounded.

Let H be a complex Hilbert space and let $A : H \to H$ be a bounded selfadjoint operator. Then (Au, u) is real for every $u \in H$. Let A_1 and A_2 be two such operators. On the set of the bounded selfadjoint operators we have the order relation $A_1 \leq A_2$ if and only if $(A_1 u, u) \leq (A_2 u, u)$ for every $u \in H$. A bounded selfadjoint operator is *positive*, and we write $A \geq 0$, if $(Au, u) \geq 0$ for every $u \in H$. (In this case $A_2 = 0$, whence the writing.) A bounded selfadjoint operator $A_1 : H \to H$ is called the *square root* of a positive bounded selfadjoint operator A on H if $A_1^2 = A$. If, in addition, $A_1 \geq 0$ then it is called the *positive square root* of A and it is denoted by $A^{1/2}$. Given A, its positive square root $A^{1/2}$ exists and is unique [Kre]. See also [Kat].

Hilbert space framework and unbounded operators [Kre]. Let H be a real or complex Hilbert space and let $A : \mathcal{D}(A) \subset H \to H$ be a densely defined unbounded operator. Then the *Hilbert-adjoint* of A is an operator $A^* : \mathcal{D}(A^*) \to H$, where $\mathcal{D}(A^*) \subset H$, and $\mathcal{D}(A^*)$ consists of all those $v \in H$ such that there exists a unique v^* such that $(Au, v) = (u, v^*)$ for all $u \in \mathcal{D}(A)$. The form of A^* follows from the relation $u^* = A^* u$. In other words, $v \in H$ belongs to $\mathcal{D}(A^*)$ if and only if the inner product (Au, v) considered as a function of v can be represented in the form $(Au, v) = (u, v^*)$. Therefore, this definition is that of the case of Banach spaces with the difference that here $\mathcal{D}(A)$ is a dense subset of H and the dual pairing relation is replaced by a corresponding relation in H, since H is isomorphic to H^*. We say that A is *symmetric* if $(Au, v) = (u, Av)$ $\forall u, v \in \mathcal{D}(A)$ and *selfadjoint* if $A = A^*$. A (densely defined) operator in a complex Hilbert space is symmetric if and only if $A \subset A^*$ (here $\subset$ stands for restriction). Hence every selfadjoint operator is symmetric. This statement is taken for a definition in [Mikh2]. A selfadjoint operator is maximally symmetric, *i.e.* it has no proper extensions. A densely defined operator A is symmetric if and only if $(Au, u) \in \mathbf{R}$ for every $u \in \mathcal{D}(A)$.

If $\mathcal{D}(A) = H$, A is symmetric if and only if it is selfadjoint. By Hellinger-Toeplitz theorem, such an operator is bounded. This is why for bounded operators, only the notion of selfadjoint was defined. In [Mikh3] the symmetric operator A is not required to be densely defined but it is required that $\mathcal{D}(A)$ be closed and the elements of $\mathcal{R}(A)$ have a finite norm. Then Mikhlin defines the positive operator using this definition of the symmetric operator. However, during the proof of the minimum of a functional he uses the density of $\mathcal{D}(A)$ in H, therefore, in fact, its definition is the same as that in [Kre].

Taking into account that $\mathcal{D}(A) \subset \mathcal{D}(A^*)$, we say that the unbounded operator A is *skew-symmetric* if $(Au, v) = -(u, A^*v)$, $\forall u, v \in \mathcal{D}(A)$. Thus, every densely defined operator A can be uniquely splitted as a sum of a symmetric operator A_s and a skew-symmetric operator A_{ss}, where $A_s = (A + A^*)/2$, $A_{ss} = (A - A^*)/2$, where by A^* here we understand the restriction of A^* to $\mathcal{D}(A)$. Obviously, $(A_{ss}u, u) = 0$, while $(A_s u, u) = (Au, u)$, $\forall u \in \mathcal{D}(A)$.

If $H = L^2(a, b)$, the operator $A^*_{|\mathcal{D}(A)}$ is called the *Lagrange* or *formal adjoint* of A and it is denoted by A^+ (Appendix 3).

Let H be a complex Hilbert space and let $A : \mathcal{D}(A) \to H$ be a densely defined unbounded operator. We say that A is *positive* if $(Au, u) \geq 0$ for every $u \in \mathcal{D}(A)$ and $(Au, u) = 0$ if and only if $u = 0$ [Mikh2]. The operator A is *positive definite* if there exists a positive constant γ such that $(Au, u) \geq \gamma^2 \parallel u \parallel^2$, $\forall u \in \mathcal{D}(A)$ [Mikh2]. In [Kre] for the positivity and positive definiteness is required (in addition) that A be symmetric. We adopt this definition.

Obviously, an unbounded positive definite operator is positive, while the inverse assertion is not true. In [Mikh3] the names positive and positive definite were inverted and some special requirements to $\mathcal{D}(A)$ were imposed. The positive square root of the positive definite operator A was discussed in Sections 3.2.4.2, 3.2.4.3. In the general case for the construction of the positive square root of an operator we recommend [LySo]. It is yielded as early as 1961.

Properties of unbounded operators A depend on $\mathcal{D}(A)$ and can change under extensions and restriction (Section 3.2.4.2).

The spectrum of the (linear) operator $A : \mathcal{D}(A) \subset X \to X$, where X is a complex normed space, is the set $\sigma(A) \subset \mathbf{C}$, $\sigma(A) = \mathbf{C} \setminus \rho(A)$, where the set $\rho(A) \subset \mathbf{C}$, called the *resolvent set* of A, consists of points $\lambda \in \mathbf{C}$, called *regular values* of A, such that the operator $R_\lambda(A) = (A - \lambda I)^{-1}$, called the *resolvent* of A, exists, is bounded and $\overline{\mathcal{D}(R_\lambda(A))} = X$. Usually, instead of $R_\lambda(A)$ we write R_λ. For every regular value λ, the equation $Au = \lambda u$ possesses the null solution only, while the equation $(A - \lambda I)u = g \in X$ possesses the unique solution $u = R_\lambda g$. The spectrum consists of points $\lambda \in \mathbf{C}$, called the *spectral values*, and it splits in three mutually disjoint spaces, namely $\sigma(A) = \sigma_p(A) \cup \sigma_c(A) \cup \sigma_r(A)$. The set $\sigma_p(A)$, called the *point* (or *discrete) spectrum* (or *eigenvalue spectrum* [Mikh5]) of A consists of all points $\lambda \in \mathbf{C}$ for which R_λ does not exist. The elements λ of $\sigma_p(A)$, are called the *eigenvalues* of A. The nontrivial solutions v of the equation $Av = \lambda v$, are called the *eigenvectors* (or *eigenfunctions* if X is a function space) *of A corresponding to λ*. To an eigenvalue λ several eigenvectors may correspond. The space of all eigenvectors of A corresponding to λ is called the *geometric eigenspace of A corresponding to* λ. Some authors omit "geometric". The maximum number of linearly independent eigenvectors corresponding to the same eigenvalue λ is called the *geometric multiplicity* of λ and it is denoted by $m_g(\lambda)$. In Russian literature, *e.g.* Mikhlin's book and Yudovich papers, $m_g(\lambda)$ is called the *rank* of λ. Hence $m_g(\lambda)$ is the dimension of the geometric eigenspace of A corresponding to λ. By the Gram-Schmidt procedure, every set of $m_g(\lambda)$ linearly independent eigenvectors of A corresponding to λ can be orthonormalized. Usually, it is understood that this procedure was already applied. Let λ be an eigenvalue. The non-vanishing vectors $w \in \mathcal{D}(A)$ such that there exists $k \in \mathbf{N}$, $k > 1$, such that $(A - \lambda I)^k w = 0$, where w are not eigenvectors of λ, are called the *generalized* or *associated eigenvectors of A corresponding to λ*. Since in Russian literature the generalized eigenvector means a solution in the generalized (weak) sense of the equation $(A - \lambda I)u = 0$, we use only the label of associate eigenvector. The space of all associated eigenvectors corresponding to some eigenvalue λ is called the *generalized eigenspace of A corresponding to λ*. Sometimes [Kat] the eigenvector is considered as an associated eigenvector too. Therefore we must beware the meaning of this notion before reading a paper. The dimension of the union of the eigenspace and generalized eigenspace of A corresponding to λ, called the *algebraic eigenspace* of A corresponding to λ, is called the *algebraic multiplicity* of λ and it is denoted by $m_a(\lambda)$. Of course, $m_g(\lambda) \leq m_a(\lambda)$. If $m_g(\lambda) = m_a(\lambda) = 1$, then λ is called a *simple eigenvalue*. The set $\sigma_c(A)$, called the *continuous spectrum of A*, consists of points $\lambda \in \mathbf{C}$ such that R_λ exists, $\overline{\mathcal{D}(R_\lambda)} = X$, but R_λ is an unbounded operator. The set $\sigma_r(A)$, called the *residual spectrum of A*, consists of points $\lambda \in \mathbf{C}$ such that R_λ exists, may be bounded or unbounded, and $\overline{\mathcal{D}(R_\lambda)} \neq X$. The above-mentioned splitting of $\sigma(A)$ is not accepted by all authors [Yos]. In particular, the discrete spectrum is not always a discrete set. However, we adopt the terminology in [Kre]; the most detailed spectral analysis is contained in [EdE]. See also [Yos]. The characteristics of spectral values and eigenvectors are different for different types of operators and spaces X, but some common features exist for large classes of operators, *e.g.* in finite-dimensional spaces, bounded, compact, selfadjoint, positive definite. These features appear significantly in compact operators. We present them separately, after introducing this concept.

Suppose that X is a finite-dimensional Hilbert space with $\dim X = n$. We choose an orthonormal basis in X and identify every operator A with the corresponding $n \times n$ matrix. Then $\sigma_c(A) = \sigma_r(A) = \emptyset$, $\sigma_p(A)$ consists of exactly n eigenvalues $\lambda_1, \ldots, \lambda_n$, some of them equal. The eigenvalues of A are the roots of the characteristic equation $\det(A - \lambda I) = 0$, which is an n-th degree polynomial equation in λ. Here I is the identity matrix. In this case $m_a(\lambda)$ is equal to the multiplicity of λ as a root of the characteristic equation. We have $\lambda_1 + \ldots + \lambda_n = \mathrm{Tr}A$, $\lambda_1 \cdots \lambda_n = \det A$, even if some eigenvalues are multiple. The matrix A is called Hermitian if $\overline{A}^T = A$, where the bar stands for the complex conjugation and ^{T} for the transposition. The operator corresponding to a Hermitian matrix is selfadjoint and its eigenvalues are real. Let $\lambda_1 \geq \lambda_2 \geq \ldots \geq \lambda_n$ be the eigenvalues of such a matrix A and denote $q(u) = \overline{u}^T A u / \overline{u}^T u$. Then $\lambda_1 = \max_{u \neq 0} q(u)$, $\lambda_n = \min_{u \neq 0} q(u)$,

$\lambda_i = \max_{u \in Y_i, \, u \neq 0} q(u)$, where Y_i is the subspace of $\mathbf{C}^n$ consisting of all vectors which are orthogonal to the eigenvectors corresponding to $\lambda_1, \ldots, \lambda_{i-1}$. If X is real and $A^T = A$, the matrix A is called a *real symmetric matrix* and its spectrum is real.

Theorem 2.1 [Kre]. *Let A be a (linear) operator on a vector space V and let $\lambda_1, \ldots, \lambda_n$ be n mutually distinct eigenvalues of A. Then n corresponding eigenvectors $u_1, \ldots, u_n$ form a linearly independent set.*

This theorem allows one to use this set as a basis for V or for a dense subspace of V. In particular, if V is finite-dimensional of dimension n (*e.g.* $\mathbf{R}^n$) and has n eigenvalues λ_i, $i = 1, \ldots, n$ with $m_g(\lambda_i) = 1$, $i = 1, \ldots, n$, then the corresponding set $\{u_1, \ldots, u_n\}$ spans V (in particular $\mathbf{R}^n$ or $\mathbf{C}^n$) and form a basis of V. If $\dim V = \infty$, the set of eigenvalues must be discrete and the set of the corresponding eigenvectors must be complete, which takes place for special operators and spaces V, as we see further. Due to its exceptional importance in applications, allowing one to solve a nonlinear problem in the space of its linearized problem, to this topic several books were dedicated, *e.g.* [Bere], [Tit], [Col1].

Let $\mathcal{B}$ be a complex Banach space and let $A : \mathcal{B} \to \mathcal{B}$ be a bounded operator. Then $\sigma(A)$ is a compact subset of $\mathbf{C}$ situated in the disk of radius $\|A\|$ and centered at the origin of $\mathcal{B}$, $|\lambda| \leq \|A\|$ for every $\lambda \in \sigma(A)$. Let $r_\sigma(A) \equiv \sup_{\lambda \in \sigma(A)} |\lambda|$ be the *spectral radius* of A. Then $r_\sigma(A) = \lim_{n \to \infty} \|A^n\|^{1/n}$; if B is complex, then $\sigma(A) \neq \emptyset$; if $A^2 = I$, then $\sigma(A) = \{0, 1\}$.

Let H be a complex Hilbert space and let $A : H \to H$ be a bounded selfadjoint (and, so, symmetric) operator. Then: $\sigma_r(A) = \emptyset$; possibly $\sigma_p(A) \neq \emptyset$; $\sigma(A) \subset \mathbf{R}$, namely $\sigma(A) \subset [m, M]$, where $m = \inf_{u \in H, \|u\|=1}(Au, u)$, $M = \sup_{u \in H, \|u\|=1}(Au, u)$ and m and M are spectral values of A; $\|A\| = \max\{|m|, |M|\} \equiv \sup_{\|u\|=1} |(Au, u)|$; the eigenvectors corresponding to numerically different eigenvalues are orthogonal; $\lambda \in \rho(A)$ if and only if there exists $c > 0$ such that $\|(A - \lambda I)u\| \geq c\|u\|$, for every $u \in H$; if, in addition, A is compact, and $H \neq \{0\}$, then A has at least one eigenvalue; let $q(u) = (Au, u)/(u, u)$ be the *Rayleigh quotient*. Then: $\sigma(A) \subset [\inf_{u \neq 0} q(u), \sup_{u \neq 0} q(u)]$; A is positive and selfadjoint if and only if its spectrum consists of nonnegative real values only; A^2 is a positive operator because $(A^2 u, u) = (Au, Au)$ and it cannot have a negative spectral value; AA^* and A^*A are selfadjoint and positive operators, so their spectra are real and cannot contain negative values; if $H \neq 0$, we say that A has a *purely discrete spectrum* or a *pure point spectrum* if A has an orthonormal set of eigenvectors total in H. The second name is not appropriate because there are concrete examples where $\sigma_c(A) \neq \emptyset$, hence $\sigma(A)$ does not reduce to $\sigma_p(A)$.

Let H be a complex Hilbert space and let $A : \mathcal{D}(A) \subset H \to H$ be an unbounded operator. In this case: if A is selfadjoint, then $\sigma(A)$ is real and closed and $\sigma_r(A) = \emptyset$; if A is symmetric then $\sigma_p(A)$ is real but not bounded; if H is separable then $\sigma_p(A)$ is at most countable (perhaps finite or empty). If $\lambda \in \sigma_r(A)$, then $\overline{\lambda} \in \sigma_p(A^*)$. If

A is symmetric, then $\sigma_p(A) \subset \sigma_p(A^*)$. The spectral properties of many classes of unbounded or bounded operators are proved by means of the properties of closed operators. Let us introduce them.

Let H be a complex Hilbert space and let $A : \mathcal{D}(A) \subset H \to H$ be an unbounded operator. We saw that A is a *closed operator* if its graph $\{(u,v) \mid u \in \mathcal{D}(A), v \in \mathcal{R}(A)\}$ is closed in $H \times H$. If A is closed and $\mathcal{D}(A)$ is closed then A is bounded. If A is bounded, then A is closed if and only if $\mathcal{D}(A)$ is closed. The Hilbert-adjoint A^* of an unbounded operator A is closed. The closure $\overline{A}$ of A is by definition the minimal closed linear extension of A (*i.e.* all other closed linear extensions of A is a closed linear extension of $\overline{A}$). Let $\overline{\mathcal{D}(A)} = H$ and let A be a symmetric operator. Then $\overline{A}$ exists and is unique. In addition, $(\overline{A})^* = A^*$. If A is closed and $\mathcal{D}(A) = H$, then A is bounded and $H = \mathcal{R}(A) \oplus \mathcal{N}(A^*)$. If A is closed and $\overline{\mathcal{D}(A)} = H$, then $\overline{\mathcal{D}(A^*)} = H$, and $A^{**} = A$. If A is symmetric, then A^{**} is a closed symmetric linear extension of A. If A is closed and injective, then A^{-1} is closed. Let H be a Hilbert space and let $A : H \to H$ be a bounded operator. We say that A is a *unitary operator* if A is bijective and $A^* = A^{-1}$. An unitary operator is closed. If A is a closed operator, then $\sigma(A) = \sigma_p(A)$.

Spectral properties of positive and positive definite operators. Let H be a complex Hilbert space and let $A : \mathcal{D}(A) \subset H \to H$ be an unbounded (linear) operator. In particular, the bounded from below operator, *e.g.* positive or positive definite, occur in calculus of variations. Of a special interest are those operators the eigenvalues of which can be used to construct minimizing sequences of the functionals associated with a Euler problem. As a consequence, in this framework, the spectral analysis concerns primarily the study of eigenvectors, the eigenvalues resulting as minima of the functionals. Since, in general, the eigenvectors are generalized solutions of the associated Euler equations, they are called the *generalized eigenvectors* or *generalized eigenelements* and the corresponding eigenvalues are called the *generalized eigenvalues*. The set of all generalized eigenvalues is the *generalized spectrum* [Mikh5], [Mikh3]. For the sake of brevity, we omit "generalized". This sense is different from that one related to the algebraic multiplicity [Kre], [Kat], [EdE]. Therefore, in presenting this topic we adopt the terminology from the Mikhlin's books and Yudovich's papers on hydrodynamic stability, more suitable to our book. Remark that the eigenelements belong, in general, to energy space H_A (Section 3.2.4.2) and eigenvalues to $\mathbf{C}$. They do not satisfy the equation $Au = \lambda u$ for $u \in \mathcal{D}(A)$ but the equation $[u, \eta]_A = \lambda(u, \eta)$ for every $\eta \in H_A$. These eigenvalues and eigenelements are ordinary eigenvalues and eigenvectors for the Friedrichs extension $\tilde{A}$ of A. If A is symmetric, then its eigenvalues λ_i are real; the eigenelements corresponding to distinct eigenvalues are orthogonal; if, in addition, H is separable, A has at most a countable set of eigenvalues. Let A be a positive definite operator. Then: its eigenelements are orthogonal in H_A as well as in H; every eigenvalue λ of A (in fact, of $\tilde{A}$ from Section 3.2.4.2) is defined by the Rayleigh quotient $\lambda = \dfrac{(\tilde{A}u, u)}{(u, u)}$

and satisfies the inequality $\lambda \geq \gamma^2$, where γ^2 is the constant of definition of A, *i.e.* it is the lower bound of A. In addition, if we denote $(\tilde{A}u, u) = |u|_A^2$, then $\lambda = \frac{|u|_A^2}{\|u\|^2}$ or, equivalently, $\lambda = |u|_A^2$, $\|u\|^2 = 1$ ($\|\cdot\|$ being the norm of H). Formally the construction of the eigenvalues λ_i proceeds similarly to the case of operators in $\mathbf{C}^n$: suppose that there exists a minimizing sequence of the Rayleigh quotient functional containing a subsequence converging in the sense of H to $v_1 \in H_A$. Then λ_1 is the minimum of this functional while the limit v_1 is the first eigenelement. In addition, $\lambda_1 = \gamma^2$. The other λ_{n+1} and v_{n+1} can be determined recurrently as minima and corresponding extremals of the same functional in the space $H_A^{(n)}$, consisting in the vectors of the form $\eta^{(n)} = \eta - \sum_{k=1}^n (\eta, v_k) v_k$, with $\eta \in H_A$. This construction is conditioned by the existence of that subsequence for which the infimum of the functional is attended, *i.e.* it is a minimum. A sufficient condition for this is that A has a discrete spectrum. Here this notion has a sense distinct from σ_p given in the above [Kre]. This is why, for some specific purposes, the splitting of $\sigma(A)$ is different from that we gave in the above [Kre].

From now on H is an infinite-dimensional complex Hilbert space and $A : \mathcal{D}(A) \subset H \to H$, an operator.

In [Mikh5] the following definition is used: let X_1 and X_2 be two Banach spaces. A densely defined operator $A : \mathcal{D}(A) \subset X_1 \to X_2$ is *completely continuous* if it transforms every bounded set of $\mathcal{D}(A)$ into a compact set of X_2. Such an operator is bounded and it can be extended by continuity to X_1.

An operator with a discrete spectrum is, by definition, a symmetric operator which has an infinite sequence of eigenvalues $\lambda_1, \lambda_2, \ldots, \lambda_n, \ldots$ and this sequence has a unique limit point at infinity, while the sequence $\{v_n\}$ of the corresponding eigenelements is complete in H [Mikh5], [Mikh3]. If A is positive definite and transforms every bounded set of H_A into a compact set in H (remark that this operator is more particular than a compact operator according to definition in [Kre] but it is compact according to definition used in [Mikh5]), then the (generalized) spectrum of A is discrete in the sense of this definition and $0 < \lambda_1 \leq \lambda_2 \leq \ldots \leq \lambda_n \leq \ldots, \lambda_n \to \infty$. Therefore for this subclass of positive definite operators the formal construction in the above is substantiated. The reciprocal assertion is also true: if A is positive definite and has a discrete spectrum then the embedding of H_A in H is compact and A^{-1} is compact and positive.

If A is positive definite and selfadjoint and γ^2 is its lower bound, then $(-\infty, \gamma^2) \subset \rho(A)$. As a consequence, $\sigma(A) \subset [\gamma^2, \infty)$ and, so, for every eigenvalue λ we have $\lambda \geq \gamma^2$. Let $\mathcal{D}(A) = H$ and assume that A is compact and positive definite (therefore selfadjoint). Then there exists an infinite orthogonal set of eigenelements v_i, $i = 1, 2, \ldots$, total in H, corresponding to the eigenvalues $\lambda_1 \geq \lambda_2 \geq \ldots \geq \lambda_n \ldots, \lambda_n \to 0$, where $\lambda = 0$ is the only possible accumulation point. Whence the possibility to expand the solutions of a nonlinear equation upon this total set of eigensolutions of the linearized equation.

If A is positive definite and has a discrete spectrum then the inverse operator A^{-1} exists and it is compact. In addition, there exists the positive square root operator $A^{1/2}$ such that if $Au = f$, then $A^{1/2}u = A^{-1/2}f$. Then $A^{1/2}u = \sum_{n=1}^{\infty} \sqrt{\lambda_n}(u, \phi_n)\phi_n$, where λ_n and ϕ_n are the eigenvalues and the corresponding eigenvectors of A. Another definition useful in numerical approximations [Mikh3] is: let H be a Hilbert space. An operator A' on a Hilbert space H, $A' : H \to H$ is called a *degenerated* or *finite* operator if $\mathcal{R}(A')$ is finite [Kat], *i.e.* it can be represented as $A'u = \sum_{k=1}^{n}(u, \psi_k)\phi_k$, where ψ_k, $\phi_k \in \mathrm{H}$ and $\{\phi_k\}$ and $\{\psi_k\}$ are linearly independent sets. If, instead of H, we have a Banach space B, then $\psi_k \in B^*$. The operator $A : H \to H$ is called a *completely continuous operator* if for every $\epsilon > 0$, $Au = A'_\epsilon u + A''_\epsilon u$, where A'_ϵ is a degenerated operator and A''_ϵ can be done as small as we wish. In the Banach space setting, $A' : X \to Y$, $\psi_k \in X^*$ and $\{\phi_k\}$ is a basis of $\mathcal{R}(A) \subset Y$ [Kat]. A completely continuous operator, therefore a compact operator, was expressed as a sum of an operator defined on a finite-dimensional space and an invertible operator defined on the quotient of H by that space. Therefore this alternative definition for compact operators is useful in approximating a solution of a nonlinear equation by its projection on the finite-dimensional spaces generated by n eigensolutions of the linearized equation. This idea is basic in Galerkin-Faedo-Hopf method. It was extensively applied to Navier-Stokes equations by Foias and Prodi [FoiP].

Since the product of a compact operator by a bounded operator is compact, it follows the important property that a compact operator may be written as a sum of a closed operator (*i.e.* with a closed graph) and an identity operator. Then the splitting of H by means of the range of a closed operator and the finite-dimensional null space of its adjoint enables one to prove the above splittings of H by means of the null and range of some powers of $A - \lambda I$, where A is compact and λ is an eigenvalue of A. On a more general plane, this splitting leads to the approximation of the dynamics of the phase space with the dynamics on a finite-dimensional inertial manifold of the phase space [ConsF], [ConsFNT].

Completely continuous (compact) operator. We saw that these two notions coincide while completely continuous mapping and compact mapping are two different concepts. Let X_1 and X_2 be two normed spaces. If the operator $A : X_1 \to X_2$ is compact then $R(A)$ is separable. If A is a compact operator so is its adjoint operator. If $\dim A(X_1) < \infty$ and the operator A is bounded, then A is compact. If A is a (linear) operator and $\dim X_1 < \infty$, then A is compact. The compact operators are completely continuous, therefore they transform the weak convergence (*i.e.* in the generalized sense) in the convergence in norm. Whence their basic role in generalized setting of differential equations. A compact operator A for $X_1 = X_2 = X$ has good spectral properties: the set of its eigenvalues is at most countable (in particular, it can be finite or empty) and its possible accumulation point is the origin; every non-vanishing spectral value is an eigenvalue; if X is infinite-dimensional, then 0 belongs to the spectrum of A; for every non-vanishing eigenvalue λ the geometric eigenspace of A is finite, *i.e.* $m_g(\lambda) < \infty$, the null spaces of $(A - \lambda I), (A - \lambda I)^2, \dots$

are finite-dimensional [Kre] implying that $m_a(\lambda) < \infty$; there exists $r \geq 1$ such that $X = \mathcal{N}(A - \lambda I)^r \otimes \mathcal{R}(A - \lambda I)^r$ for $\lambda \neq 0$, where r depends on λ; for every $\lambda \neq 0$, $\mathcal{N}(A - \lambda I)$ is finite-dimensional and so $\dim \mathcal{N}(A - \lambda I)^n < \infty$ for any $n \in \mathbf{N}$; for every $k > 0$, there are at most finitely many linearly independent eigenvectors of A corresponding to eigenvalues of absolute value greater than k.

Orthogonal projection. Let H be a Hilbert space, let Y be a closed subspace of H and let $H = Y \oplus Y^\perp$ be its splitting. This enables one to define the *projection operator* $P : H \to Y$, $P(u) = w \in Y$ and the *projection operator* $Q : H \to Y^\perp$, $Q(u) = z \in Y^\perp$. Obviously $Q = I - P$, where I is the identity operator on H. In [Kre] P is called the *orthogonal projection on* Y or *orthogonal projection of* H *onto* Y. The projection P has the following properties [Kre]: $(Pu, u) = \| Pu \|^2$, $P^2 = P$, $P \geq 0$, $\| P \| \leq 1$, $\| P \| = 1$ if $P(H) \neq \{0\}$. Moreover: a bounded linear operator $P : H \to H$, where H is a Hilbert space, is a projection if and only if P is selfadjoint and $P^2 = P$. Consider the particular case where H is a separable Hilbert space. Denote by H_1 a Hilbert subspace of H and by $H_1^\perp$ its orthogonal complement. Then $H_1^\perp$ is a Hilbert subspace of H too and H_1 and $H_1^\perp$ are separable. Let $\{\phi_n\}_{n \in \mathbf{N}}$ be an orthonormal total set in H_1. Then the projection $P : H \to H_1$, is defined by $Pu = \sum_{n=0}^\infty (u, \phi_n)\phi_n$ and the projector $Q : H \to H^\perp$, is $Qu = Iu - Pu$ [EdE], [Kre]. The projection of a Hilbert space H onto a finite-dimensional subspace of H is compact.

The projections are important in spectral theory, where they are analyzed in depth [EdE]. The method of orthogonal projections is also an important variational method [Mikh3]. In the energy method for evolution equations the projection on orthogonal subspaces represent the first step [GeoPalR00].

Basics on operator theory and spectral analysis and nonlinear mappings can be found in [AkG], [DunS], [CouH], [Fri2], [FucNSS], [EdE], [Day], [Yos], [Kre], [Mau1], [Kat], [Goldb].

Functionals [We1]. Let $V/\mathbf{K}$ be a linear space over the field $\mathbf{K}$. A *linear functional* is a linear transformation of V to $\mathbf{K}$. Let $V/\mathbf{K}$ and $W/\mathbf{K}$ be two linear spaces over the same field. Denote by $V \times W$ the space of ordered pairs (v, w) with $v \in V$ and $w \in W$. A mapping $\mathcal{F}(v, w) : V \times W \to \mathbf{K}$ which for every $w \in W$ fixed is linear in v and for every $v \in V$ fixed $\overline{\mathcal{F}}(v, w)$ is linear in w (or, equivalently, $\mathcal{F}(v, w)$ is conjugate linear in w [Kre]) is called a *sesquilinear functional (form)*. Sometimes it is called *semilinear functional* [Kre]. If $\mathbf{K} = \mathbf{R}$, then the sesquilinear form $\mathcal{F}(v, w)$ is linear in v and in w. In this case it is called a *bilinear functional*. The inner product on real spaces is a bilinear functional. A *Hermitian functional* on $V/\mathbf{K}$, where $\mathbf{K} = \mathbf{C}$, is a sesquilinear functional $\mathcal{F} : V \times V \to \mathbf{K}$, such that $\mathcal{F}(u, v) = \overline{\mathcal{F}}(v, u)$. The inner product in complex Hilbert spaces is a Hermitian functional. If $K = \mathbf{R}$, then $\mathcal{F}(u, v) = \mathcal{F}(v, u)$, i.e. $\mathcal{F}$ is a *symmetric functional*.

A *quadratic functional* is a mapping $\mathcal{F}$ from the linear space $V/\mathbf{R}$ to $\mathbf{R}$ such that $\mathcal{F}(u) = \mathcal{F}(u, u)$, where $\mathcal{F}(u, v)$ is Hermitian. Note that in this case $\mathcal{F}(u, u)$ is a real functional. In [Mikh3] this $\mathcal{F}(u, u)$ is referred to as a *homogeneous quadratic functional.* Obviously,

$$\mathcal{F}(\alpha u + \beta v) = \mid \alpha \mid^2 \mathcal{F}(u) + \alpha\overline{\beta}\mathcal{F}(u,v) + \overline{\alpha}\beta\mathcal{F}(v,u) + \mid \beta \mid^2 \mathcal{F}(v),$$

for every $u, v \in V$ and $\alpha, \beta \in \mathbf{C}$. In addition, for *quadratic functional* Mikhlin took every functional of the form $\mathcal{F}(u) + A_1 u + \overline{A}_2 u + C$, where $\mathcal{F}(u)$ is a homogeneous quadratic functional, A_1 and A_2 are operators and C is a constant. In the presentation of the variational principle 3.2.2 we used these last definitions. In [Mikh2] is introduced the *positive definite functional,* which is a homogeneous quadratic functional $\mathcal{F}(u)$ satisfying the inequality $\mathcal{F}(u) \geq \gamma^2 \parallel u \parallel^2$, for every $u \in H$, where $\overline{\mathcal{D}(\mathcal{F})} = H$ and γ is a strictly positive constant. This type of functional occurs in the theory of linear differential equations subject to nonhomogeneous boundary conditions. In this case, the solution of these problems is the minimum of the functional $\tilde{\mathcal{F}}(u) = \mathcal{F}(u) - Lu - \overline{Lu}$, where $L : \mathcal{D}(L) \to H$ is a linear functional (not necessarily bounded) such that $\mathcal{D}(\mathcal{F}) \subset \mathcal{D}(L)$.

Appendix 3

Differential operators in $L^2(a,b)$

The relationship between an operator $\mathcal{A}$, its formal adjoint $\mathcal{A}^+$ and its adjoint A^* and the role played by the boundary conditions are analyzed.

Let $C^n(a,b) \subset L^2(a,b)$ and define the unbounded differential operators $\mathcal{A}$, $\mathcal{A}^+ : C^n(a,b) \to C(a,b)$

$$\mathcal{A}u = \sum_{k=0}^{n} a_k D^k u, \quad \mathcal{A}^+ u = \sum_{k=0}^{n} (-1)^k a_k D^k u, \tag{3.1}$$

where $D^k = \dfrac{d^k}{dx^k}$, $u : (a,b) \to \mathbf{R}$, $u = u(x)$, the operator $\mathcal{A}^+$ is the adjoint of $\mathcal{A}$ in the Lagrange sense. Define the mutually distinct boundary conditions

$$B_k u(b) - B_k u(a) \equiv [B_k u(x)]_a^b = \left[\sum_{j=1}^{n-1} b_k^j D^j u(x) \right]_a^b = 0, \quad k = 1,\ldots,n, \quad b_k^j \in \mathbf{R} \tag{3.2}$$

and let A be the restriction of $\mathcal{A}$ to the linear space

$$\mathcal{D}_A = \{ u \in C^n(a,b) \cap C^{n-1}[a,b] \mid u \text{ satisfies } (3.2) \}.$$

We have

$$(\mathcal{A}u, v) = \sum_{k=0}^{n} a_k \left[\sum_{s=0}^{k-1} (-1)^s D^{k-s-1} u D^s v \right]_a^b + (u, \mathcal{A}^+ v), \quad \forall u, v \in C^n(a,b), \tag{3.3}$$

where the form of $\mathcal{A}^+$ depends only on $\mathcal{A}$ (through the coefficients a_k) and not on the boundary conditions imposed to the elements of $\mathcal{A}$.

Denote

$$[B(u,v)]_a^b = \sum_{k=0}^{n} a_k \left[\sum_{s=0}^{k-1} (-1)^s D^{k-s-1} u D^s v \right]_a^b = [B_A(u,v)]_a^b + [B_{A^*}(u,v)]_a^b \tag{3.4}$$

and let $[B_A(u,v)]_a^b$ stay for the sum of those terms of $[B(u,v)]_a^b$ which vanish for u satisfying (3.2) and let $[B_{A^*}(u,v)]_a^b$ stay for the sum of those terms of $[B(u,v)]_a^b$ which do not. Taking into account the boundary conditions (3.2), for which $[B_A(u,v)]_a^b = 0$, from (3.3) we get

$$(Au, v) = [B_{A^*}(u,v)]_a^b + (u, \mathcal{A}^+ v), \quad \forall u \in \mathcal{D}(A), \forall v \in C^n(a,b), \tag{3.5}$$

implying

$$(Au, v) = (u, A^*v), \quad \forall u \in \mathcal{D}_A, \quad \forall v \in \mathcal{D}_{A^*}, \tag{3.6}$$

where the definition of the operator A^*, adjoint of A, was taken into account. By definition of A^*, in order to obtain (3.6), we perform (3.3), then we take into account the boundary conditions (3.2) satisfied by the functions in $\mathcal{D}_A$, the remaining conditions being those which must be satisfied by $v \in \mathcal{D}_{A^*}$. It also follows that $\mathcal{A}^+_{|\mathcal{D}_{A^*}} = A^*$ and that the boundary conditions

$$[B_{A^*}(u, v)]_a^b = 0 \tag{3.7}$$

hold. In general, the boundary conditions satisfied by $v \in \mathcal{D}_{A^*}$ are different from those satisfied by $u \in \mathcal{D}_A$. Indeed, in order to impose the boundary conditions satisfied by u, we put together all terms in (3.4) which have B_* as common factor, the other factor being a sum of terms in v and its derivatives. It is not necessary that the remaining terms admit the same decomposition in sum of products.

Case of symmetric A. In this case we have

$$(Au, v) = (u, Av), \quad \forall u, v \in \mathcal{D}_A. \tag{3.8}$$

Remark 3.1. Since $C_0^\infty \subset \mathcal{D}_A$ it follows that $\overline{\mathcal{D}_A} = L^2(a, b)$. Then, letting $v \in \mathcal{D}_A$, from (3.5) we have $\mathcal{A}^+_{|\mathcal{D}_A} = A$ and the boundary condition (3.7) is satisfied.

Remark 3.2. The equality $\mathcal{A}^+_{|\mathcal{D}_A} = A$ implies the fact that $a_{2l+1} = 0$, $0 \leq 2l + 1 \leq n$, therefore, in the expression of A, only even order derivatives occur.

Remark 3.3. The form (Bu, v) in (3.4) cannot contain derivatives in u and v of the same order, *i.e.* $s = k - s - 1$. Indeed, by Remark 3.2, in (3.3) k must be an even number, while $s = k - s - 1$ implies $s = (k - 1)/2$, which is not an integer.

Consider the linear equation $Au = 0$, *i.e.*

$$\sum_{k=0}^{n} a_k D^k u = 0, \tag{3.9}$$

where $a_n \neq 0$, and let u be one solution of (3.9). Since $u, Du, \ldots, D^{k-1}u \in C^1(a, b)$, and taking into account that $D^n u = -\dfrac{1}{a_n}\left[\sum_{k=0}^{n-1} a_k D^k u\right]$, it follows that $D^n u \in C^1(a, b)$, therefore $u \in C^{n+1}(a, b)$. Then $u, Du, \ldots, D^{k-1}u \in C^2(a, b)$, and a similar reasoning shows that $D^n u \in C^2(a, b)$, implying $u \in C^{n+2}(a, b)$. Repeating this reasoning *ad infinitum* it follows that $u \in C^\infty(a, b)$. In addition, the solutions u of (3.9) are polynomials multiplied by exponentials, therefore, they belong to $C^\infty(\mathbf{R})$. In particular, $u \in C^\infty[a, b]$.

Consequence 3.1. We can redefine the domain of definition of A, $\mathcal{D}_A = \{u \in C^\infty(a, b) \mid u \text{ satisfies } (3.2)\}$.

Remark 3.4. Sometimes, variational or/and numerical reasons force us to introduce another inner product defined with the aid of A (Section 3.2.4.2). Denote it by $(\cdot,\cdot)_A$ and let $|\cdot|_A$ be the corresponding norm. Define

$$\overline{H_A} = \overline{\mathcal{D}_A}^{\,|\cdot|_A}. \tag{3.10}$$

Assume that $\overline{H_A} \subset H \subset L^2[a,b]$. Then $\mathcal{D}_A$ is dense in $L^2(a,b)$ as well as in $\overline{H_A}$ but with respect to different norms.

Remark 3.5. Since $\overline{\mathcal{D}_A} = L^2(a,b)$, it follows that if A is symmetric, then $A \subset A^*$. But the form of A and A^* are the same and the boundary conditions from $\mathcal{D}_{A^*}$ are the same as those for A^+, which, since $A = A^+$, are those from $\mathcal{D}_A$. Hence $\mathcal{D}_A = \mathcal{D}_{A^*}$, therefore A is selfadjoint.

Consequence 3.2. We can deduce additional boundary conditions from equation (3.9) and its derivatives. (We recall that A is symmetric.) The following analysis is aimed to help us in writing (Au,u) in a form more convenient for the associated variational problem. The boundary conditions satisfied by $v \in \mathcal{D}_A$ differ from those satisfied by $v \in \mathcal{D}_{A^*}$, and this holds in spite of the fact that the sum (3.7) vanishes for $v \in \mathcal{D}_A$ as well as for $v \in \mathcal{D}_{A^*}$.

In the case of $v \in \mathcal{D}_{A^*}$, the vanishing of (3.7) is a result of the fact that the boundary values of the derivatives of u other than in (3.2) must be arbitrary.

For $v \in \mathcal{D}_A$ the vanishing of (3.7) is related to the symmetry of the derivatives in u, resulted in the process of by-parts integrations when going from (u, Av) towards (Au, v). Indeed, we have

$$(Au, v) = [{}_1 B(u,v)]_a^b + \sum_{k=0}^{[n/2]} (-1)^k a_k \int_a^b D^k u D^k v\, dx, \tag{3.11}$$

$$(u, Av) = (Av, u) = [{}_1 B(v,u)]_a^b + \sum_{k=0}^{n/2} (-1)^k a_k \int_a^b D^k v D^k u\, dx, \tag{3.12}$$

where ${}_1 B(u,v)$ is a functional resulted by by-parts integrations, implying $[{}_1 B(u,v)]_a^b = [{}_1 B(v,u)]_a^b$, i.e. the announced symmetry of the sesquilinear form ${}_1 B(u,v)$. Thus, going on with the transfer of the derivatives of u on v we have

$$\sum_{k=0}^{[n/2]} (-1)^k a_k \int_a^b D^k u D^k v\, dx = [B_{A^*}(u,v)]_a^b - [{}_1 B(u,v)]_a^b + (u, Av),$$

where the first term on the right-hand side vanishes in view of (3.7). Therefore, as expected, this equality reads

$$(u, Av) = \sum_{k=0}^{[n/2]} (-1)^k a_k \int_a^b D^k u D^k v\, dx + [{}_1 B(u,v)]_a^b. \tag{3.13}$$

In general $[{}_1 B(u,v)]_a^b \neq 0$.

Example 3.1 [Mikh5]. Consider the operator A, where $Au = -D\big(p(x)(Du)(x)\big) + r(x)u$, $u : [a,b] \to \mathbf{R}$, $u = u(x)$, $\alpha u'(a) - \beta u(a) = 0$, $\gamma u'(b) + \delta u(b) = 0$, where α, β, γ, δ, ≥ 0, are constants such that $\alpha\beta \neq 0$, $\gamma\delta \neq 0$, $p(x)$, $p'(x)$ and $r(x)$ are given continuous functions on $[a,b]$, $p(x)\geq 0$, $r(x)\geq 0$, $\forall x \in [a,b]$, $p(x)$ can vanish at certain points in $[a,b]$ but $\int_a^b \frac{dx}{p(x)} < \infty$. $\mathcal{D}(A) = \{u \in C^2(a,b) \cap C^1[a,b] \mid \text{the above boundary conditions hold}\}$. A by-parts integration yields

$$(Au, v) = \int_a^b [p(x)\frac{du}{dx}\frac{dv}{dx} + r(x)u(x)v(x)]dx + \frac{\beta}{\alpha}p(a)u(a)v(a) + \frac{\delta}{\gamma}p(b)u(b)v(b)$$

for $u, v \in \mathcal{D}(A)$, therefore A is symmetric but

$$[_1 B(u,v)]_a^b \equiv \frac{\beta}{\alpha}p(a)u(a)v(a) + \frac{\delta}{\gamma}p(b)u(b)v(b) \neq 0.$$

Remark that the boundary conditions involve derivatives of order $n/2$, where $n = 2$ is the order of the differential operator A.

An important class of symmetric operators in $L^2(a,b)$ consists of operators A such that, after $[n/2]$ by-parts integrations, we have

$$[_1 B(u,v)]_a^b = 0. \tag{3.14}$$

Assume that (3.14) holds. Then, from $(3.11) - (3.13)$ it follows

$$(Au, v) = \sum_{k=0}^{[n/2]} (-1)^k a_k \int_a^b D^k u D^k v\, dx = (u, Av), \tag{3.15}$$

i.e. the associated sesquilinear form $\mathcal{F}(u,v) = (Au,v)$ is symmetric. For $u = v$, (3.15) takes the form

$$(Au, u) = \sum_{k=0}^{[n/2]} (-1)^k a_k \int_a^b (D^k u)^2\, dx = (A^{1/2}u, A^{1/2}u), \tag{3.16}$$

which is of primary interest in variational problems.

Case of nonsymmetric A. In this case, at least one odd-order derivative of u occurs in (3.1) and/or (3.14) does not hold. As a consequence, (3.15) and (3.16) are no longer valid. However, for some of these operators, it is possible to obtain a relation quite close to (3.15). This is possible if the boundary conditions (3.2) are such that the sum of terms, resulting by by-parts integration in (Au,v), vanishes. A modern and extensive treatment of non-selfadjoint eigenvalue boundary value problem can be found in [MenM].

Appendix 4

Differential operators in $(L^2(a,b))^n$. Symmetrization of operators

Consider a system of ode's with constant coefficients containing unknown functions

$$\mathbf{AU} = \mathbf{0}, \qquad \mathbf{U} \in \mathcal{D}(A), \tag{4.1}$$

where $\mathbf{A}$ is a k-th order $n \times n$ matricial differential operator of entries a_{ij}, $\mathbf{U} = (u_1, \ldots, u_n) \in C^{k_1, \ldots, k_n}(a,b) \subset \left(L^2(a,b)\right)^n$ is the unknown vector function, $u_i : (a,b) \to \mathbf{R}$, $u_i = u_i(x)$. Here k_i is the maximum order of derivatives of u_i occurring in $\mathbf{A}$ and $k = k_1, \ldots, k_n$. Each u_i satisfies one or more boundary conditions at $x = a$ and $x = b$, therefore $u_i \in C^{q_i}[a,b]$, where q_i is the maximum order of derivatives of u_i occurring in these boundary conditions.

All these smoothness requirements and boundary conditions for $\mathbf{u}$ are contained in $\mathcal{D}(\mathbf{A})$. Some results valid for the case $n = 1$ (*i.e.* a single unknown function) hold for $n > 1$ too, some others, not.

For the case $n > 1$ we are first interested in the form of symmetric $\mathbf{A}$; skew-symmetric $\mathbf{A}$; symmetric part $\mathbf{A}_s$ and skew-symmetric part $\mathbf{A}_{ss}$ of $\mathbf{A}$, $\mathbf{A}^+$, $\mathbf{A}^*$ and then in the boundary conditions. Due to the theoretical importance of symmetric operators in finding variational principles, the problem of symmetrization of $\mathbf{A}$ is our main concern in view of associating with (4.1) the most convenient variational problems. Examples are given especially for the four problems from Section 3.3.1.

Lemma 4.1. *The unknown functions of* (4.1) *are indefinitely differentiable on* (a,b), *i.e.* $u_1, \ldots, u_n \in C^\infty(a,b)$.

Proof. We show how this lemma can be proved for problems $(3.3.1) - (3.3.7)$. Similar proofs hold for (4.1). Consider the system $(3.3.1) - (3.3.3)$. Since $v \in C^4(-0.5, 0.5)$ and $\theta \in C^2(-0.5, 0.5)$, from (3.3.1) it follows that $D^2\theta \in C^2(-0.5, 0.5)$ as a difference of two functions of $C^2(-0.5, 0.5)$, that is $\theta \in C^4(-0.5, 0.5)$. Again from (3.3.1), since $v, \theta \in C^4(-0.5, 0.5)$, it follows, by the same argument, that $D^2\theta \in C^4(-0.5, 0.5)$, whence $\theta \in C^6(-0.5, 0.5)$.

Then $D\theta \in C^5(-0.5, 0.5)$ and since $\phi \in C^2(-0.5, 0.5)$, from (3.3.3) it follows that $D^2\phi \in C^2(-0.5, 0.5)$, that is $\phi \in C^4(-0.5, 0.5)$. From the fact that $D\theta, \phi \in C^4(-0.5, 0.5)$, from (3.3.3) we have $D^2\phi \in C^4(-0.5, 0.5)$, that is $\phi \in C^6(-0.5, 0.5)$.

Again from (3.3.3), since $D\theta, \phi \in C^5(-0.5, 0.5)$, it follows that $D^2\phi \in$

353

$C^5(-0.5, 0.5)$, which means that $\phi \in C^7(-0.5, 0.5)$. In a similar way, from (3.3.2) it follows that $v \in C^{10}(-0.5, 0.5)$, which, together with (3.3.1), implies $\theta \in C^{12}(-0.5, 0.5)$, and the reasoning will continue with equation (3.3.3) and so on, which proves Lemma 4.1 for $(3.3.1) - (3.3.3)$. For any other system the proof proceeds similarly.

In a general setting, $\mathcal{D}(\mathbf{A})$, $\mathcal{D}(\mathbf{A}^*) \subset \mathbf{L}^2[-0.5, 0.5]$ where $\mathbf{L}^2[-0.5, 0.5]$ here stands for $\left[L^2[-0.5, 0.5]\right]^3$, and $v, \theta, \phi, v^*, \theta^*, \phi^*$ belong to various Sobolev spaces $W^{l,2}(-0.5, 0.5)$.

However, by Lemma 4.1 and Sobolev embedding theorems, it follows that l can be as large as we want and, so, the generalized setting becomes equivalent to the classical one.

We adopted the classical setting only for the brevity to express the boundary conditions which are fundamental for our study.

Let us show how to construct $\mathbf{A}^*$ for $\mathbf{A}$ in (4.1). Let $\mathbf{A}$ be a matricial differential symmetric operator in $L^2(a, b)$. The form (expression) of the adjoint operator $\mathbf{A}^*$ is equal to that of $\mathbf{A}$ (Appendix 3). Moreover, the boundary conditions (B), obtained as a result of the by-parts integrations performed in passing from $(\mathbf{AU}, \mathbf{U}^*)$ to $(\mathbf{U}, \mathbf{AU}^*)$, must be fulfilled. The condition (B) is expressed in terms of $\mathbf{U}$ and $\mathbf{U}^*$ at $x = a$ and $x = b$. In (B) we require $\mathbf{U} \in \mathcal{D}(\mathbf{A})$ and $\mathbf{U}^* \in \mathcal{D}(\mathbf{A}^*)$ and take into account the boundary conditions for $\mathbf{U}$ occurring in $\mathcal{D}(\mathbf{A})$. Then in (B) remain some components of $\mathbf{U}$ and their derivatives the value of which are arbitrary on boundary, *i.e.* at $x = a$ and $x = b$. Next we put together all terms in the same arbitrary component of $\mathbf{U}$ or some of its derivatives. Imposing their coefficients to vanish we obtain exactly the boundary conditions for $\mathbf{U}^*$ occurring in $\mathcal{D}(\mathbf{A}^*)$. If they are the same as the boundary conditions satisfied by $\mathbf{U}$, then the smoothness of $\mathbf{U}$ and $\mathbf{U}^*$ is the same. It follows that $\mathcal{D}(\mathbf{A}) = \mathcal{D}(\mathbf{A}^*)$, *i.e.* $\mathbf{A} = \mathbf{A}^*$ and, therefore, $\mathbf{A}$ is selfadjoint.

Lemma 4.2. *The unknown functions of* (4.1) *are indefinitely differentiable on* $[a, b]$, i.e. $u_1, \ldots, u_n \in C^\infty[a, b]$.

Proof. Consider first the two-point problem

$$D^2 w(x) - k^2 w(x) = h(x), \qquad x \in (-0.5, 0.5) \qquad (4.2)$$

$$w(-0.5) = w(0.5) = 0, \qquad (4.3)$$

where k is a non-vanishing constant and $h(x) \in C^0[-0.5, 0.5]$ is a known function. It has a unique solution in the class $C^2(-0.5, 0.5) \cap C^0[-0.5, 0.5]$. Indeed, let w_1 and w_2 be two solutions of (4.2), (4.3). Then their difference $u = w_1 - w_2$ satisfies the problem

$$D^2 u(x) - k^2 u(x) = 0, \qquad x \in (-0.5, 0.5) \qquad (4.2')$$

$$u(-0.5) = u(0.5) = 0. \qquad (4.3')$$

For $k \neq 0$ the general solution of (4.2') is $u = A\sinh(kx) + B\cosh(kx)$, which reduces to $u = 0$ for the boundary conditions (4.3'). Therefore $w_1 = w_2$. It is understood that we are concerned only with the classical solutions of (4.2), (4.3). This is why we supposed that h is a continuous function. The solution of (4.2), (4.3) can be written as $w(x) = r(x) - p(x)$ where r and p are the solutions of the following problems

$$D^2 r(x) - k^2 r(x) = h(x), \qquad x \in (-0.5, 0.5) \qquad (4.2'')$$

$$r(0.5) = Dr(0.5) = 0, \qquad (4.3'')$$

and

$$D^2 p(x) - k^2 p(x) = 0, \qquad x \in (-0.5, 0.5) \qquad (4.2''')$$

$$p(0.5) = 0, \qquad p(-0.5) = r(-0.5) = -\frac{1}{k}\int_{0.5}^{-0.5} h(\xi)\sinh(k/2 + k\xi)d\xi. \qquad (4.3''')$$

The Cauchy problem in r consists of a nonhomogeneous equation and two homogeneous initial conditions, while the two-point problem in p consists of a homogeneous equation and two conditions: one homogeneous at $x = +0.5$ and the other nonhomogeneous at $x = -0.5$.

By the variation of coefficients method it follows that they have the following unique solutions

$$r(x) = \frac{1}{k}\int_{0.5}^{x} h(\xi)\sinh[k(x - \xi)]d\xi, \qquad (4.4)$$

$$p(x) = \frac{\sinh[k(x - 0.5)]}{k\sinh k}\int_{0.5}^{-0.5} h(\xi)\sinh[k(1/2 + \xi)]d\xi, \qquad (4.4')$$

whence the expression for the solution of (4.2), (4.3)

$$w(x) = \frac{1}{k}\int_{0.5}^{x} h(\xi)\sinh[k(x - \xi)]d\xi$$
$$- \frac{\sinh[k(x - 0.5)]}{k\sinh k}\int_{0.5}^{-0.5} h(\xi)\sinh[k(1/2 + \xi)]d\xi. \qquad (4.4'')$$

From (4.4'') it can be seen that if $h(x) \in C^k[-0.5, 0.5]$, then $w(x) \in C^{k+2}[-0.5, 0.5]$.

Equation (3.3.1) and the boundary conditions (3.3.4) or (3.3.7) give rise to a problem of the form (4.2), (4.3) where $w = \theta$ and $h = -v$. It follows that $\theta \in C^2[-0.5, 0.5]$, whence we have $D\theta \in C^1[-0.5, 0.5]$. Then we can say that equation (3.3.3) is of the form (4.2) where $w = \phi$ and $h = -D\theta$. In the conditions (3.3.4) or (3.3.7) we deduce that $\phi \in C^2[-0.5, 0.5]$ and, consequently, $D\phi \in C^1[-0.5, 0.5]$. In this way equation (3.3.2), in the condition (3.3.4), can be put in the form (4.2), where $w = (D^2 - k^2)v$ and $h = k^2 R(1 + El)\theta + k^2 RElD\phi$, which implies $(D^2 - k^2)v \in C^2[-0.5, 0.5]$. But $v \in C^0[-0.5, 0.5]$, whence $D^2 v \in C^0[-0.5, 0.5]$, which gives $v \in C^2[-0.5, 0.5]$ and, again, by the same reasoning, it follows that $D^2 v \in C^2[-0.5, 0.5]$, which implies $v \in C^4[-0.5, 0.5]$. Coming back to equation (3.3.1), we can continue the above reasonings. Finally we obtain the result stated by Lemma 4.2 for the problem $(3.3.1) - (3.3.4)$. In the case of the conditions (3.3.7), a direct reasoning, similar to that for (4.2), (4.3) on a fourth order equation in v, gives $v \in C^4[-0.5, 0.5]$.

Consider now the problem (3.3.1), (3.3.6). In view of Lemma 3.3.1, we can differentiate (4.2) to obtain an equation of the form (4.2), where $w = D\theta$ and $h = -Dv$. Again we may conclude, exactly as in the above, that $D\theta \in C^2[-0.5, 0.5]$ and finally $v, \theta, \phi \in C^\infty[-0.5, 0.5]$.

In the case of the boundary conditions (3.3.5), instead of the problem (3.3.3), (3.3.5) we consider the problem corresponding to (3.3.5) and to the equation obtained by differentiating (3.3.3); now $D\phi$ stands for w and $-D^2\theta$ for h etc.

Thus Lemma 4.2 is proved for the four problems in Section 3.3.1. Similar reasonings hold for (4.1).

Remark 4.1. By Lemma 4.2, from equations $(3.3.1) - (3.3.3)$, we can deduce some additional boundary conditions taking into account the given boundary conditions. These extra conditions appear as consistency conditions between the equations and the boundary conditions, since, owing to the smoothness on $[-0.5, 0.5]$, the limits as $x \to \pm 0.5$ of the terms of the equations are just the values of these terms at $x = \pm 0.5$. If the terms in equations would have been only from $C^k(-0.5, 0.5)$, then from the equations we could deduce only the limit of some combinations of the solution and some of its derivatives and not their values at $x = \pm 0.5$.

Remark 4.2. By appropriate differentiations, Lemma 4.2 enables us to reduce the system $(3.3.1) - (3.3.3)$ to an equation in v, or θ, or ϕ. This equation has the same form for v, θ and ϕ, however the boundary conditions satisfied by each of these functions differ from the boundary conditions satisfied by the others. Part of these boundary conditions are deduced (see Remark 4.1) directly from the equations and/or their derivatives.

In the case of more complicated problems (4.1), the boundary conditions must also be accounted for, and the solutions $(4.4'')$ of problems $(4.2) - (4.3)$ must be used if the obtained equation for one unknown function is of an order higher than for other unknown functions.

Lemma 4.3. *Splitting the unknown functions in even and odd parts, the problem (4.1) splits into two other distinct problems: one problem for some among these parts and the other problem for other parts. If derivatives in the equations and boundary conditions in (4.1) are only even or only odd, then (4.1) splits in an even problem, for the even parts, and an odd problem, for the odd parts.*

Proof. This lemma follows immediately remarking that every function can be uniquely written as a sum of two functions, one odd and the other even, and taking into account the fact that an even function is equal to an odd one if and only if they are equal to the null function.

More precisely, consider a two-point problem $B_e f + B_o f = 0$ at $x = \pm 0.5$ for a linear vector ode $L_e f + L_o f = 0$, for $x \in (-0.5, 0.5)$, where B_e, B_o, L_e and L_o are linear matricial differential operators with constant coefficients and f is the unknown vector function, such that B_e and L_e contain even-order derivatives only and B_o and L_o contain odd-order derivatives only. Let $f = f_e + f_o$, where $f_e = 0.5[f(x) + f(-x)]$ and $f_o = 0.5[f(x) - f(-x)]$. Hence f_e is an even function and f_o is an odd function. Since an even-order derivative of an even function is an even function, an odd-order derivative of an even function is an odd function, an even-order derivative of an odd function is an odd function and an odd-order derivative of an odd function is an even function, it follows that $0 = L_e f + L_o f = L_e f_e + L_o f_o + L_e f_o + L_o f_e$, where $L_e f_e + L_o f_o$ is an even function and $L_e f_o + L_o f_e$ is an odd function. As their sum is the null function, it follows that each of these functions is null. A similar result holds for the boundary conditions. Consequently, the given problem reduces to the two-point problem

$$B_e f_e + B_o f_o = 0, \qquad B_e f_o + B_o f_e = 0, \qquad \text{at } x = \pm 0.5$$

for the ode's

$$L_e f_e + L_o f_o = 0, \qquad L_e f_o + L_o f_e = 0, \qquad \text{for } x \in (-0.5, 0.5),$$

respectively.

These equations and boundary conditions are of the same order as those in the given problem but their number doubled. However, symmetry properties make the obtained problem of the same difficulty as the given problem. The advantage of the obtained problem is that it allows easier and somehow formal computations. This is due to the fact that in each equation and boundary condition the even and odd parts of the components of f occur in a special way, leading to the splitting of the given problem to two distinct problems.

For instance, the problem $(3.3.1) - (3.3.3)$, $(3.3.4)$ reads

$$\begin{cases} (D^2 - k^2)\theta_e + v_e = 0, \\ (D^2 - k^2)^2 v_e - k^2 R(1 + El)\theta_e - k^2 RElD\phi_o = 0, \quad x \in (-0.5, 0.5) \\ (D^2 - k^2)\phi_o + D\theta_e = 0, \\ v_e = D^2 v_e = \theta_e = \phi_o = 0, \quad \text{at } x = \pm 0.5 \end{cases}$$

$$\begin{cases} (D^2 - k^2)\theta_o + v_o = 0, \\ (D^2 - k^2)^2 v_o - k^2 R(1 + El)\theta_o - k^2 RElD\phi_e = 0, \quad x \in (-0.5, 0.5) \\ (D^2 - k^2)\phi_e + D\theta_o = 0, \\ v_o = D^2 v_o = \theta_o = \phi_e = 0, \quad \text{at } x = \pm 0.5 \end{cases}$$

hence it consists of two problems: one in v_e, θ_e and ϕ_o and the other in v_o, θ_o and ϕ_e as stated in Lemma 4.3.

Remark 4.3. In fluid dynamics, the relation between boundary conditions and the equations via the class of the solution is extremely important since it concerns, in particular, the connection between the motion inside a domain and the motion of the boundaries. The motion of the boundaries is sometimes the cause of the motion inside the domain, on one hand, and on the other one the adherence at the wall decelerates the fluid flow by friction. From the mathematical point of view this is reflected by the following two cases:

a) due to the corresponding smoothness of the solution, the differential equations are valid only inside the domain $\Omega \subset R^n$, while on its closure $\overline{\Omega} = \Omega \cup \partial\Omega$ ($\partial\Omega$ stands for the boundary of Ω) the solution together with some of its derivatives is only continuous. As a consequence, we can say nothing about the value of the other derivatives of the solutions at the wall. In particular, these derivatives cannot exist on $\partial\Omega$;

b) if the solution and its derivatives occurring in the equations are continuous on $\overline{\Omega}$, then we know the value of more linear combinations of the derivatives of the solution on $\partial\Omega$. Then the equations and the conditions are more closely linked than in the case a). Introducing the boundary conditions in the equations we get consistency conditions which express the assigned values on $\partial\Omega$ of combinations of the derivatives of the solution other than those given by the boundary conditions.

In this way we have extra conditions which diminish the set of solutions. Unlike the case a) now the solutions are determined not only by the equations and the boundary conditions but also by these extra conditions.

In the case of the nonuniqueness, the solutions from b) will be those solutions from a) which satisfy these additional boundary conditions and, therefore, correspond to a more particular physical situation or to none. If in both settings a) and b) of the same physical situation, the solution exists and is unique, then the second setting will give *a priori* behavior of the solution, expressed by the above mentioned additional boundary conditions.

Remark 4.4. The infinite regularity of u_i on $[a, b]$ allows one to write (4.1) in the form of a two-point problem for a single equation, say in u_1, $L_1 u_1 = 0$. The lacking boundary conditions are obtained from equations. The same equation is satisfied by all u_i, $i = 1, \dots, n$, *i.e.* $L_1 u_i = 0$, $i = 1, \dots, n$, but the associated boundary conditions are different.

Sometimes, for some u_s, it may appear that the equation has a higher-order than for other u_i, *e.g.* $L_2 L_1 u_s = 0$. However, in this case, the use of the boundary conditions and of the form (4.2)–(4.3) show that $L_2(L_1 u_s) = 0$ has a unique solution $L_1 u_s = 0$, *i.e.* u_s satisfies the same equation like all other unknown functions.

Remark 4.5. Assume that the entries a_{ij} of the operator A in (4.1) are $a_{ij} = \sum_{s=1}^{k} a_{ij}^s D^s$, where $D^s = \dfrac{d^s}{dx^s}$ and a_{ij}^s are constants. Assume that the vector functions $\mathbf{U} = (u_1, \dots, u_n)$ of $\mathcal{D}(A)$ are defined on $[-0.5, 0, 5]$ (*i.e.* $a = -0.5$,

$b = 0.5$). The operator $A^T : \mathcal{D}(A) \rightarrow H$, the entries a_{ij}^T of which are defined by $a_{ij}^T = a_{ji}$ is referred to as the *transpose* of A. If A contains only even-order derivatives and the sum (B)

$$\left[\sum_{i=1}^{n}\sum_{j=1}^{n}\sum_{s=0}^{k-1}(-1)^s a_{ij} D^s u_j D^{k-s-1} u_j^*\right]_{-0.5}^{0.5} = 0, \tag{B}$$

of the terms occurring as a result of the by-parts integration and taken at the boundaries $x = \pm 0.5$ vanish, then A is symmetric if and only if $A = A^T$. In this case the matrix A is symmetric. The entries a_{ji}^* of the adjoint operator A^* and, therefore, of A^+, are $a_{ji}^* = \sum_{s=1}^{n}(-1)^s a_{ij}^s D^s$. Therefore A is selfadjoint if and only if $a_{ij} = a_{ij}^*$ and $\mathcal{D}(A) = \mathcal{D}(A^*)$, i.e if and only if $\mathcal{D}(A) = \mathcal{D}(A^*)$ and the following *rule* holds: *the coefficients of the even-order derivatives in a_{ij} and a_{ji} are the same while the coefficients of odd order derivatives in a_{ij} and a_{ji} are opposite.* Since a selfadjoint operator is symmetric and its form is equal to that of A^*, it follows that this rule holds for symmetric operators too.

This rule must be supplemented with the boundary condition (B); otherwise natural conditions occur and with the given eigenvalue problem, defined by A, we cannot associate a variational principle in the simplest way. If the operator A contains only even-order derivatives and it is symmetric then $A = A^T = A^*$. In this case the associated differential matrix is symmetric.

In general, the symmetry of the operator A does not imply the symmetry of the associated matrix A. More exactly, if the operator A is symmetric, then the associated matrix is symmetric in the even derivatives and skew-symmetric in the odd derivatives.

If the operator A is symmetric and contains only odd-order derivatives, then the corresponding matrix A is skew-symmetric, *i.e.* $a_{ij} = -a_{ji}$. In particular, all entries of the main diagonal are vanishing. On the contrary, a skew-symmetric operator containing only odd-order derivatives is associated with a matrix symmetric in these derivatives. Finally, a skew-symmetric operator containing only even-order derivatives is associated with a skew-symmetric matrix.

Remark 4.6. Sometimes, a nonsymmetric matricial ordinary differential linear operator can become symmetric by multiplying one or more equations in its corresponding system of equations by some constants. This can be seen from the expressions of a_{ij} (Remark 4.5).

Remark 4.7. Unlike the case $n = 1$, a matricial differential operator A can contain even- as well as odd-order derivatives.

Appendix 5

Fourier series expansions

Total sets in $L^2(a,b)$. Backward integration technique. As we saw, the spaces $L^2(a,b)$ and $L^2(\Omega)$, in which we frequently work in this book, are separable Hilbert spaces, therefore the entire theory of Fourier series applies.

Fourier series are not termwise differentiable in general. However, if a Fourier series is uniformly convergent, then it is termwise integrable. If the separable Hilbert space is $H = L^2(-0.5, 0.5)$, this enables one to determine the Fourier coefficients of the derivative in terms of the Fourier coefficients of the function. Thus, let $u = \sum_{n=1}^{\infty} u_n e_n$ be an expansion in Fourier series of a function u with respect to the total orthonormal set $\{e_n\}$, $e_n \in H$, therefore $u_n = (u, e_n)$. Let $\{g_n\}$ be another total set in H such that, eventually, except for the first term and up to some constant factor $g_n = De_n$, where D stand for the derivative with respect to the independent variable x. Obviously, at least for total set of sines and cosines, the only one of interest for us, with the same possible exception, $Dg_n = e_n$. Then $Du = \sum_{n=1}^{\infty} u_n^{(1)} g_n$, where $u_n^{(1)} = (Du, g_n)$. But $\int_{-0.5}^{0.5} Du(x) g_n(x) dx = [u(x)g_n(x)]_{-0.5}^{0.5} - \int_{-0.5}^{0.5} u(x) Dg_n(x) dx = [u(x)g_n(x)]_{-0.5}^{0.5} - \int_{-0.5}^{0.5} u(x)e_n(x) dx = [u(x)g_n(x)]_{-0.5}^{0.5} - u_n$, i.e. $u_n^{(1)} = [u(x)g_n(x)]_{-0.5}^{0.5} - u_n$. This method of obtaining $u_n^{(1)}$ in terms of u_n and the boundary conditions for u is referred to as the *backward integration technique* [DiP61]. It is fundamental for the following because it reduces an ode to an algebraic equation in the Fourier coefficients of the solution only.

The unknown functions $\mathbf{U}$ and their derivatives $D^k \mathbf{U}$, $k = 1, 2, \ldots, l$, occurring in the boundary-value problem in Sections 3.3 and 3.4, belong to $\mathbf{L}^2(-0.5, 0.5)$, therefore $\mathbf{U} \in W^{l,2}(-0.5, 0.5)$. In addition, $\mathbf{U}$ and $D^k \mathbf{U}$ are smooth on $[-0.5, 0.5]$ and certain of their linear combinations are supposed to vanish at $x = \pm 0.5$. The existing Fourier series theory [Fich], [Fri2], [How], [ButzN] concerns real periodic functions on $\mathbf{R}$. Hence the first step in applying this theory is to extend $\mathbf{U}$ from $[-0.5, 0.5]$ to $\mathbf{R}$ by periodicity up to $\mathbf{U}$. To each total set of expansion functions a specific periodicity and symmetry properties correspond. We consider two types of symmetries: with respect to some points $\bar{x}$ and with respect to some straight lines $x = \bar{x}$. Namely, a function $f : \mathbf{R} \to \mathbf{R}$ $f = f(x)$ is skew-symmetric with respect to the point $\bar{x}$ (respectively symmetric with respect to a straight line $x = \bar{x}$) if the

361

new function $h(x) \equiv f(x - \bar{x})$ is odd with respect to the origin (respectively even with respect to the axis $x = 0$). For instance, in using the total set $\{E_{2n-1}\}_{n=1}^{\infty}$, where $E_{2n-1}(x) = \sqrt{2}\cos[(2n - 1)\pi x]$, the extended periodic function is even with respect to x and its graph is symmetric with respect to the points $(0.5 + k, 0)$, $k \in \mathbf{Z}$ and has the (main) period equal to 2. Similarly, the total set $\{F_{2n-1}\}_{n=1}^{\infty}$, where $F_{2n-1}(x) = \sqrt{2}\sin[(2n - 1)\pi x]$, is used for functions which are symmetric with respect to the straight lines $x = 0.5 + k$, $k \in \mathbf{Z}$ and which are odd functions of x. Other total sets involving expansion functions $\cos\frac{(2n-1)\pi}{4}x$ can lead to periods equal to 8 etc.

As a consequence of the *continuation by periodicity* implying certain symmetries, the obtained function $\hat{\mathbf{U}}$ defined on $\mathbf{R}$ may happen to be no longer continuous even if the function defined on $[-0.5, 0.5]$ was smooth. For instance, the even functions on $[-0.5, 0.5]$ extended up to $\mathbf{R}$ and expanded on the functions E_{2n-1} undergo jumps at $x = 0.5 + k$, $k \in \mathbf{Z}$ and, so, their first derivative is no longer a Lebesgue integrable function but it can be associated with a singular distribution, *e.g.* the Dirac δ. In fact, if the function f has a jump, then its derivative will be $\frac{df}{dz} = \left(\frac{df}{dz}\right) + \sum_{j=1}^{k} S_j \delta x_j$, where $\left(\frac{df}{dz}\right) \in L^2$. In this way, the theory of Fourier series goes beyond the Sobolev spaces theory because the derivatives are no longer regular functionals, *i.e.* of function type as in this last theory.

Every such continuation must not change the smoothness and symmetry properties of $\mathbf{U}$ *in* $[-0.5, 0.5]$; *if it does, then the chosen total set is not appropriate.*

The sum $\mathbf{S}$ of the Fourier series is defined at every $x \in \mathbf{R}$. The relationship between it and the function $\mathbf{U}$ itself is $\mathbf{S}(x) = \frac{\hat{\mathbf{U}}(x-0)+\hat{\mathbf{U}}(x+0)}{2}$ and it is called the Dirichlet sum. If $\hat{\mathbf{U}}$ is continuous at x, then $\hat{\mathbf{U}}(x)$ is equal to $\mathbf{S}(x)$. If $\hat{\mathbf{U}}$ is not continuous at x, then several situations can occur. For instance, in the case of the skew-symmetry of $\hat{\mathbf{U}}$ with respect to the points $x = \pm0.5$, the sum $\mathbf{S}$ can be equal to 0 while $\mathbf{U}(\pm0.5)$ is non-vanishing. It is only if $\hat{\mathbf{U}}(\pm0.5) = 0$ that this value coincides with $\mathbf{S}(\pm0.5)$ and so, $\mathbf{S}$ yields the true value of $\mathbf{U}$ at the boundary. On the contrary, if $\hat{\mathbf{U}}(\pm0.5) = 0$ and $\hat{\mathbf{U}}$ is an odd function such that the expansion functions are $F_{2n-1}(x)$, which are not vanishing at $x = \pm0.5$, then we must impose the constraint that the sum of the series vanishes even if every expansion function $F_{2n-1}(x)$ does not vanish at $x = \pm0.5$. This constraint reads $\Gamma_1 \equiv \sum_{n=1}^{\infty} \hat{\mathbf{U}}_n F_{2n-1}(\pm0.5) = 0$. If $\hat{\mathbf{U}}$ belongs to the domain of definition of a functional, in a variational problem, the constraint $\Gamma_1 = 0$ turns it into an isoperimetric problem. This shows that in handling with Fourier series in variational principles we must be careful just at their sensitive points: the boundary conditions.

It also follows that there are two types of methods based on Fourier series. The first concerns total sets of expansion functions satisfying all boundary conditions of the given problem [HarR]. Basically they are Galerkin-Ritz procedures. When applied to hydromagnetic and hydrodynamic stability they are known as *Chandrasekhar-Galerkin method* (C-G). In this case all boundary conditions are satisfied termwise by the series. The second type concerns expansion functions satisfying only part of the boundary conditions and, so, they lead to isoperimetric problems. The numerical advantage of this method have been pointed out by DiPrima [DiP61]. Starting with 1948, it was applied to problems of elastic stability in pioneering works by B. Budiansky [BudHC], [BudK], [BudDiP]. This is why we shall refer to this technique as the *Budiansky-DiPrima* (B-D) *method.* In complicated problems, the B-D method is preferable to C-G methods. Among others, we proved this, in particular for problems $(3.3.1) - (3.3.7)$, in [Geo87], where the convergence of the involved series was proved too.

In variational problems, suitable total sets can be constructed by taking into account that, by Theorem 1.1, if a linear subset M of $L^2(a,b)$ contains $C_0^\infty(a,b)$ (which is dense in $L^2(a,b)$) then M is dense in $L^2(a,b)$ too and, therefore, total in $L^2(a,b)$ [Mikh5]. This M may consist in smooth functions satisfying certain boundary conditions.

Expansion functions E_{2n-1} and F_{2n-1}. Among the various possibilities of extending an even function $w \in C^\infty[-0.5, 0.5]$ up to a periodic function $\hat{w} : \mathbf{R} \to \mathbf{R}$, we choose the one which leads to Fourier series each term of which vanishes at $x = \pm 0.5$. This comes to define $\hat{w}$ as a function even about the straight line $x = 0$, odd about the point $x = -0.5$, and with period equal to 2. By Fourier series theory [Fikh] we have, in the L^2 sense, the equality

$$\hat{w}(x) = \sum_{n=1}^{\infty} \hat{w}_{2n-1} E_{2n-1}(x), \tag{5.1}$$

where $E_{2n-1} = \sqrt{2}\cos[(2n-1)\pi x]$, $\hat{w}_{2n-1} = \frac{1}{2}\int_\alpha^{\alpha+2} \hat{w}(x)E_{2n-1}(x)dx$, $n = 1, 2, \ldots, \alpha \in \mathbf{R}$; this series converges on $\mathbf{R}$. Denoting by $S(x)$ its Dirichlet sum, we have $S(x) = \frac{\hat{w}(x-0)+\hat{w}(x+0)}{2}$, $\forall x \in \mathbf{R}$. Owing to the symmetry properties specified in the above we obtain $\hat{w}_{2n-1} = \int_{-0.5}^{0.5} \hat{w}(x)E_{2n-1}(x)dx$, $w(x) = \hat{w}(x) = S(x)$ for $x \in (-0.5, 0.5)$ and $S(\pm 0.5) = 0$. This last equality follows from the fact that at $x = \pm 0.5$ the function $\hat{w}$ has jumps passing from a value to its opposite (for instance $\hat{w}(0.5-0) = w(0.5)$, $\hat{w}(0.5+0) = -w(0.5)$). Hence, generally, $\hat{w} \in C^\infty(\frac{2n-1}{2}, \frac{2n+1}{2})$, $n \in \mathbf{Z}$, while at the points $x = \frac{2n-1}{2}$, it has jumps. If $w(0.5) = 0$, then $\hat{w} \in C(R)$.

The same properties of symmetry and periodicity are valid for $D^{2l}w$ and $\widehat{D^{2l}w}$, $l = 0, 1, \ldots$. Consequently, the functions w and $D^{2l}w$, which are continuous on $[-0.5, 0.5]$, can be expressed as

$$D^{2l}w(x) = \begin{cases} \widehat{D^{2l}w}(x) = \sum_{n=1}^{\infty} \hat{w}_{2n-1}^{(2l)} E_{2n-1}(x), & \text{for } x \in (-0.5, 0.5), \\ \\ D^{2l}w(0.5), & \text{at } x = \pm 0.5, \end{cases} \tag{5.2}$$

when $D^{2l}w(0.5) \neq 0$, and

$$D^{2l}w(x) = \widehat{D^{2l}w}(x) = \sum_{n=1}^{\infty} \hat{w}_{2n-1}^{(2l)} E_{2n-1}(x), \quad x \in [-0.5, 0.5], \tag{5.2'}$$

when $D^{2l}w(0.5) = 0$, $l = 0, 1, 2, \ldots$.

Similarly, if $u \in C^{\infty}[-0.5, 0.5]$ is an odd function we can extend it to R such that its extension $\hat{u}$ is an odd function about the straight line $x = 0$ and an even function about the point $x = -0.5$ and periodic with period 2. An analogous reasoning leads us to

$$\hat{u}(x) = \sum_{n=1}^{\infty} \hat{u}_{2n-1}(x) F_{2n-1}(x), \tag{5.3}$$

where $F_{2n-1} = \sqrt{2} \sin[(2n-1)\pi x]$, $\hat{u}_{2n-1} = \int_{-0.5}^{0.5} \hat{u}(x) F_{2n-1}(x) dx$, $n = 1, 2, \ldots$. In addition, $\hat{u}(x) \in C(\mathbf{R})$, $u(x) = \hat{u}(x) = S(x)$ on $[-0.5, 0.5]$, hence we have

$$D^{2l}u(x) = \widehat{D^{2l}u(x)} = \sum_{n=1}^{\infty} \hat{u}_{2n-1}^{(2l)}(x) F_{2n-1}(x), \quad \text{on} \quad x \in [-0.5, 0.5]. \tag{5.4}$$

Remark that in this case jumps could occur only at the points $x = n$, $n \in \mathbf{Z}$, about which the function $\hat{u}$ (and also $\widehat{D^{2l}u}$) is odd; but, due to the fact that $u(0) = 0$ (u is a continuous function on $[-0.5, 0.5]$ and odd about $x = 0$) and that $\hat{u}(n) = u(0)$ (by construction), it follows that the jumps are equal to zero.

The derivatives $D^{2l+1}w$, $l = 0, 1, 2, \ldots$, have the same properties as u, so from the above analysis it follows that $D^{2l+1}w \in C^{\infty}(\mathbf{R})$ and the formulae

$$D^{2l+1}w(x) = \widehat{D^{2l+1}w}(x) = \sum_{n=1}^{\infty} \hat{w}_{2n-1}^{(2l+1)}(x) F_{2n-1}(x), \quad x \in [-0.5, 0.5], \tag{5.5}$$

hold. The derivatives $D^{2l+1}u$, $l = 0, 1, 2, \ldots$, have the same properties as w, so $D^{2l+1}u \in C^{\infty}(\frac{2n-1}{2}, \frac{2n+1}{2})$, $n \in \mathbf{Z}$ and

$$D^{2l+1}u(x) = \begin{cases} \widehat{D^{2l+1}u}(x) = \sum_{n=1}^{\infty} \hat{u}_{2n-1}^{(2l+1)} E_{2n-1}(x), & \text{for} \quad x \in (-0.5, 0.5), \\ D^{2l+1}u(0.5), & \text{at} \quad x = \pm 0.5 \end{cases} \tag{5.6}$$

for $D^{2l+1}u(0.5) \neq 0$, and

$$D^{2l+1}u(x) = \widehat{D^{2l+1}u}(x) = \sum_{n=1}^{\infty} \hat{u}_{2n-1}^{(2l+1)} E_{2n-1}(x), \quad x \in [-0.5, 0.5], \tag{5.6'}$$

for $D^{2l+1}u(0.5) = 0$, $l = 0, 1, 2, \ldots$.

The above-analyzed continuous functions w, $D^k w$, u, $D^k u$, $k = 0, 1, 2, \ldots$, have bounded derivatives on $[-0.5, 0.5]$ hence they have bounded variation. It follows that [Fikh] their Fourier series are uniformly convergent on $[-0.5, 0.5]$ and, thus, these series may be integrated termwise. By backward integration technique we have

$$
\begin{cases}
\hat{w}^{(2k)}_{2n-1} = \int_{-0.5}^{0.5} \widehat{D^{2k}w}(x)E_{2n-1}(x)dx = \int_{-0.5}^{0.5} D^{2k}w(x)E_{2n-1}(x)dx \\[2mm]
= [D^{2k-1}wE_{2n-1}]_{-0.5}^{0.5} - \int_{-0.5}^{0.5} \widehat{D^{2k-1}w}(x)DE_{2n-1}(x)dx = (2n-1)\pi\hat{w}^{(2k-1)}_{2n-1}, \\[2mm]
\hat{u}^{(2k+1)}_{2n-1} = \int_{-0.5}^{0.5} \widehat{D^{2k+1}u}(x)E_{2n-1}(x)dx = \int_{-0.5}^{0.5} D^{2k+1}u(x)E_{2n-1}(x)dx \\[2mm]
= [D^{2k}uE_{2n-1}]_{-0.5}^{0.5} - \int_{-0.5}^{0.5} \widehat{D^{2k}u}(x)DE_{2n-1}(x)dx = (2n-1)\pi\hat{u}^{(2k)}_{2n-1},
\end{cases}
\tag{5.7}
$$

hence the termwise differentiation holds for $\widehat{D^{2k-1}w}$ and for $\widehat{D^{2k}u}$.

In the same way, we get

$$
\begin{aligned}
\hat{w}^{(2k+1)}_{2n-1} &= \int_{-0.5}^{0.5} \widehat{D^{2k+1}w}(x)F_{2n-1}(x)dx = \int_{-0.5}^{0.5} D^{2k+1}w(x)F_{2n-1}(x)dx \\[2mm]
&= [D^{2k}wF_{2n-1}]_{-0.5}^{0.5} - \int_{-0.5}^{0.5} \widehat{D^{2k}w}(x)DF_{2n-1}(x)dx \\[2mm]
&= 2\sqrt{2}(-1)^{n+1}D^{2k}w(0.5) - (2n-1)\pi\hat{w}^{(2k)}_{2n-1},
\end{aligned}
\tag{$5.8)_1$}
$$

$$
\begin{aligned}
\hat{u}^{(2k)}_{2n-1} &= \int_{-0.5}^{0.5} \widehat{D^{2k}u}(x)F_{2n-1}(x)dx = \int_{-0.5}^{0.5} D^{2k}u(x)F_{2n-1}(x)dx \\[2mm]
&= [D^{2k-1}uF_{2n-1}]_{-0.5}^{0.5} - \int_{-0.5}^{0.5} \widehat{D^{2k-1}u}(x)DF_{2n-1}(x)dx \\[2mm]
&= 2\sqrt{2}(-1)^{n+1}D^{2k-1}u(0.5) - (2n-1)\pi\hat{u}^{(2k-1)}_{2n-1},
\end{aligned}
\tag{$5.8)_2$}
$$

therefore we cannot differentiate termwise the derivatives of $\widehat{D^{2k}w}$ and $\widehat{D^{2k-1}u}$.

The above analysis can be done in terms of $\widehat{D^{k}\hat{w}}$ instead of $\widehat{D^{k}w}$; only slight differences would appear.

Expansion functions E_{2n} and F_{2n}. Here we follow the same presentation lines as in the previous point: first we mention symmetry and periodicity properties of the extended function, then we relate the expanded function and its derivatives to the sum of the corresponding series and, finally, we provide the relationships between the Fourier coefficients of consequent derivatives, separately for the odd and the even functions and for their odd and even derivatives.

In our computations we use the above analyzed Fourier expansions for functions vanishing at $x = \pm 0.5$, while for functions which are non-vanishing at these points we take

$$
\hat{w} = \hat{w}_0 + \sum_{n=1}^{\infty} \hat{w}_{2n}E_{2n}(x),
\tag{5.9}
$$

for even functions $w \in C^\infty[-0.5, 0.5]$, by extending them to an even function $\hat{w} \in C(\mathbf{R})$, symmetric with respect to the straight lines $x = 0.5 \cdot k$, $k \in \mathbf{Z}$, and periodic with period 1, and

$$\hat{u} = \sum_{n=1}^{\infty} \hat{u}_{2n} F_{2n}(x), \tag{5.10}$$

for odd functions $u \in C^\infty[-0.5, 0.5]$, by extending them to an odd function

$$\hat{u} \in C^\infty\left(\frac{2n-1}{2}, \frac{2n+1}{2}\right), \qquad n \in \mathbf{Z},$$

symmetric with respect to the points $x = k$, $k \in \mathbf{Z}$ and periodic with period 1. By the same arguments as above we can write

$$w = \begin{cases} \hat{w} = \hat{w}_0 + \displaystyle\sum_{n=1}^{\infty} \hat{w}_{2n} E_{2n}(x), & x \in (-0.5, 0.5) \\[2mm] w(0.5), & \text{at} \quad x = \pm 0.5 \end{cases} \tag{5.11}$$

$$Dw = \begin{cases} \widehat{Dw} = \displaystyle\sum_{n=1}^{\infty} \hat{w}_{2n}^{(1)} F_{2n}(x), & x \in (-0.5, 0.5) \\[2mm] Dw(0.5), & \text{at} \quad x = 0.5 \\[2mm] -Dw(0.5), & \text{at} \quad x = -0.5 \end{cases} \tag{5.12}$$

where $\hat{w}_0 = \int_{-0.5}^{0.5} \hat{w}(x)dx$, $\hat{w}_{2n}^{(1)} = -2n\pi\hat{w}_{2n}$. Imposing to w to satisfy the condition $w(\pm 0.5) = 0$ we obtain the constraint

$$\hat{w}_0 + \sum_{n=1}^{\infty} \hat{w}_{2n}(-1)^n \sqrt{2} = 0, \tag{5.13}$$

since, on one hand, $w(-0.5 + 0)$ is equal to $w(-0.5) = 0$ by the continuity of w and, on the other hand, it is equal to the right-hand side of (5.13) (this follows from (5.11)).

The relationship between the Fourier coefficients of the successive derivatives,

$$\begin{cases} D^{2k+1}w(x) = \widehat{D^{2k+1}w}(x) = \displaystyle\sum_{n=1}^{\infty} \hat{w}_{2n}^{(2k+1)} F_{2n}(x), \\[2mm] D^{2k}w(x) = \widehat{D^{2k}w}(x) = \hat{w}_0^{(2k)} + \displaystyle\sum_{n=1}^{\infty} \hat{w}_{2n}^{(2k)} E_{2n}(x), \end{cases} \tag{5.14}$$

are

$$\begin{cases} \hat{w}_0^{(2k)} = 2D^{2k-1}w(0.5), & \hat{w}_{2n}^{(2k+1)} = -2n\pi\hat{w}_{2n}^{(2k)}, \\[2mm] \hat{w}_{2n}^{(2k)} = (-1)^n 2\sqrt{2}D^{2k-1}w(0.5) + 2n\pi\hat{w}_{2n}^{(2k-1)} \end{cases} \tag{5.15}$$

for even functions $\hat{w}$ and

$$\begin{cases} D^{2k+1}u(x) = \widehat{D^{2k+1}u}(x) = u_0^{(2k+1)} + \displaystyle\sum_{n=1}^{\infty} \hat{u}_{2n}^{(2k+1)} E_{2n}(x), \\[2mm] D^{2k}u(x) = \widehat{D^{2k}u}(x) = \displaystyle\sum_{n=1}^{\infty} \hat{u}_{2n}^{(2k)} F_{2n}(x), \end{cases} \tag{5.16}$$

where

$$\hat{u}_0^{2k+1} = 2D^{2k}u(0.5), \quad \hat{u}_{2n}^{(2k+1)} = (-1)^n 2\sqrt{2}D^{2k}u(0.5) + 2n\pi\hat{u}_{2n}^{(2k)}, \quad \hat{u}_{2n}^{(2k)}$$
$$= -2n\pi\hat{u}_{2n}^{(2k-1)}$$

(5.17)

for odd functions $\hat{u}$.

Remark 5.1. In the computations we take into account that $\{E_1, E_3, \dots\}$ and $\{F_1, F_3, \dots\}$ are orthonormal sets on $[-0.5, 0.5]$; $\{1, E_1, E_3, \dots\}$ is an orthonormal set on $[-1, 1]$; $\{1, E_2, E_4, \dots\}$ and $\{F_2, F_4, \dots\}$ are orthonormal sets on $[-0.5, 0.5]$, $\{1, E_1, E_2, \dots\}$, and $\{F_1, F_2, \dots\}$ are orthonormal sets on $[0, 1]$; $\{1, E_{1/2}, E_{3/2}, \dots\}$, and $\{F_{1/2}, F_{3/2}, \dots\}$ are orthonormal sets on $[0, 1]$. In fact, the above analyzed expansions are based on the completeness of these sets in the class of even (E_i) or odd (F_i) functions of $L^2(-0.5, 0.5)$ [Mikh5], [Kre]. In [Chan] the complete sets on $[-0.5, 0.5]$ $\{C_1, C_2, \dots\}$, and $\{S_1, S_2, \dots\}$ of functions $C_n(x) = \frac{\cosh(\lambda_n x)}{\cosh(\lambda_n/2)} - \frac{\cos(\lambda_n x)}{\cos(\lambda_n/2)}$ and $S_n(x) = \frac{\sinh(\mu_n x)}{\sinh(\mu_n/2)} - \frac{\sin(\mu_n x)}{\sin(\mu_n/2)}$, where λ_n and μ_n are the roots of the equations $\tanh(\lambda/2) + \tan(\lambda/2) = 0$, $\coth(\mu/2) - \cot(\mu/2) = 0$, vanishing with their first derivatives at $x = \pm 0.5$ are described.

Appendix 6

The direct method based on the characteristic equation

After more than hundred years of its existence, the linear theory of hydrodynamic and hydromagnetic stability is still of interest mainly due to two facts: this theory provides the necessary conditions for instability and it is much simpler than the nonlinear theory.

The eigenvalue problems governing the linear stability of certain fluid flows, and consisting in two-point problems for systems of ode's with constant coefficients, were solved by the direct method and four methods based on Fourier series. Here we deal with the simplest direct method based on the characteristic equation. We point out its main steps, the advantages over other methods, its drawbacks, and the main tricks used to simplify the computations. Some open problems are revealed too. The secular manifolds, the characteristic manifolds and their bifurcation sets, called the false neutral manifolds, are described.

Main steps of the method. The eigenvalue problems we are dealing with are written either in the form of an ode

$$\sum_{k=0}^{n} a_{n-k} D^k u = 0, \quad x \in (-0.5, 0.5) \tag{6.1}$$

and n homogeneous boundary conditions

$$B_r u = 0, \quad r = \overline{1, n}, \quad \text{at } x = \pm 0.5, \tag{6.2}$$

where $D = \frac{d}{dx}$, $u : [-0.5, 0.5] \to \mathbf{R}$, $u \in C^\infty[-0.5, 0.5]$ is the unknown function and the constant coefficients a_i depend on m physical parameters $\mathcal{R}_{\overline{1,m}}$, or in the form of a system of ode's

$$\mathbf{AU} = \mathbf{0}, \quad x \in (-0.5, 0.5) \tag{6.3}$$

and the boundary conditions

$$B_r U_i = 0, \quad r = \overline{1, n}, \quad i = \overline{1, s} \quad \text{at } x = \pm 0.5, \tag{6.4}$$

where $\mathbf{A}$ is an $s \times s$ differential matrix the entries of which are polynomials, with constant coefficients depending on $\mathcal{R}_{\overline{1,m}}$, in the derivative D. The order of this system is n.

369

In applications, the eigenvalue problems are given in the form (6.3), (6.4) as a result of simplifications operated in the perturbed mathematical models (Section 1.4). This form is more appropriate to the application of Fourier series methods, while the form (6.1), (6.2) is the starting point for the direct method. If the involved equations are complicated, in order to deduce (6.3), (6.4) to (6.1), (6.2) the variation of coefficients method, the inverse operator method (Section 3.4.3.1) and some tricks are used. Among the tricks we mention: performing linear combinations of the equations and their convenient differentiation. In the last case, in order to keep the order of the obtained ode (6.1) as low as possible, supplementary reasonings are necessary. For instance, the operator defining the obtained equation is written as a composition of two operators, and, thus, the equation reads, say, $\left((D^2 - k^2) \circ L_2\right)(u) = 0$. If $L_2(u) = 0$ at $x = \pm 0.5$, then it follows the equation $L_2(u) = 0$, the order of which is two units less than the order of the obtained equation (Remark 3.4.4).

By an eigenvalue of the problem (6.1), (6.2) or (6.3), (6.4) we understand a value of the chosen parameter, say $\mathcal{R}_1$, to which nontrivial solutions (called eigenvectors or eigenfunctions) u of the problem correspond. Each eigenvalue is a root of the secular equation, obtained by replacing the general solution of (6.1) into (6.2) or (6.3) into (6.4). In this way the eigenvalue depends on all other parameters. Therefore, the secular equation defines some manifolds. The most convenient (physically) *secular manifold* is called the *neutral manifold* (NM). In the parameter space $\mathbf{R}^m$ it separates the domain of linear and nonlinear stability. Consequently, first our aim is to determine the secular equation. In general, its solution is possible to get only numerically.

Remark that, in general, the parameters are positive and must satisfy some other restrictions too. As a consequence, the parameter space of physical interest is a subset of $\mathbf{R}^m$, *e.g.* $\mathbf{R}_+^{*\,m}$. However, for the sake of simplicity, we continue to denote this space by $\mathbf{R}^m$.

First, consider the problem (6.1), (6.2). In order to determine the general solution of (6.1) we formally look for it in the form $u = e^{\lambda x}$ and replace this in (6.1) to obtain the characteristic equation

$$f(\lambda) = 0, \tag{6.5}$$

where $f(\lambda)$ is an n degree polynomial in λ, the coefficients of which depend on $\mathcal{R}_{\overline{1,m}}$.

Up to a null measure set, of points $(\mathcal{R}_1, \ldots, \mathcal{R}_m) \in \mathbf{R}^m$, the roots of (6.5) are simple. Denote by

$$g(\mathcal{R}_1, \ldots, \mathcal{R}_m) = 0, \tag{6.6}$$

the equations defining the manifolds the union of which is that null measure set. They consist of points to which multiple roots of (6.5) correspond. These manifolds have various topological dimensions smaller than m and we call them *the false neutral manifolds* (FNM). This label is appropriate when the secular manifold computed for the points of FNM is void.

Let $\lambda_1, \ldots, \lambda_p$ be the distinct roots of (6.5) and let m_j, $j = 1, \ldots, p$ be their multiplicities, with $\sum_{j=1}^{p} m_j = n$. These multiplicities are deduced by taking into account the Viète relations and the fact that most physical parameters are real and positive. Corresponding to $\lambda_1, \ldots, \lambda_p$ a basis for the vector space consisting of the nontrivial solutions of (6.1) is $e^{\lambda_1 x}$, $x e^{\lambda_1 x}$, $\ldots$, $x^{m_1-1} e^{\lambda_1 x}$, $e^{\lambda_2 x}$, $x e^{\lambda_2 x}$, $\ldots$, $x^{m_2-2} e^{\lambda_2 x}$, $\ldots$, $e^{\lambda_p x}$, $x e^{\lambda_p x}$, $\ldots$, $x^{m_p-1} e^{\lambda_p x}$.

Thus, in the case of the multiple roots of (6.5), the general solution of (6.1) reads

$$u(x) = \sum_{j=1}^{p} \sum_{k=0}^{m_j-1} A_k^{(j)} x^k e^{\lambda_j x}, \tag{6.7}$$

while in the case of the simple roots of (6.5), *i.e.* $m_1 = \ldots = m_p = 1$, $p = n$, it reads

$$u(x) = \sum_{i=1}^{n} A_i e^{\lambda_i x}. \tag{6.8}$$

Introducing (6.8) and (6.7) into the boundary conditions (6.2) and taking into account (6.6), we obtain the secular equation

$$F^*(\mathcal{R}_1, \ldots, \mathcal{R}_m) = 0, \tag{6.9}$$

for the case of multiple roots of (6.5), and

$$F(\mathcal{R}_1, \ldots, \mathcal{R}_m) = 0, \tag{6.10}$$

for the case of simple roots of (6.5).

Characteristic manifolds and their bifurcation sets. Secular and neutral manifolds. Involving only $\sinh(\lambda_i/2)$ and $\cosh(\lambda_i/2)$, $i = 1, \ldots, p$, and powers of λ_i, in general, the secular equation is transcendental. It yields the dependence of $\mathcal{R}_1$ on $\mathcal{R}_2, \ldots, \mathcal{R}_m$. This represents the aim of the method.

Equation (6.10) defines a (secular) manifold in the m-dimensional space of parameters. Usually this manifold has an infinity of *sheets*. From the physical point of view, the most convenient sheet is just the neutral manifold.

In (6.10) F is a determinant the columns of which have the same form in λ_i, i corresponding to the i-th column.

Formally, (6.10) is satisfied on FNM because for each point of (6.6) at least two roots λ_i and λ_j coincide, and, therefore, the columns i and j of (6.10) are identical. Thus, formally, every FNM is a secular manifold. In fact, this is not true because (6.10) is not defined on FNM and, therefore, (6.10) is not entitled to serve as a secular equation for the points of FNM. Certain concrete examples [OpG], [GeoO90] show that FNM could be physically more convenient (if it would be a secular manifold) than the true neutral manifold given by (6.10). Whence the name of *false* beard by the manifolds defined by (6.6). In these cases the direct numerical computations are invalid. In other examples, parts of FNM proved to be limits of the secular manifolds of (6.10), or even of the neutral manifolds defined by (6.10). This is the reason why, apart from (6.10) we must solve all secular equations (6.9), corresponding to all multiplicities m_j and, so, to all manifolds (6.6). It is only in this way that we can deduce which points of $\mathbf{R}^m$ are secular points, indeed.

Formally, the secular equations (6.9) are deduced from (6.10), namely writing the column j for λ_j, while the columns $j + k$, $k = 1, \ldots, m_j - 1$ are obtained by differentiating k times the $j + k$-th column of (6.10) with respect to λ_{j+k} and then replacing λ_{j+k} by λ_j.

Equations (6.9) are valid only on the manifolds (6.6). Consequently, if some manifold defined by (6.6) is q-dimensional, then the secular manifold of (6.9), when it exists, is $q - 1$-dimensional. In this way, the secular manifolds are: those of dimension $m - 1$ (corresponding to (6.10)) and those of smaller dimension (corresponding to (6.9) and situated on the manifolds (6.6)). Whence the complicated problem concerning the relative position, intersection and geometric structure of these manifolds arises [Col3].

Recall that the *bifurcation set* B of a manifold M is the projection on the parameter space of the set B_M of the bifurcation points of M, corresponding to the points where at least two sheets of M coalesce. Hence B is the set of bifurcation values corresponding to B_M. The characteristic manifold (6.10) is m-dimensional and it belongs to the space $(\lambda_1, R_1, \ldots, R_m)$, the dimension of which is $m + 1$ if $\lambda_1, \ldots, \lambda_n \in \mathbf{R}$ and $m + 2$ if some $\lambda_i \in \mathbf{C} \setminus \mathbf{R}$. The false neutral manifolds defined by (6.6) belong to the m-dimensional parameter space and they are bifurcation sets for the characteristic equation (6.10). Some among the FNM consist of bifurcation points for some other FNM or are the bifurcation sets of these ones.

More exactly, let M_λ be the characteristic manifold (6.5). Its bifurcation set B_λ is defined by eliminating the common solutions of (6.5) and a few first derivatives of (6.5) with respect to λ. Alternatively B_λ can be determined by using the Viète relations for (6.5). Therefore B_λ is the union of the sets of parameters corresponding to double, triple, etc. roots of (6.5). Let B_λ^i be those parameters corresponding to roots λ multiple of order i. Obviously, $B_\lambda^j \subset B_\lambda^i$ for $i < j$; in particular, all B_λ^i belong to the set B_λ^2 of parameters corresponding to double roots λ of (6.5). However, roots of multiplicity equal to 2 correspond only to points of $B_\lambda^2 \setminus \bigcup_{j>2} B_\lambda^j$, roots of multiplicity equal to 3 correspond only to points of $B_\lambda^3 \setminus \bigcup_{j>3} B_\lambda^j$ a.s.o. The dimension of B_λ^j is equal to $m - j + 1$. It is possible that $B_\lambda^j = \emptyset$ for $j < k$, therefore only roots λ of multiplicity at least k exist. Moreover, B_λ^j is a bifurcation set for B_λ^{j+1}. In this way, in the parameter space $\mathbf{R}^m$, B_λ determines some regions of dimension m referred to as *strata* and B_λ itself decomposes into strata (B_λ^i). The so-structured parameter space is called *the parameter portrait*. Thus $\mathbf{R}^m$ decomposes into strata of dimensions ranging from m, and corresponding to non-multiple λ, up to, possibly, zero, corresponding to a single root λ of multiplicity n (this takes place if the algebraic system of $n-1$ equations obtained by eliminating λ has a unique solution $(R_1^0, \ldots, R_m^0)$. To each such stratum a specific general solution of (6.1), and thus, a specific secular equation corresponds. A false secular manifold is that stratum to which the void secular manifold correspond.

Remark 6.1. For more than two physical parameters, the characteristic, secular and some of the false neutral manifolds cannot be represented in $\mathbf{R}^3$. This is why, a perturbed (imperfect) bifurcation study [GoluS] is necessary. In particular, it yields the deformation of the sections in the secular manifolds for various values of some parameter, all others (except the wave number) being kept fixed. This allows us to deduce the influence of that parameter, and the corresponding physical effect, on the neutral stability curve (and domain).

Remark 6.2. The form in the parameter space of equations (6.9) and (6.10) of the secular manifolds is not appropriate for the numerical study. Indeed, in general, the closed forms of the roots λ of (6.5) and of the derivatives of (6.5) are not known. In this case, instead of the understood elimination of λ between the characteristic equation (6.5) (in the simple roots case or the derivatives of (6.5) in the multiple roots case) and the secular determinant, from numerical point of view it is more convenient to solve simultaneously the secular determinant equation, equation (6.5) and, in the multiple case, the derivatives of (6.5). In the $(m+1)$- or $(m+2)$-dimensional $(\lambda, R_1, \ldots, R_m)$-space this means to intersect the corresponding manifolds [GeoC].

Advantage and drawbacks of the direct method and tricks to its easier application. Open problems. Among the advantages of the direct method we quote: 1) the very simple general form (6.7) of the solution of (6.1) and the corresponding secular equations have only a finite number of terms, therefore these solutions are *exact*. The method based on Fourier and asymptotic series involve an infinite number of terms of the solution representation and of the secular equations. Therefore, they are approximate, even if they are called exact; 2) among all methods we know, it is the only one which provides the *false secular points*. It shows how dangerous it is to apply numerical methods without a theoretical support; 3) this method is the *simplest* among all methods used for problems (6.1), (6.2); 4) the direct method applies *irrespective of the form of the boundary conditions*. In the case of the problem (6.3), (6.4) the coefficients of the various functions are related, such that the boundary conditions can be easily written for a single component of the solution $\mathbf{U}$, which in other methods is generally impossible [Geo77]. In this way, by the direct method we can treat a lot of cases untractable by other methods; 5) in the direct method, in order to get a simpler form of the secular equation, the columns of F in (6.10) are divided by $\cosh(\lambda_i/2)$. If λ_i are purely imaginary, then this division is forbidden but the condition $\cosh(\lambda_i/2) = 0$ yields secular points valid for *every boundary conditions* [Geo82a]. This is a striking property with basic implications in applications. When applying the Fourier series based methods the expressions of coefficients have as denominator just the expression $f(\lambda)$ from (6.5). Therefore they ceased to be valid for $\lambda = \lambda_i$ and, so, the direct method must be applied in order to complete this study.

Often, instead of the given problem (6.1), (6.2) or (6.3), (6.4), the problems for the even and odd parts of the solution are solved. In spite of the fact that the transcendental secular equation has a finite number of terms containing the powers of λ_i and hyperbolic sine and cosine of $\lambda_i/2$, the solution of this equation is practically impossible to obtain. This the reason why the numerical computations are done as shown in Remark 6.2.

In the case of ode's containing only even-order derivatives, a suitable change of variables, *e.g.* $\mu = \lambda^2 - a^2$, leads to a characteristic equation the degree of which is half of the initial degree. For it, usually, the closed-form root is immediate. Thus, the root of the secular equation can be deduced without any other consideration of the characteristic equation.

The direct method was applied to hydrodynamic and hydromagnetic stability theory by us and our collaborators starting with the year 1977 [Geo77] (Section 3.4.3.1). Its more detailed description can be found in [Geo 85], [GeoPalR05].

Apart from very simple situations, a systematic theoretical investigation of the bifurcation of the involved manifolds is a difficult open problem. This is the case especially when more than three parameters occur. This was also remarked by Collatz in [Col3]. It is also in these cases that the determination of the multiplicity of the characteristic roots is another open problem. The separate numerical solution of the characteristic equation and of the secular equations require numerical methods specific to bifurcation theory. Therefore, we can avoid this by numerically solving both these equations simultaneously [GeoC].

Appendix 7

First and second order differential matricial operators

Tensor algebra. Let V be a finite dimensional Euclidean vector space. Any element of V is called a *vector*. Let $\mathbf{u}, \mathbf{v}, \mathbf{w} \in V$ be some vectors and let $\{\mathbf{e}_i\}$ be an orthonormal basis of V. Denote by v_i the *coordinates* of $\mathbf{v}$ with respect to this basis. Therefore $v_i \mathbf{e}_i$ are the *components* of $\mathbf{v}$ with respect to the same basis. The *scalar* (or *inner*) *product* $\mathbf{v} \cdot \mathbf{w}$ (or $\mathbf{w} \cdot \mathbf{v}$) of $\mathbf{v}$ by $\mathbf{w}$ is a scalar, namely $\mathbf{v} \cdot \mathbf{w} = v_i w_i$, where here and below we use the summation convention on repeating subscripts. Two vectors $\mathbf{v}$ and $\mathbf{w}$ are orthogonal if $\mathbf{v} \cdot \mathbf{w} = 0$.

Correspondingly $\frac{\mathbf{v} \cdot \mathbf{w}}{\|\mathbf{w}\|^2} \mathbf{w}$ is called the *projection* of $\mathbf{v}$ on $\mathbf{w}$. Sometimes, by projection of $\mathbf{v}$ on $\mathbf{w}$ we understand the *intensity* of the projection vector, *i.e.* $\frac{\mathbf{v} \cdot \mathbf{w}}{\|\mathbf{w}\|}$.

A linear operator from V to V is referred to as a *second order tensor*. Denote by $LinV$ the vector space of all second order tensors $\{\mathbf{T}, \mathbf{U}, \mathbf{V}, \mathbf{W}, \dots\}$. Every two vectors $\mathbf{v}$ and $\mathbf{w}$ define a second order tensor $\mathbf{v} \otimes \mathbf{w}$ by the formula $(\mathbf{v} \otimes \mathbf{w})(\mathbf{u}) = (\mathbf{w} \cdot \mathbf{u})\mathbf{v}$. This is called the *tensor* (or *dyadic*) *product* of $\mathbf{u}$ and $\mathbf{v}$. It is easy to check that $\{\mathbf{e}_i \otimes \mathbf{e}_j\}$ is a basis of $LinV$. Denote by T_{ij} *the scalar coordinates* of $\mathbf{T}$ with respect to this basis. The tensor product $\mathbf{u} \otimes \mathbf{v}$ of two vectors $\mathbf{u}$ and $\mathbf{v}$ has scalar coordinates $u_i v_j$ with respect to the basis $\{\mathbf{e}_i \otimes \mathbf{e}_j\}$, *i.e.* $\mathbf{u} \otimes \mathbf{v} = u_i v_j \mathbf{e}_i \otimes \mathbf{e}_j$. A tensor $\mathbf{T}$ is called *symmetric* if $T_{ij} = T_{ji}$ and *skew-symmetric* if $T_{ij} = -T_{ji}$. Then $\mathbf{T} \cdot \mathbf{v}$ is, by definition, a vector $\mathbf{w}$ such that $w_i = T_{ij} v_j$. Therefore $\mathbf{w}$ is the image of $\mathbf{v}$ through $\mathbf{T}$ and it is called the *right scalar product* of the tensor $\mathbf{T}$ by the vector $\mathbf{v}$. The *left scalar product* of $\mathbf{v} \cdot \mathbf{T}$ of $\mathbf{T}$ by $\mathbf{v}$ is the vector $\mathbf{w}$ of coordinates $w_j = v_i T_{ij}$. Denote by $\mathbf{T}_i$ the product $\mathbf{T} \cdot \mathbf{e}_i$. Then $\mathbf{T} \cdot \mathbf{v} = \mathbf{T}_i v_i$, where $\mathbf{T}_i$ are the *vector coordinates* of $\mathbf{T}$ with respect to $\{\mathbf{e}_i\}$. If $\mathbf{T}$ is the rate of strain tensor and $\mathbf{v}$ is the normal $\mathbf{n}$ to some internal surface, the Cauchy continuous media are defined by $\mathbf{T} \cdot \mathbf{n} = \mathbf{T}_1 n_1 + \mathbf{T}_2 n_2 + \mathbf{T}_3 n_3$, expressing the fact that the surface force is a linear function of $\mathbf{n}$.

A *contraction product* $\mathbf{U} \cdot \mathbf{V}$ of two second order tensors $\mathbf{U}$ and $\mathbf{V}$ is the composition $\mathbf{U} \circ \mathbf{V}$, with components $U_{ik} V_{kj} \mathbf{e}_i \otimes \mathbf{e}_j$, also referred to as a contraction product of $\mathbf{U}$ by $\mathbf{V}$.

A *scalar product* $\mathbf{U} \mathbf{V}$ associates with two tensors $\mathbf{U}$ and $\mathbf{V}$ the scalar $U_{ik} V_{ki}$, also referred to as the scalar product, which is invariant to the choice of the basis

375

$\{\mathbf{e}_i\}$.

Given a basis $\{\mathbf{e}_i\}$, a tensor $\mathbf{T}$ is defined by the matrix T_{ik}, also denoted by $\mathbf{T}$, of its scalar coordinates. In terms of matrices, the contraction products of two tensors is described by the product of the corresponding matrices while the scalar product of two tensors is the trace of the product of the first corresponding matrix by the transpose of the second. In other words, writing (only here) the matrices in parentheses, we have

$$\mathbf{U} \cdot \mathbf{V} = (\mathbf{U}) \cdot (\mathbf{V}); \quad \mathbf{U}\mathbf{V} = \mathrm{tr}\Big[(\mathbf{U}) \cdot (\mathbf{V}^T)\Big],$$

i.e. the coordinates of $\mathbf{U} \cdot \mathbf{V}$ are the corresponding entries of the product of the two corresponding matrices; $\mathbf{U}\mathbf{V}$ is the trace of the product of the matrix corresponding to the first tensor by the transpose of the matrix corresponding to the second tensor.

A *third order tensor* $\mathbf{T}$ is defined as a linear operator from V to $LinV$. The third order tensors form a linear space denoted by Lin^2V, where a basis is $\{\mathbf{e}_i \otimes \mathbf{e}_j \otimes \mathbf{e}_k\}$, with respect to which the scalar coordinates of $\mathbf{T}$ are T_{ijk}, namely $\mathbf{T}\cdot\mathbf{v} = \mathbf{T}_{ijk}v_k\mathbf{e}_i \otimes \mathbf{e}_j$.

If V is a Euclidean space of dimension 3 and $\{\mathbf{e}_1, \mathbf{e}_2, \mathbf{e}_3\}$ is an orthonormal basis, the Ricci tensor is defined by

$$R_{ijk} = \begin{cases} 0 & \text{if at least two among} \quad i,j,k \quad \text{coincide} \\ 1 & \text{if} \quad (ijk) \quad \text{is an even permutation of } (1,2,3) \\ -1 & \text{if} \quad (ijk) \quad \text{is an odd permutation of } (1,2,3) \end{cases}$$

This tensor, independent of the choice of the basis, is used to define the *vector product* $\mathbf{u} \times \mathbf{v}$ of two vectors $\mathbf{u}$, $\mathbf{v}$ as $(\mathbf{u} \times \mathbf{v})_k = u_i v_j R_{ijk}$.

By definition a *tensor product* $\mathbf{T} \otimes \mathbf{v}$ of a second order tensor $\mathbf{T}$ by a vector $\mathbf{v}$ is a third order tensor, namely $\mathbf{T} \otimes \mathbf{v} = T_{ij}v_k\mathbf{e}_i \otimes \mathbf{e}_j \otimes \mathbf{e}_k$. The tensor product of three vectors is a third order tensor, *i.e.* $\mathbf{u} \otimes \mathbf{v} \otimes \mathbf{w} = u_i v_j w_k \mathbf{e}_i \otimes \mathbf{e}_j \otimes \mathbf{e}_k$. Higher order tensors can be defined similarly as linear operators defined on V to Lin^nV. The higher is the order of the tensors, the larger is the number of products we can define with them.

We also recall the *mixed product* of three vectors $(\mathbf{u}, \mathbf{v}, \mathbf{w}) = \mathbf{u} \cdot (\mathbf{v} \times \mathbf{w}) = u_i v_j w_k R_{ijk}$, which is a scalar.

Here are the symmetries and the skew-symmetries of some of these products

$$\mathbf{u} \cdot \mathbf{v} = \mathbf{v} \cdot \mathbf{u}; \quad \mathbf{u} \times \mathbf{v} = -\mathbf{v} \times \mathbf{u}; \quad (\mathbf{u}, \mathbf{v}, \mathbf{w}) = (\mathbf{v}, \mathbf{w}, \mathbf{u}) = (\mathbf{w}, \mathbf{u}, \mathbf{v}) = -(\mathbf{u}, \mathbf{w}, \mathbf{v})$$

$$= -(\mathbf{v}, \mathbf{u}, \mathbf{w}) = -(\mathbf{w}, \mathbf{v}, \mathbf{u});$$

$$\mathbf{T}\cdot\mathbf{u} = \mathbf{u}\cdot\mathbf{T} \quad \text{if and only if} \quad T_{ij} = T_{ji}; \quad \mathbf{T}\cdot\mathbf{u} = -\mathbf{u}\cdot\mathbf{T} \quad \text{if and only if} \quad T_{ij} = -T_{ji}$$

Since $V = Lin^oV$, a (formally) unitary treatment of vectors and tensors is usually done [Aris], [Goo], [Youn].

Linear differential operators defined on linear spaces of tensors fields. Let t be the time and $\mathbf{x}$ stand for the space variable. By a *tensor field* (in particular *vector field*) we mean a function of t and $\mathbf{x}$, the image of which are tensors. Further assume that the tensor fields are as smooth as necessary such that the operators defined on linear spaces of vector fields have a classical meaning. The *gradient operator* ∇ transforms tensors of order n into tensors of order $n+1$, namely $\nabla \mathbf{T} = \frac{\partial T_{i_1,\ldots,i_n}}{\partial x_j}\mathbf{e}_j \otimes \mathbf{e}_{i_1} \otimes \cdots \otimes \mathbf{e}_{i_n}$.

The *divergence operator* $\nabla\cdot$ transforms tensors of order n into tensors of order $n-1$, namely $\nabla \cdot \mathbf{T} = \frac{\partial T_{i_1,\ldots,i_n}}{\partial x_{i_1}}\mathbf{e}_{i_2} \otimes \cdots \otimes \mathbf{e}_{i_n}$. In particular $\nabla \cdot \mathbf{v} = \frac{\partial v_i}{\partial x_i}$. If $\nabla \cdot \mathbf{v} = 0$, then $\mathbf{v}$ is called a *solenoidal* vector.

The *curl (rotor) operator* $\nabla\times$ in $\mathbf{R}^3$ transforms tensors of order n into tensors of order n. For instance, for $n=1$ and $n=2$ we have

$$\nabla \times \mathbf{v} = R_{ijk}\frac{\partial}{\partial x_j}v_k\mathbf{e}_i,\ \nabla \times \mathbf{T} = R_{ijk}\frac{\partial}{\partial x_j}T_{kl}\mathbf{e}_l \otimes \mathbf{e}_i$$

All these operators are of the first order. They lower by one the order of differentiation of the tensor field. A second order differential linear operator is the Laplacian, transforming tensors of order n into tensors of order n and lowering by two the order of differentiation, namely

$$\Delta \mathbf{T} = \frac{\partial^2 T_{i_1,\ldots,i_n}}{\partial x_i \partial x_i}\mathbf{e}_{i_1} \otimes \cdots \otimes \mathbf{e}_{i_n}.$$

The introduction of (vectors and) tensors as linear functions endow them with the independence of the basis and allow the formulation of physical laws and corresponding equations in an invariant form. However, since the coordinates of tensors depend on the basis, the components of the equations in various directions depend on the basis too. The association of scalar components of the second order tensors with matrices, provides to matrices a mathematical definition. The converse is also true: the presence of a matrix indicates a linear operator. As the coordinates of a tensor changes with the basis, its representation by the associated matrix changes too. In other words, in different bases, a tensor is represented by different matrices.

Here are some identities for scalar functions f, g and vector fields $\mathbf{u}$, $\mathbf{v}$, $\mathbf{w}$, involving these operators

$$\nabla(f+g) = \nabla f + \nabla g,$$

$$\nabla \cdot (\mathbf{u}+\mathbf{v}) = \nabla \cdot \mathbf{u} + \nabla \cdot \mathbf{v},$$

$$\nabla \times (\mathbf{u}+\mathbf{v}) = \nabla \times \mathbf{u} + \nabla \times \mathbf{v},$$

$$\nabla(fg) = f\nabla g + g\nabla f,$$

$$\nabla \cdot (f\mathbf{u}) = f\nabla \cdot \mathbf{u} + \mathbf{u} \cdot \nabla f,$$

$$\nabla \times (f\mathbf{u}) = f\nabla \times \mathbf{u} + \nabla f \times \mathbf{u},$$

$$\nabla \cdot (\mathbf{u} \times \mathbf{v}) = \mathbf{v}\nabla \times \mathbf{u} - \mathbf{u}\nabla \times \mathbf{v},$$

$$\nabla \times (\mathbf{u} \times \mathbf{v}) = \mathbf{v} \cdot \nabla \mathbf{u} - \mathbf{v}\nabla \cdot \mathbf{u} - \mathbf{u} \cdot \nabla \mathbf{v} + \mathbf{u}\nabla \cdot \mathbf{v},$$

$$\nabla(\mathbf{u} \cdot \mathbf{v}) = \mathbf{v} \cdot \nabla \mathbf{u} + \mathbf{u} \cdot \nabla \mathbf{v} + \mathbf{v} \times \nabla \times \mathbf{u} + \mathbf{u} \times \nabla \times \mathbf{v},$$

$$\nabla(\mathbf{u} \cdot \mathbf{v}) = \nabla \mathbf{u} \cdot \mathbf{v} + \nabla \mathbf{v} \cdot \mathbf{u},$$

$$\nabla \times \mathbf{u} \times \mathbf{u} = \mathbf{u} \cdot \nabla \mathbf{u} - \nabla \frac{\mathbf{u}^2}{2},$$

$$\nabla \cdot \nabla f = \Delta f,$$

$$\nabla \times \nabla f = \mathbf{0},$$

$$\nabla \cdot \nabla \times \mathbf{u} = 0,$$

$$\nabla \times (\nabla \times \mathbf{u}) = \nabla\nabla \cdot \mathbf{u} - \Delta \mathbf{u}.$$

Notice that whenever the vector product $\times$ or the curl operator $\nabla\times$ occur in the above formulae, it is understood that the vector space is three dimensional.

Bibliography

[AbS] Abani, K., K. M. Srivastava, *Rayleigh-Taylor instability of a viscous plasma in the presence of Hall current*, Il nuovo Cimento **26**, 2 (1975), 419–432.

[Ad] Adams, R. A., *Sobolev spaces*, Academic, New York, 1978. (first ed. 1975)

[Ak] Akhiezer, N. I., *The calculus of variations*, Blaisdell, New York, 1962. (first Russian ed. was published at Gostehizdat, Moscow, 1955)

[AkG] Akhiezer, N. I., I. M. Glazman, *Theory of linear operators in Hilbert space*, Moscow, 1950. (English translation at Ungar, New York, 1962).

[AlAlm] Allard, W. K., F.J. Almgren Jr. (eds.), *Geometric measure theory and the calculus of variations*, AMS Summer Institute, Proceedings of Symposia in Pure Mathematics, **44**, AMS, Providence, R. I., 1986.

[An] de Angelis, M., *On universal energy stability of fluid motions in unbounded domains*, Accad. Sci. Fisiche Matem. Napoli, Ser. IV, **57** (1990).

[AraL] Aramanovich, I. G., V. I. Levin, *Equations of mathematical physics*, Nauka, Moscow, 1964. (Russian)

[Aris] Aris, R., *Vectors, Tensors and the Basic Equations of Fluid Mechanics*, Dover Publications, Inc., New York, 1989. (first ed. in 1962)

[AsE] Asimov, L., A. J. Ellis, *Convexity theory and its applications in functional analysis*, London Mathematical Society Monographs Series, 1981.

[AthMR] Athanasopoulos, I., G. Makrakis, J. F. Rodrigues (eds.), *Free boundary. Theory and applications*, CRC, London, 1999.

[Aub] Aubin, T., *Pròblemes isopérimétriques et espaces de Sobolev*, C. R. Acad. Sc. Paris, **280 A**, 279, (1975); J. Diff. Geom. **11** (1976), 533–598.

[Aub1] Aubin, T., *Nonlinear analysis on manifolds. Monge-Ampère equations*, Grundlehren **252**, Springer, New York, 1982.

[Aub2] Aubin, J.-P., *Analyse fonctionnelle appliquée*, I, II, Presses Universitaires de France, Paris, 1987.

[Aus] Auslender, A., *Problèmes de minimax via l'analyse convexe et les inégalités variationnelles; théorie et algorithmes*, Springer, Berlin, 1972.

[AxB] Axelson, D., V. A. Burker, *Finite element solution of boundary value problems*, Academic, New York, 1984.

[AzB] Aziz, A. K., I. Babuška, *Survey lectures on the mathematical foundations of the finite element method*, in: *The mathematical foundations of the finite element method with applications to partial differential equations*, (A. K. Aziz ed.), Academic, New York, 1973, 5–359.

[BaC] Baiocchi, C., A. Capelo, *Disequazioni variazionali e quasivariazonali. Applicazioni ai problemi di frontiera libera*, Pitagora, Bologna, 1978. (English ed. Wiley, New York, 1984)

[Bal] Balakrishnan, A. V., *Applied functional analysis*, 2-nd ed., Applications of Mathematics **3**, Springer, New York, 1981 (first published in 1973).

[Ban] Bandle, C., *Isoperimetric inequalities and applications*, Pitman, London, 1980.

[Baz] Bazley, N. W., *Lower bounds for eigenvalues*, J. Math. Mec. **10**, 2 (1961), 289–307.

[BazF] Bazley, N. W., D. W. Fox, *Lower bounds to eigenvalues using operator decompositions of the form B^*B*, ARMA **10** (1962), 352–360.

[BelMP] Belleni-Morante, A., M. Maiellaro, *On the stability of Couette-Bénard M.H.D. flows of binary mixtures*, Annali di Mat. (**IV**), **106** (1975),

[Bena] Bénard, H., *Les tourbillons cellulaires dans une nappe liquide*, Rev. Gén. Sci. Pure Appl. **11** (1900), 1261–1271.

[BensL] Bensoussan, A., J.-L. Lions, *Impulse control and quasivariational inequalities*, Gauthier - Villars, Paris, 1984.

[Berd] Berdichevskii, V. L., *Variational principles of mechanics of continua*, Nauka, Moscow, 1982, 1983. (Russian)

[Bere] Berezanskii, Yu. M., *Expansion on eigenfunctions of selfadjoint operators*, AMS, Providence, 1965.

[Berg] Berger, M. S., *Nonlinearity and functional analysis*, Academic, New York,1977.

[Berk] Berker, R., *Inégalités vérifiées par l'énergie cinétique d'un fluide visqueux incompressible occupant un domaine spatial borné*, Bull. Tech. Univ. Istanbul, **2** (1949), 41–51.

[BernFKK] Bernstein, I. B., E. A. Frieman, M. D. Kruskal, R. M. Kulsrud, *An energy principle for hydromagnetic stability problems*, Proc. Roy. Soc. London **A 244** (1958), 17–40.

[BesIN] Besov, O. A., V. P. Il'in, S. M. Nikol'skii, *Integral representations of functions and embedding theorems*, 2 vols., Wiley, New York, 1978.

[Bic] Bichir, C.-L., *Theoretical and numerical hydrodynamic stability*, Seria de Matematică Aplicată şi Industrială, Universitatea din Piteşti, **11**, Piteşti, 2002. (Romanian)

[BicGP] Bichir, C-L., A. Georgescu, L. Palese, *A nonlinear hydrodynamic stability criterion derived by a generalized energy method*, Bul. Acad. St. Rep. Moldova, Mat., **1** (47), (2005), 85–91.

[Bio] Biot, M. A., *Variational principles in heat transfer. A unified Lagrangian analysis of dissipative phenomena*, Oxford Mathematical Monographs, Clarendon, Oxford, 1970.

[BirH] Birkhoff, G. D., M. R. Hestenes, *Natural isoperimetric conditions in the calculus of variations*, Duke Math. J. **1**, 2 (1935), 198–286.

[Bl] Bliss, G. A., *Lectures on calculus of variations*, Izd. Innostrannoi Literatury, Moscow, 1950. (Russian) (first English ed. at University of Chicago Press, 1946)

[Bol] Bolza, O., *Lectures on the calculus of variations*, 2nd edition, Chelsea Publishing Co., New York, 1961. (first ed. 1946)

[BomW] Bompiani, E., V. V. Wagner (eds.), *Geometria del calcolo delle variazioni*, 2 vols., Cremonese, Roma, 1961.

[BraP1] Bramble, J. H., L. E. Payne, *Bounds for solutions of second order elliptic partial differential equations*, Contributions to Differential Equations, **1**, 1 (1963), 95–127.

[BraP2] Bramble, J. H., L. E. Payne, *Some inequalities for uniformly elliptic operators*,

Contributions to Differential Equations, **1**, 1 (1963), 129–135.

[BreS] Breuner, S. C., L. R. Scott, *The mathematical theory of finite element methods*, Texts in Applied Mathematics **15**, Springer, New York, 1994.

[Bru] van Brunt, B., *The calculus of variations*, Universitext, Springer, Berlin, 2004.

[BudDiP] Budiansky, B., R. C. DiPrima, *Bending vibrations of uniform twisted beams*, J. Math. Phys., **39**, 4 (1960), 237–245.

[BudHC] Budiansky, B., P. C. Hu, R. W. Connor, *Notes on the Lagrangian multiplier method in elastic-stability analysis*, NACA TN **1558** (1948).

[BudK] Budiansky, B., E. T. Kruszewski, *Transverse vibrations of hollow thinwalled cylindrical beams*, NACA TR **1129** (1953).

[Bus1] Busse, F. H., *The stability of finite amplitude cellular convection and its relation to an extremum principle*, J. Fluid Mech. **30** (1967), 625–649.

[Bus2] Busse, F. H., *On the optimum theory of turbulence*, in Energy stability and convection, Pitman Research Notes in Mathematics **168**, Longman, Harlow, 1988.

[But] Buttazzo, G., *Semicontinuity, relaxation and integral representation in the calculus of variations*, Pitman Research Notes in Mathematics Series **207**, Longman Scientific and Technical, New York, 1989.

[ButH] Buttazzo, G. S., S. Hildebrandt, *One-dimensional variational problems*, Oxford Lecture Series in Mathematics and its Applications **15**, Oxford University Press, 1999.

[ButzN] Butzer, P. L., R. J. Nessel, *Fourier analysis and approximation theory*, I, Birkhäuser, Basel, 1977.

[Cal] Caltagirone, J. P., *Stability of a saturated porous layer subject to a sudden rise in surface temperature : comparison between the linear and energy methods*, Q. J. Mech. Appl. Math. **33** (1980), 47–58.

[Cara] Carathéodory, C., *Calculus of variations and partial differential equations of the first order*, Chelsea, New York, 1982. (first ed. in 1964-1965)

[Carba1] Carbonaro, B., *Generalità e limiti del principio di linearizazione*, in *Problemi di Meccanica, Statistica e Fisica del Plasma*, Atti Sem. Fis. Mat., Trieste (1979), 23–42.

[Carba2] Carbonaro, B., *On the linearization principle in magnetohydrodynamics*, Ric. di Mat. **28**, 2 (1979), 425–437.

[CarboD] Carbone, L., R. De Arcangelis, *Unbounded functionals in the calculus of variations, representation, relaxation and homogenization*, Monographs and Surveys in Pure and Applied Mathematics, **125**, Chapman & Hall/CRC, Boca Raton, FL, 2002.

[Carm] Carmi, S., *Energy stability of modulated flows*, Phys. Fluids **17** (1974), 1951–1955.

[Cat] Cattabriga, L., *Su un problema al contorno relativo al sistema di equazioni di Stokes*, Rend. Sem. Mat. Padova, **31**, (1961), 308–340.

[Cea] Céa, J., *Optimization : théorie et algorithmes*, Dunod, Paris, 1971.

[Ces1] Cesari, L., *Optimization - theory and applications. Problems with ordinary differential equations*, Applications of Mathematics **17**, Springer, Berlin, 1983.

[Ces2] Cesari, L. (ed.), *Contributions to modern calculus of variations*, Pitman Research Notes in Mathematics **148**, Longman Scientific and Technical, New York, 1986.

[Chad] Chadam, J. M., H. Rasmussen (eds.), *Emerging applications in free boundary problems; free boundary problems involving solids; free boundary problems in fluid flow with applications*, Pitman Research Notes in Mathematics Series,

280, **281**, **282**, Longman Scientific and Technical, Burnt Mill, Harlow, 1993.

[Chan] Chandrasekhar, S., *Hydrodynamic and hydromagnetic stability*, Clarendon, Oxford, 1961, 1968; Dover, New York, 1981.

[Chav] Chavel, J., *Isoperimetric inequalities: differential geometric and analytical perspectives*, Cambridge University Press, 2001.

[CheY] Cheung, Y. K., M. F. Yeo, *A practical introduction to finite element analysis*, Pitman, Boston, 1979.

[ChoI] Chossat P., G. Iooss, *The Couette-Taylor problem*, Inst. Non-Linéaire de Nice, Univ. de Nice, 1993; Applied Mathematical Sciences **102**, Springer, Berlin, 1994.

[Chu] Chudinovich, I., C. Constanda, *Variational and potential methods in the theory of bending of plates with transverse shear deformation*, Monographs and Surveys in Pure and Applied Mathematics Series, Chapman & Hall/ CRC, Boca Raton, FL, 2000.

[Col1] Collatz, L., *Eigenwertaufgaben mit technischen Anwendungen*, 2nd ed., Geert und Portig, Leipzig, 1963. (Russian ed. at Nauka, Moscow, 1968).

[Col2] Collatz, L., *Functional analysis and numerical mathematics*, Academic, New York, 1966.

[Col3] Collatz, L., *Remark on bifurcation problems with several parameters*, Springer LNM **846** (1981), 82–87.

[ConsF] Constantin, P., C. Foias, *Global Lyapunov exponents, Kaplan - York formulas and the dimension of the attractors for two-dimensional Navier-Stokes equations*, Comm. Pure Appl. Math., **38** (1985), 1–27.

[ConsFNT] Constantin, P., C. Foias, B. Nicolaenko, R. Temam, *Integral manifolds and inertial manifolds for dissipative partial differential equations*, Applied Mathematical Sciences Series, **70**, Springer, New York, 1988.

[Cont] Conti, R., *Calculus of variations, classical and modern*, CIME **1** (1966).

[Cou1] Courant, R., *Calculus of variations*, Courant Institute of Mathematical Sciences, New York, 1962.

[Cou2] Courant, R., *Variational methods for the solution of problem in equilibrium of vibrations*, Bull. AMS, **49**, 1943.

[CouH] Courant, R., D. Hilbert, *Methoden der mathematischen Physik*, I, II, Springer, Berlin, 1937. (English version issued at Interscience, New York, 1953, 1962).

[CriM] Crisciani F., R. Mosetti, *Stability criteria independent from the perturbation wave-number for forced zona flows*, J. Phys. Oceanogr., **21** (7) (1991), 1075–1079.

[Dac1] Dacorogna, B., *Direct methods in the calculus of variations*, Applied Mathematical Sciences **78**, Springer, Berlin, 1989.

[Dac2] Dacorogna, B., *Introduction to the calculus of variations*, World Scientific, Singapore, 2004.

[Daug] Dauge, M., *Elliptic boundary value problems on corner domains*, LNM **1341**, Springer, Berlin, 1988.

[DautL] Dautray, R., J.-L. Lions, *Analyse mathématique et le calcul numérique pour les sciences et les techniques*, Masson, Paris, 1988.

[Dav] Davis, S. H., *On the possibility of subcritical instabilities*, in Proc. IUTAM Symp., Springer, Berlin, 1971.

[DavK] Davis, S. H., C. von Kerczek, *A reformulation of energy stability theory*, ARMA **52**, 2 (1973), 112–117.

[Day] Day, M. M., *Normed linear spaces*, 3-rd ed., Springer, New York, 1973.

[Dea] Dean, W. R., *Fluid motion in a curved channel*, Proc. Roy. Soc. **A**, **121**, (1928),

402–420.

[DenS] Dennis, J. E., R. B. Schnabel, *Numerical methods for unconstrained optimization and nonlinear equations*, Prentice-Hall, Englewood Cliffs, 1983.

[DeoR] Deo, B. J. S., A. T. Richardson, *Generalized energy methods in electrohydrodynamic stability theory*, J. Fluid Mech., **137** (1983), 131–151.

[Dia] Diaz, J. B., *Upper and lower bounds for eigenvalues*, Proc. Symposia in Appl. Math. **8**, AMS, Providence, 1958, 53–78.

[DidG] Didwania, A. K., J. D. Goddard, *A note on the generalized Rayleigh quotient for nonselfadjoint linear stability operators*, Phys. Fluids **A 5** (1993), 1269–1271.

[DimPF] Dimitrinovič, D. S., J. E. Pečaris, A. M. Fink, *Inequalities for functions and their integrals and derivatives*, Kluwer, Dordrecht, 1994.

[DiP55] DiPrima, R. C., *Application of the Galerkin method to problems in hydrodynamic stability*, Quart. Appl. Math. **13** (1955), 55–62.

[DiP59] DiPrima, R. C., *The stability of viscous flow between rotating concentric cylinders with a pressure gradient acting round the cylinders*, J. Fluid Mech., **6** (1959), 462–468.

[DiP61] DiPrima, R. C., *Some variational principles for problems in hydrodynamic and hydromagnetic stability*, Quart. Appl. Math. **18**, 4 (1961), 375–385.

[DiPS65] DiPrima, R. C., R. Sani, *The convergence of the Galerkin method for the Taylor - Dean stability problem*, Quart. Appl. Math., **23**, 2 (1965), 183–187.

[DoPH] Donelly, R. J., I. Prigogine, R. Herman (eds.), *Nonequilibrium thermodynamics, variational techniques and stability*, University of Chicago Press, 1965.

[Drag05a] Dragomirescu, I., *Particular Georgescu - Palese - Redaelli stability criterion for a binary mixture problem*, Iaşi, 2005; *On a particular stability criterion in a binary mixture problem*, Bul. St. Univ. Politeh., Timişoara, **50** (64), 1 (2005), 53–61.

[Drag06] Dragomirescu, I., *Approximate neutral surface of a convection problem for a variable gravity field*, Rend. Sem. Mat. Univ. Pol. Torino, **64**, 3 (2006), 331–342.

[Drag07] Dragomirescu, I., *Rayleigh number in a stability problem for a micropolar fluid*, Turkish J. Math., **31**, 2 (2007), 123–137.

[Drag07a] Dragomirescu, I., *Spectral problems in hydrodynamic stability*, Seria Mat. Apl. Ind. **26**, Mirton, Timişoara, 2007. (Romanian)

[DrazR] Drazin, Ph. G., W. H. Reid, *Hydrodynamic stability*, Cambridge University Press, 1981.

[Dre] LeDret, H., *Problèmes variationnels dans les multi-domaines*, Masson, Paris, 1991.

[DroR] Drobot, S., A. Rybarski, *A variational principle of hydrodynamics*, ARMA **2** (1959), 393–410.

[DuD] Dudis, J. J. K., S. H. Davis, *Energy stability of the buoyancy boundary layer*, J. Fluid Mech. **47** (1971), 381–403.

[DunS] Dunford, N., J. T. Schwartz, *Linear operators*, 3 vols, Interscience, Wiley, New York, 1958, 1963, 1971.

[DuvL] Duvaut, G., J.-L. Lions, *Les inéquations en mécanique et en physique*, Dunod, Paris, 1972.

[Ea] Eaves, R.E., *A sufficient condition for the convergence of an infinite determinant*, SIAM J. Appl. Math. **18**, 3 (1970), 652–657.

[EbS1] Ebel, D., M. C. Shen, *On the linear stability of a toroidal plasma with resistivity, viscosity and Hall currents*, J. Math. Anal. Appl., **125** (1987), 81–103.

[EbS2] Ebel, D., M. C. Shen, *Linearization principle for a toroidal Hall current plasma with viscosity and resistivity*, Ann. Mat. Pura Appl., **150** (1988), 39–65.

[EdE] Edmunds, D. E., W. Evans, *Spectral theory and differential operators*, Oxford Mathematical Monographs, Clarendon, Oxford, 1987.

[EkG] Ekeland, J., N. Ghoussoub, *Selected new aspects of the calculus of variations in the large*, Bull. (New Series) of the AMS, **39**, 2 (2002), 207–265.

[EkT] Ekeland, J., R. Temam, *Analyse convexe et problèmes variationnels*, Dunod-Gauthier Villars, Paris, 1974; *Convex analysis and variational problems*, Classics in Applied Mathematics **28**, SIAM, Philadelphia, 1999.

[El] Elsgolc, L. E., *Calculus of variations*, Pergamon, Oxford, 1961. (first Russian ed. Differential equations and calculus of variations, Fizmatlit, Moscow, 1965)

[Er] Errafiy, M., R. Kh. Zeytounian, *The Bénard problem for deep convection: routes to chaos*, Int. J. Engng. Sc., **29**, (1991), 1363–1373.

[Fab] Fabrizio, M., *Introduzione alla mecanica razionale e ai suoi metodi matematici*, 2nd ed., Zanichelli, 1998. (first ed. 1994)

[Fai] Faierman, M., *Two parameter eigenvalue problems in ordinary differential equations*, Pitman Research Notes in Mathematics Series **205**, Longman, Wiley, New York, 1991.

[Fich] Fichera, G., *Il calcolo degli autovalori*, Boll. UMI, **1** (4) (1968), 33–95.

[Fikh] Fikhtengol'ts, G. M., *Course in differential and integral calculus*, Nauka, Moscow, 1970. (Russian)

[Fil] Filipov, V. M., *Variational principles for nonpotential operators*, Transl. Math. Monographs **77**, AMS, Providence, 1989.

[Fin] Finlayson, B. A., *The method of weighted residuals and variational principles*, Mathematics in Science and Engineering **87**, Academic, New York, 1992. (first published in 1972).

[FinS] Finlayson, B. A., L. E. Scriven, *On the search for variational principles*, Int. J. Heat and Mass Transfer, **12** (19), 799–821.

[FlaR] Flavin, J. N., S. Rionero *Qualitative estimates for partial differential equations. An introduction*, CRC Press, Boca Raton, Fl., 1996.

[Fle] Fletcker, R., *Practical methods in optimization*, Wiley, New York, 1987.

[Foi73] Foiaş, C., *Statistical study of Navier-Stokes equations*, Rend. Sem. Mat. Padova, I: **48** (1972), 219–348; II: **49** (1973), 9–123.

[FoiMT] Foiaş, C., O. Manley, R. Temam, *Attractors for the Bénard problem: Existence and physical bounds on their dimension*, J. Non-linear Anal.- T.M.A., **11** (1987), 939–967.

[FoiP] Foiaş, C., G. Prodi, *Sur le comportement global des solutions nonstationnaires des équations de Navier-Stokes en dimension 2*, Rend. Sem. Mat. Padova, **39** (1967), 1–34.

[FoiST] Foias, C., G. Sell, R. Temam, *Turbulence in fluid flows: A dynamical system approach*, IMA Volumes in Mathematics and Its Applications, **55**, Springer, New York, 1993.

[FoiT] Foias, C., R. Temam, *Remarques sur les équations de Navier-Stokes stationnaires et les phénomènes successifs de bifurcation*, Ann. Scuola Norm. Sup. Pisa, Ser. IV, **5** (1978), 29–63.

[Fom] Fomenko, A. T., *Variational principles in topology*, Mathematics and its Applications **42**, Kluwer, Dordrecht, 1989. (first Russian ed. was published by Nauka, Moscow, 1982)

[For] Förster, W., *On certain parabolic differential equations and an equivalent variational problem*, LNM **280**, Springer, Berlin, 1972, 268–272.

[FrA] Friedman, A., *Variational principles and free-boundary problems*, Wiley, New York, 1987. (first ed. 1982)

[FrAS] Friedman, A., J. Spruck, *Variational problems*, The IMA Volumes in Mathematics and its Applications **53**, Springer, New York, 1993.

[FrB] Friedman, B., *Principles and techniques of applied mathematics*, John Wiley & Sons, New York, 1956.

[Fri1] Friedrichs, K. O., *Spectraltheorie halbbeschränkter Operatoren und Andwendung der Spektralzerlegung von Differenzialoperatoren*, Math. Ann., **109**, 1 (1934), 465–487 ; An inequality for potential functions, Amer. J. Math. **68** (1946), 581–592.

[Fri2] Friedrichs, K. O., *Spectral theory of operators in Hilbert spaces*, Applied Mathematical Sciences **9**, Springer, New York, 1973.

[FucNS] Fucik, S., J. Nečas, V. Souček, *Einführung in die Variationsrechnung*, Teubner Texte zur Mathematik, Teubner, Leipzig, 1977.

[FucNSS] Fucik, S., J. Nečas, J. Souček, V. Souček, *Spectral analysis of nonlinear operators*, Springer, Berlin, 1973.

[Fun] Funk, P., *Variationsrechnung und ihre Anwendung in Physik und Technik*, Die Grundlehren der mathematischen Wissenschaften **94**, Springer, Berlin, 1970.

[Gald85] Galdi, G. P., *Nonlinear stability of the magnetic Bénard problem via a generalized energy method*, ARMA **87** (1985), 167–186.

[Gald94] Galdi, G. P., *An introduction to the mathematical theory of the Navier-Stokes equations*, 2 vols., Springer Tracts in Natural Philosophy **38**, **39**, New York, 1994.

[GaldP] Galdi, G. P., M. Padula, *A new approach to energy theory in the stability of fluid motion*, ARMA **110** (1990), 187–286.

[GaldR] Galdi, G. P., S. Rionero, *Weighted energy methods in fluid dynamics and elasticity*, Springer LNM **1134**, 1985.

[GaldS82] Galdi, G. P., B. Straughan, *An immediate connection between linear and nonlinear stability via an appropriate choice of measure: application to convection problems*, Proc. Symp. on Waves and Stability in Continua, G.N.F.M. of Italian C.N.R., Catania, Italy, 1982, 65–100.

[GaldS85] Galdi, G. P., B. Straughan, *A nonlinear analysis of the stabilizing effect of rotation in the Bénard problem*, Proc. R. Soc. London **A 402** (1985), 257–283.

[GaldS89] Galdi, G. P., B. Straughan, *Exchange of stabilities, symmetry and nonlinear stability*, ARMA **89** (1985), 211–228.

[GapZ] Gaponenko, Yu. A., V. E. Zakhvataev, *Microconvection in a binary system*, Fluid Dynamics **38**, 1 (2003), 57–68.

[Gar] Gardner, R. J., *The Brunn-Minkowski inequality*, Bull. (New Series) of the AMS, **39**, 3 (2002), 355–405.

[GelF] Gelfand, I. M., S. V. Fomin, *Calculus of variations*, Prentice-Hall, Englewood Cliffs, N. J., 1963.

[Geo73] Georgescu, A., *Stability of the Couette flow of a viscoelastic fluid*, Rev. Roum. Math. Pures Appl. **18**, 9 (1973), 1371–1374.

[Geo76] Georgescu, A., *Universal criteria of hydrodynamic stability*, Rev. Roum. Math. Pures Appl., **20** (1976), 287–302.

[Geo77] Georgescu, A., *Variational formulation of some nonselfadjoint problems occurring in Bénard instability theory*, I, Preprint Series in Mathematics **35**/1977, INCREST & Institute of Mathematics, Bucharest, 1977.

[Geo82a] Georgescu, A., *Characteristic equations for some eigenvalue problems in hydromagnetic stability theory*, Mathematica, Rev. d'Anal. Num. Théorie de l'Approx. **24** (**47**), 1-2 (1982), 31–41.

[Geo82b] Georgescu, A., *Bifurcation (catastrophe) surfaces for a problem in hydromag-*

netic stability, Rev. Roum. Math. Pures Appl., **27**, 3 (1982), 335–337.

[Geo85] Georgescu, A., *Hydrodynamic stability theory*, Kluwer, Dordrecht, 1985.

[Geo87] Georgescu, A., *Exact solutions for some instability of Bénard type,* Rev. Roum. Phys., **32**, 4 (1987), 391–397.

[Geo89] Georgescu, A., *Lagrange and the calculus of variations*, Noesis **15** (1989), 29–35.

[Geo92] Georgescu, A., *Synergetics, solitons, fractals, deterministic chaos, turbulence,* University of Timişoara Press, 1992. (Romanian)

[Geo95] Georgescu, A., *Asymptotic treatment of differential equations*, Chapman & Hall, London, 1995.

[GeoC] Georgescu, A., V. Cardoş, *Neutral stability curves for a thermal convection problem*, Acta Mechanica **37** (1980), 165–168.

[GeoLP] Georgescu, A., A. Labianca, L. Palese, *A linear instability analysis of the Bénard problem for deep convection*, Proc. 'WASCOM 2005', of the 13th Conference on Waves and Stability in Continuous Media, R. Monaco, S. Pennisi, S. Rionero, T. Ruggeri (eds.), World Scientific, Singapore, 2006, 256–261.

[GeoM99] Georgescu, A., D. Mansutti, *Coincidence of the linear and nonlinear stability bounds in a horizontal thermal convection problem*, Int. J. Non-Linear Mech., **34** (1999), 603–613.

[GeoO88] Georgescu, A., I. Oprea, *The bifurcation curve of the characteristic equation provides the bifurcation point of the neutral curve of some elastic stability*, Mathematica, Rev. d'Anal. Num. Théorie de l'Approx., **17**, 2 (1988), 141–145.

[GeoO90] Georgescu, A., I. Oprea, *Neutral stability curves for a thermal convection problem, II. The case of multiple solutions of the characteristic equation*, Acta Mechanica **81** (1990), 115–119.

[GeoOO] Georgescu, A., I. Oprea, C. Oprea, *Bifurcation manifolds in a multiparametric eigenvalue problem from linear hydromagnetic stability theory*, Mathematica, Rev. d'Anal. Num. Théorie de l'Approx., **18**, 2 (1989), 123–138.

[GeoOP] Georgescu, A., I. Oprea, D. Paşca, *Methods to determine the neutral curve in the Bénard stability*, St. Cerc. Mec. Apl. **52**, 4 (1993), 267–276. (Romanian)

[GeoPal95] Georgescu, A., L. Palese, *Améliorations des estimations de Prodi pour le spectre*, C. R. Acad. Sc. Paris, 320, 1 (1995), 891–896.

[GeoPal96a] Georgescu, A., L. Palese, *Extension of a Joseph criterion to the nonlinear stability of mechanical equilibria in the presence of thermodiffusive conductivity*, Theoretical and Computational Fluid Mechanics, **8** (1996), 403–413. (An extended version was published in 1994 in Rapp. Int. Dip. Mat. Univ. Bari **12**/1994.)

[GeoPal96b] Georgescu, A., L. Palese, *A nonlinear stability criterion of binary mixture*, ZAMM, S2 **76** (1996), 529–530.

[GeoPal96c] Georgescu, A., L. Palese, *Neutral stability hypersurfaces for an anisotropic M.H.D. thermodiffusive mixture. III. Detection of false secular manifolds among the bifurcation characteristic manifolds*, Rev. Roum. Math. Pures Appl. **41**, 1-2 (1996), 35–49.

[GeoPal97] Georgescu, A., L. Palese, *Stability spectrum estimates for confined fluids*, Rev. Roum. Math. Pures Appl., **42**, 1-2 (1997), 37–51. (first published as Rapp. Int. Dipto. Mat., Univ. Bari **9**/1994)

[GeoPal00] Georgescu, A., L. Palese, *On a method in linear stability problems. Applications to natural convection in a porous medium*, Ultra Science, **12**, 3 (2000), 324–336.

[GeoPalPasB] Georgescu, A., L. Palese, D. Paşca, M. Buican, *Critical hydromagnetic stability of a thermodiffusive state*, Rev. Roum. Math. Pures et Appl. **38**, 10 (1993),

831–840.

[GeoPalR96c] Georgescu, A., L. Palese, A. Redaelli, *Linearization principle for the stability of mechanical equilibrium of a binary mixture where the Soret and Dufour effects are present*, Rapp. Int. Dip. Mat. Univ., Bari **14**/1996.

[GeoPalR00] Georgescu, A., L. Palese, A. Redaelli, *On a new method in hydrodynamic stability theory*, Math. Sc. Res. Hot-Line **4** (7) (2000), 1–16. (A first version of this paper appeared in Rapp. Int. Dip. Mat. Univ. Bari **8**/1995.)

[GeoPalR01] Georgescu, A., L. Palese, A. Redaelli, *The complete form of the Joseph extended criterion*, Ann. Univ. Ferrara, Sez. VII, Sc. Mat., **48** (2001), 9–22. (First published as Rapp. Int. Dip. Mat. Univ. Bari, **58**/2000)

[GeoPalR05] Georgescu, A., L. Palese, A. Redaelli, *A direct method and its application to a linear hydromagnetic stability problem*, ROMAI J., **1**, 1 (2005), 67–76.

[GeoPalR06] Georgescu, A., L. Palese, A. Redaelli, *A linear magnetic Bénard problem with tensorial electrical conductivity*, Bolletino U. M. I. (8)-B (2006), 197–214. (First published as *A linear magnetic Bénard problem with Hall effect. Application of Budianski-DiPrima method*, Rapp. Int. Dipto. Mat. Univ. Bari, **15**/2003).

[GeoPo] Georgescu, A., O. Polotzka, *Stability of the Couette flow of a viscoelastic fluid II.*, Rev. Roum. Math. Pures et Appl. **22**, 9 (1977), 1223–1233; *On a Bénard convection in the presence of dielectrophoretic forces*, ASME J. of Appl. Mech., **48**, 4 (1981), 980–981.

[GeoS] Georgescu, A., A. Setelecan, *Comparative study of the analytic methods to solve some problems in hydrodynamic stability theory*, Analele Univ. Bucureşti, **38**, 1 (1989), 15–20.

[GerJ] Gershuni, G. Z., E. M. Jukhovitskii, *Convective instability of incompressible fluids*, Nauka, Moscow, 1972. (Russian) (see also Gershuni, G. Z., E. M. Jukhovitskii, A. A. Nepomniatshii, *Stability of convective flows*, Moscow, Nauka, 1989. (Russian))

[Gi1] Giaquinta, M., *Multiple integrals in the calculus of variations and nonlinear elliptic systems*, Annals of Math. Studies **105**, Princeton University Press, 1984.

[Gi2] Giaquinta, M. (ed.), *Topics in calculus of variations*, Springer LNM **1365**, 1989.

[GiH] Giaquinta, M., S. Hildebrandt, *Calculus of variations, I. The Lagrangian formalism, II. The Hamiltonian formalism*, Grundlehren der mathematischen Wissenschaften **310**, **311**, Springer, Berlin, 1996.

[GiMS] Giaquinta, M., G. Modina, J. Souček, *Certain currents in the calculus of variations, I. Cartesian currents; II. Variational integrals*, Ergebnisse der Mathematik und ihrer Grenzgebiete, Series **3**, **37**, **38**, Springer, Berlin, 1998.

[GirR] Girault, V., P.-A. Raviart, *Finite element methods for Navier-Stokes equations. Theory and algorithms*, Springer, Berlin, 1986. (first ed. issued as LNM **749** in 1979)

[Gl] Glowinski, R., *Lectures on numerical methods for nonlinear variational problems*, Tata Institute of Fundamental Research Bombay, 1980, Lectures on Mathematics and Physics; Springer Surveys on Mathematics and Physics; Springer Series on Comp. Phys., Springer, Berlin, 1984.

[GlLe] Glowinski, R., P. Le Tallec, *Augmented Lagrangian and operator-splitting methods in nonlinear mechanics*, SIAM, Philadelphia, 1989.

[GlLT] Glowinski, R., J.-L. Lions, R. Trémolières, *Numerical analysis of variational inequalities*, North-Holland, Amsterdam, 1982. (first French ed. was published by Dunod, Paris, 1976)

[Goldb] Goldberg, S., *Unbounded linear operators with applications*, McGraw-Hill, New

York, 1966.

[Goldste] Goldstein, S., *The stability of viscous fluid flow under pressure between parallel planes*, Proc. Cambridge Phil. Soc., **32** (1936), 40–54.

[Goldsti] Goldstine, H. H., *A history of the calculus of variations from the 17th through 19th century*, Studies in the History of Methods and Physical Sciences **5**, Springer, New York, 1980.

[GoluS] Golubitsky, M., D. G. Schaeffer, *Singularities and groups in bifurcation theory. I*, Springer, New York, 1985.

[Goo] Goodbody, A. M., *Cartesian Tensors: with applications to mechanics, fluid mechanics and elasticity*, Wiley, New York, 1982.

[Gou] Gould, S. H., *Variational methods for eigenvalue problems*, University of Toronto Press, London, 1966.

[GrR] Gradshteyn, J. S., I. M. Ryzhik, *Table of integrals, series, and products*, Academic, New York, 1980.

[Gra] Graves, L. M. (ed.), *Calculus of variations and its applications*, Proc. Symp. Appl. Math. **8**, AMS, Providence, 1982. (first ed. in 1958).

[Gri] Griffiths, Ph. A., *Exterior differential systems and the calculus of variations*, Birkhäuser, Boston, 1983.

[Gro] Grossi, R. O., *On the role of natural boundary conditions in the Ritz method*, Int. J. Mech. Engng. Education **16**, 3 (1988), 207–210.

[GruW] Gruber, P. M., J. M. Wills (eds.), *Convexity and its applications*, Birkhäuser, Basel, 1983.

[Guil] Guiraud, J. P., G. Iooss, *Sur la stabilité des écoulements laminaires*, C. R. Acad. Sc. Paris, **266** (1960), 1283–1286.

[HarLP] Hardy, G. H., J. E. Littlewood, G. Polya, *Inequalities*, Cambridge University Press, 1934; 1952.

[HarR] Harris, D. L., W. H. Reid, *On orthogonal functions which satisfy four boundary conditions*, The Astrophys. J. **3**, 33 (1958), 429–453.

[Herm] Herman, R., *Differential geometry and the calculus of variations*, Mathematics in Science and Engineering **49**, Academic, New York, 1968.

[Hes] Hestens, M. R., *Calculus of variations and optimal control theory*, John Wiley & Sons, New York, 1966.

[HlHNL] Hlavaček, I., J. Haslinger, J. Nečas, J. Lovisek, *Solution of variational inequalities in mechanics*, Applied Mathematical Sciences **66**, Springer-Verlag, New York, 1988.

[HoMRW] Holm, D. D., J. E. Marsden, T. Ratiu, A. Weinstein, *Nonlinear stability of fluid and plasma equilibria*, Phys. Rep. **123** (1985), 1–116.

[Hop] Hopf, E., *Über die Aufgangswertaufgabe für die hydrodynamischen Grundliechungen*, Math. Nachr. **4** (1951), 213–231.

[How] Howell, K. B., *Principles of Fourier analysis*, CRC Press, Boca Raton, FL, 2001.

[HS] Hill, A.A., B. Straughan, *Linear and non-linear stability thresholds for thermal convection in a box*, Math. Meth. in Appl. Sci., **29** (2006), 2123–2132.

[Hue] Huebner, K. H., *The finite element method for engineers*, Wiley, New York, 1975.

[HurJ] Hurle, D. T. J., E. Jakeman, *Soret-driven thermosolutal convection*, J. Fluid Mech. **47** (1971), 667–687.

[IofRS1] Ioffe, A. D., S. Reich, I. Shafrir, *Calculus of variations and optimal control*, Research Notes in Mathematics series, Chapman & Hall, CRC, Boca Raton, FL, 1999.

[IofRS2] Ioffe, A. D., S. Reich, I. Shafrir, *Calculus of variations and differential equations*, Research Notes in Mathematics Series, Technion 1998; CRC, Boca Raton, FL, 2000.

[IofT] Ioffe, A. D., V. M. Tikhomirov, *Theory of extremal problems*, Studies in Mathematics and its Applications **6**, North-Holland, Amsterdam, 1979.

[Ioo1] Iooss, G., *Sur la théorie de la stabilité linéaire des écoulements laminaires*, C. R., Acad. Sc. Paris **269** (1969), 333–336.

[Ioo2] Iooss, G., *Bifurcation of maps and applications*, Math. Studies, **36**, North-Holland, Amsterdam-New York, 1979.

[IooJ] Iooss, G., D. D. Joseph, *Elementary stability and bifurcation theory*, 2nd ed., Springer, New York, 1990. (first published in 1980)

[J65] Joseph, D. D., *On the stability of the Boussinesq equations*, ARMA, **20** (1965), 59–71.

[J66] Joseph, D. D., *Nonlinear stability of the Boussinesq equations by the method of energy*, ARMA, **22** (1966), 163–184.

[J70a] Joseph, D. D., *Uniqueness criteria for the conduction-diffusion solution of the Boussinesq equations*, ARMA **35** (1970), 169–177.

[J70b] Joseph, D. D., *Global stability of a conduction-diffusion solution*, ARMA **36**, 4 (1970), 285–292.

[J76] Joseph, D. D., *Stability of fluid motions*, 2 vols., Springer, Berlin, 1976.

[J88] Joseph, D. D., *Two fluids heated from below*, in *Energy stability and convection*, Pitman Research Notes in Mathematics **168**, Longman, Harlow, 1988.

[JH] Joseph, D. D., W. Hung, *Contributions to the nonlinear theory of stability of viscous flow in pipes and between rotating cylinders*, ARMA **44** (1971), 1–22.

[JS] Joseph, D. D., C. C. Shir, *Subcritical convective instability. Part 1. Fluid layers*, J. Fluid Mech. **26** (1966), 753–768.

[Kac] Kachoyan, B. J., *Neutral curve behavior in Taylor-Dean flow*, J. Appl. Math. Phys. (ZAMP), **38** (1987), 905–924.

[Kam] Kampé de Fériet, J., *Sur la décroissance de l'énergie cinétique d'un fluide visqueux incompressible occupant un domain borné ayant pour frontière des solides fixes*, Ann. Soc. Sci. Bruxelles, **63** (1949), 35–46.

[KanK] Kantorovich, L. V., V. I. Krylov, *Approximate methods of higher analysis*, Noordhoff, Gröningen, 1958.

[KapT] Kaplan, A., R. Tischatschke, *Stable methods for ill-posed variational problems, Prox-regularization of elliptic variational inequalities and semiinfinite problems*, Mathematical Topics **3**, Wiley-VCH, Weinheim, 1994; Akademie, Berlin, 1994.

[KaX] Kaiser, R., L., Xu, *Nonlinear stability of the rotating Bénard problem, the case $P_r = 1$*, Nonlinear Diff. Eq. Appl., **5** (1998), 283–307.

[Kat] Kato, T., *Perturbation theory of linear operators*, Die Grundlehren der mathematischen Wissenschaften, **132**, Springer, New York, 1966.

[Keld] Keldysh, M. V., *On the eigenvalues and eigenfunctions of some classes of nonselfadjoint equations*, DAN SSSR, **77**, 1 (1951), 11–14. (Russian)

[Ker1] Kerr, O. S., *Heating a salinity gradient from a vertical sidewall: linear theory*, J. Fluid Mech. **207** (1989), 323–352.

[Ker2] Kerr, O. S., *Heating a salinity gradient from a vertical sidewall: nonlinear theory*, J. Fluid Mech. **217** (1990), 529–546.

[Kin] Kinderlehrer, D., *Variational inequalities and free boundary problems*, Bull. Amer. Math. Soc. **84**, 1 (1978), 7–26.

[KinS] Kinderlehrer, D., G. Stampacchia, *An introduction to variational inequalities and their applications*, Pure and Applied Mathematics, A Series of Monographs and Textbooks **87**, Academic, New York, 1980; SIAM, Philadelphia, 2000.

[Kl] Kloeden, P., R. Wells, *An explicit example of Hopf bifurcation in fluid mechanics*, Proc. Roy. Soc. London **A 390** (1983), 293–320.

[Kos] Koschmieder, E. L., *Bénard cells and Taylor vortices*, Cambridge University Press, 1993.

[Kou] Kounadis, A. N., *Variational calculus*, National Technical University, Athens, 1985.

[Kra] Kratz, W., *Quadratic functionals in variational analysis and control theory*, Academie, Berlin, 1995.

[Kre] Kreyszig, E., *Introduction to functional analysis with applications*, John Wiley & Sons, New York, 1978.

[Kryz] Kryzhanovskii, D. S., *Isoperimeters, maximum and minimum properties of geometrical figures*, 3nd ed., Fizmatghiz, Moscow, 1959. (Russian)

[Kup] Kuperschmidt, B. A., *The variational principles of dynamics*, Advanced Series in Mathematical Physics **13**, World Scientific, Singapore, 1992.

[Lad67] Ladyzhenskaya, O. A., *On new equations describing the viscous incompressible fluid flow and about solvability in the large of this boundary value problem*, Trudy Mat, Inst. Steklova, **5** (1967), 85–104. (Russian)

[Lad69] Ladyzhenskaya, O. A., *The mathematical theory of viscous incompressible flow*, 2nd revised and enlarged, Gordon and Breach, New York, 1969. (1st ed. 1963; 1st Russian ed. 1961)

[Lad91] Ladyzhenskaya, O. A., *Attractors for semigroups and evolution equations*, Accademia Nazionale dei Lincei Series, Cambridge University Press, 1991.

[LadS67] Ladyzhenskaya, O. A., V. A. Solonnikov, *The linearization principle and invariant manifolds for problems of magnetohydrodynamics*, J. Sov. Math., (4) 8 (1967), 384–422.

[LadS76] Ladyzhenskaya, O. A., V. A. Solonnikov, *On some problems of vector analysis and generalized formulations of Navier-Stokes equations*, Sem. Leningrad, **59** (1976), 81–116. (Russian)

[LadV] Ladyzhenskaya, O. A., A. M. Vershik, *Sur l'évolution des mesures déterminées par les équations de Navier-Stokes et la réduction du problème de Cauchy pour léquation statistique de E. Hopf*, Annali della Scuola Normale Superiore di Pisa, **2** (1977), 209–230.

[Lag] Lagrange, J. L., *Leçons sur le calcul des fonctions*, Coucier, Paris, 1806.

[Lak] Lakshmikantham, V., S. Leela, *Differential and integral inequalities*, Academic, New York, 1969.

[LalC] Lalas, D. P., S. Carmi, *Thermoconvective stability of ferrofluids*, Phys. Fluids **14** (1971), 436–437.

[LamL] Lambermont, J., G. Lebon, *A rather general variational principle for purely dissipative nonstationary processes*, Annalen der Physik, 7 Folge, **28**, 1 (1972), 15–30.

[Lan] Lanczos, C., *The variational principles of mechanics*, University of Toronto Press, 1949.

[Lau] Lauwerier, H. A., *Calculus of variations in mathematical physics*, Mathematical Centre Tracts **14**, Amsterdam, 1966.

[Lav] Lavrentiev, M. A., *Variational methods in the boundary value problems for systems of equations of elliptic type*, Noordhoff, Dordrecht, 1963. (First Russian ed. was published in 1962)

[LavLy] Lavrentiev, M. A., L. A. Lyusternik, *Foundations of the calculus of variations*, O.N.T.I., 1935. (Russian) (Romanian ed. Ed. Tehnica, Bucharest, 1955)

[Lebe] Lebesgue, H., *En marge du calcul des variations*, Institut de Mathématiques, Univ. de Genève, 1963.

[Leb] Lebon, G., *A new variational principle for the nonlinear unsteady heat-conduction problem*, Q. J. Mech. Appl. Math. **29**, 4 (1976), 499–509.

[LebC] Lebon, G., Cloot, A., *A thermodynamical modeling of fluid flows through porous media: application to neutral convection*, Int. J. Heat Mass Transfer, **29**, 3 (1986), 381–390.

[LebL1] Lebon, G., J. H. Lambermont, *Generalization of Hamilton's principle to continuous dissipative systems*, J. Chem. Phys. **59**, 6 (1973), 2929–2936.

[LebL2] Lebon, G., J. H. Lambermont, *A general variational criterion for chemically active continuous media*, Annalen der Physik, Series **7**, **32**, 6 (1975), 425–432.

[Lec] Lecat, M., *Bibliographie du calcul des variations I. Depuis les origines jusqu'à 1850; - De 1850 jusqu'à 1913* Ad. Hoste, Gand, Hermann, Paris, 1916.

[Leip] Leipholz, H. H. E., *Direct variational methods and eigenvalue problems in engineering*, Monographs and Textbooks on Mechanics of Solids and Fluids, Noordhoff, Leyden, 1997.

[Leit] Leitman, G., *The calculus of variations and optimal control*, Mathematical Concepts and Methods in Science and Engineering **24**, Plenum, New York, 1981.

[Ler] Leray, J., *Étude de diverses équations intégrales nonlinéares et de quelques problèmes que pose l'hydrodynamique*, J. Math. Pure Appl., Série 9, **12** (1933), 1–82.

[Lin] Lin, C. C., *The theory of hydrodynamic stability*, Cambridge University Press, 1955.

[Lind] Lindsay, K. A., B. Straughan, *Energy methods for nonlinear stability in convection problems, primarily related to geophysics*, Continuum Mech. Thermodyn. **2** (1990), 245–277.

[LiojM] Lions, J. L., E. Magenes, *Nonhomogeneous boundary value problems and applications*, Springer, New York, 1972. (First French ed. Dunod, Paris, 1968)

[Liop] Lions, P. L., *Generalized solutions of Hamilton-Jacobi equations*, Research Notes in Mathematics and its Applications **69**, Pitman Advanced Publishing Program, Boston, 1982.

[Loi] Loiseau, J., *Le principe des variations dit Hamiltoniennes est en contradiction avec la mécanique analytique*, ARMA **19** (1965), 226–230.

[LovR] Lovelock, D., H. Rund, *Tensors, differential forms and variational principles*, Pure and Applied Mathematics, Wiley-Interscience Series of Texts, Monographs and Tracts, New York, 1975.

[Lu] Luke, J. C., *A variational principle for a fluid with a free surface*, J. Fluid Mech. **27** (1967), 395–397.

[Ly] Lyusternik, L. A., *The topology of the calculus of variations in the large*, AMS, Providence, 1982. (First published in 1967)

[LySc] Lyusternik, L. A., L. Schnirelman, *Méthodes topologiques dans les problèmes variationnels*, Herman, Paris, 1934.

[LySo] Lyusternik, L. A., L. I. Sobolev, *Elements of functional analysis*, Constable, London, 1961. (F. Ungar, New York, 1961)

[MaiP84] Maiellaro, M., L. Palese, *Electrical anisotropic effects in thermal instability*, Int. J. Engng. Sc. **22**, 4 (1984), 411–418.

[MaiP89] Maiellaro, M., L. Palese, *Anisotropic thermoconvective effects on the stability of thermodiffusive equilibrium*, Proc. 1st Nat. Congress on Mechanics, Athens, 1986; Int. J. Engng. Sc. **27**, 4 (1989), 329–335.

[MaiPL] Maiellaro, M., L. Palese, A. Labianca, *Instabilizing-stabilizing effects of M.H.D. anisotropic currents*, Int. J. Engng. Sc. **27**, 11 (1989), 1353–1359.

[Mau1] Maurin, K., *Hilbert space methods*, Mir, Moscow, 1965. (Russian)

[Mau2] Maurin, K., *Calculus of variations and classical field theory*, Lecture Notes **34**, Aarhus Universitet, Mathematik Institut, 1972.

[MenM] Mennicken, R., M. Möller *Nonselfadjoint boundary eigenvalue problems*, Elsevier, Amsterdam, 2003.

[Mikh1] Mikhlin, S. G., *Direct methods in mathematical physics*, Moscow, 1950. (Russian) (see also *Variational methods to solve problems of mathematical physics*, Usp. Mat. Nauk, **5**, 6 (40) (1950). (Russian and its Romanian translation can be found in Russian - Romanian Annals)

[Mikh2] Mikhlin, S. G, *The problem of the minimum of a quadratic functional*, Holden-Day, San Francisco, 1965. (First Russian ed. Gostehizdat, Moscow-Leningrad, 1952)

[Mikh3] Mikhlin, S. G., *Variational methods in mathematical physics*, International Series of Monographs in Pure and Applied Mathematics **50**, Pergamon, Oxford, 1964. (First Russian ed., Gostehizdat, Moscow, 1957; 2nd Russian ed., Nauka, Moscow, 1970. It is a revised and enlarged version of Mikh1.)

[Mikh4] Mikhlin, S. G., *Numerical implementation of variational methods*, Nauka, Moscow, 1966. (Russian)

[Mikh5] Mikhlin, S. G., *Mathematical physics, an advanced course*, North Holland, Amsterdam, 1970. (First Russian ed. was published at Fizmatlit, Moscow, 1968.)

[Mil] Milnor, J., *Morse Theory*, Princeton University Press, 1963.

[Mitr] Mitrinović, D. S., *Analytic inequalities*, Springer, Berlin, 1970.

[Moi] Moiseiwitsch, B. I., *Variational principles*, John Wiley & Sons, New York, 1966.

[Morr] Morrey, C. B. Jr., *Multiple integrals in the calculus of variations*, Springer, Berlin, 1966.

[Mors1] Morse, M., *The calculus of variations in the large*, 6th ed., Colloquium Publications **18**, AMS, Providence, 1978. (first ed. 1934)

[Mors2] Morse, M., *Functional analysis, topology and abstract variational theory*, Mémorial des Sciences Mathématiques, fasc. XCII, 1939.

[Mors3] Morse, M., *Global variational analysis; Weierstrass integrals on a Riemannian manifold*, Mathematical Notes, Princeton University Press, 1976.

[Mors4] Morse, M., *Variational analysis: critical extremals and Sturmian extensions*, Pure and Applied Mathematics, Wiley-Interscience Series of Texts, Monographs and Tracts, New York, 1993.

[Mosc] Mosco, U., *Implicit variational problems and quasivariational inequalities*, LNM **543**, Springer, Berlin, 1976, 82–156.

[MosoM] Mosolov, P. P., V. P. Myasnikov, *Variational methods in the flow of solid-viscous-plastic media*, Izd. Moskovskogo Universiteta, 1971.

[Moss] Mossino, J., *Inéquations isopérimétriques et applications en physique*, Herman, Paris, 1984.

[Mull] Müller, I., *Thermodynamics*, Pitman, Boston, 1985.

[Mulo06] Mulone, G., *An operative method to define generalized-energy functionals in PDE's and in convection problems*, Proc. WASCOM 2005, R. Monaco, G. Mulone, S. Rionero, T. Ruggeri (eds.), World Scientific, Singapore, 2006, 390–401.

[MuloR84] Mulone, G., S. Rionero, *Existence, uniqueness and regularity theorems for stationary thermo-diffusive mixture in a mixed problem*, Proc. Workshop on Math. Aspects of Fluids and Plasma Dyn., Trieste, 1984.

[MuloR89] Mulone, G., Rionero, S., *On the nonlinear stability of the rotating Bénard problem via the Lyapunov direct method*, J. Math. Appl. **144** (1989), 109–127.

[MuloR94] Mulone, G., S. Rionero, *On the stability of the rotating Bénard problem*, Bull. Techn. Univ. Istanbul **47** (1994), 181–202.

[MuloR97] Mulone, G., S. Rionero, *The rotating Bénard problem: new stability results for any Prandtl and Taylor numbers*, Continuum Mech. Thermodyn., **9** (1997), 347–363.

[MuloR98] Mulone, G., S. Rionero, *Unconditional nonlinear exponential stability in the Bénard problem for a mixture: necessary and sufficient conditions*. Rend. Mat. Acc. Lincei **9**, 9 (1998), 221–236.

[MuloR03] Mulone, G., S. Rionero, *Necessary and sufficient conditions for nonlinear stability in the magnetic Bénard problem*, A.R.M.A., **166** (2003), 197–218.

[MuloS85] Mulone, G., F. Salemi, *On the hydrodynamic motion in a domain with mixed boundary conditions: existence, uniqueness, stability and linearization principle*, Ann. Mat. Pura Appl., **139** (1985), 147–174.

[MuloS88] Mulone, G., F. Salemi, *Some continuous dependence theorems in M.H.D. with Hall and ion-slip currents in unbounded domains*, Rend. Acad. Sci. Fis. Mat. Napoli, IV, **55** (1988), 139–152.

[Mun] Munson, B. R., D. D. Joseph, *Viscous incompressible flow between concentric rotating cylinders, Part. 2. Hydrodynamic stability*, J. Fluid Mech., **49** (1971), 305–318.

[MutWHA] Mutabazi, I., J. E. Wesfreid, J. J. Hegseth, C. D. Andereck, *Recent experimental results in the Taylor-Dean system*, Eur. J. Mech. B. Fluids, **10**, 2 (1991), 239–245.

[NanP] Naniewicz, Z., P. D. Pavagiotopoulos, *Mathematical theory of hemivariational inequalities and applications*, Pure and Applied Mathematics, A Series of Monographs and Textbooks **188**, Marcel Dekker, New York, 1995.

[Nat] Natanson, I. P., *Theory of functions of a real variable*, Constable, London, 1961. (first Russian ed. Moscow, 1950)

[Ne] Nečas, J., *Les méthodes directes en théorie des équations elliptiques*, Academia, Prague, 1967.

[NigA] Nigam, S. D., H. C. Agrawal, *A variational principle for convection of heat*, J. Math. Mech. **9**, 6 (1960), 869–883.

[Nir] Nirenberg, L, *Variational methods in nonlinear problems*, Springer LNM **1365**, 1989.

[OdR] Oden, J. T., J. N. Reddy, *Variational methods in theoretical mechanics*, Springer, Berlin, 1976.

[OpG] Oprea, I., A. Georgescu, *Bifurcating neutral surfaces in a problem of inhibition of the thermal convection due to the magnetic field*, St. Cerc. Mec. Apl. **48**, 3 (1989), 263–278. (Romanian)

[Os] Osserman, R., *The isoperimetric inequality*, Bull. AMS **84** (1978), 1182–1238.

[OtP] Ottensen, N., S. Petersson, *Introduction to the finite element method*, Prentice Hall, Englewood Cliffs, N.J., 1992.

[Pa] Pai, Sh. I., *Magnetohydrodynamics and plasma dynamics*, Springer, Berlin, 1962.

[Pal97] Palese, L., *Electroanisotropic effects on the thermal instability of an anisotropic binary fluid mixture*, J. of Magnetohydrodynamics and Plasma Research **7**, 2/3 (1997), 101–120. (first published as Rapp. Int. Dip. Mat. Univ. Bari, **31**/1990)

[Pal05] Palese, L., *On the nonlinear stability of the M.H.D. anisotropic Bénard problem*, Int. J. Engng. Sc. **43**, (2005), 1265–1282.

[Pal06] Palese, L., *On the stability of the magnetic anisotropic Bénard problem with Hall and ion-slip currents*, Proc. WASCOM 2005, R. Monaco, G. Mulone, S. Rionero, T. Ruggeri (eds.), World Scientific, Singapore, 2006, 432–437.

[PalG03] Palese, L., A. Georgescu, *A linear magnetic Bénard problem with Hall effect. Application of Budiansky-DiPrima method*, Trudy Srednevoljskogo Matematicheskogo Obshchestva, Saransk, **10**, 1 (2008), 294–306. (First published in Rapp. Int. Dipto. Mat. Univ. Bari 15/2003.)

[PalG04a] Palese, L., A. Georgescu, *Lyapunov method applied to the anisotropic Bénard problem*, Math, Sci. Res. J. **8**, 7 (2004), 196–204.

[PalG04b] Palese, L., A. Georgescu, *On instability of the magnetic Bénard problem with Hall and ion-slip effects*, Int. J. Engng. Sci., **42** (2004), 1001–1012.

[PalGM] Palese, L., A. Georgescu, S. Mitran, *Neutral curves for the MHD Soret-Dufour driven convective instability*, Rev. Roum. Sci. Tech.-Méc. Appl. **45**, 3 (2000), 265–275. (A more complete version can be found in Rapp. Dip. Mat. Univ. Bari, **38**/1995.)

[PalGPas] Palese, L., A. Georgescu, L. Pascu, *Neutral surfaces for Soret-Dufour driven convective instability*, Rev. Roum. Sci. Tech.-Méc. Appl., **43**, 2 (1998), 251–260. (First version can be found in Rapp. Int. Dip. Mat. Univ. Bari **17**/1996.)

[PalGPashB] Palese, L., A. Georgescu, D. Paşca, D. Bonea, *Thermosolutal instability of a compressible Soret-Dufour mixture with Hall and ion-slip currents through a porous medium*, Rev. Roum. Sci. Tech.-Méc. Appl. **42**, 3-4 (1997), 279–296 (First published as Rapp. Int. Dip. Mat. Univ. Bari **5**/94.)

[Pan] Panagiotopoulos, P. D., *Inequality problems in mechanics and applications. Convex and nonconvex energy function*, Birkhäuser, Boston, 1985.

[Pay1] Payne, L. E., *New isoperimetric inequalities for eigenvalues and other physical quantities*, Trans. Symp. Partial Diff. Eqs., Interscience, New York, 1956.

[Pay2] Payne, L. E., *Isoperimetric inequalities and their application*, SIAM Rev. **9**, 3 (1967), 453–488.

[Pay3] Payne, L. E., *Improperly posed problems in partial differential equations*, SIAM, Philadelphia, 1975.

[PayW] Payne, L. E., H. F. Weinberger, *New bounds for solutions of second order elliptic partial differential equations*, Pacific J. Math., **8** (1958), 551–573.

[PayW2] Payne, L. E., H. F. Weinberger, *An exact stability bound for Navier-Stokes flow in a sphere*, in *Nonlinear problems*, R. E. Langer (ed.), University of Wisconsin Press, Madison, 1963.

[Ped] Pedregal, P., *Variational methods in nonlinear elasticity*, SIAM, 1979, 1999.

[PeS] Pellew, A., R. V. Southwell, *On a maintained convective motion in a fluid heated from below*, Proc. Roy. Soc. London A, **176** (1940), 312–343.

[Pi] Pinch, E. R., *Optimal control and the calculus of variations*, Oxford University Press, 1995.

[Pn] Pnueli, D., *Maximum rate of entropy production variational formulation and preferred mode of motion in thermal instability*, Int. J. Engng. Sci. **13** (1975), 881–888.

[PolS] Polya, G., G. Szegö, *Isoperimetric inequalities in mathematical physics*, Ann. of Math. Studies **27**, Princeton University Press, 1951.

[Pot] Postnikov, M. M., *The variational theory of geodesics*, Saunders, Philadelphia, 1967.

[Pr] Prenter, P. M., *Splines and variational methods*, Pure and Applied Mathematics, Wiley-Interscience Series of Texts, Monographs and Tracts, New York, 1975.

[Pro] Prodi, G., *Teoremi di tipo locale per il sistema di Navier-Stokes e stabilità delle soluzioni stazionarie*, Rend. Sem. Mat. Padova **32** (1962), 374–397.

[Rab] Rabinowitz, P. H., *Minimax methods in critical point theory with applications to differential equations*, CBMS **65**, AMS Providence, 1986.

[Ram] Rama Rao, K. V., *Numerical solution of the thermal instability of a micropolar fluid layer between rigid boundaries*, Acta Mechanica, **32** (1979), 79–88.

[Ras] Rassoy, D. R., *A guide to mathematical models of convection processes in geothermal systems*, in Fluid mechanics in energy conversion, SIAM, Philadelphia, 1980.

[RedV] Reddy, B. D., H. F. Voyé, *Finite element analysis of the stability of fluid motion*, J. Comput. Phys. **79** (1988), 92–112.

[Rek] Rektorys, K., *Variational methods in mathematics, science and engineering*, Reidel, Dordrecht, 1977. (first Czech ed. 1975)

[Rio68] Rionero, S., *Metodi variazionali per la stabilità asintotica in media in magnetoidrodinamica*, Ann. di Mat. IV, **78** (1968), 339–364.

[Rio71b] Rionero, S., *Sulla stabilità magnetofluidoidrodinamica nonlineare asintotica in media in presenza di effetto Hall*, Ricerche Matem. **20** (1971), 285–296.

[Rio78] Rionero, S., *On the linearization principle for the Navier Stokes equations in exterior domains*, Symposium on Nonlinear Continuum Mechanics, Venice, May 23-26, 1978.

[Rio88a] Rionero, S., *On the choice of the function in the stability of fluid motions*, in *Energy stability and convection*, Pitman Research Notes in Mathematics **168**, Longman, Harlow, 1988 (presented in 1986 at the meeting in Capri).

[Rio05] Rionero, S., *Functionals for the coincidence between linear and nonlinear stability with applications to spatial ecology and double diffusive convection*, Proc. WASCOM 2005, R. Monaco, G. Mulone, S. Rionero, T. Ruggeri (eds.), World Scientific, Singapore, 2006, 461–473.

[RioG] Rionero, S., G. Galdi, *The weight function approach to uniqueness of viscous flows in unbounded domains*, ARMA, **69** (1979), 37–52.

[RioM84] Rionero, S., G. Mulone, *On the stability of a thermodiffusive mixture in a mixed problem*, Atti VII Congr. Naz. A.I.M.E.T.A., Section I, Trieste, October 2-5, 1984, 113–124.

[RioM87a] Rionero, S., G. Mulone, *On the nonlinear stability of a thermodiffusive fluid mixture in a mixed problem*, J. Math. Anal. Appl., **124** (1987), 165–188.

[RioM88a] Rionero, S., G. Mulone, *On a maximum problem governing the nonlinear stability of rotating Bénard problem*, Ricerche di Matematica **37**, 1 (1988), 177–185.

[RioM88b] Rionero, S., G. Mulone, *Existence and uniqueness and uniqueness theorems for a steady thermodiffusive mixture in a mixed problem*, Nonlinear Analysis, Methods & Applications, **12**, 5 (1988), 473–494.

[RioM88c] Rionero, S., G. Mulone, *A nonlinear stability analysis of the magnetic Bénard problem through the Lyapunov direct method*, ARMA, **103**, 4 (1988), 347–368.

[Robe1] Roberts, P. H., *An introduction to magnetohydrodynamics*, Longman, London, 1967.

[Robe1a] Roberts, P. H., *Convection in horizontal layers with internal heat generation. Theory*, J. Fluid Mech., **30**(1967), 33–49.

[Robe2] Roberts, P. H., *Dynamo theory*, in *Irreversible phenomena and dynamical systems analysis in geosciences*, C. Nicolis, G. Nicolis (eds.), Reidel, Dordrecht, 1987, 73–133.

[Roc] Rockafellar, T., *Convex analysis*, Princeton University Press, 1997.

[RocW] Rockafellar, P. T., R. J. B. Wets, *Variational analysis*, Springer, New York,

1994, 1998.

[Ros] Rossby, H. T., *A study of Bénard convection with and without rotation*, J. Fluid Mech. **36** (1969), 309–335.

[Rot] Rothe, E. H., *Some applications of functional analysis to the calculus of variations*, Proc. Symposia in Appl. Math. **8**, AMS, Providence, 1958.

[Rud] Ruddick, B. R., T. G. L. Shirtchiffe, *Data for double diffusion: physical properties of aqueous salt-sugar solutions*, Deep Sea Research **26** (1979), 775–787.

[RueT] Ruelle, D., F. Takens, *On the nature of turbulence*, Comm. Math. Phys. **20** (1971), 167–192; **23** (1971), 343–344.

[Sag] Sagan, H., *Introduction to the calculus of variations*, McGraw-Hill, New York, 1969.

[Sa] Sattinger, D. H., *The mathematical problem of hydrodynamic stability*, J. Math. Mech. **19**, 9 (1970), 197–217.

[Sa73] Sattinger, D. H., *Topics in stability and bifurcation theory*, LNM **309**, Springer, Berlin, 1973.

[Sca] Scales, L. E., *Introduction to nonlinear optimization*, Springer, New York, 1985.

[Sche] Schechter, R. S., *The variational method in engineering*, McGraw-Hill, New York, 1967.

[SelW] Seliger, R. L., G. B. Whitham, *Variational principles in continuum mechanics*, Proc. Roy. Soc. **305** (1968), 1–25.

[Ser1] Serrin, J., *On the stability of viscous fluid motions*, ARMA **3**, 1 (1959), 1–13.

[Ser2] Serrin, J., *Mathematical principles of classical fluid mechanics*, Springer, Berlin, 1959.

[Shap] Shaposhnikov, I. G., *On the theory of convective phenomena in binary mixtures*, P.M.M. **17**, 5 (1953), 604–606.

[SharR] Sharma. R. C., N. Rani, *Hall effects on thermosolutal instability of a plasma*, Indian J. of Pure Appl. Mat. **19**, 2 (1988), 202–207.

[SharS] Sharma, R. C., K. C. Sharma, *Thermal instability of compressible fluids with Hall currents through porous medium*, Instanbul Univ. Fen. Fak. Mec. A, **43** (1978), 89–98.

[SharT] Sharma. R. C., C. Trilok, *Thermosolutal instability of compressible Hall plasma in porous medium*, Astrophys. Space Sci. **155**, 2 (1989), 301–310.

[Shin] Shinbrot, M., *Lectures on fluid mechanics*, Gordon & Breach, 1973.

[ShiJ] Shir, C., D.D. Joseph, *Convective instability in a temperature and concentration field*, A.R.M.A. **30** (1968), 38–80.

[SiPM] Simo, J. C., T. A. Posbergh, J. E. Marsden, *Stability of coupled rigid body and geometrically exact rods: block diagonalization and the energy momentum method*, Phys. Rep. **193** (1990), 279–360.

[Smir] Smirnov, V. I., *A course of higher mathematics*, 5 vols., Pergamon, Oxford, 1964. (first Russian ed. Moscow, 1947)

[Smit] Smith, D. R., *Variational methods in optimization*, Prentice Hall, Englewood Cliffs, 1974.

[SmitDS] Smith, D. R., C. V. Smith, *When is Hamilton's principle an extremum principle*, AIAA J. **12**, 11 (1974), 1573–1576.

[SmitP] Smith, P., *Convexity methods in variational calculus*, Wiley, New York, 1985.

[Sob1] Sobolev, S. L., *Partial differential equations of mathematical physics*, Pergamon, Oxford, 1964. (1st Russian ed. 1947; 2nd Russian 1950; 3rd Russian ed. 1954, Gostekhizdat, Moscow)

[Sob2] Sobolev, S. L., *Some applications of functional analysis in mathematical physics*, Translations of Mathematical Monographs **7**, AMS, Providence, 1963.

(first Russian ed. Leningrad, 1950; 2nd Russian ed. Novosibirsk, 1962)

[SolM] Solonnikov, V. A., G. Mulone, *On the solvability of some initial boundary value problems in magnetofluidmechanics with Hall and ion-slip effects*, Rend. Mat. Acc. Lincei **9**, 6 (1995), 117–132.

[Soo] Soop, M. A., *A method for calculating the minimum of a one-dimensional variational integral on a infinite interval leading to Schrödinger type eigenvalue problem*, ZAMM **48**, 6 (1968), 369–376.

[Spe] Sperb, R., *Maximum principles and their applications*, Mathematics in Science and Engineering **157**, Academic, New York, 1981.

[SpV] Spiegel, E. A., G. Veronis, *On the Boussinesq approximation for a compressible fluid*, Astrophys. J. **131** (1960), 442–447.

[Strau82] Straughan B., *Instability, nonexistence and weighted energy methods in fluid dynamics and related theories*, Pitman, Boston, 1982.

[Strau] Straughan, B., *The energy method, stability and nonlinear convection*, Applied Mathematical Sciences **91**, Springer, New York, 1991; 2nd ed. 2004 (first published in 1993).

[Strau98] Straughan, B., *Explosive instabilities in mechanics*, Springer, Heidelberg, 1998.

[StriA] Strieder, W., R. Aris, *Variational methods applied to problems of diffusion and reaction*, Springer Tracts in Natural Philosophy **24**, Springer, Berlin, 1973.

[Stru] Struwe, M., *Variational methods and their applications to nonlinear partial differential equations and Hamiltonian systems*, Ergebnisse der Mathematik, 3 Folge, **34**, Springer, Berlin, 1990, 1996.

[SuS] Sutton, G. W., A. Sherman, *Engineering magnetohydrodynamics*, McGraw--Hill, New York, 1965.

[Sy] Synge, J. L., *The hypercircle in mathematical physics*, Cambridge University Press, 1957.

[Tay36a] Taylor, G. I., *Fluid friction between rotating cylinders. I. Torque measurement*, Proc. Roy. Soc. London, A **157** (1936), 546–564.

[Tay36b] Taylor, G. I., *Fluid friction between rotating cylinders. II. Distribution of velocity between concentric cylinders when outer one is rotating and inner one is at rest*, Proc. Roy. Soc. London, A **157** (1936), 565–578.

[Te] Temam, R., *Navier-Stokes equations. Theory and numerical analysis*, North-Holland, Amsterdam, 1979; *Navier-Stokes equations and nonlinear functional analysis*, 2nd ed., SIAM, Philadelphia, 1995. (first published in 1983)

[Tik] Tikhomirov, V. M., *Fundamental principles in the theory of extremal problems*, Wiley, New York, 1985. (German ed. at Teubner, Leipzig, 1982)

[TimG] Timoshenko, S., J. M. Gere, *Theory of elastic stability*, Mc Graw-Hill, New York, 1961. (first published in 1953)

[Tit] Titchmarch, E. C., *Eigenfunction expansions associated with second-order differential equations*, Part II, Clarendon Press, Oxford, 1958.

[Tri] Tritton, D. J., M. N. Zarraga, *Convection in horizontal layers with internal heat generation*, J. Fluid Mech., **30** (1967), 21–32.

[Tro] Troutman, J. L., *Variational calculus and optimal control convexity*, Undergraduate Texts in Mathematics, 2nd ed., Springer, Berlin, 1995. (First published in 1983.)

[Tuc] Tuckey, C., *Nonstandard methods in the calculus of variations*, Pitman Research Notes in Mathematics Series **297**, A Longman Scientific and Technical Publication, Wiley, New York, 1994.

[Tur] Turnbull, R. J., *Effect of dielectrophoretic forces on the Bénard instability*, Phys. Fluids **12**, 9 (1969), 1809–1815.

[Va]   Vainberg, M. M., *Variational methods for the study of nonlinear operators*, Holden-Day, San Francisco, 1964. (first Russian ed. at Gostehizdat, Moscow, 1956)

[Ve]   Velarde, M. G., *Rayleigh-Bénard convection of fluid layers: some new theoretical results*, in Les instabilities en convection naturelle and forcée, J. K. Platten & J. C. Legros, Springer-Verlag, Berlin, (1978), 89–101.

[Vel1]   Velte, W., *Über ein Stabilitätskriterium der Hydrodynamik*, ARMA, **9**, 1 (1962), 9–20.

[Vel2]   Velte, W., *Direkte Methoden der Variationsrechnung*, Teubner, Stuttgart, 1976.

[Velt]   Vel'tishchev, N. F., *Convection in a horizontal fluid layer with a uniform heat source*, Fluid Dynamics, **39**, 2 (2004), 189–197.

[Ven]   Venkatesh, Y. V., *Energy methods in time varying system stability and instability analysis*, LNPh **68**, Springer, Berlin, 1977.

[Ver]   Veronis, G., *Effect of a stabilizing gradient of solute on thermal convection*, J. Fluid Mech. **34** (1968), 315–336.

[Vert]   Vertheim, B. A., *On the conditions of convection in a binary mixture*, P.M.M., **19**, 6 (1955), 745–750.

[Vis]   Vishik, M. I., *Asymptotic behavior of solutions of evolutionary equations*, Accademia Nazionale dei Lincei, Cambridge University Press, 1992.

[Wal]   Walter, W., *General inequalities*, **5**, **6**, Birkhäuser, Basel, 1987, 1992.

[Wan]   Wan, F. Y. M., *Introduction to the calculus of variations and its applications*, Chapman & Hall, London, 1994, 1995.

[Was]   Washizu, K., *Variational methods in elasticity and plasticity*, 3rd ed., Pergamon, Oxford, 1982. (first ed. 1968).

[We1]   Weinberger, H. F., *Variational methods for eigenvalue problems*, Lecture Notes, University of Minnesota, Bookstore, 1962.

[We2]   Weinberger, H. F., *Variational methods for eigenvalue approximation*, CBMS-NSF Regional Conference Series in Applied Mathematics **15**, SIAM, Philadelphia, 1974.

[Weinste] Weinstein, A., *An invariant formulation of the new maximum-minimum theory of eigenvalues*, J. Math. Mec. **16**, 3 (1966), 213–218.

[Weinsto] Weinstock, R., *Calculus of variations*, McGraw-Hill, New York, 1952.

[Wey]   Weyl, H., *The method of orthogonal projection in partial differential equations theory*, Duke Math. J., **7** (1940), 411–444.

[Wi]   Willen, M., *Minimax theorems*, Birkhäuser, Basel, 1996.

[Yos]   Yosida, K., *Functional analysis*, Springer, Berlin, 1965.

[Youn]   Young, E. C., *Vector and tensor analysis*, Pure and Applied Mathematics, A Series of Monographs and Textbooks, **172**, 2nd ed., Marcel Dekker, New York, 1993. (first published in 1978)

[Your]   Yourgrau, W., *Variational principles in dynamics and quantum theory*, Pitman, London, 1960.

[Yu1]   Yudovich, V. I., *On the stability of stationary flow of viscous incompressible fluids*, Dokl. Akad. Nauk SSSR, **161**, 5 (1965), 1037–1040. (Russian) (translated as Soviet Physics Dokl., **10**, 4 (1965), 293–295).

[Yu2]   Yudovich, V. I., *The linearization method in hydrodynamical stability theory*, Translations of Mathematical Monographs **74**, AMS, Providence, RI, 1989. (Russian edition issued in 1984, at Rostov University Press)

[Ze]   Zeidler, E., *Nonlinear functional analysis and its applications, III. Variational methods and optimization*, Springer, Berlin, 1985.

[Zey]   Zeytounian, R. Kh., *The Bénard problem for deep convection: rigorous deriva-

tion of approximate equations, Int. J. Engng. Sci., **27**, 11 (1989), 1361–1366.

Series on Advances in Mathematics for Applied Sciences

Editorial Board

Series on Advances in Mathematics for Applied Sciences

Aims and Scope

This Series reports on new developments in mathematical research relating to methods, qualitative and numerical analysis, mathematical modeling in the applied and the technological sciences. Contributions related to constitutive theories, fluid dynamics, kinetic and transport theories, solid mechanics, system theory and mathematical methods for the applications are welcomed.

This Series includes books, lecture notes, proceedings, collections of research papers. Monograph collections on specialized topics of current interest are particularly encouraged. Both the proceedings and monograph collections will generally be edited by a Guest editor.

High quality, novelty of the content and potential for the applications to modern problems in applied science will be the guidelines for the selection of the content of this series.

Instructions for Authors

Submission of proposals should be addressed to the editors-in-charge or to any member of the editorial board. In the latter, the authors should also notify the proposal to one of the editors-in-charge. Acceptance of books and lecture notes will generally be based on the description of the general content and scope of the book or lecture notes as well as on sample of the parts judged to be more significantly by the authors.

Acceptance of proceedings will be based on relevance of the topics and of the lecturers contributing to the volume.

Acceptance of monograph collections will be based on relevance of the subject and of the authors contributing to the volume.

Authors are urged, in order to avoid re-typing, not to begin the final preparation of the text until they received the publisher's guidelines. They will receive from World Scientific the instructions for preparing camera-ready manuscript.